Experimental and Applied Immunotherapy

Jeffrey Medin · Daniel Fowler
Editors

Experimental and Applied Immunotherapy

Humana Press

Editors
Jeffrey Medin
University of Toronto
University Health Network
Toronto, Ontario
Canada
jmedin@uhnres.utoronto.ca

Daniel Fowler
National Institutes of Health
National Cancer Institute
Experimental Transplatation
and Immunology Branch
Bethesda, Maryland
USA
dhfowler@helix.nih.gov

ISBN 978-1-60761-979-6 e-ISBN 978-1-60761-980-2
DOI 10.1007/978-1-60761-980-2
Springer New York Dordrecht Heidelberg London

Printed on acid-free paper

Humana Press is part of Springer Science+Business Media (www.springer.com)

Foreword (Précis)

Immunotherapy is now recognized as an essential component of treatment for a wide variety of cancers. Established immunotherapies include bone marrow transplantation, donor leukocyte infusions, immune adjuvants, cytokines, monoclonal antibodies, and most recently, vaccines. Experimental cancer immunotherapies on the near horizon are likely to be more potent, less toxic, and more cost effective than many of the therapies that are currently in use.

The immune system is a complex and powerful defense system. The ultimate purpose of immunity is to generate responses that protect from pathogenic microorganisms. Mounting evidence, first derived from experiments in mice, indicates that the immune system also plays a role in the control and spread of a variety of cancers. Metastases account for about 90% of cancer mortality. At face value, the trafficking and highly specific tumor recognition of lymphocytes coupled with the tissue penetration of antibodies and other immune effector molecules is a promising approach to prevent and treat metastatic tumor deposits.

The realization that cancer may be regarded as a "non-healing wound" and that the development of cancer is intimately related to inflammation has led to fundamental changes in the approach to cancer immunotherapy. The immune system has evolved a large number of regulatory pathways that serve to limit inflammation and tissue damage during chronic inflammation. An improved understanding of the tumor microenvironment has led to strategies to interrupt immune suppressive regulatory circuits so that immune effector cells and cytokines can be more potent. Basic research has identified many novel strategies to reverse the immunosuppression in the tumor microenvironment that are now being translated in the clinic. Encouraging results from a number of clinical trials make it likely that this approach, often termed "check point blockade," will become routinely used in future cancer immunotherapies.

Another fundamental advance from basic research on the tumor microenvironment is the demonstration that chronic inflammation can promote or predispose to the development of cancer. From this work, strategies for cancer prevention by modulating inflammation have emerged as another promising approach to cancer immunotherapy. This field is challenged by the inherent requirement for lengthy clinical trials so that translation of preventative cancer immunotherapy into routine clinical practice is not yet on the horizon.

The primary obstacle to the incorporation of many of these potent therapies into routine clinical practice will be the occurrence of autoimmunity. The field of cancer immunotherapy is likely to face a major challenge in what is referred to as "Type II translation," a term referring to the implementation of new therapies into the community. In the case of the new potent immunotherapies, education of oncologists from widely disparate fields of medical, surgical, and radiation oncology will be required so that cancer immunotherapy can be widely and safely adopted in the community. In this regard it is instructive to recall the lessons of allogeneic bone marrow transplantation, including the development of strategies to manage graft versus host disease. In that case, a subspecialty of medical oncologists emerged with specialized training and experience. It is likely that these clinicians will lead the development of potent combination cancer immunotherapies, and that they will in turn develop the best practices to safely implement these powerful treatments into routine clinical practice.

The convergence of a number of technologies for ex vivo gene transfer in lymphocytes has generated considerable enthusiasm for cell transfer approaches using engineered T cells. Genetic reprogramming of T cells can pharmacologically enhance the function of T cells beyond their naturally evolved capacities. The use of efficient cell culture systems combined with ex vivo gene transfer provides, in principle, a unique means to circumvent the tolerance mechanisms of tumors as well as the immune escape strategies used by tumors to avoid immune elimination. Preclinical models indicate that gene-modified T cells can be used to enhance tumor specificity, improve T cell survival, modify T cell trafficking, and counteract mechanisms that promote T cell anergy. The most advanced of these approaches in terms of clinical development is currently in a pivotal clinical trial in Europe, where studies are being conducted using allogeneic T cells transduced with a suicide gene for high-risk acute leukemia. The recent success of engineered T cells in a variety of pilot cancer trials makes it conceivable that gene-modified autologous T cells will eventually exceed the potent antitumor effects of allogeneic T cells, and therefore, that allogeneic transplantation will be replaced by autologous cancer immunotherapies.

For cancer immunotherapies, the time from discovery to approval in the USA by the Food and Drug Administration tends to be longer than industry standards for other cancer immunotherapies. Monoclonal antibodies were invented in 1975, first given to patients with lymphoma in 1980, and yet Rituximab was not commercialized until 1996. Dendritic cells were observed in the nineteenth century, named in 1973, first tested in cancer trials in the early 1990s and not commercialized as a cancer therapy until 2010. Reasons for the extended period of clinical development include the inherent complexity of the immune system and a commercial reluctance by the pharmaceutical industry. In particular, cell based immune therapies have not been thought to fit into a standard business model, and therefore the delay between pilot testing and pivotal trials, often referred to as the "valley of death," is longer for cancer immunotherapies than other forms of cancer treatment.

Cancer immunotherapy was first proposed more than a century ago. With rare exceptions, the reputation of the field suffered from disappointing results. However, recent progress in translating basic findings into potent therapies has pushed the field past the tipping point. Previous setbacks were caused by an incomplete understanding of cancer immunology. Advances in our understanding of the science of the molecular interactions between tumors and the immune system have led to many novel investigational therapies and continue to inform efforts for devising more potent therapeutics. Given the major advances in the basic sciences, the development of the next generation of cancer immunotherapy has now been converted to a project in engineering the immune system. While the continued growth of sciences in the areas of cancer biology and immunology is inevitable, the principles are sufficiently understood to generate supraphysiologic immune systems that will deliver molecularly targeted cancer immunotherapies. Collectively, the chapters in this book provide a state-of-the-art road map that will lead to the creation of these "performance enhancing drugs" with the worthy destination of surmounting cancer.

Carl H. June
Director, Translational Research and Professor
Department of Pathology and Laboratory Medicine,
University of Pennsylvania,
Philadelphia, PA, USA

Contents

Part V Transplantation

Contributors

Bryan Au
Graduate Student, Robarts Research Institute and the Department of Microbiology and Immunology, The University of Western Ontario,
London ON, Canada
cau23@uwo.ca

Angela Bahlo, Ph.D.
Department of Medicine, University of Toronto, Odette Cancer Center,
Sunnybrook Health Sciences Center, Toronto, ON, Canada
spanerd@sri.utoronto.ca

John W. Barrett, Ph.D.
Robarts Research Institute, The University of Western Ontario, London, ON, Canada
jbarrett@robarts.ca

John C. Bell, Ph.D.
Senior Scientist, Center for Cancer Therapeutics,
Ottawa Hospital Research Institute, Ottawa ON, Canada
jbell@ohri.ca

Marie-Claude Bourgeois-Daigneault, B.Sc.
Laboratoire d'Immunologie Moléculaire, Département de Microbiologie et Immunologie, Université de Montréal, CP 6128 Succ Centre ville,
Montréal, QC, Canada
jacques.thibodeau@umontreal.ca

Jonathan Bramson, Ph.D.
Director, Center for Gene Therapeutics, Professor, Department of Pathology and Molecular Medicine McMaster University, Hamilton ON, Canada
bramsonj@mcmaster.ca

Malcolm K. Brenner, M.D., Ph.D.
Fayez Sarofim Chair, Professor of Medicine and Pediatrics,
Director, Center for Cell and Gene Therapy Baylor College of Medicine,
Texas Children's Hospital and The Methodist Hospital, Houston TX, USA
mbrenner@bcm.edu

Ryan Buensuceso
Graduate Student, Robarts Research Institute and the Department of Microbiology and Immunology, The University of Western Ontario, London ON, Canada
rbuensuc@uwo.ca

Sonali de Chickera
Graduate Student, Robarts Research Institute and the Department of Anatomy and Cell Biology, The University of Western Ontario, London ON, Canada
sdechick@uwo.ca

Gregory A. Dekaban Ph.D.
Scientist, Robarts Research Institute,
Professor, The Department of Microbiology and Immunology,
The University of Western Ontario, London ON, Canada
dekaban@robarts.ca

Vasiliki Economopoulos
Graduate Student, Robarts Research Institute and the Department of Medical Biophysics, The University of Western Ontario, London ON, Canada
vecon@imaging.robarts.ca

Ronan Foley, M.D.
FRCPC Director, Stem Cell Laboratory, Juravinski Hospital and Cancer Centre, Hamilton Health Sciences, Hamilton, ON, Canada
foleyr@hhsc.ca

Paula Foster, Ph.D.
Scientist, Robarts Research Institute,
Associate Professor, Department of Medical Biophysics, Robarts Research Institute, Department of Medical Biophysics, The University of Western Ontario, London ON, Canada
pfoster@imaging.robarts.ca

Daniel H. Fowler, M.D.
National Institutes of Health, National Cancer Institute, Experimental Transplatation and Immunology Branch, Bethesda, MD, USA
dhfowler@helix.nih.gov

Moïra François, Ph.D.
Department of Experimental Medicine, McGill University, Montreal, QC, Canada
moira.francois@mail.mcgill.ca

Jacques Galipeau, M.D.
Department of Experimental Medicine, McGill University, Montreal, QC, Canada
moira.francois@mail.mcgill.ca

Ulrike Gerdemann, M.D.
Center for Cell and Gene Therapy, Baylor College of Medicine, Texas Children's Hospital and The Methodist Hospital, Houston, TX, USA
uxgerdem@txccc.org

Natalie Grinshtein, Ph.D.
Department of Pathology and Molecular Medicine, McMaster University,
Hamilton, ON, Canada
ngrinshtein@hotmail.com

David A. Jaffray, Ph.D.
Professor, Departments of Radiation Oncology and Medical Biophysics,
Princess Margaret Hospital/Ontario Cancer Institute,
University of Toronto, Toronto ON, Canada
david.jaffray@rmp.uhn.on.ca

Robert Kammerer, DVM
Institute of Immunology, Friedrich-Loeffler-Institute, Tuebingen, Germany
robert.kammerer@fli.bund.de

Armand Keating, M.D.
Division of Hematology, University of Toronto, ON, Canada
armand.keating@uhn.on.ca

Yoko Kosaka, Ph.D.
Division of Hematology, University of Toronto, ON, Canada
armand.keating@uhn.on.ca

Réjean Lapointe, Ph.D.
Laboratoire d'Immunologie Moléculaire, Département de Microbiologie et
Immunologie, Université de Montréal, CP 6128 Succ Centre ville,
Montréal, QC, Canada
jacques.thibodeau@umontreal.ca

Bruce Levine, Ph.D.
Director, Clinical Cell and Vaccine Production Facility, Research Associate
Professor, Department of Pathology and Laboratory Medicine
Abramson Family Cancer Research Institute, The University of Pennsylvania
School of Medicine, Hospital of the University of Pennsylvania,
Philadelphia PA, USA
LevineBL@mail.med.upenn.edu

Megan Levings, Ph.D.
Department of Surgery, University of British Columbia
and Child Family Research Institute, Vancouver, BC, Canada
megan.levings@ubc.ca

Brian D. Lichty, Ph.D.
Center for Gene Therapeutics, Department of Pathology
and Molecular Medicine, McMaster University, Hamilton, ON, Canada
lichtyb@mcmaster.ca

Jeffrey A. Medin, Ph.D.
Senior Scientist, Ontario Cancer Institute, Professor, Department of Medical Biophysics and the Institute of Medical Science, University of Toronto, Toronto ON, Canada
jmedin@uhnres.utoronto.ca

Amanda Moretti, M.Sc.
Institute of Medical Science, University of Toronto, Toronto, ON, Canada
amanda.moretti@utoronto.ca

Takashi Murakami, M.D., Ph.D.
Division of Bioimaging Sciences, Center for Molecular Medicine, Jichi Medical University, 3311-1 Yakushiji, Tochigi, 329-0498, Japan
takmu@jichi.ac.jp

Megan Nelles, Ph.D.
Department of Medical Biophysics, University of Toronto, Toronto, ON, Canada
mnelles@uhnres.utoronto.ca

Brad H. Nelson, Ph.D.
Deeley Research Centre, BC Cancer Agency, Victoria, BC, Canada
bnelson@bccancer.bc.ca

Christopher J. Paige, Ph.D.
Vice-President Research, Senior Scientist, Ontario Cancer Institute, Professor, Departments of Medical Biophysics and Immunology, University of Toronto, University Health Network, Toronto ON, Canada
paige@uhnresearch.ca

Claude Perreault, M.D.
Department of Medicine, Université de Montréal, Maisonneuve-Rosemont Hospital, Montréal, QC, Canada
c.perreault@videotron.ca

Daniel J. Powell, Jr. Ph.D.
Pathology and Laboratory Medicine, Clinical Cell and Vaccine Production Facility, Ovarian Cancer Research Center, The University of Pennsylvania School of Medicine, Philadelphia, PA, USA
Poda@mail.med.upenn.edu

Christoph Rader, Ph.D.
Antibody Technology Section, Experimental Transplantation and Immunology Branch, Center for Cancer Research, National Cancer Institute, NIH, Bethesda, MD, USA
raderc@mail.nih.gov

Michael Ricci, BHSc.
Program, McMaster University, Hamilton ON, Canada
riccimj@mcmaster.ca

Vincenzo Salerno, Ph.D.
Ontario Cancer Institute, Toronto ON, Canada
vsalerno@uhnres.utoronto.ca

Elizabeth Scheid, B.Sc.
FRCPC Director, Stem Cell Laboratory, Juravinski Hospital and Cancer Centre, Hamilton Health Sciences, Hamilton, ON, Canada
foleyr@hhsc.ca

David Spaner, M.D., Ph.D.
Department of Medicine, University of Toronto, Odette Cancer Center, Sunnybrook Health Sciences Center, Toronto, ON, Canada
spanerd@sri.utoronto.ca

Kyle B. Stephenson, Ph.D.
Candidate, Medical Sciences, Department of Pathology and Molecular Medicine, Center for Gene Therapeutics, McMaster University, Hamilton ON, Canada
stephekb@mcmaster.ca

Jacques Thibodeau, Ph.D.
Laboratoire d'Immunologie Moléculaire, Département de Microbiologie et Immunologie, Université de Montréal, CP 6128 Succ Centre ville, Montréal, QC, Canada
jacques.thibodeau@umontreal.ca

Adele Y. Wang, Ph.D.
Student, Experimental Medicine Program, University of British Columbia, Vancouver BC, Canada
adele@interchange.ubc.ca

John R. Webb, Ph.D.
Deeley Research Centre, BC Cancer Agency, Victoria, BC, Canada
bnelson@bccancer.bc.ca

Yixin Xu, Ph.D.
Candidate, Immunology and Reproductive Biology Lab, State Key Laboratory of Pharmaceutical Biotechnology, School of Life Science, Nanjing University, Nanjing, China
xuyixin39@gmail.com

Wolfgang Zimmermann, Ph.D.
Director, LIFE Center, Tumor Immunology Laboratory, University Clinic, Ludwig-Maximilians-University Munich, Munich, Germany
wolfgang.zimmermann@med.uni-muenchen.de

Part I
T Cell Therapy: State-of-the-Art

Chapter 1
Extending the Use of Adoptive T Cell Immunotherapy for Infections and Cancer

Ulrike Gerdemann and Malcolm K. Brenner

Abstract Adoptive transfer of antigen-specific T cells has proven to be an effective and powerful therapeutic tool in the prevention and treatment of viral infections (e.g., cytomegalovirus [CMV], Epstein-Barr virus [EBV], and adenovirus) and virus-associated diseases, such as EBV-associated lymphoproliferative disease (LPD), that arise in the immunocompromised host. This therapeutic approach has also been extended to the treatment of cancer and has shown some success in patients with melanoma and EBV-associated malignancies such as Hodgkin's lymphoma and nasopharyngeal carcinoma. However, this strategy has been less successful in other malignancies. To improve the efficacy of adoptively transferred tumor-reactive T cells, a number of groups have sought to identify better immunotherapeutic target antigens and to design protocols for the optimal *in vitro* propagation of tumor-reactive T cells, which are often otherwise anergized or tolerized. Another approach that has recently come to fore involves the genetic modification of T cells using genes that confer properties such as new antigen specificity, improved homing to tumor sites, or increased resistance to tumor immune evasion. This chapter evaluates recent advances in tumor immunotherapy, including T cell engineering, and speculates on the future potential of adoptive T cell transfer in the field of cancer therapy.

Keywords Cancer immunotherapy • Adoptive T cell transfer • Tumor immunology • Gene therapy • T cells

---

U. Gerdemann (✉)
Center for Cell and Gene Therapy, Baylor College of Medicine, Texas Children's Hospital and The Methodist Hospital, Houston, TX, USA
e-mail: uxgerdem@txccc.org

J. Medin and D. Fowler (eds.), *Experimental and Applied Immunotherapy*,
DOI 10.1007/978-1-60761-980-2_1, © Springer Science+Business Media, LLC 2011

Current State of Translational T Cell Therapy

Immunotherapy for Viral Infections Post-HSCT

The infusion of *ex vivo*-expanded virus-specific cytotoxic T lymphocytes (CTLs) has been shown to both prevent and treat viral infections that arise in the immunocompromised host, such as recipients of T cell-depleted hematopoietic stem cell transplants (HSCT) [1]. In such individuals, T cell function can be impaired for at least 12 months posttransplant, thereby increasing host susceptibility to viral infections [1].

Cytomegalovirus (CMV)

Cytomegalovirus (CMV) is a latent herpes virus that is frequently reactivated after allogeneic HSCT; less frequently, CMV is acquired as a primary infection post-transplant. Primary CMV infection or CMV reactivation can result in severe pneumonitis and colitis that account for a significant number of post-transplant fatalities. Available pharmacologic therapies to treat CMV may lack efficacy and may also cause serious toxicity, such as bone marrow, renal, and hepatic impairment. As such, a clinical need exists to develop alternative therapies, such as T cell therapy, to prevent and treat CMV infection. In initial research, Riddell and colleagues infused *ex vivo*-expanded CMV-reactive $CD8^+$ T cell clones into 14 allogeneic HSCT patients in an attempt to prevent CMV reactivation; CMV-specific T cell therapy was both safe and effective in terms of restoring antiviral immunity *in vivo*. T cell receptor (TCR) clonotyping experiments determined that the transferred T cells persisted for at least 8 weeks; importantly, continued persistence of CMV-specific immunity was associated with the development of a concomitant CMV-specific $CD4^+$ T helper response [2]. In subsequent research, Einsele and colleagues generated polyclonal CMV-specific CTL lines containing both $CD4^+$ and $CD8^+$ T cells and transferred such cells to patients with CMV viremia that was resistant to antiviral chemotherapy. The clinical results were impressive: that is, infusion of a relatively small number of cells (10^7 cells/m^2) significantly reduced CMV viral load in each of 7 evaluable patients. The antiviral effect was sustained in five patients and transient in two patients who had the highest virus load. Importantly, one of these latter patients cleared the CMV virus completely after receiving a second T cell infusion. However, the remaining patient succumbed to fatal CMV encephalitis after refusing a second CTL infusion [3].

Epstein-Barr Virus (EBV)

Reactivation of the gamma-herpes virus EBV may cause a lethal lymphoproliferative disorder (post-transplant lymphoproliferative disease; PTLD) after HSCT or solid organ grafting [4]. Although many patients with PTLD may respond to withdrawal of immunosuppression or infusion of the anti-B cell antibody rituximab, the disease may progress in other individuals and result in lethality [4]. T cell therapy has been highly successful when used as prophylaxis or treatment of PTLD. Initially, Rooney and colleagues generated EBV-specific CTL *ex vivo* by using EBV-transformed lymphoblastoid cell lines (EBV-LCL) as a stimulator cells; resultant T cells were then adoptively transferred to immunocompromised patients at risk of developing EBV-associated PTLD [5]. Since 1993, these investigators have infused over 100 transplant recipients with donor-derived polyclonal T cell lines; through these efforts, it has been established that a dose of 2×10^7 CTL/m^2 is safe and effective for both prophylaxis and treatment of EBV-related PTLD. The first 26 patients enrolled in this study received CTLs that were genetically marked with an oncoretroviral vector containing the neomycin resistance gene (*neo*). Long-term follow-up showed that the marked cells could be detected for as long as 9 years post-infusion [6].

Other Viruses

More recent research has demonstrated the safety and efficacy of CTL lines that simultaneously target multiple viruses including EBV, CMV and adenovirus (Adv). Specifically, Leen and colleagues infused such trivirus-specific CTL at a dose of 2×10^5/kg into allogeneic HSCT recipients receiving a graft from a donor who was EBV and CMV seropositive. Infusion of this small number of virus-specific CTL was associated with T cell expansion *in vivo* and appeared to protect hosts against all three viruses, thereby indicating that broad spectrum treatment might be provided from a single T cell infusion [7] (Table 1.1).

Immunotherapy for Virus-Associated Malignancies

EBV Lymphoma

In the immunocompromised host, EBV PTLD can develop into frank lymphoma; in addition, even in immune competent individuals, EBV infection has been associated with both Hodgkin's lymphoma (HL) and non-Hodgkin's lymphoma (NHL) [4]. Lymphomas that arise in immunosuppressed individuals have the type 3 pattern of viral latency, in which almost all latent-cycle EBV genes are expressed, some of which are typically highly immunogenic [4]. It is therefore not surprising that administration of EBV-specific CTL can successfully treat lymphomas that emerge during EBV-PTLD following HSCT and solid organ transplants. Indeed, EBV-specific CTL were used to treat 12 patients with EBV-related lymphoma that developed after allogeneic HSCT: complete remission was observed in 10 of 12 patients. One patient with very advanced disease died within a week of receiving the CTLs; a second patient initially responded to CTL therapy but then developed progressive disease. In this latter individual, *in vitro* tumor characterization revealed that a virus deletion mutant emerged following CTL therapy; that is, the tumor target epitope recognized

Table 1.1 Clinical studies: Virus-specific CTL therapy

Virus-specificity	Expansion protocol	Infused cells	Antigen source	Advantages	Disadvantages
CMV [2]	*In vitro* expanded	$CD8^+$ T cell clones	CMV-infected fibroblasts	Effective prophylaxis	Prolonged culture Single peptide specificity Limited *in vitro* persistence
CMV [3]	*In vitro* expanded	Polyclonal CTL	CMV lysate	Effective treatment $CD4^+$/CD8 CTL	Prolonged culture
CMV [8]	*In vitro* expanded	Polyclonal CTL	Inactivated CMV virus	Effective treatment $CD4^+$/CD8 CTL	2–3 weeks *in vitro* culture
CMV [9]	Tetramer selected	Selected peptide specific $CD8^+$ T cells	CMV peptide	Rapid selection Effective treatment	Expensive Requires large starting blood volumes Limited to specific class I HLA types and viruses with high frequency of circulating specific cells
CMV [10]	*In vitro* expanded	A2 peptide specific $CD8^+$ T cells	A2 CMV-NLV peptide	Effective prophylaxis	Prolonged culture Single peptide specificity Limited *in vivo* persistence?
CMV [11]	*In vitro* expanded	Polyclonal, pp65 specific	Ad5f35pp65 vector	Effective prophylaxis $CD4^+$/CD8 CTL	Prolonged culture Expensive clinical grade vector production
EBV [5, 6, 12–15]	*In vitro* expanded	Polyclonal	EBV-LCL	Effective prophylaxis $CD4^+$/CD8 CTL	Prolonged culture (LCL + CTL) Infectious EBV Viral escape mutants

EBV [16, 17]	*In vitro* expanded	Partly matched, polyclonal	EBV-LCL	Immediate availability (banked CTL) CD4⁺/CD8 CTL	Prolonged culture Infectious EBV Limited persistence depending on the level of HLA match
Adv [18]	IFN-γ selected	Selected, polyclonal T cells	Adenoviral antigen	Rapid selection Effective treatment CD4⁺/CD8 CTL	Expensive Requires large starting blood volumes Limited in vivo persistence? Selection of terminally differentiated CTL?
Multivirus (EBV, CMV, ADV) [7]	*In vitro* expanded	Polyclonal	Ad5f35pp65-transduced LCL	Effective prophylaxis/treatment CD4⁺/CD8 CTL Multivirus specificity	Prolonged culture Infectious EBV virus Expensive clinical grade vector production

by the transferred CTL line had been deleted, thereby rendering the CTL line ineffective as an antitumor effector [12, 13]. This study demonstrates the safety and efficacy of virus-specific CTLs as treatment for immunogenic tumors that arise in the immunocompromised host and additionally highlights a concern for adoptive immunotherapy protocols: that is, even in a setting that utilizes polyclonal CTL, a mutation in a tumor-specific antigen can ultimately result in tumor escape (Table 1.2).

The success and safety profile of adoptively-transferred T cells in immunocompromised patients has prompted the extension of this modality for the treatment of tumors that arise in immunocompetent individuals. EBV is present in virtually all undifferentiated nonkeratinizing nasopharyngeal carcinomas (NPCs) and in 40–50% of HL. EBV-related lymphomas, HL, and NPC express a restricted pattern of weakly immunogenic EBV antigens, namely EBV nuclear antigen 1 (EBNA1) and latent membrane proteins LMP1 and LMP2, which are not typically immunodominant components of polyclonal EBV-specific CTL lines [4]. In spite of this restricted pattern of tumor antigen expression, EBV-specific CTL therapy of EBV-associated HL resulted in complete remission in 5 of 14 patients, with partial remission observed in 1 patient and stable disease observed in 5 patients [19]. EBV-specific CTLs have also been evaluated in 10 patients with NPC. Patients who were in remission from NPC at the time of CTL therapy remained disease free; however, in patients with bulky disease, CTL therapy resulted in limited responses that were sometimes transient [20].

To optimize the antigenic targeting of CTLs directed against HL/NHL in immunocompetent patients, Bollard and collaborators prepared CTLs whose specificity was skewed toward the more restricted array of weak tumor antigens expressed by the malignancy. Specifically, tumor-reactive T cells were manufactured *ex vivo* using stimulator cells comprised of dendritic cells (DCs) and EBV-LCL that were genetically modified by transduction with an adenoviral vector to overexpress the otherwise weak EBV antigen LMP2, and more recently, both LMP1 and LMP2 [21]. Importantly, this cell manufacturing protocol substantially increased the frequency of LMP2-specific tumor reactive cells within the resultant CTL lines. Clinical results were impressive: nine of ten patients (5HL/5NHL) with high-risk lymphoma treated in remission have remained in remission after CTL therapy. In addition, five of six patients (3HL/ 2NHL/ 1SCAEBV) with active relapsed disease sustained complete tumor responses after CTL therapy [21]. Since publication of this study (performed between 2003 and 2007), an additional five patients with active disease have been treated; three of these patients have achieved complete remission (Dr. Bollard, personal communication of unpublished data). Taken together, these results indicate that adoptive T cell therapy is safe and can also be effective for the treatment of virus-associated malignancies in the immunocompetent host (Table 1.2).

Immunotherapy for Melanoma

T cell immunotherapy has also been used to treat melanoma, with promising clinical results. In a pilot study, Rosenberg and colleagues reported that infusion

Table 1.2 Clinical studies: Tumor-specific CTL therapy

Tumor-specificity	Expansion protocol	Infused cells	Antigen source	Advantages	Disadvantages
EBV-HL [19]	*In vitro* expanded	Polyclonal	LCL	Objective responses even in relapsed patients No toxicity Persistance >12 month	Prolonged culture Infectious EBV HL LMP1-2 antigens under represented
NPC [20]	*In vitro* expanded	Polyclonal	LCL	Significant antitumor response No toxicity	Prolonged culture Infectious EBV NPC LMP 1-2 antigens under represented Antitumor response temporary in patients with bulky disease
EBV-HL [21]	*In vitro* expanded	Polyclonal	Ad5f35LMP-2 transduced DCs and LCLs	Increased expansion of LMP2 specific CTL Objective response even in relapsed patients No toxicity	Prolonged culture Infectious EBV Expensive clinical grade vector production
Melanoma [22–26]	*In vitro* expansion	Polyclonal tumor infiltrating lymphocytes	Autologous tumor-cells	Objective responses rate: 49–72% Durable response in patients with lymphodepletion	Prolonged culture Autologous tumor material required IL-2 toxicity Limited persistence in patients without lymphodepletion

of melanoma-specific tumor-infiltrating lymphocytes (TILs) and co-administration of high-dose interleukin 2 (IL-2) produced clinical responses in ~50% of patients with metastatic melanoma [22]. However, a later clinical trial from the same group using T cell clones directed against the melanoma-associated antigen, gpl00, either with or without IL-2 co-administration, reported poor clinical responses with only one minor response and one mixed response. Importantly, in this study, it was determined that the adoptively transferred cells failed to engraft or persist *in vivo* [27]. Subsequently, improved clinical results were achieved using a modified treatment protocol that incorporated a host lymphodepletion step prior to CTL infusion; host lymphodepletion was performed in an attempt to improve the *in vivo* expansion and persistence of the adoptively-transferred cells. Specifically, 13 patients with metastatic melanoma refractory to standard therapies received immunodepleting chemotherapy consisting of a combination of fludarabine plus cyclophosphamide followed by adoptive transfer of highly selected, TIL-derived, tumor-reactive T cells and high-dose IL-2. Six of 13 patients achieved an objective clinical response and four others demonstrated a mixed response; in some cases, even large bulky tumors regressed [23]. Although these results are certainly impressive and an example of effective immunotherapy, the overall approach is somewhat lacking in feasibility because the collection of autologous TILs for individual therapeutic use is restricted to patients whose tumors are amenable to surgical resection. Furthermore, the *ex vivo* expansion of large cell numbers of CTL (>10^{10} cells) is a relatively complex and expensive procedure; besides, there is concern that infusion of large numbers of activated T cells can be associated with significant toxicity, such as pulmonary impairment (Table 1.2) [28, 29].

T Cells Directed Against Nonviral Antigens

One of the challenges of adoptive immunotherapy for nonviral cancers remains the identification of strongly immunogenic target antigens. Model tumor antigens should be specifically and universally expressed on tumor cells in order to both focus antitumor immunity and limit collateral tissue damage, and ideally should be essential for the maintenance of the oncogenic phenotype of the tumor. However, the majority of known tumor antigens do not meet these criteria. That is, tumor antigens typically are not neo-antigens uniquely present in cancer cells but rather are antigens that are also expressed in normal cells; in such cases, peripheral blood T cells are tolerized to the antigens in an attempt to prevent auto-immunity. This current understanding that the majority of tumor antigens can be classified as self-antigens indicates that further efforts should seek to discover novel tumor targets; and, in cases where truly tumor specific antigens cannot be identified, there exists a need to optimize cell culture protocols for the generation of CTL endowed with a capacity to overcome the mechanisms that establish T cell tolerance against "self" antigens expressed on tumors.

Classification of Tumor Antigens

Tumor Antigens and T-Cell Immunogenicity

Tumor-associated antigens (TAA) can be classified into four main groups based on expression and tissue distribution patterns: (1) Unique antigens (e.g., MUM1) result from single mutations that are tumor- and patient-specific and therefore are only expressed in neoplastic cells [30]. Unique antigens, which may be relatively immunogenic, are often considered ideal for immunotherapy because the possibility exists that tumor cells might be specifically targeted without the destruction of nearby normal tissue [31]. However, because unique antigens are also typically patient-specific, identification of the mutated gene and subsequent generation of an individualized CTL cell product targeting the identified antigen can be highly labor- and cost-intensive. (2) A second antigen category consists of the shared lineage-restricted antigens, such as melanoma cell and normal melanocyte expression of the antigens MART, gp100, or Melan-A. Such antigens are also strongly immunostimulatory, with a potency somewhat equivalent to weak viral antigens; as such, the generation of tumor-specific T cells against these antigens can be accomplished from healthy donors and patients with minimal *ex vivo* manipulation [32, 33]. Because of the shared antigenic expression, adoptive cell therapy using T cells specific for melanoma-specific CTL or TILs has been associated with T cell-mediated destruction of normal melanocytes, thereby resulting in vitiligo as well as ocular and systemic auto-immunity [34]. (3) A third group of antigens consists of the shared tumor-specific TAA, which includes the cancer testis antigens [CTA] such as MAGE, BAGE, GAGE, NY-ESO-1 and PRAME; such CTA are expressed in multiple tumors but not in healthy organs, with the exception of germ-line tissue [35]. CTLs specific for CTA may represent an optimal approach because the T cells can be produced on a large-scale to provide broad-spectrum protection against a variety of tumors. Indeed, CTAs have been targeted in both vaccine and T cell therapy protocols, with evidence of clinical efficacy [34–37]. (4) Finally, a fourth group of antigens consists of shared TAA that are highly overexpressed in multiple different tumors and expressed at low levels in healthy tissue; antigenic members of this group include CEA, hTERT and SURVIVIN. There exists limited clinical data regarding the safety of targeting these antigens *in vivo*; however, SURVIVIN- and CEA-specific T cells have been isolated from the peripheral blood of patients who have cleared their tumors, thereby suggesting that such antigens can be effectively targeted *in vivo* [36]. It should be noted that T cells targeted to these antigens may carry the risk of inducing collateral damage to normal tissues that co-express the antigen (e.g., CEA expression on both tumor and normal biliary epithelium) [37].

Identification of Novel Tumor Antigens

Two main approaches have been used to identify novel tumor antigens. One approach starts with a pre-existing TAA-specific T cell clone with *unknown* specificity (direct immunology), whereas the second approach starts with a *predicted* tumor-associated antigen or epitope ("reverse" immunology). In the direct immunology approach, a patient-derived tumor-specific T cell clone is used to screen a tumor-derived cDNA library, synthetic peptide library, or peptides that have been eluted from the tumor cell surface to define the minimal epitope peptide recognized by the T cells. The unknown source of antigen (i.e., the novel tumor antigen) may also then be simultaneously discovered [38]. By comparison, the "reverse" immunology technique starts with a putative TAA that is predicted from expression profiling and data mining. Subsequently, the antigen is validated by assessing its immunogenicity using algorithms that predict HLA-restricted epitope peptides; then, functional *in vitro* assays and *in vivo* studies are performed to confirm the target antigen as a genuine TAA [39]. Although both approaches have been successfully used, they are labor- and cost-intensive strategies; in addition, due to internal bias, many tumor antigens are not identified by these conventional approaches.

Optimizing Cell Culture Protocols for Tumor-Specific CTL Generation

In addition to choosing the optimal tumor antigen for T cell stimulation, successful adoptive immunotherapy relies on the availability of effective protocols for activating and expanding tumor-specific CTLs *ex vivo*. The generation of an effective antigen-specific immune response requires professional antigen-presenting cells (APCs), which not only present antigen but also provide costimulation and polarizing cytokines such as IL-12 that drive T cell differentiation down the Th1/Tc1 effector pathway that has been associated with effective antitumor immunity [33]. Culture supplementation with additional cytokines, such as IL-7 and IL-21 [32], which are primarily produced by non-T cell populations, further facilitates the activation of naïve or tolerized T cells.

Antigen-Presenting Cells (APCs)

Different sources of APCs have been used in previous tumor immunotherapy studies. Although myeloid dendritic cells (DCs) are most likely the most potent APC, widespread use of DCs is limited by their restricted numbers, which in turn impedes large-scale T cell production. Isolation of sufficient numbers of DCs is especially problematic in cancer patients who have received immune-depleting intensive chemotherapy. In contrast, use of immortalized EBV-LCL represent an unlimited source of APC; however, EBV-LCL expression of endogenous EBV antigens interferes with expansion of antigen-specific T cells that recognize subdominant tumor antigens. Thus, other sources of autologous APCs that are available in large quantities and are free of viral antigens are currently being evaluated as APCs for both *ex vivo* and *in vivo* CTL stimulation protocols; such alternative APCs include activated T-APCs and CD40L-activated B cell blasts [33, 40, 41]. Several groups have also improved the stimulatory capacity of available APC sources through genetic modification. Such modifications include forced expression of co-stimulatory molecules like CD40, OX40, CD70, B7-1, ICAM-1, and LFA-3, or induction of stimulatory cytokine secretion such as IL-12 or IL-7 to induce more efficient T cell stimulation for viral or tumor antigens [33, 42–44].

Despite improvements in the efficient generation of autologous APCs for CTL stimulation, existing approaches still need to be individually prepared for each patient. The availability of "off-the-shelf" APCs to initiate and/or expand tumor-specific CTLs would simplify and abbreviate the CTL generation process, thereby significantly enhancing cost effectiveness. Toward this aim, artificial antigen presenting cells (aAPCs) and their cell-free substitutes have been developed to stimulate the *ex vivo* expansion of T cells without the need for autologous APCs.

Cellular aAPCs have been created from human leukemia cell lines, insect cells, or mouse fibroblasts [45–49]. These systems, however, often require genetic modification to effectively present antigen. For example, the leukemia cell line, K562, has been modified to express costimulatory molecules such as CD137 or $CD8^+$, to secrete a range of cytokines, and to express HLA genes [48, 49]. However, such cell-based systems may potentially carry the risk of infection or tumorigenicity. On the other hand, cell-free aAPC platforms, including micron-size latex, polyglycolide, magnetic beads, or lipid-based vesicles eliminate this risk of infection; however, such approaches in some cases are limited by lack of biocompatibility [50–53]. Ultimately, the optimal aAPC must be GMP-compliant, potent, and able to reproducibly support the efficient expansion of antigen-specific T cells *ex vivo*. As cell therapy becomes more widely used, this area of development is likely to be of intense interest.

Cytokines

Effective induction of cellular antitumor immunity also relies on immune-modulating and growth promoting cytokines. Tumor-specific T cells isolated from whole blood or tumor biopsy samples are often anergized or tolerized, and possess poor proliferative capabilities. Addition of exogenous cytokines to overcome this deficiency may be counter-productive if inhibitory T regulatory cells (Tregs) present in the culture are preferentially expanded. For example, IL-2, a cytokine typically used for T cell expansion, expands both antigen-specific T cells and Tregs. Alternative cytokines are required to selectively expand tumor-specific effector T cells without promoting the expansion of Tregs. IL-15, which like IL-2 is a cytokine that signals through a common gamma chain expressing receptor, overcomes T cell tolerance of tumor-specific CTL without promoting Treg growth [54]; on the other hand, IL-7 improves the survival of naive, memory and activated tumor-specific T cells [55]. In combination with IL-12, which is required for T cell polarization to the Th1/Tc1 differentiation pathway, cytokines such as IL-7 and IL-15 act in an additive or synergistic manner to enhance IFN-γ production, proliferation, and cytotoxic function of antigen-specific T cells [31, 33, 54–56]. IL-6 may also benefit TAA CTL generation due to its ability to skew naive CD4+ T cells to a Th17 phenotype while preventing Treg formation [57]. Similar characteristics have been attributed to the IL-2 family member cytokine, IL-21, which is involved in the differentiation of naïve T cells into Th17 cells [32, 58] Tumor antigen-specific Th17 cells have been shown to control the growth of established B16 tumors in a mouse model. Martin-Orozco *et al.* demonstrated that this antitumor activity was due to enhanced CCL20 chemokine production by tumor tissues, recruitment of dendritic cell into tumor sites, and activation of tumor-specific CD8+ T cells [59, 60]. It should be noted, however, that combining individually effective cytokines may simply produce antagonistic or even paradoxical effects; as such, the combinations used, and the sequence of their introduction, needs careful analysis for each type of tumor specific T cell culture.

Genetic Modification of T Cells

Redirecting T-cell Specificity (Genetic Modification)

Because most TAA are either "self" antigens or "naïve" targets for the immune system, the isolation and expansion of tumor-reactive T cells from cancer patients and healthy donors has proven problematic. To circumvent this practical limitation of tumor immunotherapy, investigators have genetically modified T cells to render them capable of recognizing TAAs. The two most common approaches are: (1) gene modification with TCR variable α and β chains cloned from high-affinity TAA-specific T cell clones; and (2) expression of chimeric antigen receptors (CARs) that typically recognize tumors through single-chain variable fragments (scFv) isolated from TAA specific antibodies (Fig. 1.1).

TCR Gene Transfer

T cells can be genetically modified to express transgenic α and β chains of the rare tumor-reactive TCR that can be obtained from T cells isolated from patients. Transfer of the tumor-specific TCR to autologous, mitogen-activated T cells allows the rapid production of large numbers of tumor peptide-specific T cells. *In vitro* characterization of the re-directed T cells has shown that the cells acquire the antigenic specificity of the parent T cell clone, and can respond with IFN-γ production and specific tumor cell lysis when exposed to their cognate antigen. Such redirected T cells targeting the leukemia antigens WT1 or the melanoma antigen gp100 eliminate tumors in murine tumor models [61, 62]. To date, this promising approach has been applied to the generation of T cells directed against melanoma antigens, minor histocompatibility antigens, and common oncoproteins [63].

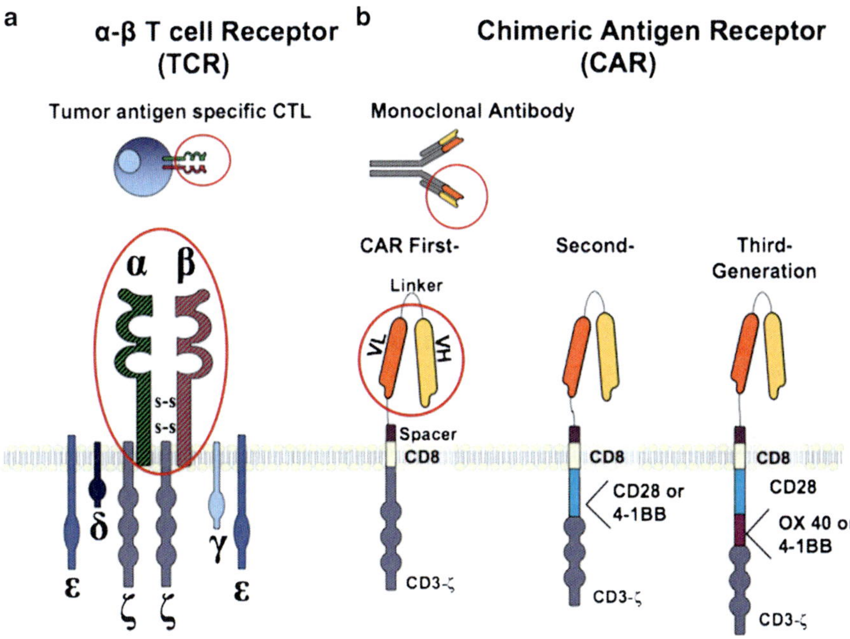

Fig. 1.1 Structure of transgenic T cell receptors (*TCR*) and chimeric antigen receptors (*CAR*). (**a**) Transgenic TCR are amplified from tumor antigen-specific T cell lines. Inclusion of an additional disulfide bond in the transgenic TCR reduces risk of misspairing with endogenous TCR. (**b**) Schematic illustrates first, second and third generation CAR receptors. The ectodomain of the CAR receptors are derived from monoclonal tumor antigen-specific antibodies. Second and third generation CAR include one (second) or more (third) costimulatory endodomains, which enhances Th1 cytokine production (IL-2, TNF-α, IFN-γ), proliferation, and survival by upregulation of anti-apoptotic molecules

However, current strategies for TCR gene transfer possess an inherent biology that may limit extensive clinical application. Specifically, transferred α and β chains can cross-pair with endogenous TCR chains, thereby forming hybrid TCRs with either no activity or with new and unwanted autoimmune reactivity [63]. The frequency of this cross-pairing problem can be reduced by modification of trans-membrane-association domains through the introduction of additional cysteines, which form additional disulfide bonds that minimize dimerization with endogenous α and β chains [64]. Also, the use of γδ-T cells as a platform for αβ transgenic TCR transduction may prevent this problem [65].

Importantly, clinical trials evaluating the adoptive transfer of TCR-transgenic T cells are currently being implemented; Table 1.3 summarizes completed and ongoing trials. The first human clinical trial using TCR-transgenic T cells was reported by the Rosenberg group, who used re-directed T cells to treat metastatic melanoma [78]. In that study, T cells were genetically modified with a TCR recognizing the melanoma antigen MART-1, and infused into 17 patients after nonmyeloablative host conditioning

Table 1.3 Completed/ongoing clinical trials of using chimeric antigen receptors (CARs)

First generation CAR		
Tumor entity	Antigen	Receptor type/T cell source
Ovarian cancer [3, 6]	α-Folate receptor	ScFv-Fcε Rly activated T cells
Renal cell carcinoma [37, 66]	CAIX	ScFv-CD4-Fcε Rly activated T cells
Refractory follicular lymphoma [67]	CD19	ScFv-CD3ζ activated T cells
Mantle cell lymphoma [68]	CD20	ScFv-CD3ζ activated T cells
Neuroblastoma [69]	GD2	ScFv-CD3ζ activated T cells; EBV CTL
Neuroblastoma [70, 71]	L1 cell adhesion molecule	ScFv-CD3ζ activated T cells
Second generation CAR		
Tumor entity	Antigen	Receptor type
CLL [28, 72]	CD19	ScFv-CD28-CD3ζ activated T cells
B-Cell NHL and chronic lymphocytic leukemia [73, 74][a]	CD19	ScFv-CD28-CD3ζ ScFv-CD3ζ activated T cells
B-Cell NHL and chronic lymphocytic leukemia [73, 74][a]	CD19	ScFv-CD28-CD3ζ ScFv-CD3ζ activated T cells; EBV CTL
Lung malignancies [75–77][a]	Her2 TGF-β	EBV CTL
Her-2 expressing metastatic	Her2	Activated T cells

[a]Ongoing studies, clinical results not yet published

and exogenous administration of 2–12 doses of IL-2 (720,000 international units/kg) every 8 h. The transgenic cells persisted long-term *in vivo* (from 2 months to 1 year) and objective regression of metastatic lesions was observed in two patients (13%) [78]. As such, this study demonstrated the safety and feasibility of this immunotherapeutic approach. However, the clinical effectiveness was clearly reduced relative to this group's previous study using adoptively transferred TILs [22, 24, 25].

To improve the clinical efficacy of this form of immuno-gene therapy, the same investigators substituted high-affinity melanoma-specific TCRs for the lower-affinity receptors that were used in the original study. This study, which was published in 2009, found that adoptive transfer of such high-affinity TCR-expressing T cells to 36 patients with metastatic and refractory melanoma produced objective cancer regressions in 30% of recipients of MART-1 specific T cells and 19% of recipients of gp-100 specific T cells. Also, unlike the prior clinical trial, several patients developed autoimmune symptoms in the skin, eyes and ears [79]. These autoimmune toxicities were not correlated with antitumor responses, thereby raising the question of whether the effects represented collateral damage (T cell targeting of normal antigen-expressing tissue) or "off-target" toxicity; such off-target toxicity may result either from cross reactivity with another self antigen or from cross-pairing of native and transgenic α and β chains to form new, self-reactive specificities [79]. Hence, this study demonstrates the importance of selecting high affinity TCRs to induce effective clinical responses and also highlights a potential safety concern in terms of autoimmune toxicity.

One major logistical limitation of expressing transgenic TCR to target cancers is that each TCR usually only binds one peptide in association with one HLA Class I/II polymorphism. This has two undesirable consequences. First, this biology increases the risk of tumor escape by mutation or downregulation of the targeted epitope; and second, a multiplicity of TCRs must be made and validated to offer the possibility of matching at least one of the patient's HLA polymorphisms.

Genetic Modification with Chimeric Antigen Receptors

T cell specificity can also be altered by expressing CARs. These CARs are composed of two regions. First, the extracellular domain (ectodomain) is responsible for antigen recognition and usually contains a scFv that incorporates the heavy and light variable chains (V_H and V_L, respectively) of a monoclonal antibody joined by a flexible linker. Second, the intracellular signaling domain (endodomain) is linked to the scFv; in first generation CARs, this domain consisted of either the TCRζ chain (CD3-ζ) or the IgE high-affinity receptor (FcεRIγ) motifs (Fig. 1.1). CAR expression allows tumors to be targeted in an HLA-unrestricted manner, thereby increasing patient eligibility. Furthermore, whereas endogenous T cell receptors bind only short peptides derived from protein antigens, CARs extend the range of antigens that can be recognized to include carbohydrates and glycolipids.

Preclinical studies have demonstrated that T cells expressing CARs can eliminate tumors in murine models. However, clinical trials with "first-generation" CAR receptors, containing only the CD3-ζ signaling domain, were disappointing (see next section). Following engagement of tumor antigens, these endodomains did not provide sufficient co-stimulatory signaling to induce T cell activation, proliferation, and cytokine production. In studies involving second-generation CARs, additional intracellular endodomains were added to enhance their *in vivo* function. These endodomains are derived from costimulatory molecules such as CD28, and their addition to the CD3-ζ chain enables cytokine production in the absence of costimulation, thereby promoting the proliferation of transduced T cells [24, 73, 80, 81]. Inclusion of endodomains from other costimulatory molecules such as 4–1BB, OX40, and ICOS (inducible T cell costimulator) further enhances cytokine production and effector function of CAR modified T cells *in vitro* [82, 83]. A schematic of first, second, and third generation CARs is shown in Fig. 1.1 An alternative approach uses viral-antigen-specific T cells that are modified to express CARs; such T cells receive co-stimulation when they encounter the viral antigens presented on professional APCs.

Clinical Studies

Clinical studies using first-generation CARs identified two limitations. First, most of the used receptors were derived from murine antibodies that were likely immunogenic, thereby limiting the *in vivo* persistence of the transgene-expressing T cells. For example, Kershaw *et al.* performed adoptive transfer therapy of autologous T cells modified with a CAR directed against the ovarian-associated α-folate receptor (α-FR); in this study, an antibody response against the chimeric receptor was identified [84]. Similar results were seen by Lamar and colleagues, who adoptively transferred T cells that expressed a CAR targeting carbonic anhydrase (for therapy of renal cell carcinoma): all three treated patients developed an anti-scFv antibody response directed against the murine portion of the CAR. The immunogenicity of recombinant receptors can be reduced by using human antibody fragments as recognition domains. Currently available murine hybridoma antibodies can be humanized by replacing murine framework regions; in addition, fully human recombinant single-chain antibodies can be generated by phage display technology [85–87]. A second limitation identified in the first-generation CAR clinical trial performed by Lamar and colleagues was the development of cholestasis in all patients after adoptive T cell transfer; this toxicity prompted the closure of further accrual to the clinical trial. Importantly, because carbonic anhydrase is expressed on biliary epithelium, this adverse event of cholestasis was considered an "on-target" toxicity [37, 66]. In sum, these findings illustrate both the potential importance and efficacy of using humanized CARs *in vivo* and the potential dangers of targeting antigens expressed on normal tissue.

First reports of patients treated with second-generation CARs that target CD19, which is expressed on B cell leukemia and lymphoma cells, showed moderate clinical effects in nonlymphodepleted patients. However, Sadelain *et al.* reported the development of a fatal cytokine storm that was associated with renal and respiratory failure in a patient who received lymphodepleting chemotherapy followed by a single dose of T cells (3×10^7 cells/kg) modified with a second generation CAR [28]. Similarly, in a separate clinical trial performed by Rosenberg *et al.*, a fatality was observed when a patient first received lymphodepleting chemotherapy followed by the adoptive transfer of T cells expressing HER-2 specific CAR that incorporated CD28 and 4-1BB endodomains [29]. Further investigation will be needed to delineate whether these adverse events were related to CAR-mediated tumor lysis or to excessive activation of the transferred T cells. If the former is true, then perhaps dose de-escalation and infusion of smaller cell numbers will be sufficient to achieve antitumor activity without toxicity. However, if the latter is true, then perhaps future investigations may need to revert to the use of second or first generation CARs.

Genetic Modification of T Cells to Improve *in vivo* Proliferation and Survival

T-cell Persistence and Survival in vivo

Once tumor-specific T cells are adoptively transferred, they must expand *in vivo* and persist for a sufficient interval to ensure clearance of all clonogenic tumor cells. However, both preclinical and clinical studies have shown that *ex vivo* manipulated T cells have limited *in vivo* proliferation and persistence. T cell proliferation *in vivo* requires continued antigenic stimulation, either in response to tumor cells or to professional APCs cross-presenting tumor antigens. Following *in vivo* expansion, a proportion of the tumor-specific T cells should ideally enter the memory T cell compartment to provide long-term protection; such memory T cells must retain the ability to reactivate and proliferate on antigenic challenge. To circumvent the host immune response, tumors can possess potent and multi-faceted mechanisms to inhibit effective antigen presentation, T cell proliferation, and T cell entry into the memory compartment [88]. In response, investigators have developed several countermeasures designed to protect against such tumor evasion. These therapeutic strategies include; (a) the infusion of T cells with memory-type characteristics, (b) genetic modification of T cells to improve survival, and (c) genetic modification of infused cells to withstand the tumor microenvironment.

T-cell Sources for Genetic Modification

Several studies have attempted to characterize the optimal T cell population to ensure long-term T cell survival and immunity after adoptive transfer. Using a nonhuman primate model of CMV infection, Berger and colleagues isolated and infused effector memory ($CD62L^-CD28^-CD8^+Fas^{hi}$) and central memory ($CD62L^+CD28^+CD8^+Fas^{hi}$) CMV-specific T cell clones; the authors found that the antigen-specific central memory-derived T cells survived longer *in vivo* than antigen-specific effector memory T cells [89]. Additional studies have evaluated T cell telomere length and expression of molecules preferentially expressed on T central memory cells such as costimulatory molecules (CD27, CD28) and homing receptors (CCR7); such studies also support the hypothesis that central memory type T cells persist for longer periods of time relative to effector memory T cells [89–92]. It is therefore possible that interventions such as selection and infusion of cells on the basis of central memory phenotype may enhance the *in vivo* life-span of adoptively-transferred cells and thus improve clinical efficacy.

Pule and colleagues took a different and novel approach to this question, as they compared the longevity of OKT3-activated T cell blasts and polyclonal EBV-specific CTLs; each of these cell populations were modified with a CAR targeting the GD2 antigen, which is expressed on neuroblastoma. Importantly, the EBV-specific CTLs survived substantially longer than their OKT3-activated counterparts, thereby demonstrating that infusion of genetically-modified polyclonal virus-specific CTLs (which are derived from both central memory and effector memory T cells) may overcome the need for time-consuming and expensive up-front selection of T cells with a central memory phenotype [69].

Gene Modification to Enhance T-Cell Proliferation

T cell growth and survival factors, such as IL-2, IL-15 and IL-7, are also crucial for *in vivo* persistence and survival of adoptively-transferred T cells. As such, it is possible that genetic modification of T cells to overexpress these growth promoting cytokines might induce autocrine proliferation *in vivo*. Using an oncoretroviral vector, Liu and colleagues overexpressed IL-2 in T cells, and found enhanced T cell proliferation without reduction in tumor killing *in vitro* [93]. Alternatively, other groups have genetically modified T cells to secrete IL-15, which may be advantageous relative to IL-2 because of its reduced capacity to promote the expansion of Tregs. For example, in a comparative study, Quintarelli and colleagues genetically modified EBV-specific T cells with oncoretroviral vectors encoding either IL-15 or IL-2; both cytokines promoted *ex vivo* and *in vivo* T cell expansion without affecting CTL antigen specificity or effector function [94]. The improved survival and expansion of IL-15 modified T cells was associated with increased antitumor activity in an *in vivo* murine model, and was not associated with the promotion of Treg cells.

T cell growth and survival can also be increased by engineering T cells to respond to cytokines already present in the tumor environment. For example, Vera and colleagues have shown that transgenic expression of the IL-7 receptor by antigen-specific CTLs restores their responsiveness to IL-7 and sustains their expansion *in vitro* and *in vivo* without affecting their antigen specificity or dependency [95]. Finally, overexpression of the human telomerase reverse transcriptase (hTERT) gene by oncoretroviral transduction has been investigated. Although this approach greatly increases the number of population doublings of transduced T cells by preventing telomere erosion, it has also been associated with genomic instability, which may limit its safety and hence clinical application [96–98].

Manipulating the Infused T Cells to Counteract Tumor Evasion Strategies

Genetic modification of T cells can also be used to counteract the inhibitory tumor microenvironment. Tumor evasion strategies include the secretion of TGF-β, which is a multifunctional cytokine that promotes tumor growth through angiogenesis; TGF-β also limits T cell proliferation and effector function, and induces T cell tolerance. Additionally, TGF-β enhances the induction and expansion of Tregs [99, 100]. Thus, it is possible that prevention of TGF-β modulation of adoptively transferred T cells may improve T cell efficacy. We have shown that mature, antigen-specific effector T cells modified to express a dominant-negative TGF-β receptor type II (dnTGF-RII) are resistant to the antiproliferative effects of TGF-β *in vitro*, thereby prolonging their *in vivo* persistence and enhancing tumor elimination in mice bearing TGF-β-expressing tumors [75]. Furthermore, the gene-modified T cells persisted only as long as the mice were vaccinated with antigen; as such, the gene-modified T cells safely maintained their antigen dependence. In a long-term safety study, no lymphoproliferation or autoimmunity was observed after transfer of dnTGF-RII-modified, antigen-specific murine splenocytes into immune competent mice [75]. Bollard and colleagues are currently assessing the *in vivo* safety and efficacy of dnTGF-RII-modified T cells in a Phase I clinical trial in patients with EBV positive HD.

Indoleamine 2, 3-dioxygenase (IDO), which is secreted by tumor cells or by anergized APCs, also induces T cell inhibition or apoptosis *in vivo*. IDO mediates its T cell immunosuppressive effects both directly by depleting the amino acid tryptophan and indirectly by increasing the levels of metabolic waste products, such as kynurenine. In an attempt to overcome this immunosuppression, investigators have used IDO inhibitors, such as MT-1 or prostaglandins; however, these efforts have not yielded remarkable benefit [101]. An alternative approach may be to modify T cells for the purpose of downregulating GCN2, which is a major component of IDO-mediated suppression in T cells. In preliminary studies, Munn *et al.* showed that GCN2-knockout cells were refractory to IDO-induced anergy; however, this approach has not yet been tested in T cells exposed to the tumor environment [102]. Still yet another approach is to antagonize enforced T cell apoptosis that occurs in the tumor environment; toward this aim, investigators have transduced T cells with antiapoptotic genes, including Bcl-2 and Bcl-xL, which increases T cell resistance to death and IL-2 cytokine withdrawal [103, 104]. Dotti *et al.* took the opposite approach to preventing apoptosis, which was to downregulate pro-apoptotic genes in T cells; T cells were transduced with an oncoretroviral vector encoding a small interfering RNA (siRNA) targeting Fas, thus making the T cells resistant to Fas/FasL-mediated apoptosis [105].

Genetic Modification of T Cells to Improve Safety

Suicide Genes

Genetic strategies that enhance the life span of T cells, interferes with their homeostasis, or changes their antigen specificity to recognize "self" antigens, carry a safety risk due to excessive lymphoproliferation or toxicity to normal organs. Furthermore, integrating viral vectors used for gene modification have the risk of insertional mutagenesis; this risk could be obviated by the rapid and complete elimination of infused cells [106, 107]. Therefore, careful risk evaluation may dictate the need to integrate a safety switch or suicide gene into T cell therapy approaches. The best characterized approach utilizes the thymidine kinase gene from herpes simplex virus I (HSV-tk). TK phosphorylates the relatively nontoxic prodrug ganciclovir (GCV), which then becomes phosphorylated by endogenous kinases to GCV-triphosphate; then, chain termination and single-strand breaks occur on drug incorporation into DNA, thereby leading to T cell death. Several phase I–II studies, and a more recent phase III study, have shown that ganciclovir administration can be used to largely eliminate transferred TK-modified cells *in vivo* [108–111]. Unfortunately, the TK gene product can be immunogenic; indeed, specific immune responses directed against this transgenic protein have been detected in recipients of transgene-expressing T cells. Such an immune response leads to the premature and unintentional elimination of infused T cells, thereby compromising the persistence and hence efficacy of the transferred T cells [112]. To overcome this problem, Sato *et al.* developed a "nonimmunogenic" suicide system based on the use of a humanized variant of the thymidylate kinase (tmpk) gene. Tmpk converts the prodrug AZT into the toxic form AZT-triphosphate, which induces potent cell cytotoxicity. In pre-clinical studies these authors demonstrated that application of AZT induced apoptosis of both dividing and nondividing cell lines and primary human and mouse cells that had been transduced with Tmpk using a lentiviral-vector [113].

Transgenic human CD20, which can be activated by a monoclonal chimeric anti-CD20 antibody, has been proposed as a nonimmunogenic safety system; however, this approach would result in the unwanted loss of normal B cells for 6 months or more [114]. A further alternative suicide gene strategy utilizes human pro-apoptotic molecules fused with an FKBP variant; this FKBP variant is optimized to bind a chemical inducer of dimerization (CID), AP1903, which is a synthetic drug that has proven safe in healthy volunteers [115]. Administration of this small molecule results in cross-linking and activation of the proapoptotic target molecules. This inducible system has been explored in human T cells using Fas or the death effector domain (DED) of the Fas-associated death domain–containing protein (FADD) as proapoptotic molecules. Up to 90% of T cells transduced with these inducible death molecules underwent apoptosis after administration of CID [116, 117]. Although these experimental results are promising, it is possible that elimination of only 90% of the transduced cells may be insufficient to ensure safety of genetically modified cells *in vivo*. Moreover, death molecules that act upstream of most apoptosis inhibitors may be ineffective for apoptosis induction in other cell types. As a step toward overcoming this potential obstacle, Straathof and colleagues modified a late-stage apoptosis pathway molecule, caspase 9, and showed that this suicide gene could be stably expressed in human T cells without compromising their functional and phenotypic characteristics. The T cells demonstrated acute sensitivity to a chemical inducer of dimerization, which caused apoptosis in 99% of transduced cells [94, 118].

Targeted Integration

Reports of leukemia development caused by insertional mutagenesis using retroviral vectors for stem cell modification have spurred development of safer techniques for modifying cell DNA sequences; such alternatives seek to either pre-ordain the integration site of the transgene or use modified vectors that possess a safer profile [106, 107]. Zinc-finger nucleases, which are artificial restriction enzymes generated by fusing a zinc-finger DNA-binding domain to a DNA-cleavage domain, are currently being evaluated. These nucleases cleave specific genomic sequences, thereby allowing safe insertion of the gene of interest [119]. Bonini and colleagues recently showed the targeted integration and stable expression of a GFP gene into the CCR5 gene region in central memory T lymphocytes [120]. Initial clinical trials to evaluate the efficacy and specificity of using zinc-finger nucleases for gene integration are being planned. Researchers are also evaluating lentiviral vectors (LVs) as a potentially safer alternative for genetic modification. LVs have been shown to be less prone to integrate near transcription start sites and therefore may reduce the risk of oncogenesis [120, 121].

Counteracting the Tumor Microenvironment

Apart from strategies to modify T cells, a number of groups have manipulated the tumor microenvironment to make it less suppressive of effector T cell function.

Nonspecific Lymphodepletion

Early trials using TILs in melanoma patients showed that a combination of lymphodepletion followed by TIL infusion increased the objective response rate from 31% (using TILs alone) to between 49% and 72%; the higher response rates were observed with more intensive conditioning, which included both fludarabine plus cyclophosphamide chemotherapy and total body irradiation [26]. The investigators attributed the enhanced efficacy to two factors: (1) the elimination of $CD4^{+}$ $CD25^{+}$ Tregs and other cells with suppressor function *in vivo*; and (2) the increased availability of T cell growth-promoting cytokines, such as IL-15 and IL-7. Although the clinical results were impressive, the conditioning regimens were associated with significant toxicity.

We have investigated a less toxic lymphodepletion strategy that uses leucocyte-depleting antibodies to induce lymphopenia prior to T cell infusion [122]. An ideal depleting antibody should spare stem cells, allow myeloid cells to recover rapidly, and have a short half-life, thereby allowing the immediate infusion of *ex vivo*-expanded T cells; of course, monoclonal antibodies used for such interventions must also be clinically available. A pair of rat lympholytic monoclonal antibodies directed against the human CD45 molecule fulfills these requirements. Although CD45 is ubiquitously expressed by hematopoietic cells, expression is highest on T cells; clinical studies have shown that T cells are depleted by the antibody, whereas hemopoietic stem cells are spared [122]. Anti-CD45 has recently been used as a lymphodepleting agent in NPC patients who subsequently received EBV-CTL. This study showed that anti-CD45 was safe *in vivo* and produced transient lympohodepletion. The adoptively transferred EBV-CTL expanded significantly more and had increased tumor-antigen responsiveness in those patients who received anti-CD45 relative to recipients of T cells alone [122].

Specific Treg Depletion

A number of research groups have specifically depleted regulatory T cells by targeting Treg-associated molecules such as the glucocorticoid-induced TNF receptor family molecule (GITR), CD25, and CTLA-4 [123–125]. Preliminary mouse experiments indicated that a single administration of agonistic anti-GITR monoclonal antibody to tumor-bearing mice provoked potent tumor-specific immunity and eradicated established tumors; these studies suggested that interventions that target GITR-expressing cells may enhance T cell therapy strategies [124]. Alternatively, Attia and colleagues used a fusion protein of IL-2 and diphtheria toxin (Denileukin Diftitox, DAB389IL-2,

ONTAK) to selectively eliminate Tregs, which express high-affinity IL-2 receptors [123]. Thirteen patients were treated; however, no objective clinical responses were observed, and the drug showed a limited capacity to deplete Tregs. The same group treated 14 patients with metastatic melanoma by administration of a fully human anti-CTLA-4 antibody (MDX-010) in conjunction with vaccination using two gp100 melanoma-associated $CD8^+$ peptides. CTLA-4 is not found on the surface of most resting T cells but is transiently upregulated after T cell activation; in contrast, naturally occurring immunosuppressive regulatory T cells constitutively express surface CTLA-4. Blockade of CTLA-4 induced grade III/IV autoimmune manifestations in six patients (43%), but also mediated objective cancer regression in three patients (21%; two complete responses and one partial response) [125]. A later report from Ribas and colleagues used a CTLA-4-blocking antibody as therapy in 39 patients with solid malignancies; complete or partial responses were observed in 10% of patients and stable disease was observed in a further 23% of patients [126]. However, this study and others have determined that prolonged administration or high doses of anti-CTLA-4 can result in grade III/IV autoimmune toxicity. These approaches to anti-CTLA therapy do not appear to increase objective tumor response rates; as such, low-dose anti-CTLA-4 (3 mg/kg) may more safely break tolerance to human tumors [126–128].

Scale-Up of Tumor CTL Therapy

Despite increasing numbers of reports highlighting the promising clinical results of tumor-specific T cell immunotherapy trials, broader implementation of this therapy is limited by the cost and complexity of CTL manufacture. The cells must be cultured for weeks or months to produce sufficient cell numbers for infusion (up to 1×10^9 cells/m^2); such manufacturing requires highly specialized facilities, infrastructure, and cell culture technologists. In addition, viral gene transfer vectors are expensive to produce; furthermore, because of the prolonged and intensive patient monitoring required, clinical trial evaluation of these cell therapies is expensive. Broader implementation of T cell therapies is also limited by the fact that products must be generated on an individual patient basis; unlike pharmaceuticals, current T cell therapies cannot be used for immediate "off-the-shelf" use. Hence, we need radical solutions not just to reduce the cost, complexity and time of CTL manufacture, but also to address their lack of immediate availability.

Simplify Large-Scale CTL Production

Unfortunately, scale-up is hindered by the cumbersome, labor-intensive, and inefficient methods currently used to grow T cells. To date, most groups have expanded T cells for clinical use in a variety of culture vessels including 2 cm^2 wells in 24-well plates, tissue culture-treated flasks, or gas-permeable tissue culture bags. Many of these vessels are not suitable for routine production of large cell numbers. Furthermore, in standard static culture vessels, the volume of medium used for cell culture is restricted by gas diffusion. For example, in the 2 cm^2 wells of 24-well plates, the volume of media is restricted to 1 mL/cm^2, which represents a volume that in turn limits the supply of nutrients that are rapidly consumed by proliferating T cells. Acidic pH and waste build-up further impedes cell growth and survival; as such, the maximum cell density that can typically be achieved is about 2 x $10^6/cm^2$. Consequently, a skilled GMP technician must frequently manipulate cultures to replenish media and growth factors in order to sustain the expansion of large T cell numbers.

To overcome these limitations, a number of closed-system bioreactors have been developed to improve cell output with minimal cell handling. Mechanical rocking, stirring, or a pump system can be used to increase the availability of O_2 in the culture; in addition, media and nutrients can be exchanged by perfusion. Examples of such bioreactors include stirred tank bioreactors and static hollow fiber bioreactors. Stirred bioreactors allow excellent gas exchange and can readily be scaled-up. Unfortunately, shear stress associated with the stirring rate reduces cell viability, and cultures require frequent medium sampling to evaluate growth-limiting factors such as glucose and waste metabolites [129, 130]. In contrast, constant medium perfusion in hollow fiber bioreactors results in the dilution of metabolites without shear stress; however, cell sampling during the culture is not possible, thereby making it difficult to assess T cell status [131, 132]. High cell densities can also be achieved in culture bags on rocking platforms; for example, the Wave Bioreactor has been used by Jensen and colleagues for therapeutic T cell production [133]. Although all of the above-mentioned methods are GMP applicable and can produce large numbers of cells, their disadvantages are the cost of purchase of specialized equipment for media and oxygen perfusion and nutrient control, as well as the complexity of running and maintaining the equipment [133]. Moreover, although genetically engineered T cells and TILs can be cultured in these bioreactors, such equipment has proven inefficient for tumor antigen-specific CTL production; that is, the generation of CTL has a strict requirement for T cell interaction with antigen-presenting-cells and feeder cells that cannot be disrupted by mechanical agitation.

Alternatively, Vera and co-workers have described a simple, fast and low cost static culture system which allows expansion of both primary T cells as well as antigen-specific CTLs to large number at high cell densities; this approach thereby dramatically lowers the cost and complexity of T cell production. This novel Gas-permeable Rapid Expansion cultureware (G-Rex) is essentially a plastic-based cultureware with a semi-permeable silicon membrane at its base to allow O_2–CO_2 exchange. Gas exchange from below permits an increased depth of medium to be used above, providing more nutrients and diluting waste products [134]. Expanded use of this simple cultureware could certainly increase the feasibility of tumor-specific T cell therapies.

"Off the Shelf" CTLs Cells

Another major barrier still to be overcome is the delay and complexity of producing individual autologous T cell products for infusion. One approach that may overcome this obstacle is the use of banked samples of partially HLA-matched

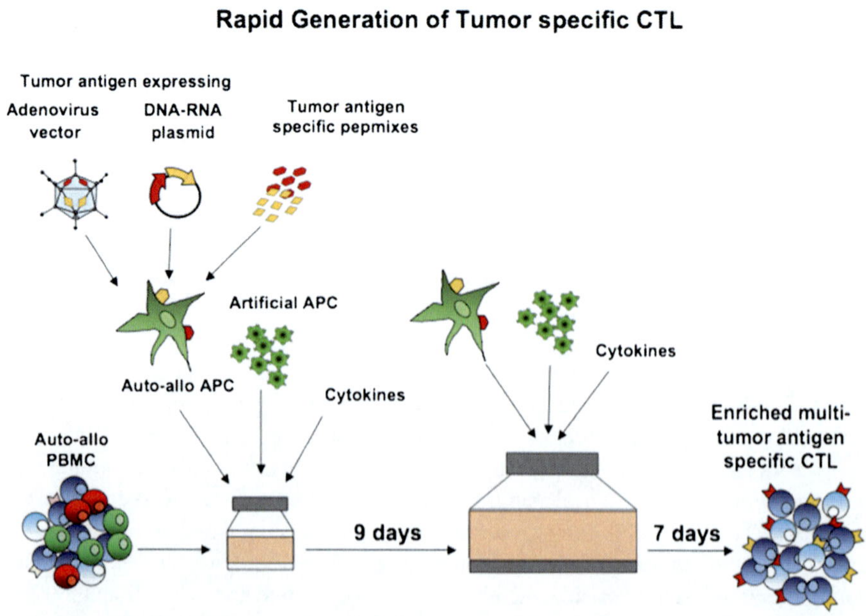

Fig. 1.2 A schematic view of the process for rapid tumor-specific cytotoxic T lymphocytes (CTL) manufacture developed by our group. APCs are modified by adenoviral transduction, plasmid transfection, or pepmix loading to express tumor antigens for stimulation of autologous or allogeneic HLA partially-matched (auto/allo) T cells. Artificial APCs are used to enhance T cell stimulation and to provide a feeder layer. Combinations of cytokines such as IL-12, IL-7, IL-21, IL-28, IL-6, IL-15, or IL-2 are added to overcome T cell anergy and inhibit proliferation of Tregs. After 9 days of culture, tumor-specific antigens are restimulated using antigen presenting autologous or allogeneic HLA partially-matched (auto/allo) APCs, of T cell donor origin, and artificial APCs as well as cytokines. T cells are expanded using novel gas permeable cell cultureware, called a G-Rex, which ensures maximal expansion of tumor-specific T cells. Seven days after the second stimulation, and following QA/QC testing, specific CTLs can be infused into patients

virus-specific CTL; this method has proven effective as therapy of EBV-LPD that developed after solid organ transplant [16, 17]. Such an approach may be adapted for therapy of other tumors; because of the lack of HLA-matching, such T cells would predictably be susceptible to a graft rejection response and therefore may have a short persistence *in vivo* that may necessitate multiple T cell infusions to achieve a therapeutic response. Figure 1.2 shows a schematic of a protocol to rapidly, cost-effectively, and simply produce tumor-specific T cells.

Cost Effectiveness of Adoptive T-Cell Therapy versus Conventional Therapies

Cancer is the second-most common cause of death in the USA. The National Institutes of Health estimates the overall economic cost of cancer at $228.1 billion in 2008: $93.2 billion for direct medical costs (total of all health expenditures);

$18.8 billion for indirect morbidity costs (cost of lost productivity due to illness); and $116.1 billion for indirect mortality costs (cost of lost productivity due to premature death). One of the major contributors to the cost is the actual cancer treatment (www.cancer.org). As an example, a conventional treatment course for stage III–IV melanoma costs about $9,756 for the initial 6 months of treatment following diagnosis, and escalates thereafter [135]. Treatment of malignancies which require an autologous or allogenic stem cell transplant can easily cost more than $150,000 [136]. A major component of these expenses includes the costs associated with treating the multiple side effects related to the use of aggressive, nonspecific therapies. Therefore, alternative treatment options that lower treatment costs are desperately required.

T cell immunotherapy may represent a more target-specific and less toxic therapy for cancer treatment, thereby potentially greatly increasing its cost effectiveness, which is the most important measure of a treatment's value. The cost of treating CMV infection by conventional therapies versus T cell immunotherapy has been compared: virus specific T cell therapy not only reduced treatment costs from $15,000 to $10,559 but also lowered morbidity. New cell growth devices such as bioreactors, use of growth promoting cytokines, and rapid generation protocols as discussed above, will further reduce the cost of CTL generation. Hence, tumor-specific T cell therapy when used in combination with conventional cancer therapies should substantially reduce direct medical costs and indirect morbidity costs, thereby relieving individual, societal, and community expenditure.

References

1. Leen AM, Heslop HE (2008) Cytotoxic T lymphocytes as immune-therapy in haematological practice. Br J Haematol 143:169–179
2. Walter EA, Greenberg PD, Gilbert MJ et al (1995) Reconstitution of cellular immunity against cytomegalovirus in recipients of allogeneic bone marrow by transfer of T-cell clones from the donor. N Engl J Med 333:1038–1044
3. Einsele H, Roosnek E, Rufer N et al (2002) Infusion of cytomegalovirus (CMV)-specific T cells for the treatment of CMV infection not responding to antiviral chemotherapy. Blood 99:3916–3922
4. Cohen JI, Bollard CM, Khanna R et al (2008) Current understanding of the role of Epstein-Barr virus in lymphomagenesis and therapeutic approaches to EBV-associated lymphomas. Leuk Lymphoma 49 Suppl 1:27–34
5. Rooney CM, Smith CA, Ng C et al (1995) Use of gene-modified virus-specific T lymphocytes to control Epstein-Barr virus-related lymphoproliferation. Lancet 345:9–13
6. Heslop HE, Ng CYC, Li C et al (1996) Long-term restoration of immunity against Epstein-Barr virus infection by adoptive transfer of gene-modified virus-specific T lymphocytes. Nat Med 2:551–555
7. Leen AM, Myers GD, Sili U et al (2006) Monoculture-derived T lymphocytes specific for multiple viruses expand and produce clinically relevant effects in immunocompromised individuals. Nat Med 12:1160–1166
8. Peggs KS, Verfuerth S, Pizzey A et al (2003) Adoptive cellular therapy for early cytomegalovirus infection after allogeneic stem-cell transplantation with virus-specific T-cell lines. Lancet 362:1375–1377

9. Cobbold M, Khan N, Pourgheysari B et al (2005) Adoptive transfer of cytomegalovirus-specific CTL to stem cell transplant patients after selection by HLA-peptide tetramers. J Exp Med 202:379–386
10. Micklethwaite K, Hansen A, Foster A et al (2007) *ex vivo* expansion and prophylactic infusion of CMV-pp65 peptide-specific cytotoxic T-lymphocytes following allogeneic hematopoietic stem cell transplantation. Biol Blood Marrow Transplant 13:707–714
11. Micklethwaite KP, Clancy L, Sandher U et al (2008) Prophylactic infusion of cytomegalovirus-specific cytotoxic T lymphocytes stimulated with Ad5f35pp65 gene-modified dendritic cells after allogeneic hemopoietic stem cell transplantation. Blood 112:3974–3981
12. Gottschalk S, Ng CY, Perez M et al (2001) An Epstein-Barr virus deletion mutant associated with fatal lymphoproliferative disease unresponsive to therapy with virus-specific CTLs. Blood 97:835–843
13. Rooney CM, Smith CA, Ng CY et al (1998) Infusion of cytotoxic T cells for the prevention and treatment of Epstein-Barr virus-induced lymphoma in allogeneic transplant recipients. Blood 92:1549–1555
14. Comoli P, Basso S, Zecca M et al (2007) Preemptive therapy of EBV-related lymphoproliferative disease after pediatric haploidentical stem cell transplantation. Am J Transplant 7:1648–1655
15. Gustafsson A, Levitsky V, Zou JZ et al (2000) Epstein-Barr virus (EBV) load in bone marrow transplant recipients at risk to develop posttransplant lymphoproliferative disease: prophylactic infusion of EBV-specific cytotoxic T cells. Blood 95:807–814
16. Haque T, Wilkie GM, Taylor C et al (2002) Treatment of Epstein-Barr-virus-positive post-transplantation lymphoproliferative disease with partly HLA-matched allogeneic cytotoxic T cells. Lancet 360:436–442
17. Haque T, Wilkie GM, Jones MM et al (2007) Allogeneic cytotoxic T-cell therapy for EBV-positive posttransplantation lymphoproliferative disease: results of a phase 2 multicenter clinical trial. Blood 110:1123–1131
18. Feuchtinger T, Matthes-Martin S, Richard C et al (2006) Safe adoptive transfer of virus-specific T-cell immunity for the treatment of systemic adenovirus infection after allogeneic stem cell transplantation. Br J Haematol 134:64–76
19. Bollard CM, Aguilar L, Straathof KC et al (2004) Cytotoxic T lymphocyte therapy for Epstein-Barr virus+Hodgkin's disease. J Exp Med 200:1623–1633
20. Straathof KC, Bollard CM, Popat U et al (2005) Treatment of nasopharyngeal carcinoma with Epstein-Barr virus – specific T lymphocytes. Blood 105:1898–1904
21. Bollard CM, Gottschalk S, Leen AM et al (2007) Complete responses of relapsed lymphoma following genetic modification of tumor-antigen presenting cells and T-lymphocyte transfer. Blood 110:2838–2845
22. Rosenberg SA, Packard BS, Aebersold PM et al (1988) Use of tumor-infiltrating lymphocytes and interleukin-2 in the immunotherapy of patients with metastatic melanoma. A preliminary report. N Engl J Med 319:1676–1680
23. Dudley ME, Wunderlich JR, Robbins PF et al (2002) Cancer regression and autoimmunity in patients after clonal repopulation with antitumor lymphocytes. Science 298:850–854
24. Dudley ME, Wunderlich JR, Yang JC et al (2005) Adoptive cell transfer therapy following non-myeloablative but lymphodepleting chemotherapy for the treatment of patients with refractory metastatic melanoma. J Clin Oncol 23:2346–2357
25. Rosenberg SA, Yannelli JR, Yang JC et al (1994) Treatment of patients with metastatic melanoma with autologous tumor-infiltrating lymphocytes and interleukin 2. J Natl Cancer Inst 86:1159–1166
26. Dudley ME, Yang JC, Sherry R et al (2008) Adoptive cell therapy for patients with metastatic melanoma: evaluation of intensive myeloablative chemoradiation preparative regimens. J Clin Oncol 26:5233–5239
27. Dudley ME, Wunderlich J, Nishimura MI et al (2001) Adoptive transfer of cloned melanoma-reactive T lymphocytes for the treatment of patients with metastatic melanoma. J Immunother 24:363–373

28. Brentjens RJ, Riviere I, Hollyman D et al (2009) Unexpected toxicity of cyclophosphamide followed by adoptively transferred CD19-targeted T cells in a patient with bulky CLL. Mol Ther 17:157–157
29. Rosenberg SA (2009) Gene Transfer Safety Assessment Board Report. National Institutes of Health (NIH) Recombinant DNA Advisory Committee (RAC) Meeting 1 December 2009. http://oba.od.nih.gov/oba/index.html.Accessed 12 January 2010
30. Coulie PG, Lehmann F, Lethe B et al (1995) A mutated intron sequence codes for an antigenic peptide recognized by cytolytic T lymphocytes on a human melanoma. Proc Natl Acad Sci USA 92:7976–7980
31. Sensi M, Anichini A (2006) Unique tumor antigens: evidence for immune control of genome integrity and immunogenic targets for T cell-mediated patient-specific immunotherapy. Clin Cancer Res 12:5023–5032
32. Kaka AS, Shaffer DR, Hartmeier R et al (2009) Genetic modification of T cells with IL-21 enhances antigen presentation and generation of central memory tumor-specific cytotoxic T-lymphocytes. J Immunother 32:726–736
33. Foster AE, Leen AM, Lee T et al (2007) Autologous designer antigen-presenting cells by gene modification of T lymphocyte blasts with IL-7 and IL-12. J Immunother 30:506–516
34. Mandelcorn-Monson RL, Shear NH, Yau E et al (2003) Cytotoxic T lymphocyte reactivity to gp100, MelanA/MART-1, and tyrosinase, in HLA-A2-positive vitiligo patients. J Invest Dermatol 121:550–556
35. Simpson AJ, Caballero OL, Jungbluth A et al (2005) Cancer/testis antigens, gametogenesis and cancer. Nat Rev Cancer 5:615–625
36. Reker S, Becker JC, Svane IM et al (2004) HLA-B35-restricted immune responses against survivin in cancer patients. Int J Cancer 108:937–941
37. Lamers CH, Sleijfer S, Vulto AG et al (2006) Treatment of metastatic renal cell carcinoma with autologous T-lymphocytes genetically retargeted against carbonic anhydrase IX: first clinical experience. J Clin Oncol 24:e20–e22
38. Kessler JH, Melief CJ (2007) Identification of T-cell epitopes for cancer immunotherapy. Leukemia 21:1859–1874
39. Viatte S, Alves PM, Romero P (2006) Reverse immunology approach for the identification of CD8⁺ T-cell-defined antigens: advantages and hurdles. Immunol Cell Biol 84:318–330
40. Quintarelli C, Dotti G, De AB et al (2008) Cytotoxic T lymphocytes directed to the preferentially expressed antigen of melanoma (PRAME) target chronic myeloid leukemia. Blood 112:1876–1885
41. Coughlin CM, Vance BA, Grupp SA et al (2004) RNA-transfected CD40-activated B cells induce functional T-cell responses against viral and tumor antigen targets: implications for pediatric immunotherapy. Blood 103:2046–2054
42. Bonehill A, Tuyaerts S, Van Nuffel AM et al (2008) Enhancing the T-cell stimulatory capacity of human dendritic cells by co-electroporation with CD40L, CD70 and constitutively active TLR4 encoding mRNA. Mol Ther 16:1170–1180
43. Hodge JW, Rad AN, Grosenbach DW et al (2000) Enhanced activation of T cells by dendritic cells engineered to hyperexpress a triad of costimulatory molecules. J Natl Cancer Inst 92:1228–1239
44. Palena C, Zhu M, Schlom J et al (2004) Human B cells that hyperexpress a triad of costimulatory molecules via avipox-vector infection: an alternative source of efficient antigen-presenting cells. Blood 104:192–199
45. Sun S, Cai Z, Langlade-Demoyen P et al (1996) Dual function of *Drosophila* cells as APCs for naive CD8⁺ T cells: implications for tumor immunotherapy. Immunity 4:555–564
46. Latouche JB, Sadelain M (2000) Induction of human cytotoxic T lymphocytes by artificial antigen-presenting cells. Nat Biotechnol 18:405–409
47. Maus MV, Thomas AK, Leonard DG et al (2002) *ex vivo* expansion of polyclonal and antigen-specific cytotoxic T lymphocytes by artificial APCs expressing ligands for the T-cell receptor, CD28 and 4-1BB. Nat Biotechnol 20:143–148
48. Suhoski MM, Golovina TN, Aqui NA et al (2007) Engineering artificial antigen-presenting cells to express a diverse array of co-stimulatory molecules. Mol Ther 15:981–988

49. Kim JV, Latouche JB, Riviere I et al (2004) The ABCs of artificial antigen presentation. Nat Biotechnol 22:403–410
50. Tham EL, Jensen PL, Mescher MF (2001) Activation of antigen-specific T cells by artificial cell constructs having immobilized multimeric peptide-class I complexes and recombinant B7-Fc proteins. J Immunol Methods 249:111–119
51. Shalaby WS, Yeh H, Woo E et al (2004) Absorbable microparticulate cation exchanger for immunotherapeutic delivery. J Biomed Mater Res B Appl Biomater 69:173–182
52. Prakken B, Wauben M, Genini D et al (2000) Artificial antigen-presenting cells as a tool to exploit the immune 'synapse'. Nat Med 6:1406–1410
53. Steenblock ER, Fahmy TM (2008) A comprehensive platform for *ex vivo* T-cell expansion based on biodegradable polymeric artificial antigen-presenting cells. Mol Ther 16:765–772
54. Teague RM, Sather BD, Sacks JA et al (2006) Interleukin-15 rescues tolerant $CD8^+$ T cells for use in adoptive immunotherapy of established tumors. Nat Med 12:335–341
55. Kaneko S, Mastaglio S, Bondanza A et al (2009) IL-7 and IL-15 allow the generation of suicide gene-modified alloreactive self-renewing central memory human T lymphocytes. Blood 113:1006–1015
56. King JW, Thomas S, Corsi F et al (2009) IL15 can reverse the unresponsiveness of Wilms' tumor antigen-specific CTL in patients with prostate cancer. Clin Cancer Res 15:1145–1154
57. Korn T, Mitsdoerffer M, Croxford AL et al (2008) IL-6 controls Th17 immunity *in vivo* by inhibiting the conversion of conventional T cells into Foxp3+ regulatory T cells. Proc Natl Acad Sci USA 105:18460–18465
58. Korn T, Bettelli E, Oukka M et al (2009) IL-17 and Th17 Cells. Annu Rev Immunol 27:485–517
59. Martin-Orozco N, Muranski P, Chung Y et al (2009) T helper 17 cells promote cytotoxic T cell activation in tumor immunity. Immunity 31:787–798
60. Muranski P, Boni A, Antony PA et al (2008) Tumor-specific Th17-polarized cells eradicate large established melanoma. Blood 112:362–373
61. Xue SA, Gao L, Hart D et al (2005) Elimination of human leukemia cells in NOD/SCID mice by WT1-TCR gene-transduced human T cells. Blood 106:3062–3067
62. Abad JD, Wrzensinski C, Overwijk W et al (2008) T-cell receptor gene therapy of established tumors in a murine melanoma model. J Immunother 31:1–6
63. Stauss HJ, Cesco-Gaspere M, Thomas S et al (2007) Monoclonal T-cell receptors: new reagents for cancer therapy. Mol Ther 15:1744–1750
64. Cohen CJ, Li YF, El-Gamil M et al (2007) Enhanced antitumor activity of T cells engineered to express T-cell receptors with a second disulfide bond. Cancer Res 67:3898–3903
65. Hiasa A, Nishikawa H, Hirayama M et al (2009) Rapid alphabeta TCR-mediated responses in gammadelta T cells transduced with cancer-specific TCR genes. Gene Ther 16:620–628
66. Lamers CH, Langeveld SC, Groot-van Ruijven CM et al (2007) Gene-modified T cells for adoptive immunotherapy of renal cell cancer maintain transgene-specific immune functions *in vivo*. Cancer Immunol Immunother 56:1875–1883
67. Jensen MC, Popplewell L, Digiusto DL et al (2007) A firts-in-human clinical trial of adoptive therapy using CD19-specific chimeric antigen receptor re-directed T cells for recurrency/refractory follicular lymphoma. Mol Ther 15:142–142
68. Till BG, Jensen MC, Wang J et al (2008) Adoptive immunotherapy for indolent non-Hodgkin lymphoma and mantle cell lymphoma using genetically modified autologous CD20-specific T cells. Blood 112:2261–2271
69. Pule MA, Savoldo B, Myers GD et al (2008) Virus-specific T cells engineered to coexpress tumor-specific receptors: persistence and antitumor activity in individuals with neuroblastoma. Nat Med 14:1264–1270
70. Gonzalez S, Naranjo A, Serrano LM et al (2004) Genetic engineering of cytolytic T lymphocytes for adoptive T-cell therapy of neuroblastoma. J Gene Med 6:704–711
71. Park JR, Digiusto DL, Slovak M et al (2007) Adoptive transfer of chimeric antigen receptor re-directed cytolytic T lymphocyte clones in patients with neuroblastoma. Mol Ther 15:825–833

72. Brentjens RJ, Hollyman D, Weiss M et al (2009) A phase I trial for the treatment of chemo-refractory chronic lymphocytic leukemia with CD19-targeted autologous T cells. Mol Ther 16:15
73. Loskog A, Giandomenico V, Rossig C et al (2006) Addition of the CD28 signaling domain to chimeric T-cell receptors enhances chimeric T-cell resistance to T regulatory cells. Leukemia 20:1819–1828
74. Rossig C, Pscherer S, Landmeier S et al (2005) Adoptive cellular immunotherapy with CD19-specific T cells. Klin Padiatr 217:351–356
75. Bollard CM, Rossig C, Calonge MJ et al (2002) Adapting a transforming growth factor beta-related tumor protection strategy to enhance antitumor immunity. Blood 99:3179–3187
76. Ahmed N, Ratnayake M, Savoldo B et al (2007) Regression of experimental medulloblastoma following transfer of HER2-specific T cells. Cancer Res 67:5957–5964
77. Ahmed N, Salsman VS, Yvon E et al (2009) Immunotherapy for osteosarcoma: genetic modification of T cells overcomes low levels of tumor antigen expression. Mol Ther 17:1779–1787
78. Morgan RA, Dudley ME, Wunderlich JR et al (2006) Cancer regression in patients after transfer of genetically engineered lymphocytes. Science 314:126–129
79. Johnson LA, Morgan RA, Dudley ME et al (2009) Gene therapy with human and mouse T cell receptors mediates cancer regression and targets normal tissues expressing cognate antigen. Blood 114:535–546
80. Haynes NM, Trapani JA, Teng MW et al (2002) Single-chain antigen recognition receptors that costimulate potent rejection of established experimental tumors. Blood 100:3155–3163
81. Maher J, Brentjens RJ, Gunset G et al (2002) Human T-lymphocyte cytotoxicity and proliferation directed by a single chimeric TCRzeta/CD28 receptor. Nat Biotechnol 20:70–75
82. Pule MA, Straathof KC, Dotti G et al (2005) A chimeric T cell antigen receptor that augments cytokine release and supports clonal expansion of primary human T cells. Mol Ther 12:933–941
83. Wang J, Jensen M, Lin Y et al (2007) Optimizing adoptive polyclonal T cell immunotherapy of lymphomas, using a chimeric T cell receptor possessing CD28 and CD137 costimulatory domains. Hum Gene Ther 18:712–725
84. Kershaw MH, Westwood JA, Parker LL et al (2006) A phase I study on adoptive immunotherapy using gene-modified T cells for ovarian cancer. Clin Cancer Res 12:6106–6115
85. Hombach A, Schneider C, Sent D et al (2000) An entirely humanized CD3 zeta-chain signaling receptor that directs peripheral blood t cells to specific lysis of carcinoembryonic antigen-positive tumor cells. Int J Cancer 88:115–120
86. Verhoeyen M, Milstein C, Winter G (1988) Reshaping human antibodies: grafting an antilysozyme activity. Science 239:1534–1536
87. Cai X, Garen A (1995) Anti-melanoma antibodies from melanoma patients immunized with genetically modified autologous tumor cells: selection of specific antibodies from single-chain Fv fusion phage libraries. Proc Natl Acad Sci USA 92:6537–6541
88. Leen AM, Rooney CM, Foster AE (2007) Improving T cell therapy for cancer. Annu Rev Immunol 25:243–265
89. Berger C, Jensen MC, Lansdorp PM et al (2008) Adoptive transfer of effector $CD8^+$ T cells derived from central memory cells establishes persistent T cell memory in primates. J Clin Invest 118:294–305
90. Klebanoff CA, Gattinoni L, Torabi-Parizi P et al (2005) Central memory self/tumor-reactive $CD8^+$ T cells confer superior antitumor immunity compared with effector memory T cells. Proc Natl Acad Sci USA 102:9571–9576
91. Powell DJ, Jr., Dudley ME, Robbins PF et al (2005) Transition of late-stage effector T cells to $CD27^+$ $CD28^+$ tumor-reactive effector memory T cells in humans after adoptive cell transfer therapy. Blood 105:241–250
92. Zhou J, Shen X, Huang J et al (2005) Telomere length of transferred lymphocytes correlates with *in vivo* persistence and tumor regression in melanoma patients receiving cell transfer therapy. J Immunol 175:7046–7052
93. Liu K, Rosenberg SA (2001) Transduction of an IL-2 gene into human melanoma-reactive lymphocytes results in their continued growth in the absence of exogenous IL-2 and maintenance of specific antitumor activity. J Immunol 167:6356–6365

94. Quintarelli C, Vera JF, Savoldo B et al (2007) Co-expression of cytokine and suicide genes to enhance the activity and safety of tumor-specific cytotoxic T lymphocytes. Blood 110:2793–2802
95. Vera JF, Hoyos V, Savoldo B et al (2009) Genetic manipulation of tumor-specific cytotoxic T lymphocytes to restore responsiveness to IL-7. Mol Ther 17:880–888
96. Dagarag M, Evazyan T, Rao N et al (2004) Genetic manipulation of telomerase in HIV-specific $CD8^+$ T cells: enhanced antiviral functions accompany the increased proliferative potential and telomere length stabilization. J Immunol 173:6303–6311
97. Hooijberg E, Ruizendaal JJ, Snijders PJ et al (2000) Immortalization of human $CD8^+$ T cell clones by ectopic expression of telomerase reverse transcriptase. J Immunol 165:4239–4245
98. Migliaccio M, Amacker M, Just T et al (2000) Ectopic human telomerase catalytic subunit expression maintains telomere length but is not sufficient for $CD8^+$ T lymphocyte immortalization. J Immunol 165:4978–4984
99. Li MO, Wan YY, Sanjabi S et al (2006) Transforming growth factor-beta regulation of immune responses. Annu Rev Immunol 24:99–146
100. Peng Y, Laouar Y, Li MO et al (2004) TGF-β regulates *in vivo* expansion of Foxp3-expressing $CD4^+CD25^+$ regulatory T cells responsible for protection against diabetes. Proc Natl Acad Sci USA 101:4572–4577
101. Zamanakou M, Germenis AE, Karanikas V (2007) Tumor immune escape mediated by indoleamine 2,3-dioxygenase. Immunol Lett 111:69–75
102. Munn DH, Sharma MD, Baban B et al (2005) GCN2 kinase in T cells mediates proliferative arrest and anergy induction in response to indoleamine 2,3-dioxygenase. Immunity 22:633–642
103. Charo J, Finkelstein SE, Grewal N et al (2005) Bcl-2 overexpression enhances tumor-specific T-cell survival. Cancer Res 65:2001–2008
104. Eaton D, Gilham DE, O'Neill A et al (2002) Retroviral transduction of human peripheral blood lymphocytes with Bcl-X(L) promotes *in vitro* lymphocyte survival in pro-apoptotic conditions. Gene Ther 9:527–535
105. Dotti G, Savoldo B, Pule M et al (2005) Human cytotoxic T lymphocytes with reduced sensitivity to Fas-induced apoptosis. Blood 105:4677–4684
106. Hacein-Bey-Abina S, von Kalle C, Schmidt M et al (2003) LMO2-associated clonal T cell proliferation in two patients after gene therapy for SCID-X1. Science 302:415–419
107. Hacein-Bey-Abina S, von Kalle C, Schmidt M et al (2003) A serious adverse event after successful gene therapy for X-linked severe combined immunodeficiency. N Engl J Med 348:255–256
108. Bonini C, Ferrari G, Verzeletti S et al (1997) HSV-TK gene transfer into donor lymphocytes for control of allogeneic graft versus leukemia. Science 276:1719–1724
109. Bonini C, Bondanza A, Perna SK et al (2007) The suicide gene therapy challenge: how to improve a successful gene therapy approach. Mol Ther 15:1248–1252
110. Ciceri F, Bonini C, Stanghellini MT et al (2009) Infusion of suicide-gene-engineered donor lymphocytes after family haploidentical haemopoietic stem-cell transplantation for leukaemia (the TK007 trial): a non-randomised phase I-II study. Lancet Oncol 10:489–500
111. Munshi NC, Govindarajan R, Drake R et al (1997) Thymidine kinase (TK) gene-transduced human lymphocytes can be highly purified, remain fully functional, and are killed efficiently with ganciclovir. Blood 89:1334–1340
112. Traversari C, Marktel S, Magnani Z et al (2007) The potential immunogenicity of the TK suicide gene does not prevent full clinical benefit associated with the use of TK-transduced donor lymphocytes in HSCT for hematologic malignancies. Blood 109:4708–4715
113. Sato T, Neschadim A, Konrad M et al (2007) Engineered human tmpk/AZT as a novel enzyme/prodrug axis for suicide gene therapy. Mol Ther 15:962–970
114. Introna M, Barbui AM, Bambacioni F et al (2000) Genetic modification of human T cells with CD20: a strategy to purify and lyse transduced cells with anti-CD20 antibodies. Hum Gene Ther 11:611–620
115. Iuliucci JD, Oliver SD, Morley S et al (2001) Intravenous safety and pharmacokinetics of a novel dimerizer drug, AP1903, in healthy volunteers. J Clin Pharmacol 41:870–879

116. Thomis DC, Marktel S, Bonini C et al (2001) A Fas-based suicide switch in human T cells for the treatment of graft-versus-host disease. Blood 97:1249–1257
117. Spencer DM, Belshaw PJ, Chen L et al (1996) Functional analysis of Fas signaling *in vivo* using synthetic inducers of dimerization. Curr Biol 6:839–847
118. Straathof KC, Pule MA, Yotnda P et al (2005) An inducible caspase 9 safety switch for T-cell therapy. Blood 105:4247–4254
119. Carroll D (2008) Progress and prospects: zinc-finger nucleases as gene therapy agents. Gene Ther 15:1463–1468
120. Provasi E, Genovese P, Magnani Z et al (2009) T cell receptor gene transfer into early differentiated lymphocytes by lentiviral vectors for safe and effective adoptive immune therapy of leukemia. Mol Ther 17:159–159
121. Schambach A, Baum C (2008) Clinical application of lentiviral vectors – concepts and practice. Curr Gene Ther 8:474–482
122. Louis CU, Straathof K, Bollard CM et al (2009) Enhancing the *in vivo* expansion of adoptively transferred EBV-specific CTL with lymphodepleting CD45 monoclonal antibodies in NPC patients. Blood 113:2442–2450
123. Attia P, Maker AV, Haworth LR et al (2005) Inability of a fusion protein of IL-2 and diphtheria toxin (Denileukin Diftitox, DAB389IL-2, ONTAK) to eliminate regulatory T lymphocytes in patients with melanoma. J Immunother 28:582–592
124. Ko K, Yamazaki S, Nakamura K et al (2005) Treatment of advanced tumors with agonistic anti-GITR mAb and its effects on tumor-infiltrating $Foxp3^+CD25^+CD4^+$ regulatory T cells. J Exp Med 202:885–891
125. Phan GQ, Yang JC, Sherry RM et al (2003) Cancer regression and autoimmunity induced by cytotoxic T lymphocyte-associated antigen 4 blockade in patients with metastatic melanoma. Proc Natl Acad Sci USA 100:8372–8377
126. Ribas A, Camacho LH, Lopez-Berestein G et al (2005) Antitumor activity in melanoma and anti-self responses in a phase I trial with the anti-cytotoxic T lymphocyte-associated antigen 4 monoclonal antibody CP-675,206. J Clin Oncol 23:8968–8977
127. Attia P, Phan GQ, Maker AV et al (2005) Autoimmunity correlates with tumor regression in patients with metastatic melanoma treated with anti-cytotoxic T-lymphocyte antigen-4. J Clin Oncol 23:6043–6053
128. Robinson MR, Chan CC, Yang JC et al (2004) Cytotoxic T lymphocyte-associated antigen 4 blockade in patients with metastatic melanoma: a new cause of uveitis. J Immunother 27:478–479
129. Carswell KS, Papoutsakis ET (2000) Culture of human T cells in stirred bioreactors for cellular immunotherapy applications: shear, proliferation, and the IL-2 receptor. Biotechnol Bioeng 68:328–338
130. Foster AE, Forrester K, Gottlieb DJ et al (2004) Large-scale expansion of cytomegalovirus-specific cytotoxic T cells in suspension culture. Biotechnol Bioeng 85:138–146
131. Malone CC, Schiltz PM, Mackintosh AD et al (2001) Characterization of human tumor-infiltrating lymphocytes expanded in hollow-fiber bioreactors for immunotherapy of cancer. Cancer Biother Radiopharm 16:381–390
132. Knazek RA, Wu YW, Aebersold PM et al (1990) Culture of human tumor infiltrating lymphocytes in hollow fiber bioreactors. J Immunol Methods 127:29–37
133. Tran CA, Burton L, Russom D et al (2007) Manufacturing of large numbers of patient-specific T cells for adoptive immunotherapy: an approach to improving product safety, composition, and production capacity. J Immunother 30:644–654
134. Vera J, Brenner L, Gerdemann U et al (2010) Accelerated production of antigen-specific T-cells for pre-clinical and clinical applications using Gas-permeable Rapid Expansion cultureware (G-Rex). J Immunother 33:305–315
135. Taylor DC, Zhou Z, Haider S et al (2006) Health-care utilization and cost for the treatment of melanoma in the six months following diagnosis. J Clin Oncol 24:18005
136. Saito AM, Cutler C, Zahrieh D et al (2008) Costs of allogeneic hematopoietic cell transplantation with high-dose regimens. Biol Blood Marrow Transplant 14:197–207

Part II
Non-T Cell Therapeutic Approaches

Chapter 2
B Lymphocytes in Cancer Immunology

David Spaner and Angela Bahlo

Abstract The role of B lymphocytes in the pathogenesis and treatment of cancer has not received as much attention as the role of T cells. However, most patients with solid tumors harbor circulating antitumor antibodies and most tumors contain a population of infiltrating B cells implying an association between oncogenic events and B-cell activation. B-cell immunity can be beneficial by providing antibody-mediated protection from oncogenic viruses or a source of recombinant tumor-specific antibodies that can be used in combination with chemotherapeutic regimens. However, activation of B cells may also be detrimental to an effective antitumor response. Tumor-reactive antibodies and B cells often recognize antigens that are generated during the unscheduled apoptotic and necrotic death processes, which accompany tumor progression and may be involved in wound-healing processes that promote tumor growth and impair protective T-cell responses. Therefore, methods to eliminate autoreactive B cells, or switch them to a B effector-1 (Be-1) phenotype that amplifies Th1/Tc1-type T-cell responses, which are typically associated with effective antitumor responses, may improve the clinical outcomes of T-cell-mediated immunotherapies. Possible strategies include the administration of B-cell-depleting monoclonal antibodies, use of targeted B-cell stimulatory agents such as Toll-like Receptor agonists, and adoptive transfer of large numbers of *ex vivo* generated tumor-reactive Be-1 cells.

Keywords B lymphocytes • Cancer vaccines • Chronic lymphocytic leukemia • Regulatory B cells • Tumor immunology

---

D. Spaner (✉)
Department of Medicine, University of Toronto, Odette Cancer Center,
Sunnybrook Health Sciences Center, Toronto, ON, Canada
e-mail: spanerd@sri.utoronto.ca

J. Medin and D. Fowler (eds.), *Experimental and Applied Immunotherapy*,
DOI 10.1007/978-1-60761-980-2_2, © Springer Science+Business Media, LLC 2011

Introduction

Current immunotherapeutic strategies in experimental cancer are focused primarily upon the augmentation of T-cell immunity. Nonetheless, recombinant antibodies, which represent a product of B cells, are playing an increasing role in current clinical cancer therapy [1]. However, B cells themselves have not been studied exhaustively in terms of their potential role in tumorigenesis or suitability as therapeutic targets. One historical reason for this "T-cell-centric" view of cancer biology was the early availability of reagents, such as CD4 and CD8 antibodies, which allowed T cells to be classified into different functional subsets, thereby facilitating detailed studies of T-cell-mediated effects. By comparison, the study of human B-cell biology was delayed for some years due to the lack of similar reagents to clearly differentiate B-cell subsets [2]. In addition, while B cells have long been known to produce antibodies, their ability to act as effector cells in an immune response has only been recognized relatively recently [3, 4]. The following emerging research findings indicate that: (1) B cells have a major impact on tumorigenesis; (2) targeting B cells may improve the efficacy of T-cell-mediated immunotherapy, and (3) B cells themselves may have important antitumor activity in some settings. The purpose of this chapter is to discuss how some of this new information might be incorporated into the design of future cancer immunotherapeutic strategies. Although B cells can clearly undergo malignant transformation into lymphomas and leukemias, the discussion here will focus on the modulatory effects of normal B cells on solid tumor biology, with an additional focus on clinical results in humans.

Peripheral Human B-cell Development

The majority of lymphocytes in the blood are T cells, making up 22–30% of total nucleated white cells. Circulating B cells represent only 7–10% of white blood cells and consist of a number of different subsets that participate in immune responses in secondary lymphoid tissues and at sites of tumor formation [5]. Approximately 75% of circulating B cells do not express CD27, indicating that they have recently emerged from the bone marrow and have not yet encountered antigen in the periphery (see Fig. 2.1). The *Ig* locus of these cells is germ-line indicating they have not yet undergone the somatic hypermutation process in germinal centers that increases the affinity of their B-cell receptors (BCRs) for specific antigens. CD27-negative B cells can be divided into transitional, prenaïve, and naïve B cells on the basis of their expression of CD38 (Fig. 2.1) [6]. Transitional B cells, which have recently emerged from the bone marrow and constitute about 2% of circulating B cells, express high levels of CD5, CD38, IgM, and IgD and are enriched for cells with autoreactive BCRs. Prenaïve B cells comprise approximately 7% of circulating B cells and have lower levels of CD38 but continue to express CD5, IgM, and IgD.

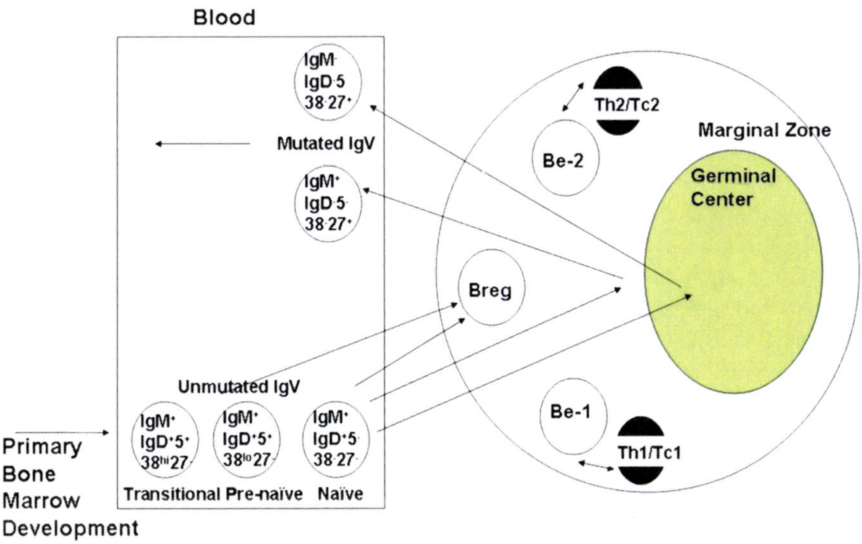

Fig. 2.1 Peripheral B-cell development. As described in the text, antigen-inexperienced primary B cells that have been selected in the bone marrow enter the blood as transitional, prenaïve, and naïve cells that undergo further differentiation in germinal centers and marginal zones of secondary lymphoid organs under the control of antigen. Memory B cells with mutated immunoglobulin variable genes then enter the recirculating pool. Possible sites of development of Bregs and effector B cells are also indicated

Transitional and prenaïve cells are thought to represent intermediate stages before B cells become naïve cells that are competent to respond to foreign antigens. Possibly because of their expression of CD5, which inhibits signaling through the BCR [7], transitional and prenaïve cells exhibit impaired calcium release and undergo activation-induced cell death in response to BCR cross-linking. In contrast, naïve B cells proliferate upon antigen activation. Unlike naïve cells, transitional and prenaïve B cells also undergo spontaneous apoptosis when placed in culture without exogenous stimulatory signals. This predisposition to die in response to antigenic signaling or absence of trophic factors is thought to ensure that transitional and prenaïve cells have a limited survival *in vivo* unless they encounter an antigen that they recognize and that the process of culling auto-reactive cells, initiated during primary development in the bone marrow, is continued in the periphery [8]. However, transitional and prenaïve cells can receive pro-survival signals via cytokines, such as IL-4, IL-10, and IL-21, and costimulatory molecules, such as CD40 [6]. Accordingly, such B cells may persist at sites of inflammation where their auto-reactivity may influence the outcome of immune responses and contribute to immunopathology [9], which may include antitumor immunity (see below).

Prenaïve cells lose expression of CD38 and CD5 and mature into naïve cells, which constitute around 65% of circulating B cells [6]. $CD38^{-}CD5^{-}CD27^{-}IgM^{+}IgD^{+}$ naïve cells acquire the ability to respond to antigenic signals through their BCR by proliferating and differentiating into short-lived plasma cells that secrete IgM antibodies. Other B cells of the activated clone mature into memory cells in the germinal centers through the processes of somatic hypermutation and class-switching, which are under the control of T cells (Fig. 2.1). Memory B cells are long-lived, respond more strongly to subsequent antigenic stimulation compared to naïve cells, are characterized by expression of CD27 in the absence of CD38 or IgD, and comprise approximately 25% of circulating B cells. Some memory cells continue to express IgM and do not undergo class-switching, despite acquiring mutations in their Ig V region genes. Such cells, which are classified as IgM^+ memory B cells [10], are thought to take part in T-cell-independent responses to polysaccharide antigens and represent circulating marginal zone B cells. By comparison, classical memory B cells undergo class-switching in the germinal center, down-regulate IgM expression, use one of the IgG subtypes, IgA, or IgE genes to form the heavy chain of their antigen receptor, and ultimately recognize protein antigens under the control of helper T cells (Fig. 2.1).

B-Cell Effector States

In addition to their well-known ability to differentiate into plasma cells and secrete antibodies, B cells also influence immunity by serving as antigen-presenting-cells (APCs). Naïve B cells are thought to represent an immunosuppressive type of APC because they have been shown to tolerize T cells that interact with them [11, 12]. However, under appropriate conditions that may involve CD40 ligation and cytokine signaling, a naïve B cell can serve as a relatively potent APC that expresses costimulatory molecules such as CD80, CD86, and ICOS, and activates both $CD4^+$ and $CD8^+$ T cells [13].

B cells exert effector functions not only through the production of antibodies, but also by making cytokines [14]. As a result of interactions with T cells, B cells can be directed to secrete polarized groups of cytokines that parallel those of the dichotomous Th1/Tc1 and Th2/Tc2 differentiation states that exist within T-cell subsets [15]. B effector 1 (Be-1) cells arise through interactions with Th1/Tc1-type T cells and secrete cytokines characteristic of this type of immune response, including IFN-γ, IL-12 and TNF-α. In contrast, B effector 2 (Be-2) cells arise through interactions with Th2/Tc2-type T cells and secrete a polarized pattern of cytokines that includes IL-2, IL-4, IL-6, IL-13, and TNF-α. Through cross-talk with interacting T cells, these polarized B effector states serve to differentially reinforce and amplify Th1/Tc1-type T cells that promote cellular immunity or Th2/Tc2-type T cells that promote humoral immunity [16].

Further research will be required to define completely the precursor cells that give rise to effector Be-1 and Be-2 cells and characterize the molecular mechanisms that drive B cells into these states. Not surprisingly, in view of the association with T-cell interactions, effector B cells are thought to originate from recently activated naïve B cells that enter germinal follicles to begin the processes of somatic hypermutation and class-switching [14]. Subsets of recirculating memory B cells may be already programmed to develop into cytokine-producing Be-1 or Be-2 cells [17]. Be-1 differentiation is thought to result from signaling through IFN-γ receptors on B cells [18] which induces the transcription factor T-bet to regulate gene expression in Be-1 cells in a manner analogous to the role it plays in regulating gene expression in Th1/Tc1-type T cells [19]. In contrast, signaling through the IL-4α receptor is thought to control B-cell differentiation towards Be-2 cells [20]. Because proinflammatory cytokine production by human B cells is enhanced by phorbol esters [13, 21], strong activation of mitogen-activated protein kinase (MAPK) signaling pathways may be needed for effector B-cell differentiation [22]. This MAPK activation may be contributed by a variety of signaling complexes on the B-cell surface, including the antigen receptor, MHC molecules [19, 20], and concomitant signaling through multiple toll-like receptors (TLRs) [23] or through a combination of TLRs and cytokine receptors [14, 21].

B cells can differentiate into regulatory cells (Bregs) that are characterized by production of immunosuppressive cytokines such as IL-10 and TGF-β [24]. In contrast to effector B cells, which amplify T-cell responses, IL-10 secreting B cells have been demonstrated to dampen effector T-cell responses in a variety of experimental situations [24], including the inhibition of immune responses against tumors [25]. The cellular origins and molecular mechanisms accounting for Breg differentiation are incompletely understood. Unlike effector B cells, which differentiate in the germinal follicle, it has been reasoned that Bregs develop from marginal zone B cells, or perhaps from $CD5^+$ transitional or prenaïve cells [26]. In mice, CD5-expressing cells of the so-called B1-B cell lineage are thought to give rise to Bregs [14]. However, the existence of the analogous cell lineage in humans remains uncertain. Production of IL-10 by some human B cells is associated with strong activation of the transcription factor, STAT-3 [21]. The tone and duration of MAPK signaling may also determine if B cells acquire regulatory functions. When B cells of marginal zone origin are treated only with IL-2 and a TLR-7 agonist, they produce little TNF-α but make the high levels of IL-10 associated with the Breg phenotype. However, if the cells are concomitantly treated with diacylglycerol mimetics, which activate ERK via Ras guanyl nucleotide-releasing proteins (RasGRPs) [27], IL-10 is "switched off," both TNF-α and-β production are increased, and the B cells acquire strong T-cell stimulatory capabilities [21].

In addition to their ability to make antibodies and cytokines and serve as APCs, activated B cells can acquire cytotoxic capabilities that may be of importance for antitumor immunity. For example, an Epstein-Barr Virus (EBV)-infected B cell line established from a breast cancer biopsy was shown to lyse breast cancer cells *in vitro* [28]. However, other activated B cells can kill activated T cells and may thereby inhibit T-cell-mediated responses [29]. Killer B cells often express molecules that are characteristic of Breg cells, including CD5, IL-10, and TGF-β. These observations suggest that Bregs may exert their inhibitory effects via both immunosuppressive cytokine secretion and direct lysis of T cells. The mechanism of killing can occur through diverse TNF and TNF receptor (TNFR) family members such as Fas ligand (CD178) and Fas, TRAIL (CD253) and its death receptors such as DR5 (TNFRSF10B or CD262), and programmed death ligands 1 and 2 (PDL1:CD274 and PDL2: CD273) [14]. In addition, some human B cells stimulated by IL-21 together with TLR or BCR agonists express granzyme B [30] and may thereby kill through perforin-mediated mechanisms typically associated with cytotoxic $CD8^+$ T cells or NK cells [31].

B Cells and Cancer

Evidence that B-cell activation is connected to cancer progression comes from an extensive literature on the presence of antibodies that recognize tumor antigens in cancer patients and a much smaller literature on the infiltration of tumors by B lymphocytes.

Serology

Circulating antibodies that recognize antigens expressed by cancer cells have been found in most patients with solid tumors [32, 33]. Using SEREX technology, where patient sera is used to screen recombinant cDNA libraries obtained from tumors, over 2,500 different proteins are listed in the Cancer Immunome database [http://ludwig-sun5.unil.ch/CancerImmunomeDB/] from breast, gastric, renal, lung, prostate, hepatic, and ovarian cancer, as well as melanoma, mesothelioma, sarcoma, neuroblastoma, lymphomas, and leukemias. Most of these antigens are ubiquitous cytoplasmic proteins such as actin, cytokeratin, DNA polymerases, and heat-shock proteins. They are not tumor-specific and would be mainly protected from circulating antibodies by their predominantly intracellular location, although such antigens can be externalized during inflammatory and apoptotic processes that accompany tumor growth (see below) [34]. Accordingly, antibodies that target these antigens would not seem capable of mediating therapeutic antitumor responses. It is possible that the relative inability to detect cell surface antigens that are more accessible to antibodies relates in part to the use of bacteria to express mammalian cDNA in SEREX assays. Bacteria lack glycosylation enzymes and are therefore unable to make glycoproteins found on the plasma membranes of eukaryotic cells [35].

Other techniques, distinct from SEREX methods, have been used to characterize naturally arising anticancer antibodies in human patients. Using a "candidate" antigen approach, antibodies to cell surface receptors, such as the HER-2/*neu* epidermal growth factor receptor (EGFR) which is overexpressed on 25–50% of breast tumors, are found in the sera of nearly a quarter of patients [36]. By making hybridomas from B cells in draining lymph nodes, or from tumor-infiltrating B cells (TIBs), antibodies that recognize cell surface glycoproteins and cytoplasmic proteins have been identified [35]. More recently, the specificities of B cells that infiltrate solid tumors have been identified by amplifying Ig V regions, cloning and sequencing these rearranged genes, constructing combinatorial libraries of single-chain variable region gene fragments (scFVs), and then selecting for tumor-binding capacity [37]. Using such approaches, it has been shown that some antitumor responses are directed against glycolipid antigens. However, even these sophisticated techniques continue to demonstrate that many antibodies made by tumor-infiltrating B cells recognize intracellular autoantigens, such as actin [38], that become externalized during apoptotic processes or are oxidized or proteolytically degraded during apoptosis [34]. Taken together, these observations suggest that intratumoral B cells, as well as B cells in organized lymphoid tissues that make circulating antibodies in cancer patients, often recognize structures associated with apoptosis and cell death which are processes that accompany tumor progression [39, 40].

Tumor-Infiltrating B Cells

Lymphocytic infiltrates are found in most solid tumors. The dominant cell population is usually T cells; in general, the more T cells that are found in a tumor, the better the prognosis [41, 42]. B cells are also a component of intratumorallymphocytic infiltrates, albeit usually a minor population compared to T cells. However, in early ductal breast carcinoma in situ, infiltrating B cells are found in excess of T cells and form the predominant intratumoral lymphocyte population [43]. It is also interesting to note that medullary breast cancer, which constitutes 3–7% of all breast cancers and has a favorable prognosis compared with other types of infiltrating ductal carcinomas, is characterized by infiltrates of B cells and plasma cells [38], along with T cells [37]. Tumor-infiltrating B cells (TIBs) are also found in other types of breast cancer [44] and other cancers including melanoma [45], lung cancer [46], and mesothelioma [47].

B cells can enter tumors in response to chemoattractants produced during the inflammation that accompanies, and may even cause, tumor progression [48]. However, by cloning rearranged immunoglobulin genes in tumor biopsies and comparing VH gene usage and the mutation status of *Ig* genes, it appears that intratumoral B cells are related and selected by antigen responses *in situ*, rather than being recruited nonspecifically from the blood into the tumor [38]. Given the antigen-specificity of the antibodies made by some TIBs described above, it seems possible that intratumoral B cells may often be responding to antigens on apoptotic bodies or to intracellular proteins that have been degraded by proteases or oxidized during the inflammatory processes inside a tumor. However, there is little information as to whether these intratumoral B cells are Be-1 cells, Be-2 cells, or Bregs (see above).

B Cells and Cancer: Friends or Foes?

While the evidence that B cells and their antibody products are associated with cancer seems clear, whether this association is protective, causal, or simply incidental has not been clarified. The answer seems to depend in part on how early the cancer is in its development. Most evidence suggests that, once the tumor is established, B cells probably have a negative effect on protective antitumor responses and may even facilitate tumor progression.

While cancers are characterized by collections of aberrant genetic events that corrupt signaling pathways and interfere with normal cell death processes [49], cancer progression is also intimately intertwined with inflammation [48]. Agents that cause cancer, such as cigarette smoke in lung cancer [50], ultraviolet light in skin cancers [51], ulcerative colitis in colon cancer [52], and micro-organisms such as *Helicobacter pylori* in intestinal cancers [53], *Hepatitis* B and C in hepatomas [54], and Human *papilloma virus* (HPV) in cervical cancer [55], are also associated with chronic inflammation. Inflammation may provide signals that promote growth of genetically-aberrant cells and may further select for more aggressive tumor cells by increasing genetic instability [48].

As cancers grow, inflammatory processes seem to become self-sustaining. Because of the break-down in control mechanisms that prevent unrestrained cellular proliferation, tumor cells continue to grow beyond the limits that are normally supported by environmental nutrients and blood supply [56]. However, even tumor cells with impaired cell death pathways cannot grow indefinitely in nutrient-poor conditions and undergo "unscheduled" apoptotic or necrotic death [39]. Apoptosis has important consequences for antitumor T-cell responses as it has been associated with peripheral tolerance mechanisms and the deviation of immune responses away from protective Th1/Tc1-type responses [57]. By comparison, necrosis causes inflammation, which leads to production of chemokines and cytokines associated with wound repair. These repair mechanisms can then be used by the tumor cells for further growth and another round of the wound-repair cycle [58]. This type of biology has led to the idea that tumors are analogous to "wounds that do not heal" [59, 60]. Although this model is clearly oversimplified, these general principles of how cancers develop are of some relevance in trying to better understand the role of B cells in tumor progression.

Evidence for a Protective Effect of B Cells in Antitumor Responses

As described above, B cells can potentially inhibit the development and progression of cancers by making antitumor antibodies or by differentiating into appropriate effector B-cell states. B-cell-derived antibodies play an essential role in protection against viral infections. In this context, B cells can protect against tumor development by helping to clear oncogenic viruses before they can become established and initiate tumor development. An excellent example of this is the use of HPV vaccines to prevent cervical cancer [61].

Recombinant antibodies have clearly been shown to contribute to the clearance of established tumors in patients. The efficacy of antibodies against CD20 (Rituximab®) in lymphoma [62] or HER-2/*neu* (Herceptin®) in breast cancer [63] have resulted in almost paradigmatic changes in treatment strategies for these cancers. Similarly, antibodies against angiogenic factors, such as VEGF, slow progression of metastatic disease [64] and antibodies against glycolipid gangliosides that are overexpressed on cancer cells, particularly melanoma, are under clinical investigation [65]. The therapeutic activity of these antibodies can be increased even further by coupling them to toxins such as radioactive isotopes or cytotoxic proteins of bacterial or plant origin [66]. Antibodies with similar specificities as the recombinant antibodies can be demonstrated to arise naturally in cancer patients [35]. The levels of naturally arising antibodies are generally very low and well below the therapeutic concentrations that can be obtained by injecting recombinant antibodies. Accordingly, it seems unlikely that naturally arising antitumor antibodies can be effective in clearing established tumors and, as discussed below, may even promote tumor growth as a result of their low concentrations. However, vaccines that enhance endogenous production of these antibodies might have therapeutic potential, a concept that has been validated in an experimental model where vaccination with a recombinant adenovirus expressing a truncated HER-2/*neu* antigen resulted in sufficient antibody production to block HER-2/*neu* function and clear subcutaneous HER-2/*neu*-expressing breast cancers in mice [67].

Several experimental models demonstrate a possible protective role for B cells against tumors that may be attributable to effector B cells. Lung metastases caused by intravenous injection of the chemically induced rat mammary adenocarcinoma, MADB106, are significantly increased when host B cells are depleted by specific antibodies [68]. The mechanism in this model seems to be a local effect of pulmonary B cells, which promote IFN-γ production and facilitate killing of tumor cells by NK cells [69]. It is possible that the protective cells in this model may represent Be-1 cells. In a mouse model, J558L plasmacytoma cells engineered to overexpress lymphotoxin (TNF-β) were cleared in syngeneic BALB/c mice through B-cell-dependent mechanisms because the tumors were significantly infiltrated with lymphocytes that expressed B220 (a B-cell marker) and failed to grow in nude mice (which lack T cells but contain B cells) but did grow in SCID mice (which lack both T and B cells) [70]. These findings are again suggestive of a role for effector B cells in tumor clearance. Similarly, a fusion of a tumor-specific antibody (directed against the human EGFR) and lymphotoxin prevented pulmonary metastases following intravenous injection of the human melanoma cell line, M24met, in nude mice (but not SCID mice). This therapeutic effect was accompanied by infiltration of B220$^+$ cells into the metastases [71]. Taken together, these observations suggest that B cells can protect against cancers under certain conditions. However, the experimental models may have limited application to the clinical setting, which typically involves treating established tumors rather than preventing tumor initiation.

Evidence for a Negative Effect of B Cells on Antitumor Responses

In principle, naturally arising antibodies against cell surface proteins, carbohydrates, and lipids might be expected to kill tumor cells by activating complement, causing antibody-mediated cellular cytotoxicity, or initiating signaling events that cause apoptosis [72]. Such scenarios may occur in the early stages of cancer but it is almost impossible to study these types of "negative" situations in the clinic without the presence of an actual tumor. In practice, tumors progress despite the presence of circulating antitumor antibodies. A simple explanation for this situation is that the cell surface structures on tumor cells that are targeted by antibodies are autoantigens and, as a result of tolerance mechanisms, high-affinity antibodies cannot be made in sufficient titers to mediate an effective antitumor response.

Weak, humoral immune responses that fail to clear tumor cells may actually have a detrimental effect on the clinical outcome by contributing to the inflammatory responses that drive tumor progression [73]. For example, transgenic mice that express HPV early region genes under the control of a human keratin 14 promoter exhibit multistage development of invasive squamous cell carcinoma of the epidermis. When they are crossed to $Rag1^{-/-}$ mice, which have a complete absence of functional B and T cells, tumorigenesis is markedly delayed and associated with reduced inflammation. Adoptive transfer of B cells or sera (which presumably contained antitumor antibodies) from the wild-type transgenic mice restored inflammatory cell infiltrates and tumor progression in premalignant lesions. These results suggest that antitumor antibodies cause inflammation that promotes the growth of cancer cells [74]. Similar concepts have been invoked to explain the role of antibodies to the foreign ganglioside, *N*-glycolylneuraminic acid (Neu5Gc), which accumulates in metabolically active cancer cells [75]. Injection of large amounts of anti-Neu5Gc antibodies slowed progression of Neu5Gc-bearing tumor cells but low amounts of antibodies promoted tumor growth. Tumor progression resulting from low levels of antibodies could be inhibited by cyclooxygenase-2 (COX-2) inhibitors, thereby suggesting that the antibodies induced an inflammatory state that promoted tumor growth.

Cell-mediated immunity, involving cytotoxic T cells, is generally thought to be the most important arm of the immune system for clearing established tumors [76]. Antibody production is primarily the result of humoral immunity that is promoted by Th2/Tc2-type T cells and B cells can promote antigen-driven responses to deviate towards Th2/Tc2-type responses [77]. Since Th2/Tc2 cells are not as efficient as Th1/Tc1 cells at clearing tumor cells, B cells are often considered detrimental to effective antitumor responses [25, 78]. However, the recent identification of Be-1 cells, which amplify Th1/Tc1-type T-cell responses [3], challenges this idea and suggests that only some B-cell effector states, presumably Be-2 cells and Bregs, are detrimental to effective antitumor immunity.

B cells have recently been found to play important roles in wound-healing [79]. Although B cells are not prominent components of cutaneous wounds, their removal by genetic means [79, 80] impeded the wound healing process by decreasing the production of cytokines, including TGF-β and IL-10. Furthermore, wound-healing was improved by adoptive transfer of IL-10 secreting B cells [80]. Given the concept of cancer as a "wound that doesn't heal" [59], these findings suggest that the small numbers of B cells found in cancer stroma might have properties of IL-10-secreting Bregs that promote tumor growth by both inhibiting local antitumor T-cell responses and promoting the processes of wound-healing [81].

B cells that participate in wound-healing are likely those that recognize antigens on apoptotic bodies and cytoplasmic proteins that have been oxidized or degraded by proteases, as such processes are associated with tissue damage. Such B cells would then secrete antibodies that cause apoptotic bodies to be cleared rapidly by monocytes and dendritic cells, limiting the presence of free autoantigens and inflammatory signals which cause immune responses and also ensuring these important APCs tolerize T cells rather than activate them [82]. A teleologic explanation for why B cells behave in this manner during normal wound healing might be to prevent toxic type 1 T-cell responses and scarring. While this behavior may preserve normal tissue functioning once a wound is repaired, analogous processes in a tumor microenvironment would inhibit clearance of tumor cells by T cells. Moreover, activation of B cells by apoptotic bodies could result in production of cytokines such as IL-10 and TGF-β that can inhibit T-cell responses and promote tumor growth. B cells can also be activated by adhesion molecules in an antigen-independent fashion [83], which could also lead to cytokine production, T-cell suppression, and tumor growth. For example, $CD5^{+}IgM^{+}$ B1-B cells that express the glycoprotein, MUC18 (also known as melanoma cell adhesion molecule), were found to bind B16 melanoma cells that also expressed MUC18 *in vivo* via MUC18/MUC18 interactions [84]. This heterotypic cell–cell interaction led to enhanced metastasis of the melanoma, perhaps by increasing ERK-signaling in the tumor cells. While the existence of B1-B cells in humans is still unclear, intriguingly, it was found that $CD5^{+}IgM^{+}$ cells (which may represent transitional or prenaïve B cells as described above) accumulated in biopsies from melanoma patients and correlated with MUC18 expression on human melanoma cells [84]. Taken together, these observations suggest that some types of intratumoral B lymphocytes may promote cancer progression by direct interactions with tumor cells.

Chronic Lymphocytic Leukemia as a Paradigm for Tumor Promotion by B Cells

Experiments in mice can be used to examine the role of B cells in tumorigenesis by removing B-cell populations via genetic or pharmacological means and adoptively transferring B-cell populations. Such experimental approaches can accentuate the typical effects of B-cell, allowing them to be uncovered in a "background" of competing physiological phenomena [79, 80]. In humans, this approach is obviously not feasible. However, a specialized clinical condition, chronic lymphocytic leukemia (CLL), may serve to illustrate some of the negative effects of B cells on solid tumors in humans.

CLL is the most common leukemia in the developed world. Chemotherapy is indicated for symptomatic disease but CLL patients are often asymptomatic and initial clinical management typically consists only of observation, sometimes for long periods of time [85]. The disease consists of an expansion of monoclonal B cells that express CD5 and low levels of IgM. The originating cell-type of CLL is not clear but the presence of somatic hypermutation in the *Ig* locus and low expression of CD38 in about half the cases suggests a postgerminal center origin, possibly in memory IgM⁺ cells (Fig. 2.1) [10]. On the other hand, the absence of somatic hypermutation and high CD38 expression in the remaining cases suggests an origin in transitional or prenaïve cells [86]. Regardless, BCRs on CLL cells-often recognize autoantigens such as rheumatoid factor, DNA, actin, and myosin, many of which are generated during inflammation and apoptosis [34, 87] and have been shown to be recognized by B cells that infiltrate solid tumors [38]. Moreover, CLL cells express high levels of IL-10 and TGF-β and characteristically suppress T-cell responses by a variety of mechanisms which include inhibiting CD40L signaling in T cells [88], killing T cells via Fas/FasL interactions [89], or dysrupting immune synapses [90]. These properties have led some scientists to speculate that CLL may be a tumor of regulatory B cells [91]. Accordingly, insights into the effects of Bregs on solid tumor progression in humans may be provided by studying the behavior of solid tumors that arise in CLL patients.

Compared to other people, CLL patients have more than double the risk of developing solid tumors. These cancers are mainly squamous cell skin cancers but also include melanoma, prostate, breast, gastrointestinal, lung, and other tumors [92, 93]. This increased risk is independent of specific treatment for CLL and solid tumors often arise in patients who are being managed only by observation, suggesting that some intrinsic property of the increased monoclonal B cell population is responsible. It is possible that the regulatory properties of the CLL B cells may be preventing effective antitumor T-cell responses or perhaps may be encouraging inflammatory processes (from uncontrolled viral infections, for example) which promote tumor progression.

However, another clinical observation is that, when solid tumors arise in CLL patients, they are often much more virulent than usual [94, 95]. The explanation for this phenomenon is also not clear but may again be related to impairment of protective antitumor T-cell responses. However, it is interesting that CD5⁺ B cells have been implicated in promoting melanoma progression in both mice and humans through direct interactions with melanoma cells [84]. CLL cells characteristically express CD5 and perhaps CLL cells use their autoreactive BCRs to bind to solid tumor cells, become activated, and produce cytokines that promote the growth of solid tumors.

Interestingly, the regulatory phenotype of CLL B cells seems to be somewhat plastic. For example, primary CLL cells can be grown in tissue culture in the presence of cytokines and TLR-agonists and maintain their suppressive features, such as IL-10 production and inability to stimulate T cells. However, in the presence of strong ERK-activation, which occurs with signaling through the BCR or with diacylglycerol agonists, the CLL cells acquire features of Be-1 cells, shut off IL-10 production, make high levels of inflammatory cytokines such as TNF-α, and strongly stimulate proliferation of Th1/Tc1-type T cells [13, 21]. Importantly, under these conditions, CLL cells are able to kill model tumors, such as MCF-7 breast cancer cells, *in vitro* (F. Wen, D. Spaner, unpublished data). Taken together, these clinical observations support the concept that B cells, particularly regulatory cells, may have a-major negative impact on the development and progression of solid tumors. However the *in vitro* results also raise the possibility that the phenotypic state of tumorigenic B cells may be manipulated to convert them into antitumor effectors.

B-Cell-Directed Cancer Immunotherapy

The above discussion suggests that B cells may play a positive role in preventing the development of cancer but have mainly negative effects on successful clearance of established tumors. These concepts suggest that depleting or enhancing specific B-cell populations may be of use in curative immunotherapy strategies.

Eliminating Negative B-Cell Effects

If B cells are inhibiting antitumor T-cell responses and promoting tumor growth, then B-cell depletion may potentially improve the results of cancer immunotherapy for established solid tumors. Interestingly, although most cancers are incurable, many are responsive to radiation therapy and chemotherapy that are highly toxic to lymphocytes, especially B cells. Although usually considered a side-effect, it is possible that depletion of B cells removes a source of trophic factors for tumor cells. As such, the B-cell depletion that occurs with these modalities may represent one mechanism that is partly responsible for their therapeutic benefits [96]. In addition, removal of B cells may promote the activity of the remaining antitumor T cells and lead to better control of the tumor, as evidenced by the abscopal effect of radiotherapy [97] or the increased activity of antigen-reactive T-cell clones injected into B-cell deficient hosts [98, 99].

The immunostimulatory properties of conventional chemotherapy are being actively investigated [22, 100] and involve other cell populations in addition to B cells. Specific depletion of B cells can be achieved with recombinant antibodies. While conventionally used to treat B-cell malignancies, these antibodies could also be used in solid tumor patients to eliminate nonmalignant B cells that produce trophic factors for tumors and immunosuppressive factors for T cells. The CD20 antibody, Rituximab®, eliminates B cells quite effectively and safely [101] and other CD20 antibodies, such as Ofatumumab® [102], CD23 antibodies such as Lumiliximab® [103], and antibodies against CD22 (Epratuzumab®) [1] are becoming available for clinical use. In an experimental murine model, CD20 antibodies slowed the growth of established $CD20^-$ solid tumors but did not induce tumor regression. However, in combination with vaccines, monoclonal antibody-mediated B-cell depletion led to enhanced antitumor responses associated with both increased numbers of activated CD8$^+$ splenic T cells and tumor regression [104]. Consistent with these findings, treatment of colorectal cancer patients with Rituximab® as a single agent led to regression of metastases in 4 of 8 evaluable patients [105]. These findings suggest that it may be advantageous to use B-cell-depleting antibodies to improve the results of cancer vaccines [106] or adoptively transferred tumor-reactive T cells [107].

A number of problems can be anticipated with these approaches. One potential limitation is that the antibodies do not readily distinguish between different effector B-cell classes. Elimination of Be-2 cells and especially Bregs are probably desirable but elimination of Be-1 cells, which amplify Th1/Tc1-type immune responses, may be detrimental to a successful T-cell-mediated antitumor response. In addition, B-cell depletion leads to increased numbers of transitional B cells that enter the circulation from the bone marrow [108]. It is not yet known if these cells might be more easily recruited into the Breg compartment and negate an otherwise therapeutic benefit.

Promoting Positive B-Cell Effects

Vaccines and Recombinant Antibodies

Vaccines that increase protective antibody titers and prevent infections with oncogenic viruses represent one modality by which B cells can be effectively manipulated for meaningful antitumor activity. The HPV vaccine, which prevents cervical cancer, is one of the best examples of this [61]. More universal use of the hepatitis B vaccine would likely prevent many cases of hepatoma [109] although vaccines capable of preventing the development of viral escape mutants in immunocompromised patients are needed to deal with the problem of HBV vaccine failure in a minority of subjects. An effective vaccine against *Helicobacter pylori* would similarly be expected to prevent the development of many gastric cancers [110]. Given that viruses have been estimated to be involved in 15–20% of cancers world-wide [111], continued development of prophylactic vaccines is likely to play an important role in cancer prevention.

Similarly, the development of recombinant antibodies to cell surface structures expressed predominantly by cancer cells will continue to be an important area for cancer therapy. The ability to generate libraries of single chain variable fragments (scFvs) overcomes many of the laborious steps associated with traditional methods of making hybridomas and offers a way to rapidly generate therapeutic antibodies to any desired antigen [37]. However, more detailed understanding of how these antibodies exert their antitumor effects *in vivo* [72] is still needed in order to develop strategies to improve the clinical results, such as increasing complement activation [112] or antibody-dependent cellular cytotoxicity [113].

Enhancing B-cell Activity In situ

Since B-cell effector states seem to be somewhat plastic (see above), an alternative approach to deleting inhibitory B cells in order to improve the therapeutic efficacy of antitumor T cells might be to turn intra-tumoral and intra-nodal B cells into Be-1 effectors in situ. The expected outcome of such an approach would be to amplify and prolong a Th1/Tc1-type of antitumor T-cell response sufficiently to clear tumor cells.

At least three different signals may be necessary to cause B cells to turn off production of immunosuppressive cytokines such as IL-10 and express the costimulatory molecule pattern required for strong stimulation of Th1/Tc1-type T cells [21]. These signals are provided by cytokines, such as IFN-γ [19] or IL-2 family members [114], TLR-agonists [115] or TNFR family members such as CD40 [116, 117], and strong MAPK activation, such as provided by diacylglycerol analogs [22] or possibly HLA-class II antibodies [83, 118]. While these reagents are not absolutely specific for B cells [119], clinical efficacy of such combinations is likely to depend on meaningful differentiation of B cells into the Be-1 phenotype *in vivo*. A more important stumbling block may be the well-known difficulties of extrapolating *in vitro* observations to *in vivo* settings [120]. Problems of hypoxia and poor vasculature with incomplete drug penetration into tumor microenvironments [121] may prevent immmunomodulatory agents from being able to increase the immunogenicity of intranodal and intratumoral B cells sufficiently to promote effective antitumor activity in situ [122].

Adoptive B-Cell Transfer

B cells turn out to be relatively easy to culture and expand to large numbers *in vitro* [123]. A relatively unexplored area of B-cell immunotherapy is "tissue engineering" with activated B cells that have been generated *in vitro*. For example, immunogenic B cells can be used as a vaccine platform to present tumor antigens to T cells [124] with significant potential advantages over dendritic cells because of the ease of generating large numbers for the repeated injections thought to be necessary for vaccine efficacy [125].

Adoptive T-cell therapy is another active area of cancer immunotherapy research [107]. As described in this chapter, B cells are capable of killing tumors [29], and may also be able to elicit antitumor responses following injection of large numbers into patients [4]. More importantly, coinjection of large numbers of Be-1 cells might amplify the effects of adoptively-transferred tumor-reactive T cells. It is not clear which peripheral blood subsets would be most suitable for initiating B-cell expansion cultures. Transitional and prenaïve cells are enriched in B cells with auto-reactive BCRs and thus may be more easily activated by cancer autoantigens to mediate killing of tumor cells. As with T cells, antigen-specificity and enhanced immunogenicity may be genetically engineered into B cells before infusion [126]. Regardless, the availability of methods to rapidly grow large numbers of B cells offers the opportunity to explore the potential benefits of adoptive B-cell therapy for cancer.

References

1. Castillo J, Winer E, Quesenberry P (2008) Newer monoclonal antibodies for hematological malignancies. Exp Hematol 36:755–768
2. Jackson SM, Wilson PC, James JA et al (2008) Human B cell subsets. Adv Immunol 98:151–224
3. Harris DP, Haynes L, Sayles PC et al (2000) Reciprocal regulation of polarized cytokine production by effector B and T cells. Nat Immunol 1:475–482
4. Li Q, Teitz-Tennenbaum S, Donald E et al (2009) *In vivo* sensitized and *in vitro* activated B cells mediate tumor regression in cancer adoptive immunotherapy. J Immunol 183:3195–3203
5. Schmidlin H, Diehl SA, Blom B (2009) New insights into the regulation of human B-cell differentiation. Trends Immunol 30:277–285
6. Lee J, Kuchen S, Fischer R et al (2009) Identification and characterization of a human CD5+ pre-naive B cell population. J Immunol 182:4116–4126
7. Dalloul A (2009) CD5: a safeguard against autoimmunity and a shield for cancer cells. Autoimmun Rev 8:349-353
8. Hardy RR, Hayakawa K (2001) B cell development pathways. Annu Rev Immunol 19:595–621
9. Ohmori H, Kanayama N (2003) Mechanisms leading to autoantibody production: link between inflammation and autoimmunity. Curr Drug Targets Inflamm Allergy 2:232–241
10. Tangye SG, Good KL (2007) Human IgM+CD27+ B cells: memory B cells or "memory" B cells? J Immunol 179:13–19
11. Ho WY, Cooke MP, Goodnow CC et al (1994) Resting and anergic B cells are defective in CD28-dependent costimulation of naive CD4+ T cells. J Exp Med 179:1539–1549
12. Fuchs EJ, Matzinger P (1992) B cells turn off virgin but not memory T cells. Science 258:1156–1159
13. Spaner DE, Shi Y, White D et al (2006) Immunomodulatory effects of Toll-like receptor-7 activation on chronic lymphocytic leukemia cells. Leukemia 20:286–295
14. Lund FE (2008) Cytokine-producing B lymphocytes-key regulators of immunity. Curr Opin Immunol 20:332–338
15. Dalgleish A (1999) The relevance of non-linear mathematics (chaos theory) to the treatment of cancer, the role of the immune response and the potential for vaccines. QJM 92:347–359
16. Wojciechowski W, Harris DP, Sprague F et al (2009) Cytokine-producing effector B cells regulate type 2 immunity to *H. polygyrus*. Immunity 30:421–433
17. Cariappa A, Boboila C, Moran ST et al (2007) The recirculating B cell pool contains two functionally distinct, long-lived, posttransitional, follicular B cell populations. J Immunol 179:2270–2281
18. Lund F, Randall T (2010) Effector and regulatory B cells: modulators of $CD4^+$ T cell immunity. Nat Rev Immunol 10:236–247
19. Harris DP, Goodrich S, Gerth AJ et al (2005) Regulation of IFN-γ production by B effector 1 cells: essential roles for T-bet and the IFN-γ receptor. J Immunol 174:6781–6790
20. Harris DP, Goodrich S, Mohrs K et al (2005) The development of IL-4-producing B cells (B effector 2 cells) is controlled by IL-4, IL-4 receptor alpha, and Th2 cells. J Immunol 175:7103–7107

21. Tomic J, White D, Shi Y et al (2006) Sensitization of IL-2 signaling through TLR-7 enhances B lymphoma cell immunogenicity. J Immunol 176:3830–3839
22. Shaha SP, Tomic J, Shi Y et al (2009) Prolonging microtubule disruption enhances the immunogenicity of CLL cells. Clin Exp Immunol 158:186–198
23. Barr TA, Brown S, Ryan G et al (2007). TLR-mediated stimulation of APC: Distinct cytokine responses of B cells and dendritic cells. Eur J Immunol 37:3040–3053
24. Mauri C, Ehrenstein MR (2008) The 'short' history of regulatory B cells. Trends Immunol 29:34–40
25. Inoue S, Leitner WW, Golding B et al (2006) Inhibitory effects of B cells on antitumor immunity. Cancer Res 66:7741–7747
26. Walker WE, Goldstein D (2007) Neonatal B cells suppress innate toll-like receptor immune responses and modulate alloimmunity. J Immunol 179:1700–1710
27. Stang SL, Lopez-Campistrous A, Song X et al (2009) A proapoptotic signaling pathway involving RasGRP, Erk, and Bim in B cells. Exp Hematol 37:122–134
28. Katano M, Kubota E, Nagumo F et al (1994) Inhibition of tumor cell growth by a human B-cell line. Biotherapy 8:1–6
29. Lundy S (2009) Killer B lymphocytes: the evidence and the potential. Inflamm Res 58:345–357
30. Jahrsdorfer B, Blackwell SE, Wooldridge JE et al (2006) B-chronic lymphocytic leukemia cells and other B cells can produce granzyme B and gain cytotoxic potential after interleukin-21-based activation. Blood 108:2712–2719
31. Spaner D, Raju K, Radvanyi L et al (1998) A role for perforin in activation induced cell death. J Immunol 160:2655–2664
32. Lee YT, Sheikh KM, Quismorio F et al (1985) Circulating anti-tumor and autoantibodies in breast carcinoma: relationship to stage and prognosis. Breast Cancer Res Treat 6:57–65
33. Preuss KD, Zwick C, Bormann C et al (2002) Analysis of the B-cell repertoire against antigens expressed by human neoplasms. Immunol Rev 188:43–50
34. Catera R, Silverman G, Hatzi K et al (2008) Chronic lymphocytic leukemia cells recognize conserved epitopes associated with apoptosis and oxidation. Mol Med 14:665–674
35. Coronella-Wood JA, Hersh E (2003) Naturally occurring B-cell responses to breast cancer. Cancer Immunol Immunother 52:715–738
36. Disis ML, Knutson KL, Schiffman K et al (2000) Pre-existent immunity to the HER-2/neu oncogenic protein in patients with HER-2/neu overexpressing breast and ovarian cancer. Breast Cancer Res Treat 62:245–252
37. Kotlan B, Simsa P, Teillaud JL et al (2005) Novel ganglioside antigen identified by B cells in human medullary breast carcinomas: the proof of principle concerning the tumor-infiltrating B lymphocytes. J Immunol 175:2278–2285
38. Hansen MH, Nielsen H, Ditzel H (2001) The tumor-infiltrating B cell response in medullary breast cancer is oligoclonal and directed against the autoantigen actin exposed on the surface of apoptotic cancer cells. Proc Natl Acad Sci USA 98:12659–12664
39. Zeh HJ, III, Lotze M (2005) Addicted to death: invasive cancer and the immune response to unscheduled cell death. J Immunother 28:1–9
40. Vakkila J, Lotze M (2004) Inflammation and necrosis promote tumour growth. Nat Rev Immunol 4:641–648
41. Dunn G, Bruce A, Ikeda H et al (2002) Cancer immunoediting: from immunosurveillance to tumor escape. Nat Immunol 3:991–998
42. Zhang L, Conejo-Garcia J, Katsaros D et al (2003) Intratumoral T cells, recurrence, and survival in epithelial ovarian cancer. N Engl J Med 348:203–213
43. Lee A, Happerfield L, Bobrow L et al (1997) Angiogenesis and inflammation in ductal carcinoma in situ of the breast. J Pathol 181:200–206
44. Pavoni E, Monteriu G, Santapaola D et al (2007) Tumor-infiltrating B lymphocytes as an efficient source of highly specific immunoglobulins recognizing tumor cells. BMC Biotechnol 7:70
45. Zhang H, Lake D, Barbuto J et al (1995) A human monoclonal antimelanoma single-chain Fv antibody derived from tumor-infiltrating lymphocytes. Cancer Res 55:3584–3591

46. Imahayashi S, Ichiyoshi Y, Yoshino I et al (2000) Tumor-infiltrating B-cell-derived *IgG* recognizes tumor components in human lung cancer. Cancer Invest 18:530–536
47. Shigematsu Y, Hanagiri T, Kuroda K et al (2009) Malignant mesothelioma-associated antigens recognized by tumor-infiltrating B cells and the clinical significance of the antibody titers. Cancer Sci 100:1326–1334
48. Colotta F, Allavena P, Sica A et al (2009) Cancer-related inflammation, the seventh hallmark of cancer: links to genetic instability. Carcinogenesis 30:1073–1081
49. Hanahan D, Weinberg R (2000) The hallmarks of cancer. Cell 100:57–70
50. Smith R, Glynn T (2000) Epidemiology of lung cancer. Radiol Clin North Am 38:453–470
51. Miller AJ, Mihm MC, Jr (2006) Melanoma. N Engl J Med 355:51–65
52. Guo H, Loeb L (2003) Tumbling down a different pathway to genetic instability. J Clin Invest 112:1793–1795
53. Suarez F, Lortholary O, Hermine O et al (2006) Infection-associated lymphomas derived from marginal zone B cells: a model of antigen-driven lymphoproliferation. Blood 107:3034–3044
54. Hayashi P, Di Bisceglie A (2005) The progression of hepatitis B- and C-infections to chronic liver disease and hepatocellular carcinoma: presentation, diagnosis, screening, prevention, and treatment of hepatocellular carcinoma. Med Clin North Am 89:345–369
55. Schiffman M, Castle PE, Jeronimo J et al (2007) Human papillomavirus and cervical cancer. Lancet 370:890–907
56. Gatenby RA, Gillies R (2004) Why do cancers have high aerobic glycolysis? Nat Rev Cancer 4:891–899
57. Steinman RM, Turley S, Mellman I et al (2000) The induction of tolerance by dendritic cells that have captured apoptotic cells. J Exp Med 191:411–416
58. Green DR, Ferguson T, Zitvogel L et al (2009) Immunogenic and tolerogenic cell death. Nat Rev Immunol 9:353–363
59. Dvorak H (1986) Tumors: wounds that do not heal. Similarities between tumor stroma generation and wound healing. N Engl J Med 315:1650–1659
60. Schafer M, Werner S (2008) Cancer as an overhealing wound: an old hypothesis revisited. Nat Rev Mol Cell Biol 9:628–638
61. Kahn J (2009) HPV vaccination for the prevention of cervical intraepithelial neoplasia. N Engl J Med 361:271–278
62. Cheson BD, Leonard J (2008) Monoclonal antibody therapy for B-cell non-Hodgkin's lymphoma. N Engl J Med 359:613–626
63. Hudis C (2007) Trastuzumab-mechanism of action and use in clinical practice. N Engl J Med 357:39–51
64. Yang JC, Haworth L, Sherry R et al (2003) A randomized trial of bevacizumab, an anti-vascular endothelial growth factor antibody, for metastatic renal cancer. N Engl J Med 349:427–434
65. Irie RF, Ollila DW, O'Day S et al (2004) Phase I pilot clinical trial of human IgM monoclonal antibody to ganglioside GM3 in patients with metastatic melanoma. Cancer Immunol Immunother 53:110–117
66. Pastan I, Hassan R, FitzGerald D et al (2007) Immunotoxin treatment of cancer. Annu Rev Med 58:221–37
67. Park J, Terabe M, Steel J et al (2008) Therapy of advanced established murine breast cancer with a recombinant adenoviral ErbB-2/neu vaccine. Cancer Res 68:1979–1987
68. Quan N, Zhang Z, Demetrikopoulos M et al (1999) Evidence for involvement of B lymphocytes in the surveillance of lung metastasis in the rat. Cancer Res 59:1080–1089
69. Jones HP, Wang YC, Aldridge B et al (2008) Lung and splenic B cells facilitate diverse effects on *in vitro* measures of antitumor immune responses. Cancer Immun 8:4
70. Qin Z, Blankenstein T (1995) Tumor growth inhibition mediated by lymphotoxin: evidence of B lymphocyte involvement in the antitumor response. Cancer Res 55:4747–4751
71. Reisfeld RA, Gillies SD, Mendelsohn J et al (1996) Involvement of B lymphocytes in the growth inhibition of human pulmonary melanoma metastases in athymic nu/nu mice by an antibody-lymphotoxin fusion protein. Cancer Res 56:1707–1712

72. Cartron G, Watier H, Golay J et al (2004) From the bench to the bedside: ways to improve rituximab efficacy. Blood 104:2635–2642
73. Houghton AN, Uchi H, Wolchok J (2005) The role of the immune system in early epithelial carcinogenesis: B-ware the double-edged sword. Cancer Cell 7:403–405
74. de Visser KE, Korets LV, Coussens L (2005) De novo carcinogenesis promoted by chronic inflammation is B lymphocyte dependent. Cancer Cell 7:411–423
75. Hedlund M, Padler-Karavani V, Varki NM et al (2008) Evidence for a human-specific mechanism for diet and antibody-mediated inflammation in carcinoma progression. Proc Natl Acad Sci USA 105:18936–18941
76. Spaner D (2004) Amplifying cancer vaccine responses by modifying pathogenic gene programs in tumor cells. J Leukoc Biol 76:338–351
77. Smith KM, Garside P, McNeil R et al (2006) Analysis of costimulatory molecule expression on antigen-specific T and B cells during the induction of adjuvant-induced Th1 and Th2 type responses. Vaccine 24:3035–3043
78. Qin Z, Richter G, Schuler T et al (1998) B cells inhibit induction of T cell-dependent tumor immunity Nat Med 4:627–630
79. Iwata Y, Yoshizaki A, Komura K et al (2009) CD19, a response regulator of B lymphocytes, regulates wound healing through hyaluronan-induced TLR4 signaling. Am J Pathol 175:649–660
80. Oliveira H, Popi A, Bachi A et al (2010) B-1 cells modulate the kinetics of wound-healing process in mice. Immunobiology 215:215–222
81. Harless W (2009) Revisiting perioperative chemotherapy: the critical importance of targeting residual cancer prior to wound healing. BMC Cancer 9:118
82. Melcher A, Todryk S, Hardwick N et al (1998) Tumor immunogenicity is determined by the mechanism of cell death via induction of heat shock protein expression. Nat Med 4:581–587
83. Scholl PR, Geha R (1994) MHC class II signaling in B-cell activation. Immunol Today 15:418–422
84. Staquicini F, Tandle A, Libutti S et al (2008) A subset of host B lymphocytes controls melanoma metastasis through a melanoma cell adhesion molecule/MUC18-dependent interaction: evidence from mice and humans. Cancer Res 68:8419–8428
85. Hallek M, Cheson BD, Catovsky D et al (2008) International workshop on chronic lymphocytic leukemia. Guidelines for the diagnosis and treatment of chronic lymphocytic leukemia: a report from the International Workshop on Chronic Lymphocytic Leukemia updating the National Cancer Institute-Working Group 1996 guidelines. Blood 111:5446–5456
86. Caligaris-Cappio F, Ghia P (2008) Novel insights in chronic lymphocytic leukemia: are we getting closer to understanding the pathogenesis of the disease? J Clin Oncol 26:4497–4503
87. Lanemo MA, Hellqvist E, Sidorova E et al (2008) A new perspective: molecular motifs on oxidized LDL, apoptotic cells, and bacteria are targets for chronic lymphocytic leukemia antibodies. Blood 111:3838–3848
88. Cantwell M, Hua T, Pappas J, Kipps T (1997) Acquired CD40-ligand deficiency in chronic lymphocytic leukemia. Nat Med 3:984–989
89. Tinhofer I, Marschitz I, Kos M et al (1998) Differential sensitivity of CD4+ and CD8+ T lymphocytes to the killing efficacy of Fas (Apo-1/CD95) ligand+ tumor cells in B chronic lymphocytic leukemia. Blood 91:4273–4281
90. Ramsay A, Johnson A, Lee A et al (2008) Chronic lymphocytic leukemia T cells show impaired immunological synapse formation that can be reversed with an immunomodulating drug. J Clin Invest 118:2427–2437
91. Stevenson F, Caligaris-Cappio F (2004) Chronic lymphocytic leukemia: revelations from the B-cell receptor. Blood 103:4389–4395
92. Tsimberidou AM, Wen S, McLaughlin P et al (2009) Other malignancies in chronic lymphocytic leukemia/small lymphocytic lymphoma. J Clin Oncol 27:904–910
93. Hisada M, Biggar R, Greene M et al (2001) Solid tumors after chronic lymphocytic leukemia. Blood 98:1979–1981
94. Mehrany K, Weenig R, Pittelkow M et al (2004) High recurrence rates of Basal cell carcinoma after Mohs surgery in patients with chronic lymphocytic leukemia. Arch Dermatol 140:985–988

95. Hartley B, Searle AE, Breach NM et al (1996) Aggressive cutaneous squamous cell carcinoma of the head and neck in patients with chronic lymphocytic leukaemia. J Laryngol Otol 110:694–695
96. Prehn RT (1994) Stimulatory effects of immune reactions upon the growths of untransplanted tumors. Cancer Res 54:908–914
97. Kaminski JM, Shinohara E, Summers J et al (2005) The controversial abscopal effect. Cancer Treat Rev 31:159–172
98. Appay V, Voelter V, Rufer N et al (2007) Combination of transient lymphodepletion with busulfan and fludarabine and peptide vaccination in a phase I clinical trial for patients with advanced melanoma. J Immunother 30:240–250
99. Spaner D, Sheng-Tanner X, Schuh A (2002) Aberrant regulation of superantigen responses during T-cell reconstitution and graft-versus-host disease in immunodeficient mice. Blood 100:2216–2224
100. Lake RA, Robinson B (2005) Immunotherapy and chemotherapy – a practical partnership. Nat Rev Cancer 5:397–405
101. Rastetter W, Molina A, White C (2004) Rituximab: expanding role in therapy for lymphomas and autoimmune diseases. Annu Rev Med 55:477–503
102. Castillo J, Milani C, Mendez-Allwood D (2009) Ofatumumab, a second-generation anti-CD20 monoclonal antibody, for the treatment of lymphoproliferative and autoimmune disorders. Expert Opin Investig Drugs 18:491–500
103. Rosenwasser L, Meng J (2005) Anti-CD23. Clin Rev Allergy Immunol 29:61–72
104. Kim S, Fridlender Z, Dunn R et al (2008) B-cell depletion using an anti-CD20 antibody augments antitumor immune responses and immunotherapy in nonhematopoetic murine tumor models. J Immunother 31:446–457
105. Barbera-Guillem E, Nelson M , Barr B et al (2000) B lymphocyte pathology in human colorectal cancer. Experimental and clinical therapeutic effects of partial B cell depletion. Cancer Immunol Immunother 48:541–549
106. Oizumi S, Deyev V, Yamazaki K et al (2008) Surmounting tumor-induced immune suppression by frequent vaccination or immunization in the absence of B cells. J Immunother 31:394–401
107. Rosenberg S, Restifo N, Yang J et al (2008) Adoptive cell transfer: a clinical path to effective cancer immunotherapy. Nat Rev Cancer 8:299–308
108. Palanichamy A, Barnard J, Zheng B et al (2009) Novel human transitional B cell populations revealed by B cell depletion therapy. J Immunol 182:5982–5993
109. Chang M (2009) Cancer prevention by vaccination against hepatitis B. Recent Results Cancer Res 181:85–94
110. Del G, Malfertheiner P, Rappuoli R (2009) Development of vaccines against Helicobacter pylori. Expert Rev Vaccines 8:1037–1049
111. Javier R, Butel J (2008) The history of tumor virology. Cancer Res 68:7693–7706
112. Klepfish A, Rachmilewitz E, Kotsianidis I et al (2008) Adding fresh frozen plasma to rituximab for the treatment of patients with refractory advanced CLL. QJM 101:737–740
113. Taylor R, Lindorfer M (2008) Immunotherapeutic mechanisms of anti-CD20 monoclonal antibodies. Curr Opin Immunol 20:444–449
114. Spaner D, Hammond C, Mena J et al (2004) Effect of IL-2R beta-binding cytokines on costimulatory properties of chronic lymphocytic leukaemia cells: implications for immunotherapy. Br J Haematol 127:531–542
115. Spaner D, Masellis A (2007) Toll-like receptor agonists in the treatment of chronic lymphocytic leukemia. Leukemia 21:53–60
116. Sotomayor E, Borrello I, Tubb E et al (1999) Conversion of tumor-specific CD4+ T-cell tolerance to T-cell priming through *in vivo* ligation of CD40. Nat Med 5:780–787
117. Khalil M, Vonderheide R (2007) Anti-CD40 agonist antibodies: preclinical and clinical experience. Update. Cancer Ther 2:61–65
118. Rech J, Repp R, Rech D et al (2006) A humanized HLA-DR antibody (hu1D10, apolizumab) in combination with granulocyte colony-stimulating factor (filgrastim) for the treatment of non-Hodgkin's lymphoma: a pilot study. Leuk Lymphoma 47:2147–2154

119. Hamzah J, Nelson D, Moldenhauer G et al (2008) Vascular targeting of anti-CD40 antibodies and IL-2 into autochthonous tumors enhances immunotherapy in mice. J Clin Invest 118:1691–1699
120. Spaner D, Foley R, Galipeau J et al (2008) Obstacles to effective Toll-like receptor agonist therapy for hematologic malignancies. Oncogene 27:208–217
121. Minchinton A, Tannock I (2006) Drug penetration in solid tumours. Nat Rev Cancer 6:583–592
122. Spaner D, White D, Shaha S et al (2010) A phase I/II trial of toll-like receptor-7 agonist immunotherapy in chronic lymphocytic leukemia. Leukemia 24:222–226
123. Kondo E, Gryschok L, Klein-Gonzalez N et al (2009) CD40-activated B cells can be generated in high number and purity in cancer patients: analysis of immunogenicity and homing potential. Clin Exp Immunol 155:249–256
124. Schultze J, Grabbe S, Bergwelt-Baildon M (2004) DCs and CD40-activated B cells: current and future avenues to cellular cancer immunotherapy. Trends Immunol 25:659–664
125. Woodland D (2004) Jump-starting the immune system: prime-boosting comes of age. Trends Immunol 25:98–104
126. Morgan R, Dudley M, Wunderlich J et al (2006) Cancer regression in patients after transfer of genetically engineered lymphocytes. Science 314:126–129

Chapter 3
Monoclonal Antibody Therapy for Cancer

Christoph Rader

Abstract Since the approval of *rituximab* (Rituxan®) for the treatment of B-cell non-Hodgkin's lymphoma (B-NHL) in 1997, nine additional monoclonal antibodies (mAbs) have been approved by the FDA for cancer therapy. Currently, more than 1,300 clinical studies registered at ClinicalTrials.gov investigate mAb therapy of cancer, including more than 150 phase III clinical trials. In concert with their clinical acceptance, mAbs in oncology have become commercially attractive. Four out of the ten approved mAbs have reached blockbuster status with annual sales exceeding $1 billion. The top three selling cancer drugs are all mAbs. These numbers indicate the potential of mAbs to play a leading role in cancer therapy for decades to come. Although mAbs provide a proven drug platform beyond the proof-of-concept stage, future success will depend on broadening and potentiating mAb therapy through antigen discovery, antibody engineering, use of mAbs in combination with chemotherapy and radiotherapy, and personalized medicine.

Keywords Antibody engineering • Cancer • Hematologic malignancy • Solid malignancy • Therapeutic monoclonal antibodies

General Considerations

Introduction

Cancer immunotherapy is based on administered or induced components of the immune system that selectively target and eradicate tumor cells. Strategies include monoclonal antibody (mAb) therapy, cytokine therapy, vaccination, hematopoietic stem cell transplantation, and adoptive cell transfer [1]. Since the approval of *rituximab*

C. Rader (✉)
Antibody Technology Section, Experimental Transplantation and Immunology Branch, Center for Cancer Research, National Cancer Institute, NIH, Bethesda, MD, USA
e-mail: raderc@mail.nih.gov

J. Medin and D. Fowler (eds.), *Experimental and Applied Immunotherapy*,
DOI 10.1007/978-1-60761-980-2_3,

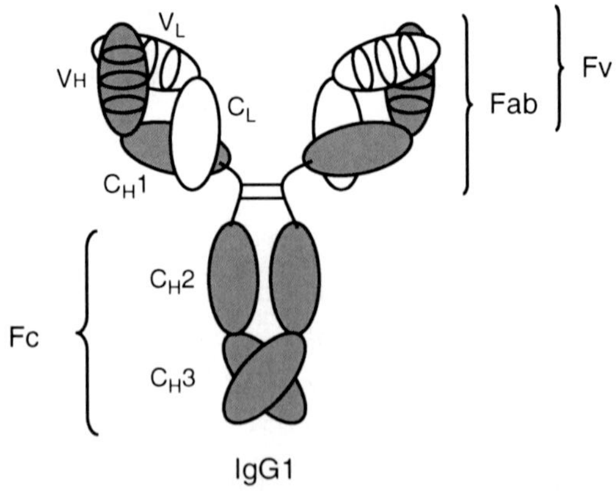

Fig. 3.1 IgG1 molecule. All approved mAbs in oncology and the vast majority of investigational mAbs in phase III clinical trials for cancer therapy are based on the IgG format. The 150-kDa IgG1 molecule is the dominantly used IgG format. It contains two identical light chains (white) and two identical heavy chains (gray). The light chains consist of an N-terminal variable domain (V_L) followed by one constant domain (C_L). The heavy chain consists of an N-terminal variable domain (V_H) followed by three constant domains (C_H1, C_H2, and C_H3). C_H1 and C_H2 are linked through a flexible hinge region that anchors four interchain disulfide bridges of the IgG1 molecule, one for each of the two light and heavy chain pairs (not shown) and two for the heavy chain pair (shown). The antigen binding site is formed by six CDRs, three provided by each variable domain. Fv, Fab, and Fc portions of the IgG1 molecule are pointed out

(Rituxan®) by the FDA in 1997, mAb therapy has become a paradigm for the success and promise of immunotherapy of cancer. In this chapter, I will examine why the antibody molecule (Fig. 3.1) is an exceptionally successful drug format for cancer therapy and how this is reflected in currently approved and investigational treatments.

MAb therapy is sometimes referred to as passive immunization, emphasizing injection of exogenous antibodies rather than induction of endogenous ones. Nonetheless, the immune system of the cancer patient can play an active role in mAb therapy by contributing cells and proteins that mediate the antitumor activity of antibodies. On the other hand, certain exogenous antibodies can eradicate tumor cells on their own by blocking survival signals, inducing apoptosis, or delivering a cytotoxic payload with independent antitumor activity. To date, 28 mAbs have achieved FDA approval, including ten mAbs for cancer therapy [2] (Fig. 3.2).[1] In addition, more than 200 different mAbs are currently in clinical trials [3]. Of 248 phase III clinical trials with mAbs registered at ClinicalTrials.gov (search terms: monoclonal antibody, interventional studies, phase III; date of search: June 5, 2009), 156 (63%) are investigated for cancer therapy. In concert with their clinical success, mAbs have become commercially viable [4–6]. More than 100 companies worldwide have mAbs in

[1] http://www.landesbioscience.com/journals/mabs/about#background. Accessed June 23, 2010.

Name	Format	Conjugation	Chemotherapy	Antigen	Indication	Approval
Rituximab (Rituxan®)	chimeric mouse/human IgG1κ	none	yes/no	CD20	B-NHL/ B-CLL	1997/ 2010
Trastuzumab (Herceptin®)	humanized IgG1κ	none	yes	HER2	breast cancer	1998
Gemtuzumab ozogamicin (Mylotarg®)	humanized IgG4κ	calicheamicin	no	CD33	AML	2000
Alemtuzumab (Campath®)	humanized IgG1κ	none	no	CD52	B-CLL	2001
Ibritumomab tiuxetan (Zevalin®)	mouse IgG1κ	^{90}Y	no	CD20	B-NHL	2002
Tositumomab (Bexxar®)	mouse IgG2aλ	^{131}I	no	CD20	B-NHL	2003
Cetuximab (Erbitux®)	chimeric mouse/human IgG1κ	none	yes/no	EGFR	colon cancer/ head and neck cancer	2004/ 2006
Bevacizumab (Avastin®)	humanized IgG1κ	none	yes	VEGF	colon cancer/ lung cancer/ breast cancer/ glioblastoma/ kidney cancer	2004/ 2006/ 2008/ 2009/ 2009
Panitumumab (Vectibix®)	human IgG2κ	none	no	EGFR	colon cancer	2006
Ofatumumab (Arzerra®)	human IgG1κ	none	no	CD20	B-CLL	2009

Fig. 3.2 FDA-approved mAbs for cancer therapy in chronological order. Since 1997, ten mAbs have been approved for cancer therapy. Six of them are indicated for hematologic malignancies and four for solid malignancies. Four mAbs are approved both as single agent and in combination with chemotherapy, depending on the specific indication. The only antibody-drug conjugate, gemtuzumab ozogamicin, was withdrawn on June 21, 2010

clinical trials. Global sales in 2008 were ~$33 billion, or ~4% of total pharmaceutical sales.[2] Notably, based on global sales in 2008, mAbs comprised the top three cancer drugs.[3]

Precision and Predictability

Key features of mAb therapy at the center of both their clinical and commercial success are precision and predictability. An analysis by the Tufts Center for the Study of Drug Development [6] revealed the approval success rates of mAbs in clinical trials are considerably higher compared to traditional small molecule drugs. This favorable difference is most pronounced in oncology with an approval success rate of ~15% for clinically investigated mAbs compared to ~5% for clinically investigated small molecule drugs; localized activity and low toxicity are the

[2] Maggon K (2009) Global monoclonal antibodies market review 2008. http://knol.google.com. Accessed January 9, 2010.

[3] Maggon K (2009) Global cancer market review 2008. http://knol.google.com. Accessed January 9, 2010.

prevailing advantages of all approved and most investigational mAbs. However, adverse events in mAb therapy can be severe [7]. In particular, the interplay with the patient's immune system can render mAbs targeting healthy lymphocytes dangerous, as recently observed for the already approved anti-integrin α_4 and α_L mAbs *natalizumab* (Tysabri®) and *efalizumab* (Raptiva®), respectively; these agents caused progressive multifocal leukoencephalopathy (PML) in a small percentage of patients. PML is caused by activation of a latent virus that infects and destroys oligodendrocytes in the white matter of the brain and, due to a lack of available treatments, is one of the most deadly opportunistic infections in AIDS and other immunocompromized patients, including those treated with certain mAbs [8]. Recently, PML has been confirmed as a rare adverse event of treatment with anti-CD20 mAb *rituximab* [9]. More dramatically, an investigational anti-CD28 mAb triggered a cytokine storm followed by multi-organ failure in 100% of healthy volunteers in a phase I clinical trial [10]. Preclinical investigations in nonhuman primates had not predicted human toxicity. In contrast with all approved mAbs and the majority of investigational mAbs, the anti-CD28 mAb was designed to activate rather than block or destroy its target cells; thus, the significant toxicity experience with the anti-CD28 mAb did not significantly alter the general perception of mAbs as relatively safe drugs, particularly in oncology. In addition, clinical trials proved the safety of immunomodulatory mAbs [11] that act through antagonizing inhibitory receptors on T cells such as CTLA4 and PD1 rather than through agonizing activating receptors such as CD3 and CD28 [12, 13].

Clinical and commercial predictability comes with molecular precision. This precision of mAb therapy is in one part founded on high specificity and strong affinity characteristic for antibody/antigen interactions and in another part due to the local confinement of the large antibody molecule (150 kDa) to the circulatory system and interstitial space. Small molecule drugs (<1 kDa), by contrast, have unlimited access to nearly all extracellular and intracellular niches, making activity and toxicity profiles less predictable.

From Hematologic to Solid Malignancies

Based on approved mAbs for cancer therapy as well as mAbs in preclinical and clinical development, hematologic malignancies, which comprise only 10% of all cancers, appear to be preferred target diseases. Currently, hematologic malignancies are indications for six out of ten (60%) approved mAbs (Fig. 3.2) and 64 out of 156 (41%) phase III clinical trials for cancer therapy registered at ClinicalTrials.gov. Why are mAbs particularly suited for the therapy of hematologic malignancies? First, the ready accessibility and availability of both target cells and effector cells and proteins in blood, bone marrow, and lymph nodes provides an optimal battle-ground for the antibody molecule. Although hematologic malignancies with bulky disease, defined as a tumor mass of >10 cm in any diameter, impose penetration challenges they still provide better access for mAbs due to their integration in circulatory and lymphatic systems. Second, the expression of cell surface proteins in hematologic malignancies is typically more homogenous and more defined than in solid malignancies. Most hematologic malignancies originate from lymphoid or myeloid cell transformations. These cells are characterized by defined combinations of cell surface proteins, also termed immunophenotypes, thereby allowing precise targeting of those hematopoietic cell lineages from which the malignant cells originated. Among the approved mAbs (Fig. 3.2), for example, is the anti-CD33 mAb *gemtuzumab ozugamicin* (Mylotarg®), which targets myeloid cells. Within the lymphoid cell lineage, the anti-CD20 mAbs *rituximab* (Rituxan®), *ibritumomab tiuxetan* (Zevalin®), *iodine I 131 tositumomab* (Bexxar®), and *ofatumumab* (Arzerra®) selectively target B cells. The anti-CD52 mAb *alemtuzumab* (Campath®), on the other hand, displays broader specificity by targeting both myeloid and lymphoid cells. It is anticipated that the discovery of cell surface proteins selectively expressed in malignant lymphoid or myeloid cells will permit further refinement of mAb therapy. Third, hematologic malignancies are often accompanied by qualitative and quantitative defects in normal leukocytes. Impaired cellular and humoral immune responses might render mAbs less immunogenic in patients with leukemia, lymphoma, and myeloma compared to cancer patients with less compromised immune systems. In addition, mAbs like rituximab and alemtuzumab may mask their intrinsic immunogenicity by targeting normal in addition to malignant leukocytes [14]. Antibody immunogenicity, which has been observed for all approved mAbs in at least a subset of patients, can decrease activity and increase toxicity profiles of mAbs [15]. Fourth, a substantial portion of hematologic malignancies, such as indolent B-cell non-Hodgkin's lymphoma (B-NHL) and B-cell chronic lymphocytic leukemia (B-CLL) progress slowly. The chronic nature of these cancers and the usually advanced age of the patients may favor gentler over aggressive treatments with the goal of disease containment rather than cure.

Based on these considerations, hematologic malignancies in general and B-cell malignancies in particular [16] have not only become dominant indications for current mAb therapy but also a preferred training ground for the development of the next generation of mAbs, such as antibody-drug conjugates [17] designed to further increase antitumor activity while maintaining low toxicity. Despite the aforementioned challenges, the knowledge gained from treating hematologic malignancies with established and investigational mAb platforms has spurred the development of mAbs for the therapy of solid malignancies. In fact, three out of four approved mAbs for the therapy of solid malignancies have already reached blockbuster drug status with annual global sales exceeding $1 billion, including: anti-HER2 mAb *trastuzumab* (Herceptin®), anti-EGFR mAb *cetuximab* (Erbitux®), and anti-VEGFA mAb *bevacizumab* (Avastin®) [3]. *Bevacizumab*, which has demonstrated therapeutic activity in a broad range of solid malignancies and is approved in combination with chemotherapy for the treatment of metastatic colorectal cancer, lung cancer, and metastatic breast cancer, in combination with interferon alpha (IFN-α) for metastatic kidney cancer, and as a single agent for the treatment of glioblastoma, is set to become the commercially most successful mAb in oncology. Its manufacturer Genentech (Table 3.2), which also makes blockbuster drugs *rituximab* and

Table 3.1 New antigens targeted by mAbs in phase III clinical trials for cancer therapy

Antigen	MAb type(s)	Indication(s)
CA9	Chimeric mouse/human IgG	Kidney cancer
CD4	Human IgG	T-NHL
CD22	Humanized IgG Humanized IgG/calicheamicin conjugate	B-NHL
CTLA4	Human IgG	Melanoma
EPCAM	Mouse IgG Bispecific antibody	Colorectal cancer Malignant ascites
GD2	Chimeric mouse/human IgG	Neuroblastoma
GD3	Mouse IgG (anti-idiotype)	Lung cancer
IGF1R	Human IgG	Lung cancer
MUC1	Mouse IgG ^{90}Y-labeled mouse IgG	Colorectal cancer Ovarian cancer
RANKL	Human IgG	Bone metastases

ClinicalTrials.gov search terms: Monoclonal antibody, interventional studies, phase III (June 5, 2009)

trastuzumab, was fully acquired by Hoffmann-La Roche in 2009, following a trend of acquisitions of biotech companies with approved mAbs or mAbs in clinical development by pharma companies. Another prominent example is ImClone Systems (Table 3.2), the manufacturer of *cetuximab*, which was acquired by Eli Lilly and Company in 2008. Fueled by the large market of patients with solid malignancies, dominated by lung cancer, colorectal cancer, breast cancer, and prostate cancer, substantial efforts by biotech and pharma companies have led to a rich and innovative pipeline of mAbs in preclinical and clinical development. In fact, a closer look at phase III clinical trials shows that mAbs for solid malignancies reveal a trend toward more innovation than mAbs for hematologic malignancies with respect to targeting new antigens. That is, out of the above-mentioned 64 phase III clinical trials in hematologic malignancies, only four (6%) investigate new mAbs to new antigens; the remaining 60 cases either investigate new indications or regimens of already-approved mAbs or investigate new mAbs to already established antigens (albeit to new epitopes). By contrast, 15 (16%) out of 92 phase III clinical trials in solid malignancies involve new mAbs to new antigens. New antigens targeted by mAbs in phase III clinical trials are shown in Table 3.1.

The trend to more innovation reflects the additional challenges that mAbs for solid malignancies face. Particularly innovative is the strategy of targeting antigens that are not expressed on tumor cells. For example, *bevacizumab* targets VEGFA, a soluble protein secreted by both tumor and normal cells that is involved in the formation of blood vessels that infiltrate tumors in a process known as tumor angiogenesis [18]. A number of drugs inhibiting tumor angiogenesis, thereby blocking the delivery of nutrients and oxygen to the growing tumor, are being investigated clinically for cancer therapy. Although *bevacizumab* is considered the first approved tumor angiogenesis inhibitor, identification of a precise mechanism of action has been clouded by the finding that the antitumor activity of chemotherapy, which also depends on blood delivery, improves in the presence of *bevacizumab* [19]. Another example of a clinically investigated tumor angiogenesis-inhibiting mAb is *bavituximab* (Table 3.2), which binds to

Table 3.2 Generic names of mAbs mentioned in this chapter

Generic name	Antigen	Manufacturer
Adalimumab	TNFα	Abbott Laboratories
Alemtuzumab	CD52	Genzyme
Bavituximab	Phosphatidylserine	Peregrine
Bevacizumab	VEGFA	Genentech
Blinatumomab	CD19/CD3 bispecific	Micromet
Catumaxomab	EPCAM/CD3 bispecific	Trion Pharma
Cetuximab	EGFR	ImClone Systems
Denosumab	RANKL	Amgen
Efalizumab	CD11A	Genentech
Figitumumab	IGF1R	Pfizer
Gemtuzumab ozugamicin	CD33	Wyeth
Ibritumomab tiuxetan	CD20	Biogen-Idec
Ipilimumab	CTLA4	Medarex
Mitumomab	GD3	ImClone Systems
Natalizumab	CD49D	Biogen Idec
Nimotuzumab	EGFR	YM BioSciences
*Ocrelizuma*b	CD20	Genentech
Ofatumumab	CD20	Genmab
Panitumumab	EGFR	Amgen
Rituximab	CD20	Genentech
Iodine I 131 tositumomab	CD20	GlaxoSmithKline
Trastuzumab	HER2	Genentech
Tremelimumab	CTLA4	Pfizer
Zalutumumab	EGFR	Genmab
Zanolimumab	CD4	Genmab

The generic nomenclature of mAbs indicates the class of pharmaceutical ("*mab*") in the suffix, preceded by animal origin (e.g., "*o*" for mouse, "*xi*" for chimeric mouse/human, "*axo*" for mouse/rat hybrid, "*zu*" for humanized, and "*u*" for human), preceded by the disease or target class (e.g., "*tu*" or "*tum*" for tumor, "*li*" or "*lim*" for immune system, "*ci*" or "*cir*" for cardiovascular, and "*os*" for bone), preceded by a unique prefix. For example, *beva-ci-zu-mab* is a humanized mAb that targets a cardiovascular antigen whereas *pani-tum-u-mab* is a human mAb that targets a tumor antigen. Antibody-drug conjugates and radioimmunoconjugates include names for the conjugated drug, chelate, or radioisotopes.

a lipid antigen, phosphatidylserine, that is selectively displayed on tumor endothelial cells [20]. Two anti-CTLA4 mAbs, *ipilimumab* and *tremelimumab*, which are in phase III clinical trials for the therapy of melanoma, exemplify a new class of immunomodulatory mAbs that target the general immune system of cancer patients [13]. Blockade of CTLA4, a T-cell surface protein that suppresses CD28-mediated activation and other inhibitory checkpoints in the immune system, is an emerging strategy for cancer immunotherapy [21], often in combination with cancer vaccines [22, 23]. Another promising mAb in phase III clinical trial that targets an antigen involved in tumor-host interactions is the anti-RANKL mAb *denosumab*. This mAb has been investigated for the treatment of osteoporosis, rheumatoid arthritis, and multiple myeloma; importantly, *denosumab* inhibits bone metastasis in solid malignancies by interfering with the interaction of RANKL and RANK that signals bone removal [24].

Direct and Indirect Mechanisms of Activity

In cancer therapy, mAbs mediate potent and selective cytotoxicity through direct or indirect mechanisms. The selective delivery of a drug as in the case of *gemtuzumab ozogamicin* or radioisotopes as in the case of *ibritumomab tiuxetan* and *iodine I 131 tositumomab* (Fig. 3.2) represents a clear therapeutic concept [17, 25]; by comparison, the mechanism of activity for unconjugated (also termed naked) mAbs, which are the dominant format among approved and investigational mAbs, is not only variable but has remained vague [26]. By binding to its antigen, which could be a receptor or ligand, a mAb can block receptor/ligand interactions that are crucial for tumor cell survival. For example, by binding to VEGFA secreted by tumor cells, *bevacizumab* blocks its interaction with VEGFR2 on endothelial cells that line tumor-infiltrating blood vessels. Antagonizing VEGFR2 leads to endothelial cell apoptosis that precedes tumor cell apoptosis [27]. *Cetuximab* has a similar mechanism of activity that targets tumor cells directly by blocking the EGF receptor (EGFR or ERBB1) [28]. In addition to the blockade of receptor/ligand interactions, receptor cross-linking is another mechanism of action that is mediated by the bivalent Fab portion of the antibody molecule (Fig. 3.1). For example, *trastuzumab*-mediated cross-linking of HER2 (ERBB2), an ERBB receptor family member without a known ligand, induces tumor cell apoptosis. Cross-linking of CD20 by *rituximab* can also activate apoptotic signaling pathways in B-NHL [29] and B-CLL [30] cells. These proapoptotic mAbs have been shown to sensitize tumor cells to chemotherapy and radiotherapy, thereby providing a strong rationale for combination treatments [31]. Furthermore, preclinical investigations have shown that chemotherapy can induce the cell surface expression of certain antigens that subsequently can be targeted by mAbs or antibody-drug conjugates [32, 33].

Completely different mechanisms of activity of some mAbs are mediated by the Fc portion (Fig. 3.1) of IgG1, which is the dominant isotype used for mAbs in oncology (for an article on mAb isotype selection, see [34]). Two effector functions termed complement-dependent cytotoxicity (CDC) and antibody-dependent cellular cytotoxicity (ADCC) define these mechanisms of activity. In CDC, the Fc portion of tumor cell surface-bound IgG1 recruits complement protein C1q, thus triggering a cytolytic response through activation of the complement cascade. CDC is thought to contribute to the activity of *rituximab* [35, 36]. In ADCC, the Fc portion of tumor cell surface-bound IgG1 activates FcμRIIIa (CD16A) and FcμRIIa (CD32A) on effector cells such as natural killer cells and macrophages. ADCC is thought to be a key mechanism of activity of several mAbs approved for the therapy of hematologic or solid malignancies [37]. In fact, the valine/phenylalanine polymorphism at position 158 in the amino acid sequence of FcμRIIIa and the histidine/arginine polymorphism at position 131 of FcμRIIa are known to modulate IgG1 binding and ADCC, and also can influence clinical responses to *rituximab* [38] and *trastuzumab* [39]. Although the Fcμ receptor (FcμR) genotype in cancer patients can predict clinical responses to conventional mAbs, a new generation of engineered mAbs has been developed to overcome these limitations and mediate more potent ADCC through improved affinity to FcμRs. This can be achieved through protein or carbohydrate engineering of the Fc portion of IgG1 [40, 41]. Collectively, ADCC is considered a key mechanism by which naked mAbs with IgG1 isotype mediate cytotoxicity in cancer therapy. Another possible mechanism by which the cancer patient's immune system contributes to the activity of mAbs is through cross-presentation of antigens released by dying tumor cells to FcμR-expressing antigen presenting cells, thus triggering T-cell responses. This vaccine effect of mAbs was recently demonstrated by the induction of idiotype-specific T cell responses in follicular lymphoma patients following treatment with *rituximab* [42]. In addition to triggering cellular immune responses, therapeutic mAbs may also initiate humoral immune responses in the cancer patient through triggering an idiotypic cascade. The idiotypic cascade can include endogenous anti-idiotypic antibodies that mimic the antigen and anti–anti-idiotypic antibodies that bind the antigen [43]. Likewise, anti-idiotypic mAbs such as the GD3-mimicking mouse mAb *mitumomab* (Table 3.1 and 3.2), which is in phase III clinical trials for lung cancer therapy, are intended to act as vaccines as they are more immunogenic than the antigen.

Whereas the utilization of mAbs for the selective delivery of cytotoxic payloads provides a more straightforward mechanism of activity that, arguably, is independent of the cancer patient's immune system and genetic background, antibody-drug conjugates and radioimmunoconjugates are lagging behind in terms of market and pipeline share. This situation may be related primarily to issues such as manufacturability, shelf-life, and administration challenges. However, in terms of potency, antibody-drug conjugates and radioimmunoconjugates are often superior to naked mAbs and do not require mAb use in combination with chemotherapy. One of several promising concepts are immunotoxins, in particular recombinant fusions of the Fv portion (Fig. 3.1) of mAbs with a truncated form of the bacterial toxin *Pseudomonas* exotoxin A [44]. For example, an anti-CD22 immunotoxin gave complete (CR) and partial responses (PR) in more than 50% of patients with refractory or relapsed hairy cell leukemia in phase I and II clinical trials at the National Cancer Institute [45, 46].

Antigen

The success of mAb therapy for cancer depends on the identification of antigens that are specifically expressed on the cell surface of tumor cells or tumor-supporting cells. In addition, growth factors that are specifically expressed by tumor or tumor supporting cells can serve as molecular targets for mAb therapy. By binding to these extracellular antigens, mAbs mediate the selective destruction of tumor cells. In contrast to conventional treatments, mAb therapy in theory does not harm healthy cells because they do not express the antigens; consequently, mAb therapy will cause fewer side effects. As candidates for mAb therapy, antigens should be expressed at high levels on the surface of tumor cells or tumor-supporting cells and should be absent from normal tissues and stem cells. Although the ideal antigen is expressed in the context of the tumor only, few truly tumor-specific antigens have been identified [47]. Nonetheless, a number of antigens with broader expression have proven to be useful for mAb therapy, as long as their expression is restricted to less critical normal tissues. The antigens CD20, CD33, and CD52, which are targeted by six out of ten approved mAbs (Fig. 3.2), are also expressed on normal blood cells of lymphoid or myeloid lineage, but not on hematopoietic stem cells. The receptor tyrosine kinase (RTK) antigens EGFR, HER2, and IGF1R [targeted by approved mAbs *cetuximab*, *trastuzumab* and *panitumumab* (Fig. 3.2) and investigational mAbs *figitumumab*, *nimotuzumab*, and *zalutumumab* (Table 3.2)] are overexpressed in some carcinomas but also expressed at lower levels in normal epithelial cells. Similarly, VEGFA, the antigen targeted by the approved mAb *bevacizumab*, is overexpressed in tumor compared to healthy tissue [27]. Nevertheless, several preclinical and clinical mAbs target antigens with more restricted expression patterns. For example, mAb L19 binds to the alternatively spliced extra-domain B of fibronectin that is expressed in the extracellular matrix of tumor tissue in solid and hematologic malignancies; this antigen is not expressed in healthy tissue. Preclinical and clinical investigations have used L19 for the selective delivery of cytotoxic payloads or cytokines [48]. A truly tumor-specific antigen is the B-cell receptor, or idiotype, expressed by malignant B cells in B-NHL and B-CLL. Custom-made mouse mAbs against individual idiotypes from lymphoma and leukemia patients were among the first mAbs that were clinically investigated in the early 1980s [49]. Anti-idiotypic antibodies also constitute the first example of personalized medicine in mAb therapy of cancer.

In addition to specific expression or overexpression on tumor cells, the suitability of cell surface antigens and epitopes as targets for mAb therapy depends on their functional implication in tumor cell proliferation and survival, their level of expression, their proximity to the cell membrane, and their stability at the cell surface. Antibody-drug conjugates rely on antigens that are internalized to intracellular compartments in order to facilitate efficient drug release and delivery.

Antibody Engineering

Structural Features

The antibody, or immunoglobulin (Ig), molecule consists of a defined covalent assembly of Ig domains that can be grouped in Fv, Fab, and Fc portions (Fig. 3.1). The most important feature of the antibody molecule is a hypervariable antigen binding site, whose diversity in humans is based on the random combination of >150 variable (V), diversity (D), and joining (J) gene segments and somatic mutations. The antigen binding site results from the convergence of six hypervariable peptide loops or complementarity determining regions (CDRs), three provided by each light and heavy chain variable domain. The six CDRs are clustered at one end of the antibody molecule (Fig. 3.1). It is primarily the variation in amino acid sequence in the CDRs that produces mAbs of differing antigen specificities. CDR1 and CDR2 of light and heavy chain are encoded within the V gene segments. The most hypervariable CDRs, CDR3 of light and heavy chain, are generated by the recombination of V and J gene segments or V, D, and J gene segments, respectively. The modular design of the antibody molecule and the stability of the Ig domain have facilitated a large variety of antibody engineering strategies based on recombinant DNA technology [50, 51].

Chimeric, Humanized, and Fully Human mAbs

The conversions of rodent mAbs derived from hybridoma technology [52] to mAbs with human constant domains were milestone achievements in the development of therapeutic antibodies. Mouse and rat mAbs are highly immunogenic in humans, triggering a human anti-mouse antibody (HAMA) or a human anti-rat antibody (HARA) response, which severely limits repeated dosing by the formation of immunocomplexes that not only prevent the therapeutic antibody from binding its antigen but are also known to induce mild to severe allergic reactions. HAMA and HARA also impair the effectiveness and safety of future treatments with other rodent-based mAbs. The first generation of mAbs with human constant domains were chimeric with rodent variable domains of light and heavy chain recombinantly fused to the corresponding human constant domains [53]. Yet, due to their remaining mouse sequences, it is possible that such chimeric mAbs could still trigger a human anti-chimera antibody (HACA) response. Therefore, the second generation of mAbs with human constant domains was further humanized by grafting the six CDRs that comprise the antigen binding site of the mouse or rat mAb into corresponding human framework regions [54]. The third generation of mAbs with human constant domains features fully human variable domains (Fig. 3.3). Although less frequent, both humanized and fully human mAbs can still be immunogenic in what is known as human anti-human antibody (HAHA) response [55, 56]. HACA and HAHA can be due to antiallotypic recognition of polymorphisms in the constant domains or anti-idiotypic recognition of the variable domains of mAbs. Nonetheless, anti-idiotypic antibody responses may be associated with beneficial immunity in cancer patients by triggering the above-mentioned idiotypic cascade [43, 57]. In addition to antibody immunogenicity that is due to the recognition of foreign amino acid sequences, anti-carbohydrate antibody responses have been reported in patients treated with *cetuximab* [58]. Finally, the nature of the antigen is likely to influence antibody immunogenicity [59].

Eight out of ten approved mAbs for cancer therapy (Fig. 3.2) have been derived from rodent mAbs generated by hybridoma technology [52]. Except for the two radioimmunoconjugates, which are given as single dose, all of these mAbs are either chimeric or humanized (Fig. 3.2). The two most recently approved mAbs,

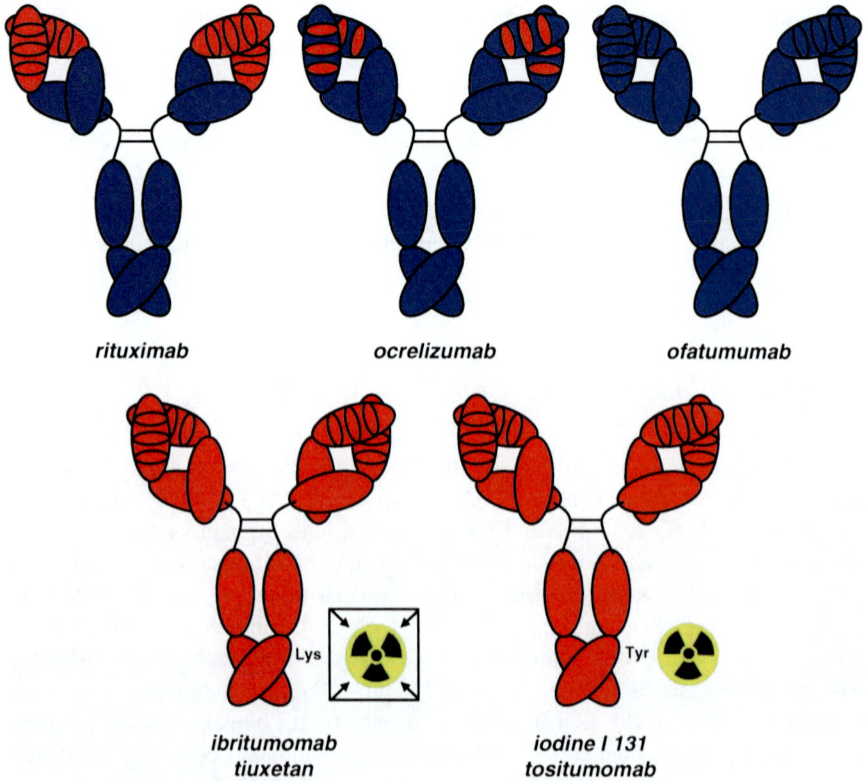

Fig. 3.3 Diversity of CD20-targeting mAbs that are approved or investigated in phase III clinical trials. *Rituximab* (*upper left*) is a chimeric mAb consisting of mouse (*red*) variable domains and human (*blue*) constant domains. Humanization, as in case of *ocrelizumab* (*upper center*), typically involves the grafting of all six CDRs of the mouse variable domains into framework regions of human variable domains. *Ofatumumab* (*upper right*) is a fully human mAb derived from transgenic mice with human light and heavy chain genes. Radioimmunoconjugates are based on antibody-chelate conjugates as in case of *ibritumomab tiuxetan* (*lower left*) which complexes radioisotope ^{90}Y or directly labeled mAbs as in case of *iodine I 131 tositumomab* (*lower right*). Lys, lysine; Tyr, tyrosine

panitumumab and *ofatumumab* are the only fully human mAbs. Both were generated by hybridoma technology using transgenic mice that express human antibodies [60]. Six additional fully human mAbs in phase III clinical trials for cancer therapy are derived from transgenic mice [61], namely: *denosumab*, *figitumumab*, *ipilimumab*, *tremelimumab*, *zalutumumab*, and *zanolimumab* (Table 3.2). Clearly, the anticipated low immunogenicity of fully human mAbs in therapeutic applications that require repeated dosing has fueled their preclinical and clinical development. Another route to fully human mAbs is through phage display of naive, immune, or synthetic antibody libraries. The 2002 FDA approval of *adalimumab* (Humira®) [62], an anti-TNFα mAb for the treatment of rheumatoid arthritis, marks the first approval of a therapeutic mAb generated by phage display. In addition to the *de novo* generation of fully human mAbs, phage display also facilitates affinity maturation and humanization as well as the selection of other therapeutically relevant features of antibodies [63, 64].

Fc Engineering

Other therapeutically relevant applications of antibody engineering have focused on the Fc rather than the Fab portion of the antibody molecule. As mentioned above, the Fc portion mediates the interaction of an antibody with FcμRIIIa and FcμRIIa in ADCC and C1q in CDC. The Fc portion also interacts with the neonatal receptor FcRn, which is responsible for the extended half-life of antibodies in circulation [65]. Fc optimization through rational design and directed evolution has allowed the tuning of effector functions as well as circulatory half-life of mAbs [41]. The Fc portion is key to the favorable and tunable pharmacokinetic and pharmacodynamic characteristics of mAbs that are based on the IgG format. Transferring these characteristics to small molecule drugs, i.e., the utilization of IgG and Fc molecules as carrier proteins, represents a new premise in preclinical and clinical investigations [66–68].

Beyond IgG

Antibody engineering has also been instrumental in the generation of mAbs that deviate from the natural IgG format of the antibody molecule [69]. Two formats of the antibody molecule that have been predominately used in antibody engineering are the 50-kD Fab fragment (Fig. 3.1) and the 25-kDa single chain Fv (scFv) fragment with two variable domains of light and heavy chain covalently linked by an artificial polypeptide. Fully human mAbs from antibody libraries are first selected as Fab or scFv formats before they are converted to IgG. In addition, due to their smaller size enabling deeper tissue penetration, both of these formats are directly investigated for the therapy of solid malignancies [70]. The same applies to the even smaller single domain antibodies or nanobodies that consist of a single human or humanized variable light or heavy chain domain [71, 72]. The modular nature of the antibody molecule has also facilitated the generation of a variety of antibody constructs with specificity for two different antigens, referred to as bispecific antibodies [73]. For the most part, bispecific antibodies have been engineered for recruiting cytotoxic T cells or NK cells to the tumor site through combining specificity for an effector cell receptor, such as CD3 or FcμRIIIa, with specificity for a tumor antigen. For example, *blinatumomab* (Table 3.2) is a bispecific T-cell engager (BiTE) antibody [74] that combines an anti-CD3 scFv and an anti-CD19 scFv; this antibody was shown to redirect cytotoxic T cells for the efficient killing of malignant B cells at very low concentrations. Objective clinical responses (PR and CR) were found at doses as low as 0.015 mg/m^2 in a phase I clinical trial for the therapy of relapsed B-NHL and B-CLL [75]. For comparison, *rituximab* is given at a standard dose of 375 mg/m^2. BiTEs and other bispecific antibody formats are also being investigated in solid malignancies. In addition, recombinant DNA technology has enabled the rational design or directed evolution of bispecific antibodies that are based on the established IgG format [76, 77]. However, not all bispecific antibodies are products of recombinant DNA technology. The original concept of bispecific antibodies involved the fusion of two hybridoma cell lines that express different mAbs to a hybrid-hybridoma cell line that expresses the corresponding bispecific antibody among other by-products. The clinically most advanced bispecific antibody based on this original concept, *catumaxomab* (Table 3.2), is a rat-mouse hybrid IgG2 mAb in phase III clinical trial for malignant ascites in epithelial cancers that combines dual specificities for CD3 and EPCAM with the Fc-mediated specificity for FcμRs [78]. The rationale is to generate tricellular complexes of tumor cells, T cells, and FcμR-expressing effector cells (including NK cells, macrophages, and dendritic cells) in order to potentiate cellular immune responses against the tumor. Other products of recombinant DNA technology with substantial promise based on clinical trial data include the above discussed immunotoxins as well as immunocytokines. An example for the latter, hu14.18-IL-2 [79] is a mAb/cytokine fusion protein in clinical development for the treatment of melanoma and neuroblastoma. It consists of a humanized anti-GD2 mAb fused to a molecule of IL-2 at the C-terminus of each heavy chain. The rationale is to selectively deliver IL-2 at the tumor site to potentiate cellular immune responses.

Clinical Performance

Overview

As a relatively young class of cancer therapeutics, initial clinical investigations of mAbs have mainly been confined to late-stage refractory or relapsed cancers following first-line and second-line standard treatments. Nonetheless, four mAbs have already been approved for initial cancer therapy. *Rituximab* in combination with cyclophosphamide/vincristine/doxorubicin/prednisone (CHOP) chemotherapy is now the first-line standard treatment of aggressive B-NHL. *Trastuzumab* in combination with paclitaxel (Taxol®) is approved for the first-line treatment of HER2-overexpressing metastatic breast cancer. *Cetuximab* in combination with radiotherapy is approved for the first-line treatment of locally or regionally advanced squamous cell carcinoma of the head and neck. Finally, *bevacizumab* is indicated for first-line treatments of (1) metastatic colorectal cancer in combination with 5-fluorouracil chemotherapy, (2) metastatic nonsquamous nonsmall cell lung cancer in combination with carboplatin/paclitaxel chemotherapy, (3) metastatic HER2-negative breast cancer in combination with paclitaxel chemotherapy, and

(iv) metastatic kidney cancer in combination with IFN-α. Notably, FDA approval of mAbs for cancer therapy does not necessarily imply that the mAb demonstrated increased overall survival (OS) in phase III clinical trials. For example, the FDA approval of *bevacizumab* for metastatic HER2-negative breast cancer in combination with paclitaxel chemotherapy was based on a progression-free survival (PFS) advantage when compared to paclitaxel chemotherapy alone; no OS benefit was found [80]. Thus, PFS has become an accepted primary endpoint in phase III clinical trials of mAbs for cancer therapy. Nonetheless, an increasing number of phase III clinical trials have demonstrated a robust OS benefit for cancer patients following mAb treatment. MAbs have prolonged the life of many patients with a variety of hematologic and solid malignancies by months to years. Prime examples from published randomized phase III clinical trials are selected in the following paragraphs. (For a more complete picture of the clinical performance of mAbs approved in oncology, see [81] and [82]).

CD20 Targeting

Rituximab has revolutionized the therapy of B-NHL [83]. Clinical responses have been demonstrated in first-line and second-line treatment of aggressive and advanced indolent B-NHL in combination with chemotherapy [84]. In a randomized phase III clinical trial for the first-line treatment of elderly patients with diffuse large B-cell lymphoma, an aggressive form of B-NHL, *rituximab* plus chemotherapy (R-CHOP) was compared to chemotherapy alone (CHOP) [85]. An OR of 82% and an OS of 70% (at 2 years) was found for the R-CHOP arm compared to 69% and 57%, respectively, for the CHOP arm. PFS and OS remained statistically significant in favor of R-CHOP at 5 years [86]. Independent R-CHOP vs. CHOP clinical trials confirmed these results in aggressive and advanced indolent B-NHL [87, 88]. In addition to B-NHL, *rituximab* in combination with chemotherapy has been approved by the FDA for the treatment of rheumatoid arthritis and B-CLL in 2006 and 2010, respectively. Despite the previously mentioned rare adverse event of PML, *rituximab* is considered a relatively safe drug with mild infusion-related toxicities as the most common side effect.

The clinical and commercial success of *rituximab* has fueled a rich pipeline of competing anti-CD20 mAbs. As shown in Fig. 3.3, these include humanized and human mAbs against the same molecule as well as those targeted against other epitopes of CD20 that may mediate more potent cytotoxicity [35, 89]. *Ofatumumab* (Figs. 3.2 and 3.3) [90] is a fully human anti-CD20 mAb that was recently approved as third-line treatment for B-CLL refractory to fludarabine and *alemtuzumab*. The only two radioimmunoconjugates among approved mAbs for cancer therapy, *iodine I 131 tositumomab* and *ibritumomab tiuxetan* (Figs. 3.2 and 3.3) both target CD20 and have demonstrated significant clinical responses in B-NHL patients. For example, a randomized phase III clinical trial for the treatment of refractory or relapsed indolent B-NHL that compared *ibritumomab tiuxetan* to *rituximab* reported an OR rate of 80% versus 56%, in favor of the radioimmunoconjugate [91]. A single dose gave durable remissions with an OS of 53% at 5 years [92]. The radioimmunoconjugate was well tolerated without revealing an increased incidence of treatment-related myelodysplastic syndrome (MDS) or AML [93]. A randomized phase III clinical trial investigated a single dose of *iodine I 131 tositumomab* as consolidation treatment after first-line treatment for advanced follicular lymphoma, an indolent form of B-NHL; the median PFS was 36.5 months at 3.5 years, which was increased relative to 13.3 months for patients who did not receive consolidation therapy. A CR of 87% was observed, including 77% of patients in PR who converted to CR after *iodine I 131 tositumomab* consolidation [94]. The radioimmunoconjugate was well tolerated and may be safer than chemotherapy, in particular for elderly patients. Despite these impressive clinical responses, radioimmunoconjugates have not been widely adopted by oncologists because the therapy must be administered in specially equipped facilities. Nonetheless, anti-CD20 radioimmunoconjugates have paved the way for the preclinical and clinical development of a broad range of radioimmunoconjugates for the therapy of hematologic and solid malignancies.

ERBB Receptor Family Targeting

With CD20 the pioneer and model antigen for mAbs in hematologic malignancies, a similar role can be attributed to antigens of the ERBB receptor family in solid malignancies [95]. ERBB receptor family members such as EGFR, HER2, and ERBB3 are implicated in the proliferation, differentiation, and survival of normal cells. Overexpression or mutation of these molecules results in dysregulated signaling, thereby promoting malignant transformation. *Trastuzumab* in combination with chemotherapy has become a standard treatment for HER2-overexpressing metastatic breast cancer. That said, less than one-third of metastatic breast cancers fulfill the requirement of HER2 overexpression as determined by immunohistochemistry (IHC) or fluorescent *in situ* hybridization (FISH). A randomized phase III clinical trial for the treatment of HER2-overexpressing metastatic breast cancer that compared *trastuzumab* plus chemotherapy to chemotherapy alone reported an OR rate of 50% versus 32% and an OS of 25.4 months versus 20.3 months, respectively [96]. An important adverse event of *trastuzumab* in combination with chemotherapy is cardiotoxicity, which is likely caused by the expression of HER2 on cardiomyocytes [97]. Cardiotoxicity may be lessened by modification of the combination drug regimen [98, 99]. *Trastuzumab* is also being investigated clinically for a variety of other HER2-overexpressing solid malignancies; results from randomized phase III clinical trials are pending.

The approved anti-EGFR mAb *cetuximab* was investigated in a randomized phase III clinical trial that compared *cetuximab* plus irinotecan (Camptosar®) chemotherapy with *cetuximab* alone for the treatment of EGFR-expressing metastatic colorectal cancer refractory to irinotecan; in this study, 82% of patient tumors expressed EGFR. This study found an advantage to the combination therapy relative to antibody therapy alone, with an OR rate of 22.9% versus 10.8% and an OS of 8.6 months versus 6.9 months [100]. In another randomized phase III clinical trial of EGFR-expressing metastatic colorectal cancer refractory to chemotherapy, *cetuximab* plus best supportive care was compared to best supportive care alone; antibody therapy appeared superior, with an OS of 6.1 months versus 4.6 months [101]. A retrospective analysis [102] revealed that the effectiveness of *cetuximab* was significantly associated with oncogenic KRAS mutations. KRAS is an intracellular protein in the EGFR signaling pathway. Oncogenic KRAS mutations are commonly found in colorectal cancer. In patients with wild-type KRAS tumors, OS in the *cetuximab* plus best supportive care arm was 9.5 months; this result compared favorably to 4.8 months in the best supportive care alone arm. By contrast, in patients whose tumors expressed mutated KRAS (43% of patients), OS in the two arms did not differ. Similar differential therapeutic responses were reported for anti-EGFR mAb *panitumumab* [103]. Anti-EGFR mAb therapy should therefore only be considered for metastatic colorectal cancer patients with wild-type KRAS tumors [104]. Oncogenic mutations of BRAF, a protein downstream of KRAS, also impair anti-EGFR mAb therapy [105].

In addition to metastatic colorectal cancer, *cetuximab* is also approved for the treatment of head and neck cancer. In a randomized phase III clinical trial for the treatment of locally or regionally advanced squamous cell carcinoma of the head and neck, *cetuximab* plus radiotherapy revealed a striking benefit when compared to radiotherapy alone; at 54 months of follow-up, the OS rates were 49.0 months versus 29.3 months, respectively [106]. As HACA has not been described as a limiting factor in repeated dosing with the chimeric mAb *cetuximab*, a clinical benefit of the potentially lower immunogenicity of fully human mAb *panitumumab* has not been established yet. However, anaphylaxis induced by *cetuximab* was found for a subset of patients with pre-existing IgE against galactose-α-1,3-galactose [58]. This carbohydrate is present on glycoproteins from nonprimate mammals and nonprimate mammalian cell lines. *Cetuximab*, which is produced by a mouse myeloma cell line, is glycosylated with a galactose-α-1,3-galactose linkage in the variable domain of the heavy chain.

The fact that oncogenic mutations in intracellular proteins downstream of EGFR can override the effectiveness of anti-EGFR mAb therapy strongly argues for direct cytotoxicity through EGFR blockade rather than indirect cytotoxicity through ADCC or CDC as the dominant mechanism of activity in the cases of *cetuximab* and *panitumumab*. Oncogenic mutations also provide a tumor-escape mechanism, resulting in resistance to mAb therapy after initial clinical responses. In fact, most cancer patients eventually progress on mAb therapy. In addition to downstream adaptations, intrinsic adaptations (e.g., epitope mutations or receptor down-regulation), lateral adaptations (e.g., up-regulation of co-receptors or alternative receptors), and upstream adaptations (e.g., up-regulation of ligands) provide possible mechanisms of resistance to anti-EGFR mAb therapy [107] and to anti-HER2 mAb therapy [108]. The up-regulation of alternative signaling pathways via other RTKs (such as

ERBB3, IGF1R, VEGFR1, and MET) in response to EGFR-directed therapy has prompted intense preclinical and clinical investigations of combination treatments that target more than one RTK signaling pathway [109]. Strategies for the simultaneous targeting of different RTK signaling pathways may include combinations of mAbs with small molecule kinase inhibitors [110], combinations of two or more monospecific mAbs, or the utilization of bispecific mAbs [77, 111].

VEGFA Targeting

One approved mAb that is being investigated in phase III clinical trials for combination with other approved mAbs is the anti-VEGFA mAb *bevacizumab*; this mAb will soon be the commercially most-successful and clinically most-broadly applied mAb in oncology. *Bevacizumab* has been clinically investigated for the therapy of a range of solid malignancies and also for certain hematologic malignancies. *Bevacizumab* is already approved for metastatic colorectal, lung, metastatic breast, and metastatic kidney cancer, all in combination with various chemotherapies or IFN-α, as well as a single agent in glioblastoma; furthermore, current phase III clinical trials include cervical cancer, head and neck cancer, gastric cancer, mesothelioma, pancreatic cancer, osteosarcoma, and ovarian cancer. Two randomized phase III clinical trials illustrate the potency of *bevacizumab* in combination with chemotherapy with respect to OS of advanced cancer patients. One trial compared *bevacizumab* plus irinotecan/fluorouracil/leucovorin chemotherapy with chemotherapy alone for the first-line treatment of metastatic colorectal cancer [112]; OS rates were 20.3 months versus 15.6 months, respectively. The other trial compared *bevacizumab* plus paclitaxel/carboplatin chemotherapy with chemotherapy alone for the treatment of late-stage lung cancer; OS rates were 12.3 months versus 10.3 months, respectively [113]. However, adverse events were significantly higher in the *bevacizumab* plus chemotherapy arm. In fact, although *bevacizumab* is set to become one of the most broadly prescribed cancer drugs, it is also the one with the most severe adverse events. Potentially fatal side effects in a small percentage of cancer patients include gastrointestinal perforation, wound-healing problems, and severe bleeding. Other current phase III clinical trials investigate combinations of *bevacizumab* with other approved mAbs for the therapy of colorectal cancer (*bevacizumab* plus *cetuximab* or *panitumumab*), breast cancer (*bevacizumab* plus *trastuzumab*), and aggressive B-NHL (*bevacizumab* plus *rituximab*). These combinations, however, are not always beneficial and were reported to be detrimental in a recently published randomized phase III clinical trial that compared *bevacizumab* plus chemotherapy plus *cetuximab* to *bevacizumab* plus chemotherapy alone for the first-line treatment of metastatic colorectal cancer [114]; reported PFS durations were 9.4 months versus 10.7 months, respectively. Significantly decreased PFS (10.0 months versus 11.4 months) and increased toxicity was also reported in a randomized phase III clinical trials that investigated the addition of *panitumumab* to *bevacizumab* plus chemotherapy [115]. This outcome is unexpected and remains unexplained mechanistically; nonetheless, it underscores the importance of randomized phase III clinical trials [116], which conclude the on-average 7.5-year long clinical development pathway of mAbs in oncology [3].

Outlook

After review of their clinical performance, it becomes clear that the currently approved mAbs are not magic bullets that cure cancer. However, mAbs can provide life-extending benefits for an increasing number of cancer patients, ranging from substantial (in certain solid malignancies) to spectacular (in certain hematologic malignancies). Perhaps more importantly, the clinical performance and the commercial viability of currently approved mAbs provide a robust platform for the development of future generations of mAbs that, as single agents or in combination, are more active and even less toxic. A principal limit of mAb therapy of cancer has clearly not been reached, providing much room for improvement.

What does it take to improve mAb therapy of cancer? The discovery of new antigens that can mediate more potent cytotoxicity without severe adverse events for mAbs, antibody-drug conjugates, or radioimmunoconjugates is well underway. Sophisticated concomitant or sequential combination treatments that address tumor heterogeneity and escape mechanisms are mandated and should be based on preclinical high-throughput screening platforms that better predict clinical performance. Targeting cancer stem cells may play a pivotal role in this effort. The success of mAb therapy will also rely on personalized medicine that takes the genetic background of patient and tumor into account, such as FcμR polymorphisms and oncogenic mutations. Successful immunotherapy of cancer will likely require the triggering and perhaps augmentation of both humoral and cellular immune components that act in concert. The recruitment and activation of T cells and NK cells through bispecific antibodies and the adoptive transfer of T cells and NK cells equipped with cell surface antibody molecules are promising strategies in this regard. Finally, like other cancer drugs and treatments, mAbs are more likely to be therapeutically successful when given earlier in the disease course. Randomized phase III clinical trials that investigate mAbs, alone or in combination, for first-line treatments of early stage cancers will take many years to provide reliable OS data, but are eagerly awaited.

Improving clinical performance is key to cost effectiveness. The flip side of the commercial success of mAbs in oncology is the fact that mAb therapy is very expensive. Depending on the indication, a 1-year treatment with *bevacizumab*, for example, costs between $50,000 and $100,000. This price reflects the high costs of preclinical and clinical development, manufacturing, and intellectual property. However, competition and generics are expected to reduce the price of mAb therapy in the near future.

Antibody engineering will continue to play a key role in developing future generations of mAbs. The fact that all approved mAbs and the vast majority of mAbs in phase III clinical trials for cancer therapy are IgG molecules suggests that this natural and first-to-market format will remain dominant for years, possibly decades. Although alternative antibody or nonantibody formats have attractive features, they lack the vast and streamlined manufacturing, regulatory, and clinical experience gained with the IgG platform. Thus, mAbs with more subtle changes that retain the overall IgG format, but gradually tune affinity, enhance effector functions, lengthen circulatory half-life, reduce immunogenicity, introduce multi-specificity, overcome tissue and tumor penetration barriers, allow site-specific drug conjugations, and can be manufactured at lower costs are likely to dominate mAb therapy of cancer in the foreseeable future.

Acknowledgements I thank members of my laboratory for comments on the manuscript, in particular Lauren R. Skeffington and Drs. Sivasubramanian Baskar, Thomas Hofer, Brian C. Shaffer, and Jiahui Yang. This work was supported by the Intramural Research Program of the Center for Cancer Research, National Cancer Institute, National Institutes of Health.

References

1. Blattman JN, Greenberg PD (2004) Cancer immunotherapy: A treatment for the masses. Science 305:200–205
2. Dimitrov DS, Marks JD (2009) Therapeutic antibodies: Current state and future trends – is a paradigm change coming soon? Methods Mol Biol 525:1–27, xiii
3. Reichert JM (2008) Monoclonal antibodies as innovative therapeutics. Curr Pharm Biotechnol 9:423–430
4. Baker M (2005) Upping the ante on antibodies. Nat Biotechnol 23:1065–1072
5. Reichert JM, Valge-Archer VE (2007) Development trends for monoclonal antibody cancer therapeutics. Nat Rev Drug Discov 6:349–356
6. Reichert JM, Rosensweig CJ, Faden LB et al (2005) Monoclonal antibody successes in the clinic. Nat Biotechnol 23:1073–1078
7. Scott CT (2005) The problem with potency. Nat Biotechnol 23:1037–1039
8. Major EO (2010) Progressive multifocal leukoencephalopathy in patients on immunomodulatory therapies. Annu Rev Med 61:35–47
9. Carson KR, Evens AM, Richey EA et al (2009) Progressive multifocal leukoencephalopathy after rituximab therapy in HIV-negative patients: a report of 57 cases from the research on adverse drug events and reports project. Blood 113:4834–4840
10. Suntharalingam G, Perry MR, Ward S et al (2006) Cytokine storm in a phase 1 trial of the anti-CD28 monoclonal antibody TGN1412. N Engl J Med 355:1018–1028
11. Waldmann TA (2006) Effective cancer therapy through immunomodulation. Annu Rev Med 57:65–81
12. Berger R, Rotem-Yehudar R, Slama G et al (2008) Phase I safety and pharmacokinetic study of CT-011, a humanized antibody interacting with PD-1, in patients with advanced hematologic malignancies. Clin Cancer Res 14:3044–3051
13. Fong L, Small EJ (2008) Anti-cytotoxic T-lymphocyte antigen-4 antibody: the first in an emerging class of immunomodulatory antibodies for cancer treatment. J Clin Oncol 26:5275–5283
14. Hwang WY, Foote J (2005) Immunogenicity of engineered antibodies. Methods 36:3–10
15. Tabrizi MA, Tseng CM, Roskos LK (2006) Elimination mechanisms of therapeutic monoclonal antibodies. Drug Discov Today 11:81–88
16. Cheson BD, Leonard JP (2008) Monoclonal antibody therapy for B-cell non-Hodgkin's lymphoma. N Engl J Med 359:613–626

17. Schrama D, Reisfeld RA, Becker JC (2006) Antibody targeted drugs as cancer therapeutics. Nat Rev Drug Discov 5:147–159
18. Folkman J (2007) Angiogenesis: an organizing principle for drug discovery? Nat Rev Drug Discov 6:273–286
19. Jain RK (2005) Normalization of tumor vasculature: an emerging concept in antiangiogenic therapy. Science 307:58–62
20. Ran S, Downes A, Thorpe PE (2002) Increased exposure of anionic phospholipids on the surface of tumor blood vessels. Cancer Res 62:6132–6140
21. Weiner LM (2007) Building better magic bullets – improving unconjugated monoclonal antibody therapy for cancer. Nat Rev Cancer 7:701–706
22. Attia P, Phan GQ, Maker AV et al (2005) Autoimmunity correlates with tumor regression in patients with metastatic melanoma treated with anti-cytotoxic T-lymphocyte antigen-4. J Clin Oncol 23:6043–6053
23. Hodi FS, Butler M, Oble DA et al (2008) Immunologic and clinical effects of antibody blockade of cytotoxic T-lymphocyte-associated antigen 4 in previously vaccinated cancer patients. Proc Natl Acad Sci U S A 105:3005–3010
24. Lipton A, Steger GG, Figueroa J et al (2007) Randomized active-controlled phase II study of denosumab efficacy and safety in patients with breast cancer-related bone metastases. J Clin Oncol 25:4431–4437
25. Milenic DE, Brady ED, Brechbiel MW (2004) Antibody-targeted radiation cancer therapy. Nat Rev Drug Discov 3:488–499
26. Weiner GJ (2007) Monoclonal antibody mechanisms of action in cancer. Immunol Res 39:271–278
27. Gerber HP, Ferrara N (2005) Pharmacology and pharmacodynamics of bevacizumab as monotherapy or in combination with cytotoxic therapy in preclinical studies. Cancer Res 65:671–680
28. Grunwald V, Hidalgo M (2003) Developing inhibitors of the epidermal growth factor receptor for cancer treatment. J Natl Cancer Inst 95:851–867
29. Stel AJ, Ten Cate B, Jacobs S et al (2007) Fas receptor clustering and involvement of the death receptor pathway in rituximab-mediated apoptosis with concomitant sensitization of lymphoma B cells to Fas-induced apoptosis. J Immunol 178:2287–2295
30. Pedersen IM, Buhl AM, Klausen P et al (2002) The chimeric anti-CD20 antibody rituximab induces apoptosis in B-cell chronic lymphocytic leukemia cells through a p38 mitogen activated protein-kinase-dependent mechanism. Blood 99:1314–1319
31. Wilson TR, Johnston PG, Longley DB (2009) Anti-apoptotic mechanisms of drug resistance in cancer. Curr Cancer Drug Targets 9:307–319
32. Rubinfeld B, Upadhyay A, Clark SL et al (2006) Identification and immunotherapeutic targeting of antigens induced by chemotherapy. Nat Biotechnol 24:205–209
33. Lane D (2006) Designer combination therapy for cancer. Nat Biotechnol 24:163–164
34. Salfeld JG (2007) Isotype selection in antibody engineering. Nat Biotechnol 25:1369–1372
35. Cragg MS, Glennie MJ (2004) Antibody specificity controls *in vivo* effector mechanisms of anti-CD20 reagents. Blood 103:2738–2743
36. Racila E, Link BK, Weng WK et al (2008) A polymorphism in the complement component C1qa correlates with prolonged response following rituximab therapy of follicular lymphoma. Clin Cancer Res 14:6697–6703
37. Clynes RA, Towers TL, Presta LG et al (2000) Inhibitory Fc receptors modulate *in vivo* cytoxicity against tumor targets. Nat Med 6:443–446
38. Weng WK, Levy R (2003) Two immunoglobulin G fragment C receptor polymorphisms independently predict response to rituximab in patients with follicular lymphoma. J Clin Oncol 21:3940–3947
39. Musolino A, Naldi N, Bortesi B et al (2008) Immunoglobulin G fragment C receptor polymorphisms and clinical efficacy of trastuzumab-based therapy in patients with HER-2/neu-positive metastatic breast cancer. J Clin Oncol 26:1789–1796
40. Satoh M, Iida S, Shitara K (2006) Non-fucosylated therapeutic antibodies as next-generation therapeutic antibodies. Expert Opin Biol Ther 6:1161–1173

41. Desjarlais JR, Lazar GA, Zhukovsky EA et al (2007) Optimizing engagement of the immune system by anti-tumor antibodies: an engineer's perspective. Drug Discov Today 12:898–910
42. Hilchey SP, Hyrien O, Mosmann TR et al (2009) Rituximab immunotherapy results in the induction of a lymphoma idiotype-specific T-cell response in patients with follicular lymphoma: support for a "vaccinal effect" of rituximab. Blood 113:3809–3812
43. Weiner LM, Dhodapkar MV, Ferrone S (2009) Monoclonal antibodies for cancer immunotherapy. Lancet 373:1033–1040
44. Pastan I, Hassan R, Fitzgerald DJ et al (2006) Immunotoxin therapy of cancer. Nat Rev Cancer 6:559–565
45. Kreitman RJ, Squires DR, Stetler-Stevenson M et al (2005) Phase I trial of recombinant immunotoxin RFB4(dsFv)-PE38 (BL22) in patients with B-cell malignancies. J Clin Oncol 23:6719–6729
46. Kreitman RJ, Stetler-Stevenson M, Margulies I et al (2009) Phase II trial of recombinant immunotoxin RFB4(dsFv)-PE38 (BL22) in patients with hairy cell leukemia. J Clin Oncol 27:2983–2990
47. Scott AM, Welt S (1997) Antibody-based immunological therapies. Curr Opin Immunol 9:717–722
48. Kaspar M, Zardi L, Neri D (2006) Fibronectin as target for tumor therapy. Int J Cancer 118:1331–1339
49. Davis TA, Maloney DG, Czerwinski DK et al (1998) Anti-idiotype antibodies can induce long-term complete remissions in non-Hodgkin's lymphoma without eradicating the malignant clone. Blood 92:1184–1190
50. Carter PJ (2006) Potent antibody therapeutics by design. Nat Rev Immunol 6:343–357
51. Demarest SJ, Glaser SM (2008) Antibody therapeutics, antibody engineering, and the merits of protein stability. Curr Opin Drug Discov Devel 11:675–687
52. Kohler G, Milstein C (1975) Continuous cultures of fused cells secreting antibody of predefined specificity. Nature 256:495–497
53. Morrison SL, Johnson MJ, Herzenberg LA et al (1984) Chimeric human antibody molecules: mouse antigen-binding domains with human constant region domains. Proc Natl Acad Sci U S A 81:6851–6855
54. Riechmann L, Clark M, Waldmann H et al (1988) Reshaping human antibodies for therapy. Nature 332:323–327
55. Ritter G, Cohen LS, Williams C, Jr. et al (2001) Serological analysis of human anti-human antibody responses in colon cancer patients treated with repeated doses of humanized monoclonal antibody A33. Cancer Res 61:6851–6859
56. Aarden L, Ruuls SR, Wolbink G (2008) Immunogenicity of anti-tumor necrosis factor antibodies-toward improved methods of anti-antibody measurement. Curr Opin Immunol 20:431–435
57. DeNardo GL, Bradt BM, Mirick GR et al (2003) Human antiglobulin response to foreign antibodies: therapeutic benefit? Cancer Immunol Immunother 52:309–316
58. Chung CH, Mirakhur B, Chan E et al (2008) Cetuximab-induced anaphylaxis and IgE specific for galactose-alpha-1,3-galactose. N Engl J Med 358:1109–1117
59. Clark M (2000) Antibody humanization: a case of the 'emperor's new clothes'? Immunol Today 21:397–402
60. Jakobovits A, Amado RG, Yang X et al (2007) From xenomouse technology to panitumumab, the first fully human antibody product from transgenic mice. Nat Biotechnol 25:1134–1143
61. Lonberg N (2005) Human antibodies from transgenic animals. Nat Biotechnol 23:1117–1125
62. Bain B, Brazil M (2003) Adalimumab. Nat Rev Drug Discov 2:693-694
63. Rader C (2001) Antibody libraries in drug and target discovery. Drug Discov Today 6:36–43
64. Hoogenboom HR (2005) Selecting and screening recombinant antibody libraries. Nat Biotechnol 23:1105–1116

65. Roopenian DC, Akilesh S (2007) FcRn: the neonatal Fc receptor comes of age. Nat Rev Immunol 7:715–725
66. Rader C, Sinha SC, Popkov M et al (2003) Chemically programmed monoclonal antibodies for cancer therapy: adaptor immunotherapy based on a covalent antibody catalyst. Proc Natl Acad Sci U S A 100:5396–5400
67. Hofer T, Thomas JD, Burke TR Jr. et al (2008) An engineered selenocysteine defines a unique class of antibody derivatives. Proc Natl Acad Sci U S A 105:12451–12456
68. Woodnutt G, Violand B, North M (2008) Advances in protein therapeutics. Curr Opin Drug Discov Dev 11:754–761
69. Hudson PJ, Souriau C (2003) Engineered antibodies. Nat Med 9:129–134
70. Nelson AL, Reichert JM (2009) Development trends for therapeutic antibody fragments. Nat Biotechnol 27:331–337
71. Holt LJ, Herring C, Jespers LS et al (2003) Domain antibodies: proteins for therapy. Trends Biotechnol 21:484–490
72. Vincke C, Loris R, Saerens D et al (2009) General strategy to humanize a camelid single-domain antibody and identification of a universal humanized nanobody scaffold. J Biol Chem 284:3273–3284
73. Chames P, Baty D (2009) Bispecific antibodies for cancer therapy. Curr Opin Drug Discov Dev 12:276–283
74. Baeuerle PA, Kufer P, Bargou R (2009) BiTE: teaching antibodies to engage T-cells for cancer therapy. Curr Opin Mol Ther 11:22–30
75. Bargou R, Leo E, Zugmaier G et al (2008) Tumor regression in cancer patients by very low doses of a T cell-engaging antibody. Science 321:974–977
76. Wu C, Ying H, Grinnell C et al (2007) Simultaneous targeting of multiple disease mediators by a dual-variable-domain immunoglobulin. Nat Biotechnol 25:1290–1297
77. Bostrom J, Yu SF, Kan D et al (2009) Variants of the antibody Herceptin that interact with HER2 and VEGF at the antigen binding site. Science 323:1610–1614
78. Shen J, Zhu Z (2008) Catumaxomab, a rat/murine hybrid trifunctional bispecific monoclonal antibody for the treatment of cancer. Curr Opin Mol Ther 10:273–284
79. Niculescu-Duvaz I (2004) Technology evaluation: EMD-273063, EMD lexigen. Curr Opin Mol Ther 6:559–566
80. Miller K, Wang M, Gralow J et al (2007) Paclitaxel plus bevacizumab versus paclitaxel alone for metastatic breast cancer. N Engl J Med 357:2666–2676
81. O'Mahony D, Bishop MR (2006) Monoclonal antibody therapy. Front Biosci 11:1620–1635
82. Rader C, Bishop MR (2007) Monoclonal antibodies in cancer therapy. In: Kaufman HL, Wolchok JD (eds) General principles of tumor immunotherapy: basic and clinical applications of tumor immunology. Springer, Dordrecht
83. Grillo-Lopez AJ (2003) Rituximab (rituxan/MabThera): the first decade (1993–2003). Expert Rev Anticancer Ther 3:767–779
84. Cvetkovic RS, Perry CM (2006) Rituximab: a review of its use in non-Hodgkin's lymphoma and chronic lymphocytic leukaemia. Drugs 66:791–820
85. Coiffier B, Lepage E, Briere J et al (2002) CHOP chemotherapy plus rituximab compared with CHOP alone in elderly patients with diffuse large-B-cell lymphoma. N Engl J Med 346:235–242
86. Feugier P, Van Hoof A, Sebban C et al (2005) Long-term results of the R-CHOP study in the treatment of elderly patients with diffuse large B-cell lymphoma: a study by the groupe d'Etude des Lymphomes de l'Adulte. J Clin Oncol 23:4117–4126
87. Habermann TM, Weller EA, Morrison VA et al (2006) Rituximab-CHOP versus CHOP alone or with maintenance rituximab in older patients with diffuse large B-cell lymphoma. J Clin Oncol 24:3121–3127
88. Hiddemann W, Kneba M, Dreyling M et al (2005) Frontline therapy with rituximab added to the combination of cyclophosphamide, doxorubicin, vincristine, and prednisone (CHOP) significantly improves the outcome for patients with advanced-stage follicular lymphoma

compared with therapy with chop alone: results of a prospective randomized study of the German Low-Grade Lymphoma Study Group. Blood 106:3725–3732
89. Teeling JL, Mackus WJ, Wiegman LJ et al (2006) The biological activity of human CD20 monoclonal antibodies is linked to unique epitopes on CD20. J Immunol 177:362–371
90. Castillo J, Milani C, Mendez-Allwood D (2009) Ofatumumab, a second-generation anti-CD20 monoclonal antibody, for the treatment of lymphoproliferative and autoimmune disorders. Expert Opin Investig Drugs 18:491–500
91. Witzig TE, Gordon LI, Cabanillas F et al (2002) Randomized controlled trial of yttrium-90-labeled ibritumomab tiuxetan radioimmunotherapy versus rituximab immunotherapy for patients with relapsed or refractory low-grade, follicular, or transformed B-cell non-Hodgkin's lymphoma. J Clin Oncol 20:2453–2463
92. Witzig TE, Molina A, Gordon LI et al (2007) Long-term responses in patients with recurring or refractory B-cell non-Hodgkin lymphoma treated with yttrium 90 ibritumomab tiuxetan. Cancer 109:1804–1810
93. Czuczman MS, Emmanouilides C, Darif M et al (2007) Treatment-related myelodysplastic syndrome and acute myelogenous leukemia in patients treated with ibritumomab tiuxetan radioimmunotherapy. J Clin Oncol 25:4285–4292
94. Morschhauser F, Radford J, Van Hoof A et al (2008) Phase III trial of consolidation therapy with yttrium-90-ibritumomab tiuxetan compared with no additional therapy after first remission in advanced follicular lymphoma. J Clin Oncol 26:5156–5164
95. Peipp M, Dechant M, Valerius T (2008) Effector mechanisms of therapeutic antibodies against ERBB receptors. Curr Opin Immunol 20:436–443
96. Slamon DJ, Leyland-Jones B, Shak S et al (2001) Use of chemotherapy plus a monoclonal antibody against HER2 for metastatic breast cancer that overexpresses HER2. N Engl J Med 344:783–792
97. Crone SA, Zhao YY, Fan L et al (2002) Erbb2 is essential in the prevention of dilated cardiomyopathy. Nat Med 8:459–465
98. Chien KR (2006) Herceptin and the hear – a molecular modifier of cardiac failure. N Engl J Med 354:789–790
99. Joensuu H, Kellokumpu-Lehtinen PL, Bono P et al (2006) Adjuvant docetaxel or vinorelbine with or without trastuzumab for breast cancer. N Engl J Med 354:809–820
100. Cunningham D, Humblet Y, Siena S et al (2004) Cetuximab monotherapy and cetuximab plus irinotecan in irinotecan-refractory metastatic colorectal cancer. N Engl J Med 351:337–345
101. Jonker DJ, O'Callaghan CJ, Karapetis CS et al (2007) Cetuximab for the treatment of colorectal cancer. N Engl J Med 357:2040–2048
102. Karapetis CS, Khambata-Ford S, Jonker DJ et al (2008) K-ras mutations and benefit from cetuximab in advanced colorectal cancer. N Engl J Med 359:1757–1765
103. Amado RG, Wolf M, Peeters M et al (2008) Wild-type KRAS is required for panitumumab efficacy in patients with metastatic colorectal cancer. J Clin Oncol 26:1626–1634
104. Allegra CJ, Jessup JM, Somerfield MR et al (2009) American Society of Clinical Oncology provisional clinical opinion: testing for KRAS gene mutations in patients with metastatic colorectal carcinoma to predict response to anti-epidermal growth factor receptor monoclonal antibody therapy. J Clin Oncol 27:2091–2096
105. Di Nicolantonio F, Martini M, Molinari F et al (2008) Wild-type BRAF is required for response to panitumumab or cetuximab in metastatic colorectal cancer. J Clin Oncol 26:5705–5712
106. Bonner JA, Harari PM, Giralt J et al (2006) Radiotherapy plus cetuximab for squamous-cell carcinoma of the head and neck. N Engl J Med 354:567–578
107. Hopper-Borge EA, Nasto RE, Ratushny V et al (2009) Mechanisms of tumor resistance to EGFR-targeted therapies. Expert Opin Ther Targets 13:339–362
108. Bedard PL, de Azambuja E, Cardoso F (2009) Beyond trastuzumab: overcoming resistance to targeted HER-2 therapy in breast cancer. Curr Cancer Drug Targets 9:148–162

109. Dancey JE, Chen HX (2006) Strategies for optimizing combinations of molecularly targeted anticancer agents. Nat Rev Drug Discov 5:649–659
110. Zhang J, Yang PL, Gray NS (2009) Targeting cancer with small molecule kinase inhibitors. Nat Rev Cancer 9:28–39
111. Lu D, Zhang H, Koo H et al (2005) A fully human recombinant IgG-like bispecific antibody to both the epidermal growth factor receptor and the insulin-like growth factor receptor for enhanced antitumor activity. J Biol Chem 280:19665–19672
112. Hurwitz H, Fehrenbacher L, Novotny W et al (2004) Bevacizumab plus irinotecan, fluorouracil, and leucovorin for metastatic colorectal cancer. N Engl J Med 350:2335–2342
113. Sandler A, Gray R, Perry MC et al (2006) Paclitaxel-carboplatin alone or with bevacizumab for non-small-cell lung cancer. N Engl J Med 355:2542–2550
114. Tol J, Koopman M, Cats A et al (2009) Chemotherapy, bevacizumab, and cetuximab in metastatic colorectal cancer. N Engl J Med 360:563–572
115. Hecht JR, Mitchell E, Chidiac T et al (2009) A randomized phase IIIb trial of chemotherapy, bevacizumab, and panitumumab compared with chemotherapy and bevacizumab alone for metastatic colorectal cancer. J Clin Oncol 27:672–680
116. Mayer RJ (2009) Targeted therapy for advanced colorectal cancer – more is not always better. N Engl J Med 360:623–625

Chapter 4
Natural Killer Cells for Cancer Immunotherapy

Yoko Kosaka and Armand Keating

Abstract Natural killer (NK) cells are immune effector cells that have long been known to possess potent cytotoxic ability. Despite this, NK cells remained relatively underrepresented in the medical literature, due in part to the strong emphasis placed on studying the mechanisms of the antigenic specificity and memory of T and B lymphocytes. Fortunately, as innate cells have gained prominence in recent years, NK cell research has blossomed and we now have a glimpse of the complexity of these cells and the potential that they have in cancer therapy. Not only do NK cells have a powerful ability to directly kill abnormal cells, they play a critical role in shaping adaptive responses by secreting a wide array of regulatory factors and interacting with multiple cell types. This chapter provides an overview of the current understanding of human NK cells, and discusses the potential of taking advantage of this knowledge to use NK cells in cancer therapy. Although much of our knowledge of NK cell biology comes from mouse studies, many of which involved models of viral or auto-immune diseases, the focus here is on observations made with human NK cells in the context of cancer.

Keywords Natural killer cells • Adoptive cell therapy • Innate immunity • Cytotoxicity • Killer cell immunoglobulin-like receptor (KIR)

NK Cell Development and Identification

NK cells were originally discovered in 1975 by two independent groups, which described cells that killed tumor cells 'spontaneously', that is, without prior exposure to the tumor [1, 2]. NK cells were also later found to be mediators of "hybrid resistance", which refers to the rejection of parental bone marrow grafts by an

A. Keating (✉)
Division of Hematology, University of Toronto, ON, Canada
e-mail: armand.keating@uhn.on.ca

J. Medin and D. Fowler (eds.), *Experimental and Applied Immunotherapy*,
DOI 10.1007/978-1-60761-980-2_4, © Springer Science+Business Media, LLC 2011

F1 hybrid, a phenomenon that had puzzled investigators [3]. It was also recognized that NK cells were distinct from B and T cells as they do not rearrange antigen-recognition receptors. These and other early studies led to the idea that NK cells utilize a novel strategy to recognize their targets.

NK cells are said to belong to the innate immune system based on their rapid response in first-line defense, the lack of classical antigen specificity, and an inability to generate memory responses [4]. Recent findings, however, suggest that this description is incomplete. Indeed, NK cells appear to undergo "priming" to mount responses and possess some memory characteristics typical of cells of the adaptive immune system [5, 6]. NK cells have direct effects on target cells that could preclude the need for a subsequent adaptive response. However, NK cells can also act indirectly as central regulators of overall innate and adaptive immune responses through cellular interactions; notably, factors produced by NK cells can influence dendritic cells (DCs). In this way, NK cells function to close the gap between innate and adaptive immunity.

The majority of NK cells reside as mature lymphocytes in peripheral blood, lymph nodes, spleen, and bone marrow. Additionally, some NK cells are found in other tissues such as lung, liver, and uterus. NK cells localized to certain sites appear to be functionally specialized; for example, NK cells found in mucosal tissues are known to secrete IL-22 [7]. In addition, decidual NK cells in the reproductive tract play a unique role in preserving tissue homeostasis during pregnancy [8]. Although much remains unclear, NK cells primarily undergo development from $CD34^+$ progenitor cells in the bone marrow. NK cell precursors have also been found in the gut. More recently, it has been shown that some NK cells, like T cells, also develop in the thymus and are dependent on IL-7 [9].

It is now evident that NK cells consist of a diverse population that can be functionally and phenotypically categorized into several subsets. In humans, NK cells are broadly identified by their expression of CD56 and lack of the pan-T cell marker CD3. NK cells can be further subdivided on the basis of the levels of CD56 expression and on CD16 (FcγRIII) expression. Cells that express high levels of CD56 have a biased propensity to respond to and produce cytokines and chemokines but have low cytolytic potential; by comparison, the $CD56^{dim}$ subset of NK cells primarily exert cytotoxic function [10]. $CD56^{dim}/CD16^{bright}$ cells make up the majority (~90%) of NK cells that circulate in peripheral blood; in contrast, the $CD56^{bright}$/CD16-negative subset is found primarily in lymph nodes, where such cells interact with antigen-presenting cells and T cells [11]. Given that roughly half of the lymphocytes in the body reside in the lymph nodes, there are much higher absolute numbers of the $CD56^{bright}$ NK cell population. Interestingly, although only a small percentage of NK cells in healthy peripheral blood are the $CD56^{bright}$ subset, in patients who have received a hematopoietic stem cell transplant these cells are the first lymphocytes to reconstitute and are present in increased numbers [12]. To date, the lineage of these NK subsets is still unclear, although there is convincing evidence that $CD56^{bright}$ cells are the precursors of $CD56^{dim}$ cells [13]. Other markers, such as CD27, have also emerged that delineate various NK cell subsets [14].

Effector Functions of NK Cells

Like other cells of the innate immune system, NK cells stand ready to fire off a rapid effector response once a danger signal is received; that is, in contrast to B and T cells, NK cells can act without the need to proliferate and without a dependence on the *de novo* production of effector molecules (Fig. 4.1). The effector functions of NK cells result from positive and negative signals received through multiple soluble and cell-bound ligands. The function most associated with NK cells is an ability to recognize and lyse abnormal target cells, such as malignant or virally infected cells. The trigger for target cell lysis occurs by stimulation through an activating receptor on NK cells, such as CD16, which enables antibody-dependent cellular cytotoxicity (ADCC) to occur. Signals derived from activating such receptors results in the release of granules containing lytic molecules such as perforin and granzymes toward the area of target cell contact, inducing target cell death [15] (Fig. 4.1a).

NK cell targets can also be induced to undergo apoptosis when the tumor necrosis factor (TNF) ligand family members TNF-α, Fas ligand, and TNF-related apoptosis-inducing ligand (TRAIL) expressed on NK cells recognize and engage their cognate receptors on target cells (Fig. 4.1b). Such interactions then induce a cascade of caspase activation that ultimately leads to apoptosis. Due to the potent apoptosis-inducing effects of TRAIL, and the observation that this mechanism can bypass regulatory MHC-KIR interactions, the TRAIL pathway has been targeted for therapeutic use [16].

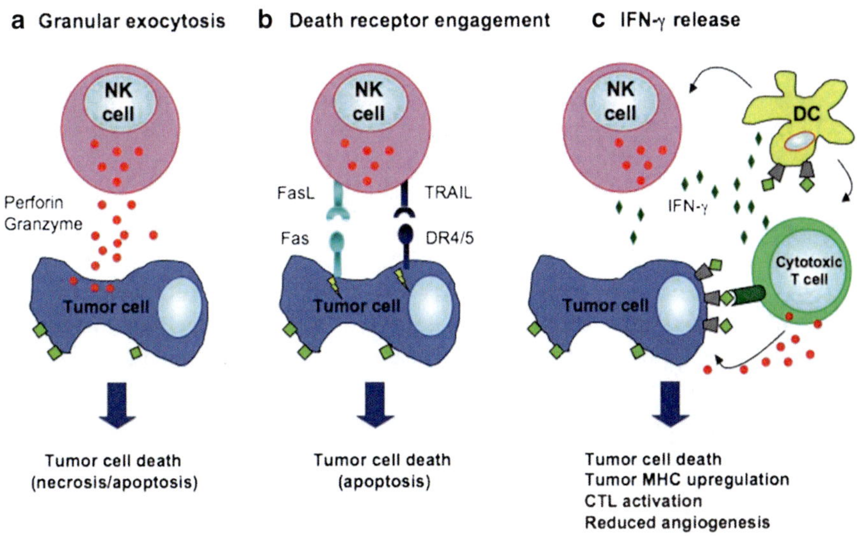

Fig. 4.1 NK cell-mediated anti-tumor responses. (**a**) Degranulation of NK cells after activation releases granzymes and perforins that kill target cells. (**b**) Expression of TNF, Fas ligand and TRAIL by NK cells bind their cognate receptors on target cells, inducing apoptosis. (**c**) IFN-γ secretion by NK cells not only has effects on tumor cells directly but also recruits and activates other immune effectors such as DCs and cytotoxic T lymphocytes

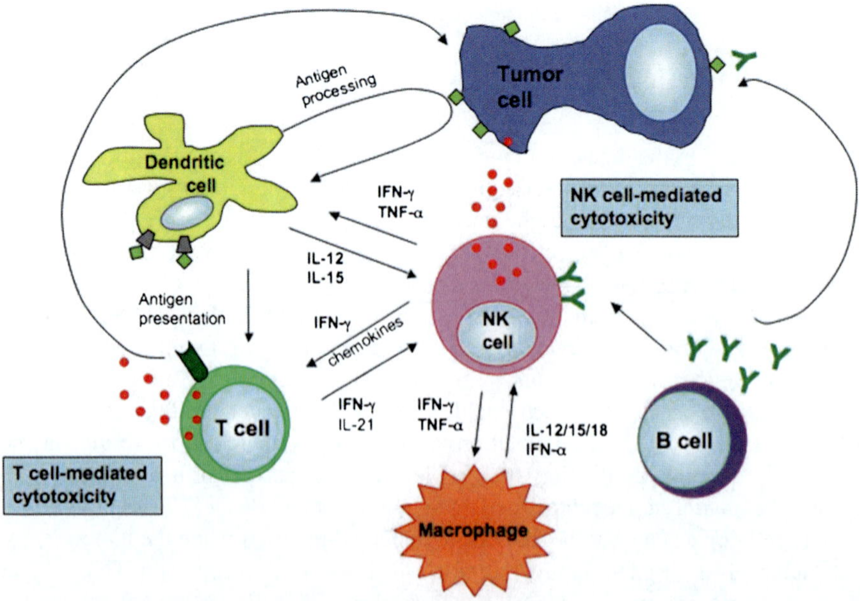

Fig. 4.2 NK cells cooperate with other cells to generate anti-tumor responses. A network of cells of the immune system participate in forming protective responses against cancer. NK cells are affected by, and in turn, affect the functional capacities of many different immune mediators, generating a regulatory feedback loop

NK cells, particularly those that are CD56^{bright} also produce many cytokines, most notably large amounts of IFN-γ thereby promoting inflammation and skewing of adaptive immune responses (Fig. 4.1c) [17]. IFN-γ not only acts directly to inhibit the proliferation of tumor cells but also contributes indirectly to the total immune response by maturing and activating antigen-presenting cells to drive effector Th1/Tc1 immune responses (Fig. 4.2). Activation of DCs through IFN-γ as well as NK cell-derived TNF-α and GM-CSF results in DC production of IL-12 and IL-15 which in turn enhances NK cell function. Such cross-talk effectively alerts other cells, such as T cells, to the presence of transformed or infected cells, thereby engendering cooperation among multiple cell types. Since chemokines are also produced in large quantities by NK cells, recruitment of cells to the microenvironment is further enhanced [18]. Importantly, IFN-γ can also directly antagonize effector Th2/Tc2 immunity and reduces regulatory T cell factors such as IL-10 and TGF-β that dampen NK cell activity and diminish antitumor responses. It is also interesting to note that NK cell-derived IFN-γ secretion might suppress tumor growth and tumor metastases by inhibiting angiogenesis [19, 20].

NK-Target Cell Recognition and Regulation by Cell Surface Receptors

Because NK cells can lyse targets without prior priming and without target cell display of specific "foreign" antigens, such formidable cells must require other modes of tight regulation to avoid harm. T cells respond to antigen presented in the context of MHC; the question remained, then, as to the stimuli that induce NK cells to respond. This question was addressed by the "missing self" hypothesis proposed by Klas Karre in 1986 as a model based on knowledge at that time that NK cells kill targets that *lack* self-MHC proteins on their surface [21]. Thus, NK cells evolved to combat aberrant cells that were coerced to down-regulate MHC class I expression. Self-MHC class I is expressed on virtually every cell in the body; loss of MHC expression is a common feature of malignant transformation or viral infection for evasion of T-cell recognition.

The concept of missing self implies that NK cells are held in check by the constant surveillance of NK cell inhibitory receptors to detect MHC Class I antigens; in this model, when a cell lacking MHC Class I is encountered, it is attacked. However, it has become apparent that regulation of NK cell activation is more complex than previously assumed. Multiple signals are received when NK cells contact other cells, and it is the balance of such positive and negative signals that dictates NK cell responsiveness (Fig. 4.3) [22]. Importantly, such contact signals are both quantitative and qualitative. Therefore, the absence of MHC alone is not sufficient for NK cells to actively engage in cytolysis of target cells: that is, there

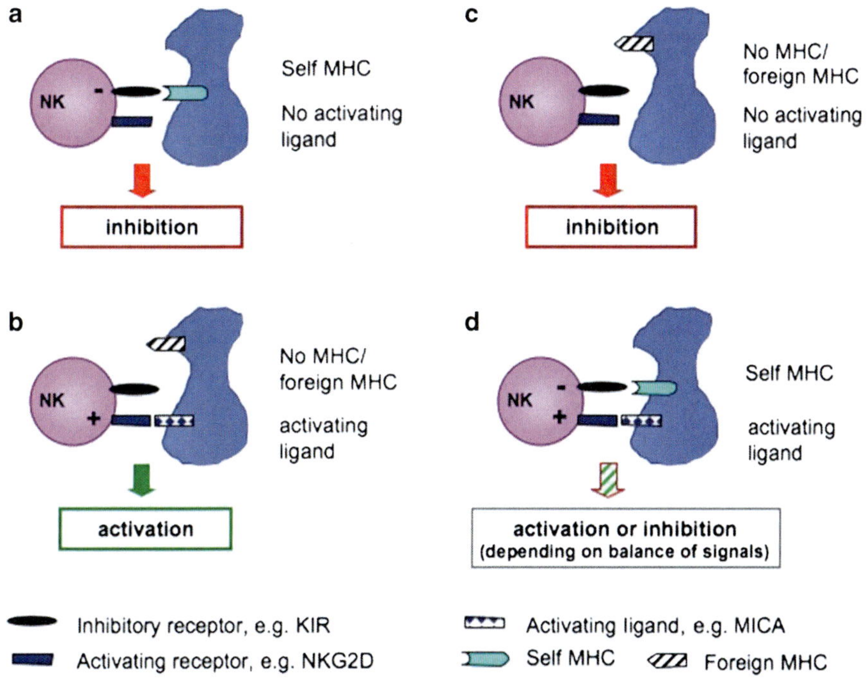

Fig. 4.3 Model of target cell recognition by NK cells. (**a**) NK cells are not activated when inhibitory receptors (such as KIRs) bind self-MHC, but do not detect ligands to activating receptors (e.g., healthy host cells). (**b**) NK cells are activated when they do not detect inhibitory KIRs but do detect ligands to activating receptors (e.g., HLA-negative malignant or pathogen-infected cells). (**c**) NK cells are not activated when neither inhibitory nor activation receptor is engaged by cognate ligands (e.g., healthy allogeneic cells). (**d**) NK cells are either activated or not activated, depending on the balance and/or strength of signals generated from both activating and inhibitory receptors

Table 4.1 Major NK cell receptors

Inhibitory	Ligand
KIR-DL (KIR2DL1, 2, 3, 5; KIR3DL1-2)	HLA-A, B, C
LIR-1/ILT2 (CD85)	Multiple HLA, HLA-G, UL18 (HCMV)
CD94-NKG2A (CD159a)	HLA-E
KLRG1	Cadherins
NKR-P1 (CD161)	LLT1
Single 7, 9 (CD328, CD329)	Sialic acid
Activating	**Ligand**
NKG2D (CD314)	MICA, MICB, ULBP1-4
Natural cytotoxicity receptors (NKp30, NKp44, NKp46)	Viral hemagglutinnins
CD16 (FcγRIIIA)	IgG
DNAM-1 (CD226), CD96	PVR, nectin-2
VLA-4, 5	VCAM-1, fibronectin
CD2	LFA-3 (CD58)
KIR (DS) (KIR2DS1-6, KIR3DS1)	HLA-C, ?
Either inhibitory or activating	**Ligand**
2B4 (CD244)	CD48
KIR2DL4	HLA-G

also appears to be a requirement for positive signals generated by an activation receptor. This biology accounts for the fact that healthy cells that express low or no MHC Class I are not killed by NK cells. For instance, erythrocytes that lack MHC class I molecules are protected against NK cell lysis; this protection is due either to an absence of an activation signal or to the presence of other non-MHC binding inhibitory ligands. In another setting, NK cells transferred to an allogeneic (non-self MHC-bearing) host might leave many healthy cells untouched due to the lack of activating receptors. Some of the major NK cell receptors that generate activating or inhibitory signals are highlighted in Table 4.1.

Inhibitory Receptors

KIRs

KIR proteins are the best-characterized family of receptors that recognizes specific alleles of the classical MHC Class Ia molecules (HLA-A, HLA-B, HLA-C) and signal NK cells to remain unresponsive to healthy host cells. Therefore, most KIRs are inhibitory, and engagement leads to negative signaling within the cell. Like MHC, the KIR genes are highly polymorphic [23]. Inheritance of specific types and numbers of KIRs is varied within the human population and continues to be defined (http://www.allelefrequencies.net). KIR expression is also heterogeneous on different populations of NK cells within an individual. Interestingly, because the KIR genes segregate independently of HLA, HLA-matched individuals do not necessarily carry the same KIRs. In addition, although most KIRs are inhibitory in nature, some are activating, such as the KIR DS group [22]. These concepts become important in situations of MHC-matched but not KIR-matched transplants.

Individual KIRs are distinguished structurally, with either two or three immunoglobulin (Ig) domains and a long or short cytoplasmic tail, and are named as such: for instance, KIR2DL4 contains two Ig domains and a long cytoplasmic tail. The KIR DL family recognizes HLA-A, B, C ligands and transduces an inhibitory signal. KIR haplotypes can be broadly categorized into two groups, A and B [24]. Group A haplotypes have a fixed number of genes while Group B haplotypes have, in addition to the Group A genes, a variable number of Group B-specific genes. The Group B haplotype contains more activating receptor genes than the Group A haplotype. These differences have been shown to influence susceptibility to preeclampsia, autoimmunity, infectious disease, and cancer [24]. In a recent study of 448 acute myeloid leukemia (AML) patients with allogeneic hematopoietic transplants, a significant benefit in overall and relapse-free survival was found when Group B haplotype donors were used, regardless of the recipient's genotype, although the underlying mechanism remains unclear [25]. Such information should increase the likelihood of successful transplants.

NKG2A/CD94 Heterodimer

NKG2A/CD94 heterodimers are similar to the KIR family, but recognize nonclassical MHC Class Ib molecules (HLA-E) and display limited polymorphisms [26]. HLA-E is widely expressed on many cell types. NKG2A/CD94 receptors are important inhibitory molecules affecting NK cell cytotoxicity; specifically, increased expression of NKG2A/CD94 correlates with impaired cytotoxicity. Importantly, the NKG2A/CD94 receptor is up-regulated on NK cells in patients with AML who received KIR-mismatched hematopoietic cell allografts but showed no graft-versus-leukemia effect [27].

Activating Receptors

There is a large collection of receptors that confer an activating signal to NK cells; this list continues to grow. Some of the best characterized in the context of cancer are described below.

NKG2D

NKG2D is nonpolymorphic and expressed as a homodimer by all NK cells. NKG2D generates critical signals to kill tumor cells; ligands are structurally related to MHC

Class I but are not involved in antigen presentation to T cells. NKG2D appears to be critical in the protective role of NK cells in tumor surveillance [28, 29]. NKG2D ligand expression is largely restricted to cells undergoing stress, such as those that are neoplastic or virally infected. Six ligands have been described so far, including MICA, MICB, and the ULBP proteins (ULBP1, ULBP2, ULBP3, and ULBP4). These ligands are expressed on many different tumor types [30]. Interestingly, soluble NKG2D ligands are shed by some tumors and suppress NK-mediated tumor cell killing [31].

Natural Cytotoxicity Receptors (NCR)

NCRs consist of the molecules NKp46, NKp30, and NKp44 and provide potent activating signals to NK cells [32]. NKp46 and NKp30 are expressed on peripheral blood NK cells, while NKp44 is up-regulated on NK cells exposed to IL-2. There appears to be a threshold expression level for NCRs to mediate cytotoxicity. For instance, NK cells from AML patients can express low levels of NCRs and correlates with a poor prognosis [33]. Several viral products, such as hemagglutinins, bind specifically to NCRs [32]. Recently, a study has identified a product released by tumor cells called HLA-B-associated transcript 3 (BAT3) as a ligand for NKp30 that triggers NK cell cytotoxicity. [34]. Although the identification of other cellular NCR ligands remains elusive, it is clear that tumors express them and that they play a role in activating NK cells. Synthetic NCR-Ig fusion proteins that target these ligands have been found to suppress tumor growth [35].

CD16 (FcγRIII)

CD16 binds to the constant region (Fc) of antibodies, thereby leading to NK cell activation and perforin-mediated target lysis. Therefore, antibody-coated cells are targets of NK cells; this mechanism plays an important role in antibody therapy, as described in later sections of this chapter.

DNAM-1 (CD226)

DNAM-1 is an activating receptor whose ligands include Nectin-2 (CD112) and poliovirus receptor (PVR, CD155). One study found that expression of DNAM-1 ligands was particularly high in myeloid leukemic cells [36]. It has also been noted that DNAM-1 expression on NK cells is decreased in myeloma patients with active disease compared with those in remission [37].

The net effect of signals received through these many complex receptor-ligand interactions appears to determine whether a particular cell is an appropriate NK cell target. This ability of NK cells to assess the strengths of simultaneous signals that determines whether or not to respond highlights their exquisite sensitivity. These requirements also appear to differ based on the NK cell subset. For instance, NK cells that have been preactivated by cytokines have a lower threshold for full activation. Evidence is mounting that 'NK synapses', similar to those that exist for T cells, can fine-tune NK cell proliferation, IFN-γ production, or cytotoxicity [38]. Coengagement of two activation receptors such as 2B4 and NKG2D has a synergistic effect in inducing NK cytotoxicity [39]. Furthermore, adhesion molecules also participate to enhance NK-target interactions, as engagement of LFA-1 on NK cells by ICAM-1 is sufficient to induce granule polarization [40]. The involvement of multiple activating and inhibiting signals affords opportunities to capitalize on the function of these receptors to undertake successful antitumor therapies.

Extrinsic Regulation of NK Cells

As with signals derived from membrane-bound ligands, many soluble factors affect the development and activation of NK cells. Stimulatory factors include, but are not restricted to: IL-1, IL-2, IL-7, IL-12, IL-15, IL-18, IL-21, IFN-α, Flt3L, and SCF. NK cells constitutively express the IL-2 receptor complex and are dependent on IL-2 for growth and activation. Low-dose IL-2 therapy in humans expands not only T cells but also endogenous NK cells [41]. Flt3L, SCF, and IL-7 are also essential for the proper development of NK cells. IL-15 is an essential survival and growth factor for NK cells and is trans-presented by DCs to induce NK cell 'priming' [42]. IL-15 can also elevate NKG2D expression on NK cells, thereby promoting optimal effector function. IL-1 and IL-18 up-regulate the expression of the IL-12 receptor on NK cells, and in turn, IL-12 and IL-18 induce NK cell cytotoxicity. IL-21 has been shown recently to support the expansion of the cytotoxic $CD56^{dim}/CD16^{+}$ population and also promotes cytotoxicity; hence, therapy with this cytokine is currently being pursued [43].

Generally, factors that are immunosuppressive to other cells, such as TGF-β and IL-10, also inhibit NK cells. Regulatory T cells that express these cytokines are known to attenuate NK cell responses by affecting NK receptor expression [44]. Tumors can also express TGF-β and induce NK hypo-responsiveness by down-regulating NKG2D ligands or NKG2D on NK cells [45]. Interestingly, NK cells can also be negatively influenced by IDO and PGE_2 derived from mesenchymal stromal cells (MSCs) *in vitro* [46]. Such findings showing an immunosuppressive effect of MSCs against NK cells and T cells are worth considering given the increased interest in pursuing adoptive therapy using NK cells or MSCs in cancer patients.

Role of NK Cells in Cancer

The concept of tumor immunosurveillance is widely accepted, but it is still debated as to what extent NK cells participate. Several lines of evidence suggest that NK cells play a key role. In murine studies, depletion of NK cells can enhance tumor growth *in vivo* [28, 47]. In humans, selective NK cell deficiencies are rare; thus, it is difficult to determine whether a clear relationship exists between NK cells and cancer in humans. However, an epidemiological survey found that individuals with low NK cell activity had a higher risk of developing cancer [48].

Although most studies focus on the effectiveness of NK cells against hematological malignancies, as described in more detail below, it is clear that NK cells are also able to efficiency lyse solid tumors [49]. The presence of NK cells infiltrating the tumor tissue is associated with a favorable prognosis in many cases [50–54]. Some recent studies have found, however, that not all tumor-infiltrating NK cells are cytotoxic [55, 56]. Therefore, even after recruitment to the tumor site, NK cells may have diminished capacity to suppress tumor growth.

Clinical use of NK Cells

Adoptive Immunotherapy

Adoptive immunotherapy refers to the clinical administration of cells that have been cultured *ex vivo* in an effort to boost antitumor immunity in the patient (Fig. 4.4a). In pioneering studies initiated in the 1980s by Rosenberg and colleagues, patient autologous cells treated *ex vivo* with IL-2, called lymphokine-activated killer (LAK) cells, were infused with IL-2 into patients with advanced metastatic renal cell cancer and melanoma [57]. This approach yielded an overall response rate of 15–20% but was found to be no more beneficial than IL-2 therapy alone. Modifications to the original protocol, including using larger numbers of LAK cells or selecting activated NK cells also did not improve disease outcome. This type of LAK therapy was also tried in a wide variety of malignancies, also with disappointing results. We now know that LAK cells include many different effector cells, among them, a proportion of T cells that are not tumor-specific. Additionally, as mentioned, T cells are MHC-restricted and are therefore only able to recognize antigens expressed in MHC-bearing tumor cells. Moreover, unknown at the time, coadministration of IL-2 with the LAK cells, which was performed to support the survival of the adoptively transferred cells, not only expands and activates "beneficial" T cells but also promotes the proliferation of immunosuppressive regulatory T cells [58, 59]. Importantly, the increased presence of regulatory T cells, which are known to inhibit NK cells, correlates with a poor disease outcome in cancer patients [60–62].

The most significant example of a potential therapeutic role for NK cells is in the setting of allogeneic hematopoietic cell transplantation. Groundbreaking studies showed that patients with AML who received haploidentical allografts had increased disease-free survival [63]. *In vitro* studies demonstrated that the leukemia-specific killing was predominantly an effect of NK cells in the graft that were activated due to a mismatch of donor NK inhibitory receptors with host MHC Class I. In these studies, it was also remarkable that the patients experienced less graft-versus-host disease (GVHD), likely due to the depletion of T cells in the graft as well as NK cell-mediated killing of host DCs, which limited host antigen presentation.

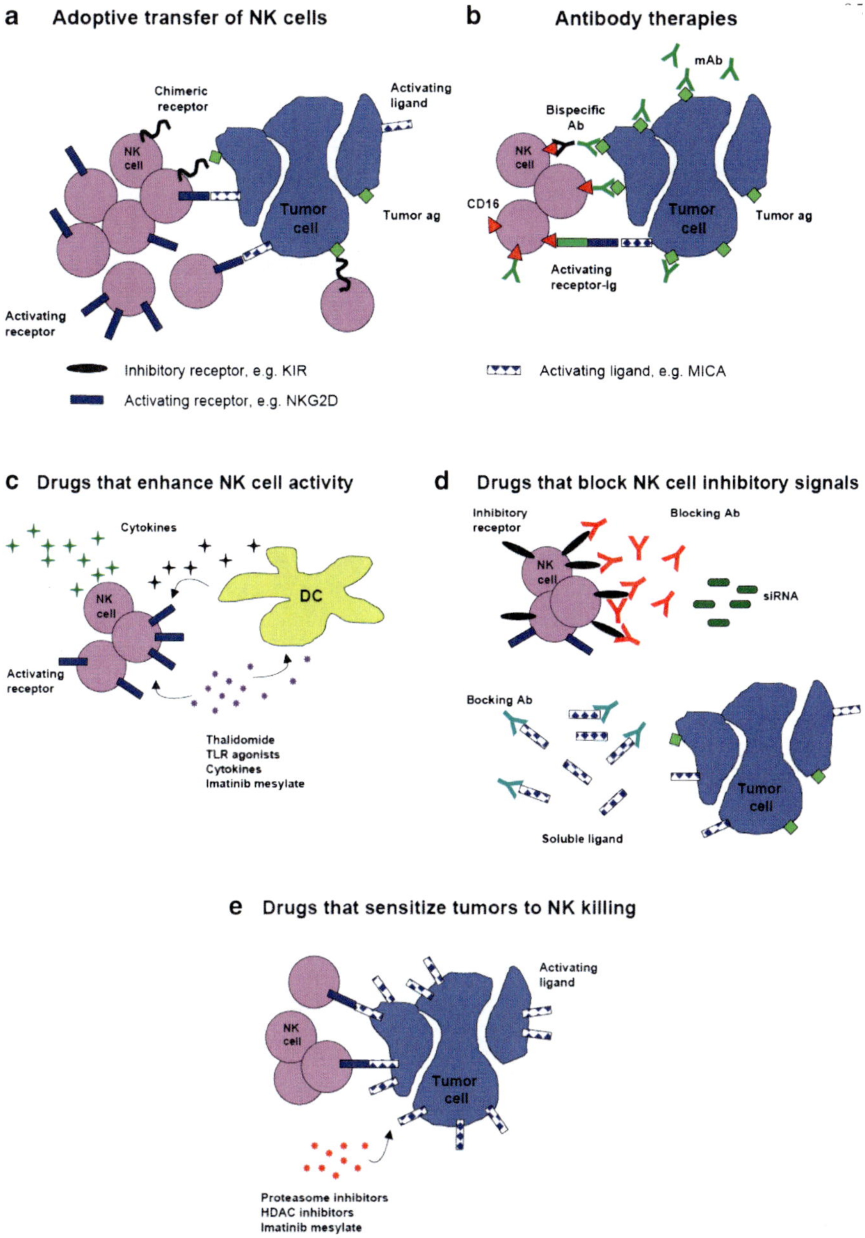

Fig. 4.4 Strategies for NK cell-based therapy in cancer. (**a**) Autologous or allogeneic NK cells or cell lines expanded/activated *in vitro* can then be administered, with an aim to increase NK cell numbers and anti-tumor activity. (**b**) Anti-tumor antibody therapy can make use of the ADCC function of NK cells by binding CD16. To enhance this linkage, bispecific antibodies can be engineered that possess both CD16-binding and tumor-antigen-binding Fab regions. (**c**) Chemotherapeutic drugs can have activating effects on NK cells, directly or indirectly through stimulating the functions of third party cells, such as DCs. (**d**) Inhibitory receptors that decrease/inactivate NK cell functions can be blocked by antibodies or siRNA. (**e**) Some anti-cancer drugs that are directly cytotoxic to tumor cells can also sensitize tumors to NK-mediated killing by upregulating ligands to NK cell receptors

Recognizing the potential of using enriched NK cells in cancer therapy, Miller and colleagues conducted a study using infusions of haploidentical NK cells following immune-depleting chemotherapy in patients with advanced melanoma, renal cell carcinoma, Hodgkin's disease, and AML [64]. In this trial, 5 of 19 patients with AML achieved complete remission. This therapy was relatively well-tolerated and did not result in GVHD. Moreover, the study found that NK cells were detectable long-term; increased levels of IL-15, which resulted from chemotherapy administration, appeared to enable persistence of the adoptively transferred NK cells.

Following those initial studies, there have been multiple reports that either support or are in conflict with the original findings of Ruggeri *et al.* [65]. Due to the complexity of NK cell recognition and differences in donor–recipient matching, it is not surprising that disparate results have been reported. Differences in study design and clinical protocols in the context of different cancers make these studies difficult to compare. Thus far, evidence supports the idea that the highest likelihood of a successful graft-versus-tumor response with NK cell therapy occurs in the setting of haplo-identical T cell-depleted hematopoietic cell transplantation in patients with AML. While donor selection is critical, KIR phenotying to select the most appropriate donor cells may also enhance the overall success, as mentioned above [25].

Although adoptive immunotherapy with *ex vivo*-expanded primary NK cells has met with some success [64], protocols to generate such cells still need considerable optimization. Several protocols have been reported, but it is not yet clear as to the extent that the expansion process alters NK cell function [66–68]. Refining culturing methods, including exposure of NK cells to various stimulatory factors such as IL-15 and/or using feeder cells, might augment cytolytic activity. Furthermore, a major challenge of adoptive cell therapy remains the ability of transferred cells to traffic to and penetrate the immunosuppressive microenvironment that surrounds solid tumors [69]. Therefore, a primary goal of adoptive immunotherapy with NK cells should be to target/redirect these cells to infiltrate tumors using specific tumor antigens and/or chemokines [70]. It is likely that the route of NK cell delivery will also be important. For example, autologous NK cells infused in colon cancer patients with liver metastases showed NK cell persistence only when injected regionally by the intra-arterial route [71].

While the clinical use of permanently transformed NK cell lines derived from patients with NK cell malignancies may at first appear to be counter-intuitive, there are several advantages in taking this approach to adoptive immunotherapy. First, some of the lines are considerably more cytotoxic than primary NK cells; irradiation of the transformed NK cells, which is utilized to prevent proliferation and potential tumor formation *in vivo*, need not reduce their cytotoxic potential [72, 73]. Importantly, because such cells are an allogeneic and readily available source, GMP-compliant manufacturing is simplified [74]. NK-92 is the prototype of a well-characterized highly cytotoxic NK cell line. NK-92 cells do not kill normal human bone marrow cells but kill T-ALL and leukemic cell lines better than LAK, normal NK cells, and T cells *in vitro* [75]. The effects of NK-92 cells are broad, with killing of approximately half of 45 primary leukemia cells tested. Preclinical studies performed in xenograft mouse models implanted with leukemia and melanoma cells showed improved survival of tumor-bearing mice treated with an NK-92 cell infusion [76]. Due to these promising results, a clinical study in children with advanced cancers was conducted in Germany [77]; in addition, phase I trials are underway in the USA and by our group in Canada. Infusion of NK-92 cells into patients with advanced disease appears to be safe and is well-tolerated [78]. In an effort to further improve the cytotoxic capacity of NK-92 cells against malignant cells, several manipulations have been attempted. One approach is to generate variants that express cytokines such as IL-2 and IL-15 to enhance cytotoxicity [79, 80]. Another strategy is to engineer NK-92 cells that bear chimeric receptors, such as single chain antitumor antibody attached to the intracellular signaling domain of the TCR zeta chain. In preclinical animal models, HER2-redirected NK-92 cells were found to lyse otherwise resistant breast, ovarian and squamous cell cancer cells [81]. Other antigens that have been targeted in this way are CD19 and CD20 [82, 83].

Another NK cell line that has shown promise for therapeutic purposes is KHYG-1 [73, 84, 85]. KHYG-1 is more cytotoxic than NK-92 *in vitro* against several leukemic targets and like NK-92, retains its cytolytic capacity following irradiation. Interestingly, KHYG-1 is characterized by constitutively polarized granules and cleaved perforin, which likely contribute to its enhanced cytotoxicity. Also, KHYG-1 expresses high levels of the activating receptors NKp44 and NKG2D, which would be expected to enhance target recognition and lytic responses. Such characteristics and further evaluation of the mechanisms involved in tumor recognition and killing by this cell line (and others) should help to broaden and identify the tumors that are susceptible to the action of NK cell lines.

Strategies that Target NK Cells

Antibody Therapies

Monoclonal antibodies, such as those targeting CD20 (rituximab) or HER2 (trastuzumab), are currently used therapeutically. NK cell-mediated ADCC plays a major role in the effectiveness of antibody-based therapies (Fig. 4.4b) [86]. Consistent with this, polymorphisms can affect the affinity of CD16 for antibodies and such genetic variation has been found to influence rituximab responsiveness in non-Hodgkin's lymphoma [87]. Consequently, novel antibody therapies currently underway take advantage of NK cell-mediated ADCC and include the use of antibodies designed to improve binding to CD16 along with bispecific antibodies that target both NK cells and tumor cells.

Combining antibody therapy with cytokines and other drugs to enhance NK cell activity appears particularly promising. IL-2 and rituximab therapy have been combined in several trials [88–90]. One study found that a thrice weekly regimen of IL-2 with rituximab therapy increased the number of NK cells and promoted ADCC, with four/five patients responding [89]. A large phase II trial in NHL patients found that IL-2 with rituximab resulted in expansion of Fc receptor-bearing cells and ADCC, although only 5/57 patients showed a clinical response [90]. Additionally, a small trial with seven patients combining rituximab and IL-2 with LAK cells found that ADCC was improved compared to pretreatment, with two patients achieving a partial remission and five patients achieving stable disease [91]. A bispecific antibody targeting CD16 and CD30 administered with IL-2 and GM-CSF has also been tested in patients with Hodgkins lymphoma; this approach resulted in increased NK cell numbers [92].

IL-12 has also been combined with antibody therapy [93–95]. A small phase I trial combining IL-12 with trastuzumab in HER2+ malignancies found that NK-derived IFN-γ was detected in patients (3 of 15) who had a clinical response or stabilization of disease [95]. In another study, the chemokines IL-8, MDC, MIP1α, MCP1, and RANTES were produced by NK cells when exposed to IL-2, IL-12, and trastuzumab-coated tumor cells and some of these chemokines were detected in the sera of breast cancer patients undergoing trastuzumab and IL-12 combination therapy [96].

Other factors appear to hold therapeutic promise. IL-15 is a good candidate, as it is a potent NK cell-stimulating factor and protects NK cells from some of the toxic effects of IL-2 [97]. In a murine model of colon cancer, combining IL-15 with anti-CD40 antibody provided a survival benefit over each agent alone. NK cells isolated from combination-treated mice were found to have greater cytotoxicity against the target tumor cells [98]. IL-21 has been tested with antibody therapy *in vitro* and its use in combination with cetuximab (anti-HER-1) or trastuzumab-coated tumor cells results in highly elevated production of IFN-γ and chemokines by NK cells [99, 100]. Recombinant soluble Apo2L/TRAIL in combination with rituximab induced apoptosis in a synergistic fashion in NHL lines and, moreover, attenuated the tumor graft in NHL xenograft models [101]. Importantly, when NK cells were removed, this effect was diminished *in vivo*, thereby demonstrating the contribution of NK cells.

Engaging Activating Signals

In addition to cytokines, other drugs are currently being studied that directly augment NK cell activity (Fig. 4.4c). TLR agonists are an exciting avenue for cancer treatment, especially as an adjuvant to vaccination strategies [102]. NK cells themselves express TLRs, which recognize products produced by bacterial and viral pathogens. TLR9 agonists, such as oligonucleotides containing CpG motifs, induce the production of large amounts of IFN-γ by NK cells when added to antibody-coated tumor cells [103]. A phase I trial of CpG in NHL patients identified increased NK activity and ADCC in some patients [104]. In addition, due to the close cooperation between DCs and NK cells, NK cells may also act indirectly in TLR-based adjuvant therapy. CpG administration following Flt3L treatment of patients who underwent autologous HCT resulted in increased NK cell cytotoxicity; this effect was primarily attributed to an indirect mechanism via the activation of plasmacytoid DCs [105].

Immunomodulatory drugs such as lenalidomide, a thalidomide analog, have antitumor activity through a variety of mechanisms, including the activation of NK cells. For instance, lenalidomide induced the upregulation of granzyme B and FasL on NK cells and augmented ADCC when added with NK cells and rituximab-coated

NHL lines [106, 107]. The tyrosine kinase inhibitor imatinib mesylate has also been found to act on DCs to activate NK cell-derived IFN-γ production in patients with gastrointestinal stromal tumors [108].

Blocking Inhibitory Signals

Another approach to boost NK responses is to block inhibitory signals on NK cells (Fig. 4.4d) [109]. The first trial to block inhibitory KIR with a monoclonal antibody in patients with hematological malignancies is underway [110]. Eventually, it is possible that multiple antibodies may be utilized to block a panel of KIR receptors. As mentioned, some tumors produce soluble NKG2D ligands, such as MICA, which down-regulate NKG2D. Antibodies generated against MICA can diminish soluble MICA-induced immunosuppression and thus promote antitumor responses [111]. These types of therapy should mobilize the activity of endogenous NK cells that are unable to provide antitumor activity on their own. Inhibitory receptor silencing by siRNA might also become an option for patient-specific treatment [112]. To manipulate NK cell activation/inhibitory receptors, one must know whether tumor cells will be able to elicit a response based on the expression of such receptors.

Chemotherapeutic Drugs

In addition to enhancing NK cell cytotoxicity directly, tumor targets can be rendered more sensitive to NK cells by altering their phenotype (Fig. 4.4e). Anticancer drugs such as proteasome inhibitors and histone deacetylase (HDAC) inhibitors modulate the expression of NK cell receptor ligands on a wide variety of tumor cells. The proteasome inhibitor bortezomib (Velcade), approved for use in multiple myeloma and mantle cell lymphoma, and the HDAC inhibitor romidepsin (Istodax) approved for use in cutaneous T cell lymphoma, induced NK cell TRAIL-mediated lysis by upregulating the TRAIL-binding receptor DR5 on tumor cell lines [113]. Furthermore, bortezomib down-regulated the expression of MHC Class I on myeloma cells, thereby resulting in greater sensitivity to autologous and allogeneic NK cells [114]. Other studies showed that tumors treated with HDAC inhibitors up-regulate the ligands for NKG2D and DNAM-1, which induce NK cell activation [115, 116]. These observations suggest the potential efficacy of using chemotherapeutic agents, especially in conjunction with adoptive NK cell therapy.

Conclusions and Future Challenges

NK cells contribute to antitumor responses by direct tumor cell lysis as well as by mobilizing other cells to build a complex multi-pronged effort to protect the host. Various forms of immunotherapy have been shown to stimulate the action of NK cells.

Recent research has demonstrated how complex the recognition of targets by NK effectors can be, with continued characterization of additional receptors and their signaling pathways. It will be important to better understand the positive and negative regulation of NK functions to better tailor therapies that capitalize on these cells.

As cells that survey the body for transformed cells and respond rapidly, NK cells likely have the most impact at the initial stages of malignancy, when tumor burden is low. It is for this reason that innovative NK cell therapies are likely to be less successful in individuals with advanced disease. This biology may help explain why results of some clinical studies with NK cells have been disappointing. Since most patients eventually succumb to metastatic disease, NK cells should be exploited to prevent malignant cells from having the opportunity to spread.

As tumors develop ways to circumvent specific arms of the immune system, approaches that deliver multiple hits will meet with the most success. A better understanding of the mechanisms of action of tumor-killing cells and drugs will make it possible to design multimodal therapies that directly or indirectly employ the activity of NK cells. Some anticancer drugs have already been shown to sensitize malignant cells to NK cell-mediated death. For adoptive immunotherapy with NK cells, it will be important to ensure that the cells reach their destination through targeting strategies; in addition, the coordinated use of drugs and the optimizing of the route of administration will be vital. Moreover, strategies must be developed that allow NK cells to retain their cytolytic efficiency and resist the dampening effects of suppressive cells and factors. As these parameters become more amenable to manipulation, it is tempting to envisage personalized therapy with a bank of NK cells or NK cell lines that can be administered according to the specific biology and needs of an individual patient.

References

1. Herberman RB, Nunn ME and Lavrin DH (1975) Natural cytotoxic reactivity of mouse lymphoid cells against syngeneic acid allogeneic tumors. I. Distribution of reactivity and specificity. Int J Cancer 16(2):216–229
2. Kiessling R, Klein E and Wigzell H (1975) "Natural" killer cells in the mouse. I. Cytotoxic cells with specificity for mouse Moloney leukemia cells. Specificity and distribution according to genotype. Eur J Immunol 5(2):112–117
3. Kumar V, George T, Yu YY et al (1997) Role of murine NK cells and their receptors in hybrid resistance. Curr Opin Immunol 9(1):52–56
4. Caligiuri MA (2008) Human natural killer cells. Blood 112(3):461–9
5. Long EO (2007) Ready for prime time: NK cell priming by dendritic cells. Immunity 26(4):385–387
6. Sun JC, Beilke JN and Lanier LL (2009) Adaptive immune features of natural killer cells. Nature 457(7229):557–561
7. Cella M, Fuchs A, Vermi W et al (2009) A human natural killer cell subset provides an innate source of IL-22 for mucosal immunity. Nature 457(7230):722–725
8. Santoni A, Zingoni A, Cerboni C et al (2007) Natural killer (NK) cells from killers to regulators: distinct features between peripheral blood and decidual NK cells. Am J Reprod Immunol 58(3):280–288

9. Vosshenrich CA, Garcia-Ojeda ME, Samson-Villeger SI et al (2006) A thymic pathway of mouse natural killer cell development characterized by expression of GATA-3 and CD127. Nat Immunol 7(11):1217–1224
10. Poli A, Michel T, Theresine M et al (2009) CD56bright natural killer (NK) cells: an important NK cell subset. Immunology 126(4):458–465
11. Fehniger TA, Cooper MA, Nuovo GJ et al (2003) CD56bright natural killer cells are present in human lymph nodes and are activated by T cell-derived IL-2: a potential new link between adaptive and innate immunity. Blood 101(8):3052–3057
12. Carson W and Caligiuri M (1996) Natural killer cell subsets and development. Methods 9(2):327–343
13. Chan A, Hong DL, Atzberger A et al (2007) CD56bright human NK cells differentiate into CD56dim cells: role of contact with peripheral fibroblasts. J Immunol 179(1):89–94
14. Hayakawa Y and Smyth MJ (2006) CD27 dissects mature NK cells into two subsets with distinct responsiveness and migratory capacity. J Immunol 176(3):1517–15 24
15. Cullen SP and Martin SJ (2008) Mechanisms of granule-dependent killing. Cell Death Differ 15(2):251–262
16. Ashkenazi A, Holland P and Eckhardt SG (2008) Ligand-based targeting of apoptosis in cancer: the potential of recombinant human apoptosis ligand 2/Tumor necrosis factor-related apoptosis-inducing ligand (rhApo2L/TRAIL). J Clin Oncol 26(21):3621–3630
17. Schoenborn JR and Wilson CB (2007) Regulation of interferon-gamma during innate and adaptive immune responses. Adv Immunol 96:41–101
18. Robertson MJ (2002) Role of chemokines in the biology of natural killer cells. J Leukoc Biol 71(2):173–183
19. Coughlin CM, Salhany KE, Gee MS et al (1998) Tumor cell responses to IFNgamma affect tumorigenicity and response to IL-12 therapy and antiangiogenesis. Immunity 9(1): 25–34
20. Sidky YA and Borden EC (1987) Inhibition of angiogenesis by interferons: effects on tumor- and lymphocyte-induced vascular responses. Cancer Res 47(19):5155–5161
21. Karre K, Ljunggren HG, Piontek G et al (1986) Selective rejection of H-2-deficient lymphoma variants suggests alternative immune defence strategy. Nature 319(6055):675–678
22. Lanier LL (2005) NK cell recognition. Annu Rev Immunol 23:225–274
23. Norman PJ and Parham P (2005) Complex interactions: the immunogenetics of human leukocyte antigen and killer cell immunoglobulin-like receptors. Semin Hematol 42(2):65–75
24. Parham P (2005) MHC class I molecules and KIRs in human history, health and survival. Nat Rev Immunol 5(3):201–214
25. Cooley S, Trachtenberg E, Bergemann TL et al (2009) Donors with group B KIR haplotypes improve relapse-free survival after unrelated hematopoietic cell transplantation for acute myelogenous leukemia. Blood 113(3):726–732
26. Braud VM, Allan DS, O'Callaghan CA et al (1998) HLA-E binds to natural killer cell receptors CD94/NKG2A, B and C. Nature 391(6669):795–799
27. Nguyen S, Dhedin N, Vernant JP et al (2005) NK-cell reconstitution after haploidentical hematopoietic stem-cell transplantations: immaturity of NK cells and inhibitory effect of NKG2A override GvL effect. Blood 105(10):4135–4142
28. Guerra N, Tan YX, Joncker NT et al (2008) NKG2D-deficient mice are defective in tumor surveillance in models of spontaneous malignancy. Immunity 28(4):571–580
29. Smyth MJ, Swann J, Cretney E et al (2005) NKG2D function protects the host from tumor initiation. J Exp Med 202(5):583–588
30. Moretta L, Bottino C, Pende D et al (2006) Surface NK receptors and their ligands on tumor cells. Semin Immunol 18(3):151–158
31. Doubrovina ES, Doubrovin MM, Vider E et al (2003) Evasion from NK cell immunity by MHC class I chain-related molecules expressing colon adenocarcinoma. J Immunol 171(12):6891–6899
32. Arnon TI, Markel G and Mandelboim O (2006) Tumor and viral recognition by natural killer cells receptors. Semin Cancer Biol 16(5):348–358

33. Fauriat C, Just-Landi S, Mallet F et al (2007) Deficient expression of NCR in NK cells from acute myeloid leukemia: Evolution during leukemia treatment and impact of leukemia cells in NCRdull phenotype induction. Blood 109(1):323–330
34. Pogge von Strandmann E, Simhadri VR, von Tresckow B et al (2007) Human leukocyte antigen-B-associated transcript 3 is released from tumor cells and engages the NKp30 receptor on natural killer cells. Immunity 27(6):965–974
35. Arnon TI, Markel G, Bar-Ilan A et al (2008) Harnessing soluble NK cell killer receptors for the generation of novel cancer immune therapy. PLoS ONE 3(5):e2150
36. Pende D, Spaggiari GM, Marcenaro S et al (2005) Analysis of the receptor-ligand interactions in the natural killer-mediated lysis of freshly isolated myeloid or lymphoblastic leukemias: evidence for the involvement of the Poliovirus receptor (CD155) and Nectin-2 (CD112). Blood 105(5):2066–2073
37. El-Sherbiny YM, Meade JL, Holmes TD et al (2007) The requirement for DNAM-1, NKG2D, and NKp46 in the natural killer cell-mediated killing of myeloma cells. Cancer Res 67(18):8444–8449
38. Nedvetzki S, Sowinski S, Eagle RA et al (2007) Reciprocal regulation of human natural killer cells and macrophages associated with distinct immune synapses. Blood 109(9): 3776–3785
39. Bryceson YT, March ME, Ljunggren HG et al (2006) Synergy among receptors on resting NK cells for the activation of natural cytotoxicity and cytokine secretion. Blood 107(1):159–166
40. Bryceson YT, March ME, Ljunggren HG et al (2006) Activation, coactivation, and costimulation of resting human natural killer cells. Immunol Rev 214:73–91
41. Miller JS, Tessmer-Tuck J, Pierson BA et al (1997) Low dose subcutaneous interleukin-2 after autologous transplantation generates sustained *in vivo* natural killer cell activity. Biol Blood Marrow Transplant 3(1):34–44
42. Lucas M, Schachterle W, Oberle K et al (2007) Dendritic cells prime natural killer cells by trans-presenting interleukin 15. Immunity 26(4):503–517
43. Skak K, Kragh M, Hausman D et al (2008) Interleukin 21: combination strategies for cancer therapy. Nat Rev Drug Discov 7(3):231–240
44. Ghiringhelli F, Menard C, Martin F et al (2006) The role of regulatory T cells in the control of natural killer cells: relevance during tumor progression. Immunol Rev 214:229–238
45. Clayton A, Mitchell JP, Court J et al (2008) Human tumor-derived exosomes down-modulate NKG2D expression. J Immunol 180(11):7249–7258
46. Spaggiari GM, Capobianco A, Abdelrazik H et al (2008) Mesenchymal stem cells inhibit natural killer-cell proliferation, cytotoxicity, and cytokine production: role of indoleamine 2,3-dioxygenase and prostaglandin E2. Blood 111(3):1327–1333
47. Street SE, Hayakawa Y, Zhan Y et al (2004) Innate immune surveillance of spontaneous B cell lymphomas by natural killer cells and gammadelta T cells. J Exp Med 199(6):879–884
48. Imai K, Matsuyama S, Miyake S et al (2000) Natural cytotoxic activity of peripheral-blood lymphocytes and cancer incidence: an 11-year follow-up study of a general population. Lancet 356(9244):1795–1799
49. Re F, Staudacher C, Zamai L et al (2006) Killer cell Ig-like receptors ligand-mismatched, alloreactive natural killer cells lyse primary solid tumors. Cancer 107(3):640–648
50. Coca S, Perez-Piqueras J, Martinez D et al (1997) The prognostic significance of intratumoral natural killer cells in patients with colorectal carcinoma. Cancer 79(12):2320–2328
51. Hsia JY, Chen JT, Chen CY et al (2005) Prognostic significance of intratumoral natural killer cells in primary resected esophageal squamous cell carcinoma. Chang Gung Med J 28(5): 335–340
52. Ishigami S, Natsugoe S, Tokuda K et al (2000) Prognostic value of intratumoral natural killer cells in gastric carcinoma. Cancer 88(3):577–583
53. Villegas FR, Coca S, Villarrubia VG et al (2002) Prognostic significance of tumor infiltrating natural killer cells subset CD57 in patients with squamous cell lung cancer. Lung Cancer 35(1):23–28
54. Zhu LY, Zhou J, Liu YZ et al (2009) Prognostic significance of natural killer cell infiltration in hepatocellular carcinoma. Chin J Cancer 28(11):1198–1202

55. Carrega P, Morandi B, Costa R et al (2008) Natural killer cells infiltrating human nonsmall-cell lung cancer are enriched in CD56 bright CD16(-) cells and display an impaired capability to kill tumor cells. Cancer 112(4):863–875
56. Schleypen JS, Baur N, Kammerer R et al (2006) Cytotoxic markers and frequency predict functional capacity of natural killer cells infiltrating renal cell carcinoma. Clin Cancer Res 12(3 Pt 1):718–725
57. Dudley ME and Rosenberg SA (2003) Adoptive-cell-transfer therapy for the treatment of patients with cancer. Nat Rev Cancer 3(9):666–675
58. Ahmadzadeh M and Rosenberg SA (2006) IL-2 administration increases CD4$^+$ CD25(hi) Foxp3+ regulatory T cells in cancer patients. Blood 107(6):2409–2414
59. Zhang H, Chua KS, Guimond M et al (2005) Lymphopenia and interleukin-2 therapy alter homeostasis of CD4$^+$ CD25$^+$ regulatory T cells. Nat Med 11(11):1238–1243
60. Bates GJ, Fox SB, Han C et al (2006) Quantification of regulatory T cells enables the identification of high-risk breast cancer patients and those at risk of late relapse. J Clin Oncol 24(34):5373–5380
61. Curiel TJ, Coukos G, Zou L et al (2004) Specific recruitment of regulatory T cells in ovarian carcinoma fosters immune privilege and predicts reduced survival. Nat Med 10(9):942–949
62. Kono K, Kawaida H, Takahashi A et al (2006) CD4(+)CD25high regulatory T cells increase with tumor stage in patients with gastric and esophageal cancers. Cancer Immunol Immunother 55(9):1064–1071
63. Ruggeri L, Capanni M, Urbani E et al (2002) Effectiveness of donor natural killer cell alloreactivity in mismatched hematopoietic transplants. Science 295(5562):2097–2100
64. Miller JS, Soignier Y, Panoskaltsis-Mortari A et al (2005) Successful adoptive transfer and *in vivo* expansion of human haploidentical NK cells in patients with cancer. Blood 105(8):3051–3057
65. Gill S, Olson JA and Negrin RS (2009) Natural killer cells in allogeneic transplantation: effect on engraftment, graft-versus-tumor, and graft-versus-host responses. Biol Blood Marrow Transplant 15(7):765–776
66. Berg M, Lundqvist A, McCoy P, Jr. et al (2009) Clinical-grade *ex vivo*-expanded human natural killer cells up-regulate activating receptors and death receptor ligands and have enhanced cytolytic activity against tumor cells. Cytotherapy 23:1–15
67. Luhm J, Brand JM, Koritke P et al (2002) Large-scale generation of natural killer lymphocytes for clinical application. J Hematother Stem Cell Res 11(4):651–657
68. McKenna DH, Jr., Sumstad D, Bostrom N et al (2007) Good manufacturing practices production of natural killer cells for immunotherapy: a six-year single-institution experience. Transfusion 47(3):520–528
69. Albertsson PA, Basse PH, Hokland M et al (2003) NK cells and the tumour microenvironment: implications for NK-cell function and anti-tumour activity. Trends Immunol 24(11):603–609
70. Gao JQ, Okada N, Mayumi T et al (2008) Immune cell recruitment and cell-based system for cancer therapy. Pharm Res 25(4):752–768
71. Matera L, Galetto A, Bello M et al (2006) *In vivo* migration of labeled autologous natural killer cells to liver metastases in patients with colon carcinoma. J Transl Med 4:49
72. Klingemann HG, Wong E and Maki G (1996) A cytotoxic NK-cell line (NK-92) for *ex vivo* purging of leukemia from blood. Biol Blood Marrow Transplant 2(2):68–75
73. Suck G, Branch DR and Keating A (2006) Irradiated KHYG-1 retains cytotoxicity: potential for adoptive immunotherapy with a natural killer cell line. Int J Radiat Biol 82(5):355–361
74. Tam YK, Martinson JA, Doligosa K et al (2003) *Ex vivo* expansion of the highly cytotoxic human natural killer-92 cell-line under current good manufacturing practice conditions for clinical adoptive cellular immunotherapy. Cytotherapy 5(3):259–272
75. Yan Y, Steinherz P, Klingemann HG et al (1998) Antileukemia activity of a natural killer cell line against human leukemias. Clin Cancer Res 4(11):2859–2868
76. Tam YK, Miyagawa B, Ho VC et al (1999) Immunotherapy of malignant melanoma in a SCID mouse model using the highly cytotoxic natural killer cell line NK-92. J Hematother 8(3):281–290
77. Tonn T, Becker S, Esser R et al (2001) Cellular immunotherapy of malignancies using the clonal natural killer cell line NK-92. J Hematother Stem Cell Res 10(4):535–544

78. Arai S, Meagher R, Swearingen M et al (2008) Infusion of the allogeneic cell line NK-92 in patients with advanced renal cell cancer or melanoma: a phase I trial. Cytotherapy 10(6):625–632
79. Nagashima S, Mailliard R, Kashii Y et al (1998) Stable transduction of the interleukin-2 gene into human natural killer cell lines and their phenotypic and functional characterization *in vitro* and *in vivo*. Blood 91(10):3850–3861
80. Zhang J, Sun R, Wei H et al (2004) Characterization of interleukin-15 gene-modified human natural killer cells: implications for adoptive cellular immunotherapy. Haematologica 89(3):338–347
81. Demirtzoglou FJ, Papadopoulos S and Zografos G (2006) Cytolytic and cytotoxic activity of a human natural killer cell line genetically modified to specifically recognize HER-2/neu overexpressing tumor cells. Immunopharmacol Immunotoxicol 28(4):571–590
82. Boissel L, Betancur M, Wels WS et al (2009) Transfection with mRNA for CD19 specific chimeric antigen receptor restores NK cell mediated killing of CLL cells. Leuk Res 33(9):1255–1259
83. Muller T, Uherek C, Maki G et al (2008) Expression of a CD20-specific chimeric antigen receptor enhances cytotoxic activity of NK cells and overcomes NK-resistance of lymphoma and leukemia cells. Cancer Immunol Immunother 57(3):411–423
84. Suck G, Branch DR, Aravena P et al (2006) Constitutively polarized granules prime KHYG-1 NK cells. Int Immunol 18(9):1347–1354
85. Suck G, Branch DR, Smyth MJ et al (2005) KHYG-1, a model for the study of enhanced natural killer cell cytotoxicity. Exp Hematol 33(10):1160–1171
86. Clynes RA, Towers TL, Presta LG et al (2000) Inhibitory Fc receptors modulate *in vivo* cytoxicity against tumor targets. Nat Med 6(4):443–446
87. Cartron G, Dacheux L, Salles G et al (2002) Therapeutic activity of humanized anti-CD20 monoclonal antibody and polymorphism in IgG Fc receptor FcgammaRIIIa gene. Blood 99(3):754–758
88. Friedberg JW, Neuberg D, Gribben JG et al (2002) Combination immunotherapy with rituximab and interleukin 2 in patients with relapsed or refractory follicular non-Hodgkin's lymphoma. Br J Haematol 117(4):828–834
89. Gluck WL, Hurst D, Yuen A et al (2004) Phase I studies of interleukin (IL)-2 and rituximab in B-cell non-hodgkin's lymphoma: IL-2 mediated natural killer cell expansion correlations with clinical response. Clin Cancer Res 10(7):2253–2264
90. Khan KD, Emmanouilides C, Benson DM, Jr. et al (2006) A phase 2 study of rituximab in combination with recombinant interleukin-2 for rituximab-refractory indolent non-Hodgkin's lymphoma. Clin Cancer Res 12(23):7046–7053
91. Berdeja JG, Hess A, Lucas DM et al (2007) Systemic interleukin-2 and adoptive transfer of lymphokine-activated killer cells improves antibody-dependent cellular cytotoxicity in patients with relapsed B-cell lymphoma treated with rituximab. Clin Cancer Res 13(8):2392–2399
92. Hartmann F, Renner C, Jung W et al (2001) Anti-CD16/CD30 bispecific antibody treatment for Hodgkin's disease: role of infusion schedule and costimulation with cytokines. Clin Cancer Res 7(7):1873–1881
93. Ansell SM, Geyer SM, Maurer MJ et al (2006) Randomized phase II study of interleukin-12 in combination with rituximab in previously treated non-Hodgkin's lymphoma patients. Clin Cancer Res 12(20 Pt 1):6056–6063
94. Bekaii-Saab TS, Roda JM, Guenterberg KD et al (2009) A phase I trial of paclitaxel and trastuzumab in combination with interleukin-12 in patients with HER2/neu-expressing malignancies. Mol Cancer Ther 8(11):2983–2991
95. Parihar R, Nadella P, Lewis A et al (2004) A phase I study of interleukin 12 with trastuzumab in patients with human epidermal growth factor receptor-2-overexpressing malignancies: analysis of sustained interferon gamma production in a subset of patients. Clin Cancer Res 10(15):5027–5037
96. Roda JM, Parihar R, Magro C et al (2006) Natural killer cells produce T cell-recruiting chemokines in response to antibody-coated tumor cells. Cancer Res 66(1):517–526
97. Kobayashi H, Dubois S, Sato N et al (2005) Role of trans-cellular IL-15 presentation in the activation of NK cell-mediated killing, which leads to enhanced tumor immunosurveillance. Blood 105(2):721–727

98. Zhang M, Yao Z, Dubois S et al (2009) Interleukin-15 combined with an anti-CD40 antibody provides enhanced therapeutic efficacy for murine models of colon cancer. Proc Natl Acad Sci USA 106(18):7513–7518
99. Roda JM, Joshi T, Butchar JP et al (2007) The activation of natural killer cell effector functions by cetuximab-coated, epidermal growth factor receptor positive tumor cells is enhanced by cytokines. Clin Cancer Res 13(21):6419–6428
100. Roda JM, Parihar R, Lehman A et al (2006) Interleukin-21 enhances NK cell activation in response to antibody-coated targets. J Immunol 177(1):120–129
101. Daniel D, Yang B, Lawrence DA et al (2007) Cooperation of the proapoptotic receptor agonist rhApo2L/TRAIL with the CD20 antibody rituximab against non-Hodgkin lymphoma xenografts. Blood 110(12):4037–4046
102. Wolska A, Lech-Maranda E and Robak T (2009) Toll-like receptors and their role in carcinogenesis and anti-tumor treatment. Cell Mol Biol Lett 14(2):248–272
103. Roda JM, Parihar R and Carson WE, 3rd (2005) CpG-containing oligodeoxynucleotides act through TLR9 to enhance the NK cell cytokine response to antibody-coated tumor cells. J Immunol 175(3):1619–1627
104. Link BK, Ballas ZK, Weisdorf D et al (2006) Oligodeoxynucleotide CpG 7909 delivered as intravenous infusion demonstrates immunologic modulation in patients with previously treated non-Hodgkin lymphoma. J Immunother 29(5):558–568
105. Chen W, Chan AS, Dawson AJ et al (2005) FLT3 ligand administration after hematopoietic cell transplantation increases circulating dendritic cell precursors that can be activated by CpG oligodeoxynucleotides to enhance T-cell and natural killer cell function. Biol Blood Marrow Transplant 11(1):23–34
106. Reddy N, Hernandez-Ilizaliturri FJ, Deeb G et al (2008) Immunomodulatory drugs stimulate natural killer-cell function, alter cytokine production by dendritic cells, and inhibit angiogenesis enhancing the anti-tumour activity of rituximab *in vivo*. Br J Haematol 140(1):36–45
107. Wu L, Adams M, Carter T et al (2008) lenalidomide enhances natural killer cell and monocyte-mediated antibody-dependent cellular cytotoxicity of rituximab-treated CD20+ tumor cells. Clin Cancer Res 14(14):4650–4657
108. Borg C, Terme M, Taieb J et al (2004) Novel mode of action of c-kit tyrosine kinase inhibitors leading to NK cell-dependent antitumor effects. J Clin Invest 114(3):379–388
109. Koh CY, Blazar BR, George T et al (2001) Augmentation of antitumor effects by NK cell inhibitory receptor blockade *in vitro* and *in vivo*. Blood 97(10):3132–3137
110. Sheridan C (2006) First-in-class cancer therapeutic to stimulate natural killer cells. Nat Biotechnol 24(6):597
111. Jinushi M, Hodi FS and Dranoff G (2006) Therapy-induced antibodies to MHC class I chain-related protein A antagonize immune suppression and stimulate antitumor cytotoxicity. Proc Natl Acad Sci USA 103(24):9190–9195
112. Mao CP, Hung CF and Wu TC (2007) Immunotherapeutic strategies employing RNA interference technology for the control of cancers. J Biomed Sci 14(1):15–29
113. Lundqvist A, Abrams SI, Schrump DS et al (2006) Bortezomib and depsipeptide sensitize tumors to tumor necrosis factor-related apoptosis-inducing ligand: a novel method to potentiate natural killer cell tumor cytotoxicity. Cancer Res 66(14):7317–7325
114. Shi J, Tricot GJ, Garg TK et al (2008) Bortezomib down-regulates the cell-surface expression of HLA class I and enhances natural killer cell-mediated lysis of myeloma. Blood 111(3):1309–1317
115. Armeanu S, Bitzer M, Lauer UM et al (2005) Natural killer cell-mediated lysis of hepatoma cells via specific induction of NKG2D ligands by the histone deacetylase inhibitor sodium valproate. Cancer Res 65(14):6321–6329
116. Schmudde M, Braun A, Pende D et al (2008) Histone deacetylase inhibitors sensitize tumour cells for cytotoxic effects of natural killer cells. Cancer Lett 272(1):110–121

Chapter 5
Dendritic Cell-Based Cancer Vaccines: Practical Considerations

Elizabeth Scheid, Michael Ricci, and Ronan Foley

Abstract Recent advances in our understanding of the immune system and its relationship to cancer have enhanced the therapeutic potential of cancer immunotherapy. The ability to use clinical cell processing to safely and effectively generate novel autologous cell-based therapies has led to a rapid expansion in early phase clinical trials evaluating the potential of cell-based cancer immunotherapies. These studies have demonstrated both feasibility and remarkable low-level toxicity. Despite these advances, clinical efficacy has been limited. This review aims to summarize both success and current limitations of cell-based immunotherapy. Specific issues include optimization and standardization of cell specific products, identification of ideal patients and cancer subtypes as well as methods to comprehensively evaluate the host response and more fully understand underlying biological effectors that are engaged.

Keywords Cancer • Cell-based therapy • Dendritic cells • Immunotherapy • Vaccines

Introduction

Cancer vaccines aim to stimulate an immune response to selectively eradicate tumor cells by a mechanism that typically involves cytotoxic T cells (CTL). Although cancer vaccines may ultimately be used to prevent cancer, their current clinical evaluation pertains to patients with established malignancy. Immunotherapy for the treatment of cancer has been investigated for decades. In the early twentieth century, "remissions" from cancer were observed in patients who developed severe streptococcal skin infections. This observation led to further investigations that

R. Foley (✉)
FRCPC Director, Stem Cell Laboratory, Juravinski Hospital and Cancer Centre, Hamilton Health Sciences, Hamilton, ON, Canada
e-mail: foleyr@hhsc.ca

J. Medin and D. Fowler (eds.), *Experimental and Applied Immunotherapy*,
DOI 10.1007/978-1-60761-980-2_5,

involved direct injection of bacterial filtrates ("Coley's toxins")[1]. Based on these early observations, it became evident that immunity indeed played a role in host protection against cancer. It is now known that the immune system is actively involved in cancer surveillance and prevention [2]. Thus, if cancer does develop, the occurrence is due either to cancer cell evasion of a functional immune system or to a fundamental immunologic error. And, by extension, it has long been proposed that if immunity against cancer might be restored, then an effective anti-tumor effect might be realized. Clinical investigators have employed a variety of methods to modulate the immune system in an attempt to augment anticancer immunity. Therapeutic strategies have included the use of monoclonal antibodies, cytokines, autologous T cells, and allogeneic bone marrow transplantation.

In the 1980s, clinical trials using recombinant cytokines completed by Rosenberg *et al.* demonstrated therapeutic benefit when interleukin-2 (IL-2) was combined with tumor-infiltrating lymphocytes (TILs) [3, 4]. The identification of antigens recognized by TILs from a patient that responded successfully [5], and the discovery of the MAGE-1 antigen from peripheral blood mononuclear cells (PBMCs) sensitized *in vitro* with melanoma cells [6], set the stage for future clinical developments [7]. The Steinman laboratory was essential in assigning an important role to dendritic cells (DCs) in antigen presentation and T-cell stimulation [8]. Given their central role in immune activation, the concept of using autologous DCs as a therapeutic cell-based vaccine has been considered for some time. The first DC vaccine trial was completed in 1995, and since then, there have been numerous phase I and II studies completed in a variety of malignancies [9, 10]. Despite remarkable preclinical murine results, DC-based vaccines have yet to be proven clearly efficacious in humans. As a result, questions exist as to the most promising directions for current and future clinical DC-based vaccine trials. This review will focus on clinical use of cell-based therapies that employ autologous DCs and highlight specific next steps that may help move the field forward.

Dendritic Cell Biology

Dendritic cells have been extensively evaluated in experimental models, and encouraging results have provided support to move DC-based therapy into the clinical setting. Many successful immunotherapy applications of bone marrow-derived DCs in preclinical cancer models have been described [11–13]. These studies have evaluated well-characterized tumor rejection antigens and have employed a variety of methods to load such antigens onto murine DCs. Results have been impressive, with DC therapy demonstrating both protective immunity against tumor challenge and therapeutic immunity against established tumors.

The mechanism of action of DCs in the development of an immune response has also been extensively evaluated in preclinical models [14, 15]. As previously mentioned, DCs are potent antigen-presenting cells (APCs) that play an essential role in the induction of T-cell responses. DCs function as specialized leukocytes to acquire, process, and present antigens to T cells within the context of major histocompatability complexes I and II [14, 15]. DCs are present at potential pathogen entry sites, and exist in a relatively immature state in those locales. When presented with inflammatory signals, DCs rapidly mature and take up foreign antigens. In doing so, they develop into APCs capable of migrating and initiating an immune response. This presentation of antigens stimulates specific T-cells that have the capability of destroying a foreign entity [14, 15].

The antigen-presenting capability of DCs may be useful as an application in therapeutic cancer immunology. It seems logical that by introducing antigens of choice to autologous DCs, one can induce the desired immune response. This can be accomplished through the isolation of DC precursors, the *ex vivo* loading of DCs with appropriate antigens, and then the reintroduction of the modified DCs back into the patient. In theory, DCs will migrate to lymph nodes and present antigens to stimulate the immune system to induce the destruction of an antigen-bearing target cancer cell. Since tumors may possess specific antigens, it seems reasonable to speculate that therapeutic effects can be achieved if one can correctly present tumor-associated antigens to effector T cells. On the other hand, if the target antigen were a self-antigen, then the vaccination platform would need to provide sufficient stimulation to overcome self-tolerance mechanisms. In any case, the ultimate goal would be to create a sustained immune response capable of eliminating or reducing the growth of cancer cells (see Fig. 5.1).

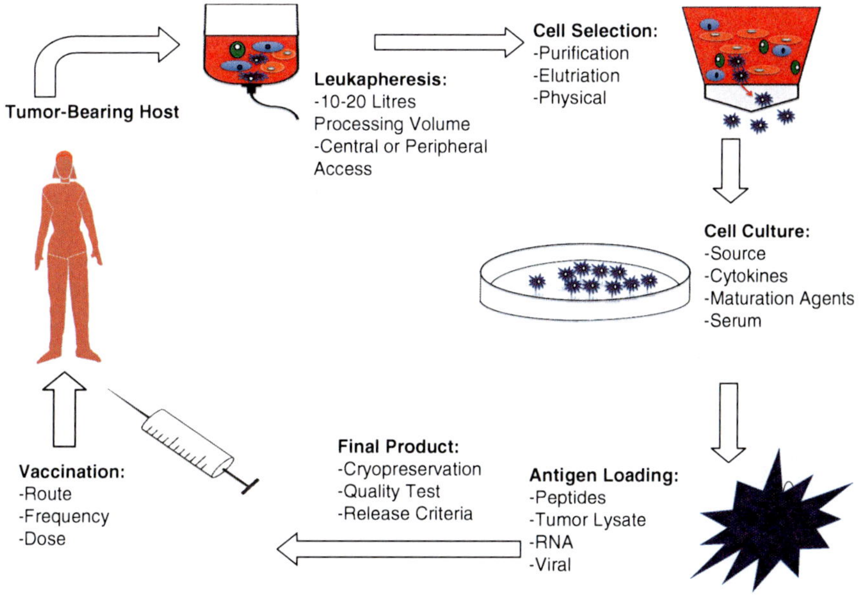

Fig. 5.1 Dendritic cell therapy overview

Autologous Clinical Vaccine Trials

Based on supportive preclinical data, translation to early phase human clinical trials occurred quickly [9]. Although technical concerns were initially raised, the ability to select specific cell populations from peripheral blood eventually made these studies feasible to implement. Isolation of peripheral blood DC precursors could be performed in hospital-based laboratories followed by the preparation of cellular vaccines that are thoroughly characterized and approved by regulatory agencies for clinical use. Collectively, these clinical trials have studied a variety of cancer types in patients at various stages of disease; the cellular vaccination platforms have typically included DCs derived from either monocyte or $CD34^+$ hematopoietic progenitors [16–18]. Such DCs have been cultured in various cytokine combinations and have been loaded with putative tumor antigens by a variety of techniques. The dose of cultured DCs, route of administration, and frequency of injections have also varied considerably. Immune outcome measures have typically involved an assessment of activation of $CD8^+$ CTLs by the release of IFN-γ (as measured by ELISPOT or intracellular flow cytometry-based assays). Clinical response described as complete response (CR), progressive disease (PD), or stable disease (SD) has in some trials been defined by RECIST (Revised Evaluation Criteria in Solid Tumour) [19]. A summary of representative clinical DC vaccine trials is shown in Table 5.1.

Clinical Dendritic Cell Studies: Issues Facing the Field

DC Source and Vaccine Manufacture

The feasibility of producing large-scale preparations of autologous DCs from cancer patients has already been established. These varied efforts have contributed to a greater knowledge of DC biology and vaccine preparation, but have not yet led to a consensus as to the best procedure for generating human DCs. There remains variability in terms of the source of cells used to generate DCs, procedures to enrich these cells, and use of final DC maturation factors. Culture conditions vary as to media type, protein source (serum, albumin, autologous plasma, or serum-free), vessel type, and length of culture [42–44].

Clinical-grade DCs can be generated *ex vivo* from autologous $CD34^+$ hematopoietic progenitor cells (HPCs) or from monocytes (MoDCs). To generate the quantities of DCs required to provide multiple vaccine doses for clinical use, high numbers of $CD34^+$ HPCs are often obtained in leukapheresis products from patients who have undergone stem cell mobilization by *in vivo* administration of GM-CSF or G-CSF [41]. An enriched population of $CD34^+$ cells is usually selected prior to culture [16], however, HPC may be cultured without $CD34^+$ enrichment and still yield enriched DCs at the end of culture [45]. In contrast with HPCs, monocytes are relatively abundant in peripheral blood, eliminating the requirement for *in vivo* cytokine mobilization

Table 5.1 Summary of clinical trials evaluating autologous dendritic cells (DCs) in the setting of cancer[a]

Cancer	Stage	DC source	Ag loading	Route/#inject/ freq.	Toxicity (grade)	Immune response (% of patients)	Tumor regression (% of evaluable patients)	References
HCC	A	Mo	lys ate– HepG2	i.v. ×6 q 3 weeks	No	50	12	[20]
Mel	A	Mo	Killed Colo829	s.c., 4–8, q 4 weeks	No	23	10	[21]
Mel	A	Mo	lys – M44, SK MEL 28, COLO 829	i.d. and s.c. ×6 biwk and 2q 6 weeks	Mild	84	6	[22]
MM	M	Mo	id. prot or pep	s.c. ×3 + i.v. ×2 biwk	No	66	6	[23]
Mel	A	Mo	lys – aut T	i.d. ×3 biwk	No	40	8	[24]
Mel	A	Mo	lys aut T	i.d. ×6 biwk	No	NA	13	[25]
CRC	A	Mo	pep- CEA, MAGE-2, HER2	i.d. ×4 q 3 weeks	Mild	27	0	[26]
MM	A	CD34	id pep	s.c. ×4 biwk	No	45	0	[27]
Breast	A	Mo	p53 pep	s.c. ×4 weeks, × 6 biwk	No	36	7	[28]
Breast	A	CD34	pep – HER2	i.v., ×3–6, q 3 weeks	No	20	20	[29]
PCa	A	Mo	prot – PAP	i.v. or i.d. or i.l., ×2, q 4 weeks	Mild	100	NA	[30]
PCa	A	Mo	pep – PSMA	i.v. ×6, q 6 weeks	No	NA	27	[31]
BCL	A	Mo	id prot	i.v. ×3, q 4 weeks + ×1 q 2–6 months	No	69	20	[32]
Mel	A	Mo	pep – MAGE-3	s.c. and i.d. ×3, biwkly + ×2 i.v. biwk	Mild	72	54	[33]
Mel	A	Mo	lys – aut T. OR pep – MAGE-3, tyro, gp100, MART1	i.n. ×4 q weeks + ×1 at 6 weeks + ×4 q weeks	No	33	12	[34]

(continued)

Table 5.1 (continued)

Cancer	Stage	DC source	Ag loading	Route/#inject/ freq.	Toxicity (grade)	Immune response (% of patients)	Tumor regression (% of evaluable patients)	References
Mel	A	Mo	lys – aut T. OR pep – MAGE-1, MAGE-3, tyro, MART1, gp100	i.n. ×4 q weeks, ×1 at week 6 + ×5 q 4 weeks	No	68	37	[35]
Mel	A	Mo	pep – MAGE-1, MAGE-3, MAGE-4, MAGE-10, tyro, gp100, melan-A	s.c. ×5 biwk + ×1 at week 14	No	75	4	[36]
Child	M	Mo	lys – aut T.	i.d. ×3 biwk	No	50	6	[37]
RCC	A	Mo	aut T. RNA	i.v. and i.d. ×3 biwk	No	85	NA*	[38]
CRC/ NSCLC	A	Mo	pep – CEA	i.v. ×2 q 4 weeks	Mild	58	25	[39]
RCC	A	Mo	lys – aut T. OR RCC cell line	i.v. or i.d. ×3 q 4 weeks	Mild	83	8	[40]
Mel	A	CD34	pep – MART1, tyro, MAGE-3, gp100	s.c. ×4 biwk	No	88	38	[41]
Mel	A	CD34	pep – MAGE-1, MAGE-3, melan-A, gp100, tyro	i.v. ×4 biwk	Mild	28	14 NED	[16]
					AV	57	15	

[a] A variety of vaccine platforms are employed. Estimates of% patients developing an immune response and % demonstrating clinical evidence of any tumor shrinkage are listed. *A* advanced, *aut T* autologous tumor, *BCL* B cell lymphoma (non-Hodgkin's), *CD34* $CD34^+$ progenitors, *Child* childhood cancers, *CRC* colorectal cancer, *HCC* hepatocellular carcinoma, *lys* lysate, *MM* multiple myeloma, *Mel* melanoma, *Mo* monocytes, *NA* not available, *NA** unable to assess CR due to treatment with other therapies/death, *NED* surgery on entry, *NSCLC* nonsmall cell lung cancer, *PCa* prostate cancer, *pep* peptide, *prot* protein, *tyro* tyrosinase

of donors before leukapheresis. Monocytes may be further enriched from apheresis products by plastic adherence to culture flasks [46], positive or negative CD14 selection by antibody beads [47], or by elutriation [48]. HPC are typically cultured for 9–28 days with the addition of a cocktail of growth factors and cytokines to promote expansion of the $CD34^+$ cells (SCF, Flt-3, IL-3, and IL-6) combined with other cytokines that promote differentiation into DC (GM-CSF and IL-4). The further addition of TNF-α matures $CD34^+$-derived DC; alternatively, DC maturation has been achieved by some groups by adding a cocktail of proinflammatory cytokines which may include IL-1β, IL-6, and PGE-2, in addition to TNF-α [49, 50]. Compared with HPC-derived DCs, monocyte-derived DCs are cultured for a shorter period; most often 7 days with some groups advocating the reduction of culture times to 2–3 days [51, 52]. For monocyte-derived DC generation, GM-CSF and IL-4 are added during the duration of the culture period [53]. DC maturation is achieved by the addition of a variety of cytokines, with a cocktail comprised of TNF-α, IL-1β, IL-6, and PGE-2 (Jonuleit Cocktail) [54] being used in the preparation of many clinical DC vaccines. However, the use of this maturation cocktail has been scrutinized in response to studies demonstrating that the addition of PGE-2 leads to reduced IL-12 secretion by DC [55]. DC secretion of IL-12 is desirable in that IL-12 polarizes T-cell responses toward $CD4^+$Th1 and $CD8^+$Tc1 CTL responses that are considered to be essential for vaccine efficacy. On the other hand, PGE-2 induces CCR7 expression, thereby reducing DC adherence to culture flasks and more critically, inducing migratory function in DC. Further studies have demonstrated that PGE-2-related reduction in IL-12 production is transitory, and that on DC-T cell contact, IL-12 production is not reduced [56].

Efforts to identify the best cytokine combination to obtain optimal DCs for clinical vaccines have included modifying the Jonuleit cocktail by eliminating PGE-2 and adding CD40L and IFN-α [57]. Other groups have induced DC maturation with LPS and IFN-γ [42], or have recommended using the TLR ligands polyI:C and R848 in addition to using PGE-2 [58]. The complexity of the observations with respect to cytokine stimuli and IL-12 production illustrates the inherent difficulty in elucidating the optimal conditions for DC maturation, and cautions against assessing one factor (e.g., IL-12 production) to determine the optimal maturation stimulus [57].

To date, monocytes have been the most common source of cells used to make clinical-grade DCs. As previously mentioned, in contrast to HPCs, monocytes can be obtained without cytokine mobilization; this feature is advantageous in situations in which heavily pretreated patients may fail to mobilize sufficient progenitor cells. Moreover, a smaller leukapheresis volume is required to obtain a monocyte-enriched product relative to the volume required to collect a HPC-enriched product. Typically, a 10 liter leukapheresis procedure will provide sufficient monocytes for a clinical-scale DC production. This factor therefore reduces the time required for apheresis, potentially reducing the need for a central intravenous access line and larger volume processing. In addition, when compared to HPC-derived DC generation, the manufacture of DCs from monocytes is generally less costly due to reduced time in culture and reduced reagent consumption. Finally, the resultant cell population in the DC vaccine product generated from monocytes is more homogenous than a DC population derived from HPCs.

Nonetheless, there has been significant debate as to the best choice of DC precursor for immunotherapy trials. Studies have described similarities between the two populations with respect to morphology, phenotype, antigen uptake, and ability to present antigen [58, 59]. $CD34^+$-derived DCs have been shown to more potent than monocyte-derived DCs for inducing allogeneic T-cell proliferation and for priming $CD8^+$ T cells [60, 61]. Procedures to optimize the large-scale generation of clinical grade moDCs [62] and CD34-DCs [45, 49, 50] have been developed. A number of clinical studies have been completed using $CD34^+$-derived DCs [29, 41, 63, 64]. Clinical and immune responses after immunization with $CD34^+$ DC in melanoma patients have been reported by Banchereau *et al.* [16]; however other studies have indicated inferior responses with $CD34^+$ DC vaccines [27, 29].

Overall, the results from human trials using monocyte or $CD34^+$ progenitors have demonstrated variable success. As such, it is currently difficult to draw definitive conclusions as to the optimal cell source for the generation of DCs. Inconsistent outcomes are more likely related to variability in the procedures used to manufacture DC vaccines and to individual patient factors rather than to significant differences in monocyte versus $CD34^+$-derived DCs. Ongoing controversies relating to DC subtype, state of maturation, and ability to migrate indicate that the DC debate is far from over. For the time being, it may be practical in designing clinical trials to consider an approach that has favorable side effect and cost profiles.

Although no standardized process exists for producing DC vaccines, standardized objectives exist with respect to the use of GMP reagents, closed systems, and the incorporation of DC vaccine release criteria. The quality of the DC product can be assessed by parameters including sterility, viability, purity, identity, stability, and potency. Typically, the expression of DC surface markers such as CD80, CD83 or CD54 [65] have been used as surrogate markers of DC potency, in addition efforts have been made to establish a standardized DC potency assay [66]. Release specifications must take into account variability arising from individual patients. The adoption of appropriate in-process controls, batch records, testing, and specifications for release may improve consistency in DC vaccine products.

Antigen Loading of DCs

A variety of methods can be used to load DCs with putative tumor rejection antigens. Ideally, the method used will provide sufficient T-cell activation to generate a clinically-effective antitumor immune response. DC vaccines may be loaded with individually defined antigens or with mixed preparations derived from tumor cells. Of course, there are advantages and disadvantages to each strategy. Clinical-grade peptides are readily available and can be simply loaded into DC preparations. In addition, posttherapy immune monitoring is relatively straightforward when such defined antigens are used. However, the list of known tumor antigens remains relatively limited and includes antigens that may have suboptimal immunogenicity. DC vaccines loaded with a blend of peptides may circumvent this. Another disadvantage of peptide loading is that peptides are restricted to specific HLA types, thereby limiting the application of this approach to only a subset of potential patients.

An interesting observation seen after administration of peptide alone, or peptide loaded DC vaccines has been the phenomenon of antigen spreading; i.e., development of immunity against antigens that are present on the tumor, but not present in the vaccine [67, 68]. Epitope spreading is seen in autoimmune disease; tissue destruction generates debris that is taken up and presented by APCs, thereby generating a diversity of autoimmune responses. Similarly, the generation of an initial antitumor response after vaccine administration can create a microenvironment amenable to the presentation of a variety of TAAs thereby generating CTL to TAA additional to those present in the vaccine. A criticism of peptide antigen loaded DC vaccines has been that the presentation of the loaded peptides may be relatively short-lived, and also not sufficient to induce an effective immune response. To improve on the efficacy of peptide loaded DC vaccines, alterations to peptides to modify MHC binding, immunogenicity, and elicit a broad repertoire of T cells have been suggested. Strategies have included the use of peptides that require processing by DC; for example long synthetic peptides comprised of both MHC-I and MHC-II epitopes and glycopeptides [69, 70].

Alternative antigen loading strategies include the use of allogeneic and autologous tumor lysates, apoptotic cells, total mRNA from tumor, or tumor-DC fusions. Such approaches may be advantageous because they provide an assortment of tumor antigens that may be more relevant to an individual patient. However, unlike the relatively straightforward immune monitoring following vaccination with defined antigens, assessment of immune responses in these alternative approaches may be more complex. Another challenge with some of these alternative approaches is the limited ability to obtain sufficient tumor tissue. This limitation may be overcome by the use of tumor mRNA; that is, tumor mRNA can be obtained from a relatively small biopsy and can then be amplified to provide material sufficient for DC vaccine manufacture. Electroporation procedures can then be used to transfect DCs [71]. Presentation of antigens by mRNA-transfected DCs has been shown to be prolonged and capable of generating effective CTL responses [71].

Viral vectors that incorporate sequences for tumor antigens may also be used to load DCs. Extensive preclinical work has established the suitability of different recombinant vectors, including adenoviridae [72–75] poxviridae [72], retroviridae [75], and rhabdoviridae [76]. In some studies, virally transduced DCs have been shown to be superior to DCs transfected by other methods [77]. When compared to peptide-pulsed DCs, adenovirus-transduced DCs have demonstrated enhanced expression of CCR7 and improved migratory capacity [77]. Moreover, viral-based genetic vaccinations have proven both clinically safe and feasible. Specifically, a clinical trial administering $CD34^+$-derived DCs modified with a tyrosinase-encoding vaccinia virus construct to six melanoma patients showed one partial response, with the majority of patients displaying an immune response [63]. Another clinical trial employing adenovirus-gp100/MART-1-transduced DC vaccines in melanoma patients resulted in the development of vitiligo or melanoma-associated hypopigmentation (MAH) in all three of the patients treated [78]; of note, MAH has been reported in other studies using DC immunotherapy for melanoma [41]. However, none of the three patients in the study reported by Tsao *et al.* [78] showed significant clinical responses.

Route of Administration

Generation of an effective T-cell response may be highly dependent on migration of antigen-loaded DCs to appropriate antigen-presentation sites. Studies to determine an optimal route of administration remain a priority. Routes for administration have included intravenous (i.v.), subcutaneous (s.c.), intradermal (i.d.), intranodal (i.n.), and intralymphatic (see Table 5.1). Some studies have randomized patients to different injection routes to allow a direct comparison of the routes using the same DC vaccine. In a pilot clinical trial completed by Fong *et al.* [30], peptide-pulsed DCs were injected by i.v., i.d., and intralymphatic routes. Immune responses were seen in all groups; however, the type of response varied among the cohorts. For example, the induction of T-cell IFN-γ production was only seen in the patients who had been injected by i.d. or intralymphatic routes. Antigen-specific antibodies developed in all treatment cohorts, however a higher frequency and titer of antigen specific antibodies were seen among the patients in the i.v. administered cohort, leading to the conclusion that i.v. administration may lead to antibody response whereas Th1 immunity is better induced by i.d. or intralymphatic injections. Another study compared melanoma patients randomly assigned to i.v., i.d., and i.n. cohorts. Results from this study also suggested that intranodal administration of DCs might be preferred for induction of T-cell activity [79].

Visualization of the trafficking and *in vivo* localization of radiolabeled DCs has shown that i.v.-infused DCs traffic through the lungs and then localize in the spleen and liver [80]. Prince *et al.* [80] reported a similar fate for i.v.-infused DCs, and also demonstrated that DC injected by s.c. and i.d. routes displayed similar but inconsistent migration to lymph nodes, with less than 2% of the cells actually reaching the regional nodes. DCs injected intranodally accumulate in the injected node, and have been shown to travel to subsequent draining nodes [81]. This observation would suggest that i.n. injection, although technically challenging, might be superior to i.d. administration. However, the further migration of DCs to other nodes may be the result of disruption of nodal architecture rather than true migration, thereby accounting for variable results between studies [81, 82]. Variable results may also be due to imprecise delivery into the node. A comparison of i.d. and i.n. injection of DCs in 22 melanoma patients demonstrated immune responses in both groups of patients. Examination of the nodes showed no disruption of the nodal architecture; in addition, there was an increase in the size of the cortex/paracortex areas, thereby suggesting the occurrence of T-cell stimulation in these areas. The frequency of response was actually greater in the i.d. group; however, the significance of this finding is not clear due to the small sample size [83].

Despite the clear research advances that are developing in this area, there is no clear answer as to the best route of DC administration, thus leading some to conclude that the best approach may be to use multiple DC administration routes [84]. It is clear, however, that with rapidly developing image technologies there is a great potential to now integrate real-time imaging. This technology will help to determine the trafficking fate of prepared, often cryopreserved DC products and therefore may help identify an optimal route of DC administration.

Cancer Type: Susceptibility to DC Therapy

The ultimate goal of DC-based vaccine trials is to eliminate tumors in all target regions, that is, to achieve a complete response (CR). One could predict that specific cancer types may be more susceptible to immune intervention than others. Studies have indicated that melanoma, colorectal, prostate, and non-Hodgkins lymphoma (NHL) may be more susceptible and applicable to DC immunotherapy (see Table 5.1). Given that patients with NHL can attain complete remission with current conventional therapies, the concept of a postinduction vaccine is appealing for the prevention of tumor relapse. As seen in Table 5.1, melanoma is also a commonly studied cancer type for DC therapy and remains an excellent disease target for immunotherapy. Several studies have demonstrated mildly encouraging results in melanoma [16, 21, 35, 36, 41, 64] and colorectal cancer [39, 85]. The colorectal studies have had some limited clinical responses in patients with advanced cancer including; tumor regression in two patients and two others experiencing stable disease [39]; reduction of CEA levels in another study [84]; stable disease and increased survival [86]); and one complete response and stable disease in another study [87].

Recently, there has been a hint of success with a DC vaccine approach in studies of antigen-loaded antigen presenting cells (Sipuleucil-T, Dendreon) in patients with hormone refractory prostate cancer. Prostate acid phosphate (PAP) is expressed in over 90% of prostate tumors. Sipuleucil-T is composed of autologous DCs pulsed with a recombinant fusion protein consisting of PAP and GM-CSF (PA2024). Phase III studies [88] of men with metastatic androgen-independent prostate cancer has recently provided positive results. An integrated analysis of data from two randomized, double-blind, placebo-controlled phase III studies showed increased overall survival in patients receiving Sipuleucil-T, with a 4.3 month survival difference and a 33% reduction in the risk of death [89]. A recent study [90] comparing Sipuleucel-T to placebo in 512 men with prostate cancer, found that median survival was extended 4.1 month by the vaccine and determined that the 3-year survival rate was increased by 38%. Adverse events of the vaccine consisted of low-grade toxicities, including fever, chills, and headache; typically, these side effects only lasted 1–2 days [90]. As such, these results using the Sipuleucil-T vaccine has restored some optimism to the use APC methods of immunotherapy, consistently showing improvement in patient survival in phase III trials [88, 89], Given the complexities of immunology in each cancer subtype, it is reasonable to expect different diseases to have better responses than others. It is also possible that other less well-studied cancer types will emerge as being immune susceptible.

Summary

The scientific rationale for using DC-based cancer vaccines is strong, and clinical trials have clearly demonstrated some success of this approach. The generation of clinical-grade cellular vaccines is clearly feasible. Moreover, administration of a wide-range of autologous DC products can occur in an outpatient setting and appears to be safe. Injection of autologous DCs often induces a measurable and predicted immune response in patients ranging from heavily pretreated individuals bearing advanced tumors to chemotherapy-naïve patients with relatively low tumor burden. In some patients, albeit at an infrequent rate, immune responses generated by DC vaccines appear to result in a definite shrinkage of one or more existing tumors. Based on these results, one could argue that the next logical step would be to move successful phase I and II trials into a phase III randomized setting in which other downstream outcome measures such as survival and quality of life can be studied. However, several challenges exist that need to be considered. Specifically, at the present time, there may be some reluctance to move toward phase III clinical trials due to both the significant cost of vaccine preparation and a minimal perceived response rate efficacy using RECIST criteria. Therefore, consideration must be made as to the next steps to be taken.

Issues in Clinical Trial Methodology

The standard process used to evaluate a cytotoxic cancer agent is to determine response rate and toxicity, disease-free survival (DFS) in responders, and then overall survival (OS). Use of these parameters is based on a premise suggesting that downstream clinical outcomes for a drug should demonstrate a significant upfront response rate (complete response CR/partial response [PR]). DC vaccines and resultant T-cell effector mechanisms may work to halt the growth, spread, and progression of cancer, which clearly could be important for downstream outcome measures; however, such immune mechanisms may not provide, or be intended to provide, a robust initial response rate. Nevertheless, how does one justify moving to the phase III clinical trial setting with less than spectacular response rate (RR) data in phase I and II clinical trials? On the other hand, could important outcomes be missed by not moving to larger randomized comparative studies capable of evaluating these later outcome measures? Since clinical responses may take longer with DC-based vaccines than with traditional cytotoxic therapies, and disease progression may occur during this time, it may be appropriate to use criteria other than RECIST to measure response endpoints [91]. Along these lines, the iSBTc Cancer Vaccine Clinical Trials Working Group has proposed guidelines in an effort to move cancer vaccine trials forward [92].

Standardization of DC Preparation

Another challenge to the field relates to a lack of standardization in DC vaccine manufacture. As previously noted, efforts to develop standard procedures may be beneficial and may facilitate more meaningful comparisons between trials.

Immunological Parameters

Although it would seem perfectly logical to conclude that measurable immune responses must be present before one might observe an effective clinical tumor response, this view is controversial and may not be valid. Efforts to improve the sensitivity and specificity of immune outcome analysis may help to ensure a tighter link between immune activation and clinical response. Not only would this be of clinical benefit, but may help guide future incremental steps to focus on and enhance immune activation. At present, simply measuring peripheral blood-activated T cells may be misleading if other immunosuppressive factors are simultaneously at play. The ability to mount an effective immune response is hindered by the presence of T-regulatory cells that exert direct immunosuppressive effects on effector T cells and other tumor derived factors present within the tumor microenvironment, such as cytokines (TGF-β and IL-10), and enzymes (IDO) which exert an immunosuppressive effect [93]. With respect to the latter, it is highly possible that even if an effective T-effector cell response was generated from a DC vaccine, immunosuppressive activity from within the local tumor environment could have a dampening or mitigating effect on any cellular responses. In this regard one may have to consider removing or temporarily disabling the tumor, perhaps by conventional therapies including radiation or cytotoxic chemotherapy.

Specific features of the individual host must also be considered. Given that some patients are capable of generating both immune and clinical responses it would seem logical, if possible, to predetermine what predictive factors are relevant in terms of patient selection. These factors may range from cancer type, stage of disease, and number of lines of prior therapy to more complex immune measures such as T-cell subset analysis and degree of response to foreign antigens. Development of a standardized immune competence assay may also be helpful for identification of those patients that are likely to benefit from this cell-based therapy.

Combination Therapy

The majority of patients treated on DC immunotherapy trials have had late-stage malignancy. Presumably, many patients harbor tumors that are too large to be susceptible to destruction by T cells alone; as such, it may be advantageous to evaluate DC vaccines in patients with early-stage disease [94]. It has also been proposed that the efficacy of DC vaccines in a metastatic setting may be increased by combining DC therapy with other therapies [95, 96]. Destruction of tumors using chemotherapy, radiotherapy, or cryo-ablation results in reduction of tumor size and the release of antigen and DC activating signals [91]. Tumor cell death induces a complicated network of reactions that can either be immunosuppressive or actively stimulate antitumor responses [97]. For example, treatment with anthracyclines such as doxorubicin induces tumor-cell apoptosis; doxorubicin-treated murine cancer cells can be taken up by DCs, thereby leading to the proliferation of $CD8^{+}$ T cells [98].

The potential for improving efficacy of DC vaccination by combining it with conventional therapies is of significant interest and has been investigated in clinical trials [99]. A clinical trial combining radiotherapy with administration of DCs in a formulation containing GM-CSF and IL-4 in refractory hepatoma patients has shown some evidence of clinical and immune responses [100]. Other potential strategies include using oncolytic viruses, either to provide oncolysates for pulsing DCs [101] or as virotherapy to reduce tumor followed by administration of DCs [102, 103]. Preclinical work combining oncolytic herpes simplex virus (HSV) followed by administration of immature DCs resulted in tumor reduction and increased survival in mice [104].

Future Directions

It may be a blend of modifications (see Table 5.2) that will strengthen the field of cell-based DC vaccines. Efforts to select patients and cancer subtypes will remain important. The ability to predict "good immune responders" based on novel assays will be beneficial. The ability to design informative comparative two-armed trials may also be of benefit. Given that primary outcome measures are largely immunobiological, there is a need to improve the frequency and level of these responses by understanding the tumor microenvironment and by improving the potency of autologous killer T cells. The ability to standardize DC preparations may be of theoretical benefit but challenging given that no one protocol has emerged as superior. Finally, combination therapy makes intuitive sense, as destruction of an initial bulky tumor followed by DC vaccine could be a strategy that is developed for a phase III clinical trial setting.

Table 5.2 Cancer dendritic cell (DC) vaccines: Future directions[a]

Disease	Susceptibility to immune intervention
Patients	Stage, age, remission, minimal residual disease immune responsiveness
Vaccine strategy	DC precursor
	Antigen loading
	Route of administration
	Novel imaging
Immune monitoring	Tests, clinically evaluable
Methodology	Two-armed studies
	Multi-institutional
	Larger studies – disease free survival/overall survival
	RECIST-other options
Combinations	Conventional therapies
	Monoclonal antibodies
	New agents
	Oncolytic vectors

[a]Summary of issues facing DC-based clinical immunotherapy studies. In reaching a position to move into phase III studies a variety of issues currently exist. Addressing these issues may help the field move beyond its current position

References

1. Davis ID, Jefford M, Parente P, et al. (2003) Rational approaches to human cancer immunotherapy. J Leukoc Biol 73(1):3–29
2. Swann JB, Smyth MJ. (2007) Immune surveillance of tumors. J Clin Invest 117(5):1137–1146
3. Lotze MT, Frana LW, Sharrow SO, et al. (1985) *In vivo* administration of purified human interleukin 2. I. Half-life and immunologic effects of the Jurkat cell line-derived interleukin 2. J Immunol 134(1):157–166
4. Rosenberg SA, Lotze MT, Muul LM, et al. (1985) Observations on the systemic administration of autologous lymphokine-activated killer cells and recombinant interleukin-2 to patients with metastatic cancer. N Engl J Med 313(23):1485–1492
5. Robbins PF, el-Gamil M, Kawakami Y, et al. (1994) Recognition of tyrosinase by tumor-infiltrating lymphocytes from a patient responding to immunotherapy. Cancer Res 54(12):3124–3126
6. van der Bruggen P, Traversari C, Chomez P, et al. (1991) A gene encoding an antigen recognized by cytolytic T lymphocytes on a human melanoma. Science 254(5038):1643–1647
7. Novellino L, Castelli C, Parmiani G. (2005) A listing of human tumor antigens recognized by T cells: March 2004 update. Cancer Immunol Immunother 54(3):187–207
8. Inaba K, Metlay JP, Crowley MT, et al. (1990) Dendritic cells as antigen presenting cells *in vivo*. Int Rev Immunol 6(2–3):197–206
9. Ridgway D. (2003) The first 1000 dendritic cell vaccinees. Cancer Invest 21(6):873–886
10. Rosenberg SA. (2001) Progress in human tumour immunology and immunotherapy. Nature 411(6835):380–384
11. Mayordomo JI, Zorina T, Storkus WJ, et al. (1995) Bone marrow-derived dendritic cells pulsed with synthetic tumour peptides elicit protective and therapeutic antitumour immunity. Nat Med 1(12):1297–1302

12. Yang S, Vervaert CE, Burch J, Jr., et al. (1999) Murine dendritic cells transfected with human GP100 elicit both antigen-specific $CD8^{(+)}$ and $CD4^{(+)}$ T-cell responses and are more effective than DNA vaccines at generating anti-tumor immunity. Int J Cancer 83(4):532–540
13. Saha A, Chatterjee SK, Foon KA, et al. (2004) Dendritic cells pulsed with an anti-idiotype antibody mimicking carcinoembryonic antigen (CEA) can reverse immunological tolerance to CEA and induce antitumor immunity in CEA transgenic mice. Cancer Res 64(14) 4995–5003
14. Thery C, Amigorena S. (2001) The cell biology of antigen presentation in dendritic cells. Curr Opin Immunol 13(1):45–51
15. Banchereau J, Steinman RM. (1998) Dendritic cells and the control of immunity. Nature 392(6673):245–252
16. Banchereau J, Palucka AK, Dhodapkar M, et al. (2001) Immune and clinical responses in patients with metastatic melanoma to $CD34^{(+)}$ progenitor-derived dendritic cell vaccine. Cancer Res 61(17):6451–6458
17. Syme R, Bajwa R, Robertson L, et al. (2005) Comparison of CD34 and monocyte-derived dendritic cells from mobilized peripheral blood from cancer patients. Stem Cells 23(1):74–81
18. Banchereau J, Schuler-Thurner B, Palucka AK, et al. (2001) Dendritic cells as vectors for therapy. Cell 106(3):271–274
19. Engell-Noerregaard L, Hansen TH, Andersen MH, et al. (2009) Review of clinical studies on dendritic cell-based vaccination of patients with malignant melanoma: assessment of correlation between clinical response and vaccine parameters 2. Cancer Immunol Immunother 58(1):1–14
20. Palmer DH, Midgley RS, Mirza N, et al. (2009) A phase II study of adoptive immunotherapy using dendritic cells pulsed with tumor lysate in patients with hepatocellular carcinoma. Hepatology 49(1):124–132
21. Palucka AK, Ueno H, Connolly J, et al. (2006) Dendritic cells loaded with killed allogeneic melanoma cells can induce objective clinical responses and MART-1 specific $CD8^{+}$ T-cell immunity. J Immunother 29(5):545–557
22. Ross M, Camacho LH, Hersh EM, et al. (2007) Clinical and immunological responses in patients with malignant melanoma treated with a dendritic cell-based vaccine. Preliminary report from a multi-institutional phase II clinical trial 1. ASCO Meeting Abstracts 25(18_Suppl):3004
23. Curti A, Tosi P, Comoli P, et al. (2007) Phase I/II clinical trial of sequential subcutaneous and intravenous delivery of dendritic cell vaccination for refractory multiple myeloma using patient-specific tumour idiotype protein or idiotype (VDJ)-derived class I-restricted peptides. Br J Haematol 139(3):415–424
24. Redman BG, Chang AE, Whitfield J, et al. (2008) Phase Ib trial assessing autologous, tumor-pulsed dendritic cells as a vaccine administered with or without IL-2 in patients with metastatic melanoma. J Immunother 31(6):591–598
25. O'Rourke MG, Johnson MK, Lanagan CM, et al. (2007) Dendritic cell immunotherapy for stage IV melanoma. Melanoma Res 17(5):316–322
26. Kavanagh B, Ko A, Venook A, et al. (2007) Vaccination of metastatic colorectal cancer patients with matured dendritic cells loaded with multiple major histocompatibility complex class I peptides. J Immunother 30(7):762–772
27. Titzer S, Christensen O, Manzke O, et al. (2000) Vaccination of multiple myeloma patients with idiotype-pulsed dendritic cells: immunological and clinical aspects. Br J Haematol 108(4):805–816
28. Svane IM, Pedersen AE, Johansen JS, et al. (2007) Vaccination with p53 peptide-pulsed dendritic cells is associated with disease stabilization in patients with p53 expressing advanced breast cancer; monitoring of serum YKL-40 and IL-6 as response biomarkers. Cancer Immunol Immunother 56(9):1485–1499
29. Dees EC, McKinnon KP, Kuhns JJ, et al. (2004) Dendritic cells can be rapidly expanded *ex vivo* and safely administered in patients with metastatic breast cancer. Cancer Immunol Immunother 53(9):777–785

30. Fong L, Brockstedt D, Benike C, et al. (2001) Dendritic cells injected via different routes induce immunity in cancer patients. J Immunol 166(6):4254–4259
31. Tjoa BA, Simmons SJ, Bowes VA, et al. (1998) Evaluation of phase I/II clinical trials in prostate cancer with dendritic cells and PSMA peptides. Prostate 36(1):39–44
32. Timmerman JM, Czerwinski DK, Davis TA, et al. (2002) Idiotype-pulsed dendritic cell vaccination for B-cell lymphoma: clinical and immune responses in 35 patients. Blood 99(5):1517–1526
33. Thurner B, Haendle I, Roder C, et al. (1999) Vaccination with mage-3A1 peptide-pulsed mature, monocyte-derived dendritic cells expands specific cytotoxic T cells and induces regression of some metastases in advanced stage IV melanoma. J Exp Med 190(11):1669–1678
34. Hersey P, Menzies SW, Halliday GM, et al. (2004) Phase I/II study of treatment with dendritic cell vaccines in patients with disseminated melanoma. Cancer Immunol Immunother 53(2):125–134
35. Nestle FO, Alijagic S, Gilliet M, et al. (1998) Vaccination of melanoma patients with peptide- or tumor lysate-pulsed dendritic cells. Nat Med 4(3):328–332
36. Schuler-Thurner B, Schultz ES, Berger TG, et al. (2002) Rapid induction of tumor-specific type 1 T helper cells in metastatic melanoma patients by vaccination with mature, cryopreserved, peptide-loaded monocyte-derived dendritic cells. J Exp Med 195(10):1279–1288
37. Geiger JD, Hutchinson RJ, Hohenkirk LF, et al. (2001) Vaccination of pediatric solid tumor patients with tumor lysate-pulsed dendritic cells can expand specific T cells and mediate tumor regression. Cancer Res 61(23):8513–8519
38. Su Z, Dannull J, Heiser A, et al. (2003) Immunological and clinical responses in metastatic renal cancer patients vaccinated with tumor RNA-transfected dendritic cells. Cancer Res 63(9):2127–2133
39. Fong L, Hou Y, Rivas A, et al. (2001) Altered peptide ligand vaccination with Flt3 ligand expanded dendritic cells for tumor immunotherapy. Proc Natl Acad Sci USA 98(15): 8809–8814
40. Holtl L, Zelle-Rieser C, Gander H, et al. (2002) Immunotherapy of metastatic renal cell carcinoma with tumor lysate-pulsed autologous dendritic cells. Clin Cancer Res 8(11):3369–3376
41. Mackensen A, Herbst B, Chen JL, et al. (2000) Phase I study in melanoma patients of a vaccine with peptide-pulsed dendritic cells generated *in vitro* from CD34$^{(+)}$ hematopoietic progenitor cells. Int J Cancer 86(3):385–392
42. Dohnal AM, Graffi S, Witt V, et al. (2009) Comparative evaluation of techniques for the manufacturing of dendritic cell-based cancer vaccines. J Cell Mol Med 13(1):125–135
43. Erdmann M, Dorrie J, Schaft N, et al. (2007) Effective clinical-scale production of dendritic cell vaccines by monocyte elutriation directly in medium, subsequent culture in bags and final antigen loading using peptides or RNA transfection. J Immunother 30(6):663–674
44. Jarnjak-Jankovic S, Hammerstad H, Saeboe-Larssen S, et al. (2007) A full scale comparative study of methods for generation of functional dendritic cells for use as cancer vaccines. BMC Cancer 7:119
45. Paczesny S, Li YP, Li N, et al. (2007) Efficient generation of CD34$^+$ progenitor-derived dendritic cells from G-CSF-mobilized peripheral mononuclear cells does not require hematopoietic stem cell enrichment. J Leukoc Biol 81(4):957–967
46. Morse MA, Zhou LJ, Tedder TF, et al. (1997) Generation of dendritic cells *in vitro* from peripheral blood mononuclear cells with granulocyte-macrophage-colony-stimulating factor, interleukin-4, and tumor necrosis factor-alpha for use in cancer immunotherapy. Ann Surg 226(1):6–16
47. Felzmann T, Witt V, Wimmer D, et al. (2003) Monocyte enrichment from leukapharesis products for the generation of DCs by plastic adherence, or by positive or negative selection. Cytotherapy 5(5):391–398
48. Berger TG, Strasser E, Smith R, et al. (2005) Efficient elutriation of monocytes within a closed system (Elutra) for clinical-scale generation of dendritic cells. J Immunol Methods 298(1–2):61–72

49. Yuan J, Kendle R, Ireland J, et al. (2007) Scalable expansion of potent genetically modified human langerhans cells in a closed system for clinical applications. J Immunother 30(6):634–643
50. Ackermann B, Troger A, Glouchkova L, et al. (2004) Characterization of CD34+ progenitor-derived dendritic cells pulsed with tumor cell lysate for a vaccination strategy in children with malignant solid tumors and a poor prognosis. Klin Pädiatr 216(3):176–182
51. Dauer M, Obermaier B, Herten J, et al. (2003) Mature dendritic cells derived from human monocytes within 48 hours: a novel strategy for dendritic cell differentiation from blood precursors. J Immunol 170(8):4069–4076
52. Tawab A, Fan Y, Read EJ, et al. (2009) Effect of *ex vivo* culture duration on phenotype and cytokine production by mature dendritic cells derived from peripheral blood monocytes. Transfusion 49(3):536–547
53. Sallusto F, Lanzavecchia A. (1994) Efficient presentation of soluble antigen by cultured human dendritic cells is maintained by granulocyte/macrophage colony-stimulating factor plus interleukin 4 and downregulated by tumor necrosis factor alpha. J Exp Med 179(4):1109–1118
54. Jonuleit H, Kuhn U, Muller G, et al. (1997) Pro-inflammatory cytokines and prostaglandins induce maturation of potent immunostimulatory dendritic cells under fetal calf serum-free conditions. Eur J Immunol 27(12):3135–3142
55. Kalinski P, Vieira PL, Schuitemaker JH, et al. (2001) Prostaglandin E(2) is a selective inducer of interleukin-12 p40 (IL-12p40) production and an inhibitor of bioactive IL-12p70 heterodimer. Blood 97(11):3466–3469
56. Boullart AC, Aarntzen EH, Verdijk P, et al. (2008) Maturation of monocyte-derived dendritic cells with Toll-like receptor 3 and 7/8 ligands combined with prostaglandin E2 results in high interleukin-12 production and cell migration 6. Cancer Immunol Immunother 57(11):1589–1597
57. Kaka AS, Foster AE, Weiss HL, et al. (2008) Using dendritic cell maturation and IL-12 producing capacity as markers of function: a cautionary tale. J Immunother 31(4):359–369
58. Chen B, Stiff P, Sloan G, et al. (2001) Replicative response, immunophenotype, and functional activity of monocyte-derived versus CD34(+)-derived dendritic cells following exposure to various expansion and maturational stimuli. Clin Immunol 98(2):280–292
59. Herbst B, Kohler G, Mackensen A, et al. (1997) CD34+ peripheral blood progenitor cell and monocyte derived dendritic cells: a comparative analysis. Br J Haematol 99(3):490–499
60. Ferlazzo G, Wesa A, Wei WZ, et al. (1999) Dendritic cells generated either from CD34+ progenitor cells or from monocytes differ in their ability to activate antigen-specific CD8+ T cells. J Immunol 163(7):3597–3604
61. Ratzinger G, Baggers J, de Cos MA, et al. (2004) Mature human Langerhans cells derived from CD34+ hematopoietic progenitors stimulate greater cytolytic T lymphocyte activity in the absence of bioactive IL-12p70, by either single peptide presentation or cross-priming, than do dermal-interstitial or monocyte-derived dendritic cells. J Immunol 173(4):2780–2791
62. Bohnenkamp HR, Noll T. (2003) Development of a standardized protocol for reproducible generation of matured monocyte-derived dendritic cells suitable for clinical application. Cytotechnology 42(3):121–131
63. Di Nicola M, Carlo-Stella C, Mortarini R, et al. (2004) Boosting T cell-mediated immunity to tyrosinase by vaccinia virus-transduced, CD34(+)-derived dendritic cell vaccination: a phase I trial in metastatic melanoma. Clin Cancer Res 10(16):5381–5390
64. Fay JW, Palucka AK, Paczesny S, et al. (2006) Long-term outcomes in patients with metastatic melanoma vaccinated with melanoma peptide-pulsed CD34(+) progenitor-derived dendritic cells. Cancer Immunol Immunother 55(10):1209–1218
65. Small EJ, Fratesi P, Reese DM, et al. (2000) Immunotherapy of hormone-refractory prostate cancer with antigen-loaded dendritic cells. J Clin Oncol 18(23):3894–3903
66. Shankar G, Bader R, Lodge PA. (2004) The COSTIM bioassay: a novel potency test for dendritic cells. J Immunol Methods 285(2):293–299
67. Disis ML, Goodell V, Schiffman K, et al. (2004) Humoral epitope-spreading following immunization with a HER-2/neu peptide based vaccine in cancer patients. J Clin Immunol 24(5):571–578

68. Wierecky J, Mueller M, Brossart P. (2006) Dendritic cell-based cancer immunotherapy targeting MUC-1. Cancer Immunol Immunother 55(1):63–67
69. Vlad AM, Muller S, Cudic M, et al. (2002) Complex carbohydrates are not removed during processing of glycoproteins by dendritic cells: processing of tumor antigen MUC1 glycopeptides for presentation to major histocompatibility complex class II-restricted T cells. J Exp Med 196(11):1435–1446
70. Tarp MA, Clausen H. (2008) Mucin-type O-glycosylation and its potential use in drug and vaccine development. Biochim Biophys Acta 1780(3):546–563
71. Met O, Eriksen J, Svane IM. (2008) Studies on mRNA electroporation of immature and mature dendritic cells: effects on their immunogenic potential. Mol Biotechnol 40(2):151–160
72. Di Nicola M, Siena S, Bregni M, et al. (1998) Gene transfer into human dendritic antigen-presenting cells by vaccinia virus and adenovirus vectors. Cancer Gene Ther 5(6):350–356
73. Kaplan JM, Yu Q, Piraino ST, et al. (1999) Induction of antitumor immunity with dendritic cells transduced with adenovirus vector-encoding endogenous tumor-associated antigens. J Immunol 163(2):699–707
74. Wan Y, Emtage P, Zhu Q, et al. (1999) Enhanced immune response to the melanoma antigen gp100 using recombinant adenovirus-transduced dendritic cells. Cell Immunol 198(2):131–138
75. Sloan JM, Kershaw MH, Touloukian CE, et al. (2002) MHC class I and class II presentation of tumor antigen in retrovirally and adenovirally transduced dendritic cells. Cancer Gene Ther 9(11):946–950
76. Boudreau JE, Bridle BW, Stephenson KB, et al. (2009) Recombinant vesicular stomatitis virus transduction of dendritic cells enhances their ability to prime innate and adaptive antitumor immunity. Mol Ther 17(8):1465–1472
77. Nakamura M, Iwahashi M, Nakamori M, et al. (2005) Dendritic cells transduced with tumor-associated antigen gene elicit potent therapeutic antitumor immunity: comparison with immunodominant peptide-pulsed DCs. Oncology 68(2–3):163–170
78. Tsao H, Millman P, Linette GP, et al. (2002) Hypopigmentation associated with an adenovirus-mediated gp100/MART-1-transduced dendritic cell vaccine for metastatic melanoma. Arch Dermatol 138(6):799–802
79. Bedrosian I, Mick R, Xu S, et al. (2003) Intranodal administration of peptide-pulsed mature dendritic cell vaccines results in superior $CD8^+$ T-cell function in melanoma patients. J Clin Oncol 21(20):3826–3835
80. Mackensen A, Krause T, Blum U, et al. (1999) Homing of intravenously and intralymphatically injected human dendritic cells generated *in vitro* from $CD34^+$ hematopoietic progenitor cells. Cancer Immunol Immunother 48(2–3):118–122
81. Prince HM, Wall DM, Ritchie D, et al. (2008) *In vivo* tracking of dendritic cells in patients with multiple myeloma. J Immunother 31(2):166–179
82. De Vries IJ, Krooshoop DJ, Scharenborg NM, et al. (2003) Effective migration of antigen-pulsed dendritic cells to lymph nodes in melanoma patients is determined by their maturation state. Cancer Res 63(1):12–17
83. Kyte JA, Mu L, Aamdal S, et al. (2006) Phase I/II trial of melanoma therapy with dendritic cells transfected with autologous tumor-mRNA. Cancer Gene Ther 13(10):905–918
84. Salcedo M, Bercovici N, Taylor R, et al. (2006) Vaccination of melanoma patients using dendritic cells loaded with an allogeneic tumor cell lysate. Cancer Immunol Immunother 55(7):819–829
85. Tamir A, Basagila E, Kagahzian A, et al. (2007) Induction of tumor-specific T-cell responses by vaccination with tumor lysate-loaded dendritic cells in colorectal cancer patients with carcinoembryonic-antigen positive tumors. Cancer Immunol Immunother 56(12):2003–2016
86. Burgdorf SK, Fischer A, Myschetzky PS, et al. (2008) Clinical responses in patients with advanced colorectal cancer to a dendritic cell based vaccine 6. Oncol Rep 20(6):1305–1311
87. Morse MA, Nair SK, Mosca PJ, et al. (2003) Immunotherapy with autologous, human dendritic cells transfected with carcinoembryonic antigen mRNA. Cancer Invest 21(3):341–349

88. Small EJ, Schellhammer PF, Higano CS, et al. (2006) Placebo-controlled phase III trial of immunologic therapy with sipuleucel-T (APC8015) in patients with metastatic, asymptomatic hormone refractory prostate cancer. J Clin Oncol 24(19):3089–3094
89. Higano CS, Schellhammer PF, Small EJ, et al. (2009) Integrated data from 2 randomized, double-blind, placebo-controlled, phase 3 trials of active cellular immunotherapy with sipuleucel-T in advanced prostate cancer. Cancer 115(16):3670–3679
90. Boyles S, Chang L. (2009) AUA 2009: prostate cancer vaccine significantly improves 3-year survival. http://www medscape com/viewarticle/702085. Accessed June 6, 2009
91. Lesterhuis WJ, Aarntzen EH, De Vries IJ et al. (2008) Dendritic cell vaccines in melanoma: from promise to proof? Crit Rev Oncol Hematol 66(2):118–134
92. Finke LH, Wentworth K, Blumenstein B, et al. (2007) Lessons from randomized phase III studies with active cancer immunotherapies – outcomes from the 2006 meeting of the Cancer Vaccine Consortium (CVC). Vaccine 25 Suppl 2:B97–B109
93. Sakaguchi S, Sakaguchi N, Shimizu J, et al. (2001) Immunologic tolerance maintained by $CD25^+$ $CD4^+$ regulatory T cells: their common role in controlling autoimmunity, tumor immunity, and transplantation tolerance. Immunol Rev 182:18–32
94. Salazar LG, Disis ML. (2005) Cancer vaccines: the role of tumor burden in tipping the scale toward vaccine efficacy. J Clin Oncol 20;23(30):7397–7398
95. Gulley JL, Madan RA, Arlen PM. (2007) Enhancing efficacy of therapeutic vaccinations by combination with other modalities. Vaccine 25(Suppl 2):B89–96. Epub 2007 June 4
96. Palucka AK, Ueno H, Fay J, et al. (2008) Dendritic cells: a critical player in cancer therapy? J Immunother 31(9):793–805
97. Fonseca C, Dranoff G. (2008) Capitalizing on the immunogenicity of dying tumor cells. Clin Cancer Res 14(6):1603–1608
98. Casares N, Pequignot MO, Tesniere A, et al. (2005) Caspase-dependent immunogenicity of doxorubicin-induced tumor cell death. J Exp Med 202(12):1691–701
99. Rozkova D, Tiserova H, Fucikova J, et al. (2009) FOCUS on FOCIS: combined chemo-immunotherapy for the treatment of hormone-refractory metastatic prostate cancer. Clin Immunol 131(1):1–10
100. Chi KH, Liu SJ, Li CP, et al. (2005) Combination of conformal radiotherapy and intratumoral injection of adoptive dendritic cell immunotherapy in refractory hepatoma. J Immunother 28(2):129–135
101. Bai L, Koopmann J, Fiola C, et al. (2002) Dendritic cells pulsed with viral oncolysates potently stimulate autologous T cells from cancer patients. Int J Oncol 21(4):685–694
102. Li QX, Liu G, Wong-Staal F. (2008) Oncolytic virotherapy as a personalized cancer vaccine. Int J Cancer 123(3):493–499
103. Xia D, Moyana T, Xiang J. (2006) Combinational adenovirus-mediated gene therapy and dendritic cell vaccine in combating well-established tumors. Cell Res 16(3):241–259
104. Farrell CJ, Zaupa C, Barnard Z, et al. (2008) Combination immunotherapy for tumors via sequential intratumoral injections of oncolytic herpes simplex virus 1 and immature dendritic cells. Clin Cancer Res 14(23):7711–7716

Chapter 6
Mesenchymal Stromal Cells: An Emerging Cell-Based Pharmaceutical

Moïra François and Jacques Galipeau

Abstract Over the course of the last decade, mesenchymal stromal cells (MSC) made a remarkable entry into the field of cellular therapy. Their ability to differentiate into several mesenchymal lineages, as well as their role in hematopoiesis, provided the basis for several clinical investigations in the field of regenerative medicine, ranging from heart repair to hematopoietic support in patients undergoing hematopoietic peripheral blood progenitor cells transplantation. In addition, MSC were also shown to modulate the immune response, either by acting as an immunosuppressant on several immune cells (T and B cells, dendritic cells, and macrophages), or on IFN-γ stimulation, as antigen presenting cells (APC) for the priming of $CD4^+$ and $CD8^+$ T cells. Although the exact mechanisms by which MSC mediate their immunosuppressive effect is not fully elucidated, several *in vivo* results from animal disease models and clinical trials in humans has proven the potential of MSC as immunosuppressive as well as anti-inflammatory agents. Conversely, MSC can also stimulate the immune system by presenting exogenously acquired antigen to T cells, a feature currently investigated in the context of cell-based vaccines for cancer immunotherapy. The mechanisms underlying the physiological roles and immuno-modulatory properties of MSC must, however, be clarified in order to optimize their beneficial impact while minimizing unwanted phenomena. The present review hereby attempts to summarize and reflect on the latest breakthroughs concerning the elucidation of MSC properties and their clinical applications with a special attention to their role in immunotherapy.

Keywords Hematopoiesis • Immunosuppression • Immunotherapy • Mesenchymal stromal cells • Regeneration

---

M. François (✉)
Department of Experimental Medicine, McGill University, Montreal, QC, Canada
e-mail: moira.francois@mail.mcgill.ca

J. Medin and D. Fowler (eds.), *Experimental and Applied Immunotherapy*,
DOI 10.1007/978-1-60761-980-2_6, © Springer Science+Business Media, LLC 2011

Introduction and Classification

In 1976, Alexander Friendenstein described a novel colony of fibroblast-like cells isolated from the bone marrow by plastic adherence with what appeared to be stem cell characteristics, as these cells could proliferate and differentiate into bone and cartilage *in vitro* [1]. Other groups later proceeded to demonstrate that these cells could differentiate into several mesenchymal lineages, such as bone, cartilage, muscle, fat, tendons, and neurons [2–6]. According to this observation, Arnold Caplan introduced the term *mesenchymal stem cells* to name this unique cell population, referring to their mesenchymal pluripotency [7]. The name mesenchymal *stem* cell was, however, later challenged by researchers in the field of mesenchymal cell therapy. Their main objection was based on results obtained by Colter *et al.* who demonstrated that early cultures of human mesenchymal stem cells (MSC), maintained at low density, contained three distinct cell populations based on their size and shape: small round cells, spindle-shaped cells, and large flat cells. Further analysis revealed that each of these cell populations possessed different levels of cell replication and differentiation potential. The small round cell population, named "RS" for its rapid self-renewing potential, was found to differentiate more extensively into osteocytes, adipocytes, and chondrocytes and to possess additional surface markers compared to the other populations [8, 9]. The members of the Mesenchymal and Tissue Stem Cell Therapy Committee of the International Society of Cellular Therapy (ISCT) arrived at the conclusion that MSC cultures may contain stem cells but most of the cells isolated did not meet the criteria characteristic of stem cells (self-renewal and mesenchymal pluripotency). Therefore, they proposed the less controversial name "multipotent mesenchymal *stromal* cells" (MSC) because MSC are consistently found to be part of the stroma, independent of the tissue from which they arise [10]. Indeed, cell populations with multipotent mesenchymal plasticity and self-renewing potential are found in various tissue, including: adipose tissue [11], umbilical cord blood [12], and placenta [13]. However, bone marrow-derived MSC are the best characterized and most used in current research. This chapter will therefore focus on adult bone marrow-derived MSC.

As MSC research has matured, different isolation methods, tissue origins, and characterization criteria have been developed. In order to standardize the elements defining MSC, the Mesenchymal and Tissue Stem Cell Therapy Committee implemented a list of criteria. First, MSC populations must be plastic-adherent; second, although many isolation methods based on surface cell markers have been used [14–16], the only markers selected for identification of MSC were positivity for CD105, CD73, CD90 expression and negative staining for the hematopoietic cell markers CD45, CD34, CD14 or CD11b, CD79a or CD19, and HLA-DR. Third, cells should be shown to differentiate into osteoblasts, adipocytes, and chondroblasts *in vitro*; and finally, the origin of the cells should always be clearly stated, such as "bone marrow-derived" or "adipose-derived" MSC [17].

Physiological Functions of MSC

MSC and Hematopoiesis

MSC are active components of the bone marrow hematopoietic niche. They are mainly localized in the endosteum of the bone where they give rise to pericytes, myofibroblasts, osteocytes, and endothelial cells [18], all functional elements of the bone marrow stroma supporting hematopoietic stem cell and progenitor cell development. MSC also express fibronectin, laminin, collagen, and proteoglycans, which are part of the extracellular matrix of the bone marrow stroma. Importantly, MSC directly interact with hematopoietic cells via an array of surface markers and cytokines, which regulate differential aspects of HSC development: quiescence, proliferation, and differentiation. Cell–cell contact between MSC and hematopoietic cells is mediated by several adhesion molecules, including ICAM-1, ICAM-2, ICAM-3, VCAM-1, LFA-3, CD44, and CD72 [19]. MSC have also been shown to express hematopoietic growth factors such as BMP4, Flt-3, LIF, OSM, SCF, SDF-1, and TGF-β along with interleukins such as IL-1, IL-6, IL-7, IL-8, IL-11, IL-14, and IL-15 [20]. In regard to their critical role in hematopoiesis, MSC have been used to maintain and expand HSC in culture [21] and to promote engraftment and hematopoietic recovery in patients receiving peripheral blood hematopoietic stem cell (PBSC) transplantations, as detailed below.

PBSC Transplantation

Chemotherapy and radiotherapy regimens given to patients prior to PBSC transplantation have been shown to not only kill host hematopoietic stem cells, but to also alter the hematopoietic niche micro-environment [22], which could comprise hematopoietic engraftment and hematopoiesis recovery in patients undergoing PBSC transplantation. Due to their essential involvement in the hematopoietic niche, it has been suggested that co-implantation of MSC along with PBSC could enhance hematopoietic recovery. Based on this hypothesis, several clinical trials have been performed [23–26]. Although no toxicity related to MSC implantation was observed, no significant improvement in hematopoietic recovery was noticed, with the exception of a study performed by Koc *et al.* in which a majority of patients presented an enhanced platelet recovery profile in comparison with historical values using standard procedure [25]. More importantly, several other clinical trials in which MSC were co-infused with PBSC revealed that despite successful donor HSC engraftment, donor MSC could not be detected [23, 24, 26]. It is therefore unclear whether cotransplantation of MSC, despite their known implication in hematopoiesis, have any positive impact on autologous PBSC transplantation. In order to properly test this hypothesis, randomized studies comparing standard PBSC transplantation with or without MSC would need to be conducted. Nonetheless, the absence of MSC's engraftment does not necessarily confirm that

MSC are not involved in hematopoietic recovery following PBSC transplantation. Co-infused MSC could transiently support the newly engrafted donor HSC through the expression of hematopoietic cytokines and matrix components until the host hematopoietic stroma has fully recovered.

MSC Homing

In vivo experiments in rodents [27] and nonhuman primates [28] have demonstrated that a large portion of MSC are trapped in the lung following intravenous administration via peripheral vein [27]. However, infused MSC may subsequently migrate to several organs, including liver, spleen, bone marrow, and kidney. The migratory potential of MSC is associated with expression of VCAM-1, which allows them to interact with endothelial cells [29] and travel throughout the body. In addition, MSC have been shown to express various chemokine receptors [30] (CCR1, CCR7, CCR9, CXCR4, CXCR5, and CXCR6) that promotes their migration to specific sites; for example, MSC migration to the bone marrow [31] and heart [32] can occur via SDF-1, whereas MSC migration to an inflammation site and tumor microenvironment may occur via other CC and CXC chemokines [33]. Their mesenchymal differentiation potential in combination with their ability to migrate throughout the body suggest that MSC could be used in regenerative medicine for the treatment of several disorders such as osteogenesis imperfecta (OI) and acute tissue injury repair response such as myocardial infarction.

Osteogenesis Imperfecta

Children with OI are born with one mutated copy of collagen type I. This protein is the primary structural element in bone formation; as a result, children with OI develop bone deformities and fragility leading to frequent fractures and a short stature. Based on the ability of MSC to migrate to the bone and differentiate into osteoblasts, Horwitz *et al.* hypothesized that the infusion of whole bone marrow, which contains mesenchymal progenitors, could attenuate if not cure OI. In 1999, a first clinical study was conducted on three infants with severe deforming OI [34]. The infants were infused with an unmanipulated bone marrow graft from HLA-identical or single antigen-mismatched siblings following a myelo-ablative conditioning regimen. Impressively, all three patients demonstrated hematopoietic engraftment, and new bone formation could be seen in bone biopsies performed 3 months after transplantation. Total body bone mineralization and growth were increased, while the fracture incidence was reduced in the first 6 months following BMT. Encouraged by these results, Horwitz and colleagues conducted a second clinical trial in which infants with OI were not only given BMT, but were also infused with *in vitro*-expanded allogeneic MSC derived from their BMT donor [35]. In order to observe donor MSC engraftment, MSC were transduced prior to the implantation using retrovectors encoding for either the neomycin phosphotransferase gene or the β-galactosidase gene. Polymerase chain reaction using primers against both these genes permitted quantification of MSC engraftment in recipients. Of the six infants treated, five demonstrated MSC engraftment in the bone or the bone marrow stroma accompanied by a net increase in growth rate. That said, only one patient demonstrated a substantial increase in total body mineralization. Since no changes were observed in the only patient in whom injected MSC did not engraft, the beneficial effects seen in the other infants can presumably be attributed to the infused MSC. The authors have not identified the reason for the quasi-absence of bone mineralization in the second study using *in vitro* expanded MSC plus BMT compared to BMT only. Since bone mineralization was assayed only once 3 month after the first MSC infusion, delay in bone mineralization is possible. The authors also suggest that bone linear growth and mineralization are two independent mechanisms in which MSC could have an effect on the former but not on the latter. Furthermore, it is possible that such MSC-based treatment of OI would benefit from the selection of a precise subpopulation of MSC; specifically, the RS cells described by Colter *et al.* may be beneficial because they are more prone to engraftment and osteoblast differentiation [8, 9].

Myocardial Infarction

The cardioprotective and regenerative properties of MSC in animal models of myocardial infarction (MI), have been extensively studied. Makino *et al.* have demonstrated *in vitro* that murine MSC treated with 5-azacytidine, a cytosine analog that regulates the expression of genes implicated in cell differentiation, could differentiate into cardiomyocytes and form myotube-like structures capable of synchronous beating [36]. In addition, *in vivo* studies performed in mouse [6], rat [37], and swine [38] models of MI demonstrated that MSC could engraft into the myocardium and express muscle-specific and endothelial cell-specific proteins, thus indicating environmental-induced differentiation of MSC into cardiomyocytes and vascular endothelial cells. Overall, administration of MSC following MI was shown to significantly improve heart function, although the underlying mechanism behind the therapeutic effect of MSC is still unclear [37–40]. MSC-induced heart repair may arise from factors secreted by MSC that exert a paracrine effect on the heart micro-environment. Indeed, under a hypoxic environment, MSCs were shown to secrete several pro-angiogenic and antiapoptotic factors (VEGF, FGF, placental growth factor, IL-6, and MCP-1) which contribute to the recruitment and proliferation of endothelial cells [41] and cardiac precursors [42]. In addition, such factors diminish the level of hypoxia-induced apoptosis in endothelial cells [43] and cardiomyocytes [44]. MSC conditioned-media was also reported to inhibit cardiac fibroblast proliferation and collagen synthesis [40], which may account for the reduction in fibrosis following MSC transplantation [45]. The release of proinflammatory cytokines by ischemic tissue plays an important role in the severity of heart remodeling events such as scarring, hypertrophy of myocytes, and apoptosis of cardiomyocytes and endothelial cells [46]. A study performed in a rat model of MI showed a reduction in the expression of pro-inflammatory cytokines

TNF-α, IL-1β, and IL-6 in the noninfarct area of the heart following MSC transplantation, thereby suggesting that MSC immunosuppressive properties can reduce cytokine-induced heart remodeling [47]. Based on these encouraging preliminary results in animals, MSC are currently under investigation in clinical trials.

MSC Immuno-regulatory Functions

Background MSC Immune Characteristics

The molecular and cellular basis of the immune regulatory properties of MSC has thus far not been fully elucidated. However, several observations indicate the direct involvement of MSC in immune-mediated mechanisms. First of all, MSC express several immune receptors that modulate their phenotype; namely, IFN-γ, TNF-α and TGF-β receptors have been found on the cell surface of MSC [19]. In addition, Toll-like receptors (TLR) have been identified on MSC. Mouse MSC were found to express all TLR with the exception of TLR9 [48]; in contrast, human MSC express only TLR3 and TLR4 at levels comparable to primary macrophages [49]. As stated earlier, MSC also express chemokine receptors that can promote cell migration to sites of injury or inflammation were they can produce chemokines, leading to the recruitment of innate immune cells such as neutrophils, macrophages, and NK cells.

In regard to antigen presentation, nonactivated, resting mouse and human MSC were shown to express variable levels of MHC class I and no MHC class II molecules. In addition, constitutive expression of co-stimulatory molecules CD80, CD86, and CD40 was not detected on human MSC, although some C57BL/6 mouse MSC clones were observed to express low level of CD80 [50]. On the other hand, activation with IFN-γ has been reported to up-regulate MHC class I and class II levels on MSC, without any effect on expression of CD80, CD86, CD40, CD28, ICOSL, and 4-1BBL costimulatory molecules [50, 51]. Notably, MSC express several interleukins, particularly high levels of IL-6 that can modulate T-cell responses. Some authors have also found that MSC produce IL-12, a key factor in activating the innate immune response [52]. However, when human or mouse MSC were carefully compared to macrophages, we observed that human and mouse MSC produced very low levels of IL-12 even after optimal stimulation with IFN-γ priming and TLR stimulation [49, 52]. Similarly, we (unpublished data) and others [53], have not detected IL-10 protein expression in mouse nor human MSC, although others have stated that they have detected IL-10 transcript in resting [54] or TLR-activated [55] mouse MSC.

MSC and Immunosuppression: Direct versus Indirect T-Cell Inhibition

Apart from being nonhematopoietic stem cell progenitors with mesenchymal plasticity and stromal properties for the support of HSC homeostasis, MSC have also been shown to exert a profound effect on immune regulation. The overall outcome of MSC-mediated immunosuppression is inhibition of T-cell activation and proliferation. This process has been shown to be mediated by MSC directly on T cells or indirectly on other immune cells, which in turn suppress T-cells activation. The immunosuppressive effect of MSC on T-cell activation has been extensively studied in past reviews [19, 56, 57] and will not be discussed in detail here. However, we will mention that MSC-mediated inhibition of T-cell activation has been tested in response to various stimuli and has been attributed mostly to secreted factors such as prostaglandin E2 (PGE2), TGF-β, IL-10, indoleamine 2,3-dioxygenase (IDO), nitric oxide (NO) and others, but also by cell–cell contact through B7.H1 and its receptor on activated T cells, PD-1. MSC have also been shown to induce T-cell differentiation into immunosuppressive regulatory T cells (Treg) [58–60] (Fig. 6.1).

Direct T-Cell Immunosuppression

Overall, most information gathered on MSC-mediated T-cell immunosuppression comes from *in vitro* studies and none of the mechanisms have been confirmed *in vivo,* except for the involvement of NO. NO produced *de novo* by NO synthase (iNOS/NOS2A) is known to suppress T-cell proliferation by inhibiting STAT5 phosphorylation downstream of the IL-2 receptor, which is essential to T-cell activation and proliferation [61]. Oh *et al.* demonstrated that IFN-γ produced by activated T cells induces the production of NO by MSC, which in turn inhibits T-cell activation and proliferation [62,63]. In addition, in a mouse model of graft-versus-host disease (GVHD) and delayed-type hypersensitivity (DTH), Ren *et al.* observed that T-cell apoptosis and cell-cycle arrest, normally seen following the injection of MSC in their 2 mouse models, was completely mitigated when MSC derived from $iNOS^{-/-}$ mice or $IFN\gamma R1^{-/-}$ mice were used. These results thus confirm the immunosuppressive role of NO produced by IFN-γ-stimulated MSC [64]. Furthermore, recent *in vivo* data strongly supports the role of MSC-derived chemokine derivatives-N-terminal truncated CCL2 in particular as a suppressor mechanism for immunoglobulin-producing B cells as well as Th1 and Th17 T-cells [65,66].

Indirect T-Cell Immunosuppression

Via Macrophages

Although MSC do not produce IL-10, they can stimulate macrophages and DC to secrete IL-10, which in turn has a profound immunosuppressive effect on T cells. A recent study (2009) by Nemeth *et al.* demonstrated that MSC can reduce mortality in a mouse model of peritonitis associated with septicemia and release of bacterial toxins in the circulation [56]. In this study, mouse MSC injected in the systemic circulation of septic mice localized in the lung, where they were found

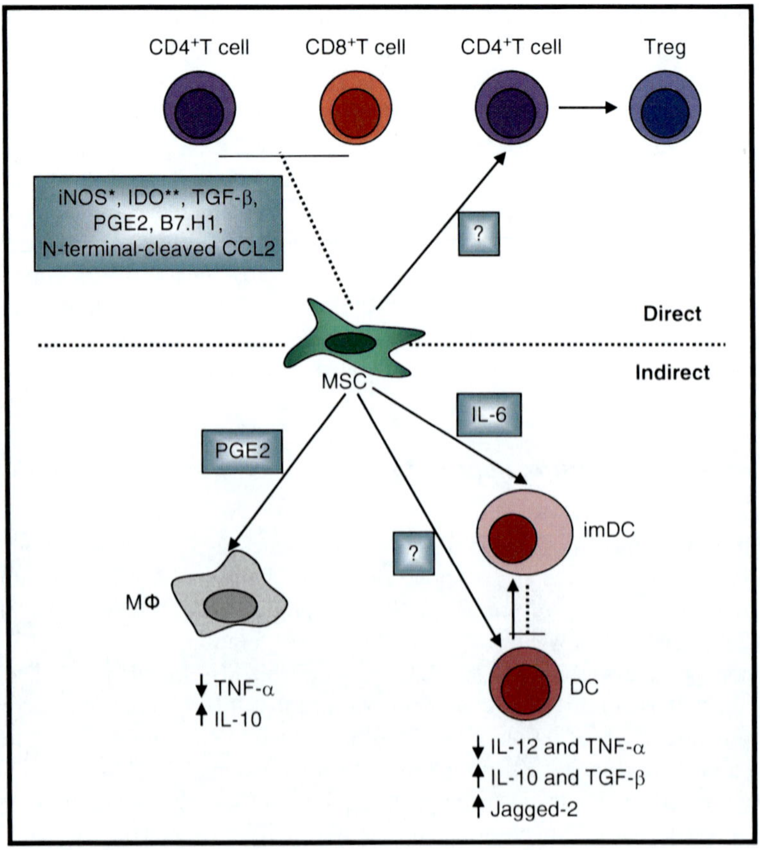

Fig. 6.1 Mesenchymal stromal cells (*MSC*)-mediated direct versus indirect T cell immunosuppression. MSC have been shown to mediate T cell immunosuppression through direct and indirect pathways. The expression of iNOS*, indoleamine 2,3-dioxygenase (*IDO***), TGF-β, PGE2 and surface expression of B7.H1 by MSC can directly inhibit T cell proliferation and activation. MSC can also induce the conversion of CD4+ T cells into regulatory T cells (*Treg*) through unknown mechanisms. Indirectly, MSC can suppress T cell activation by acting on other immune cells which in turn modify the micro-environment from inflammatory to immunosuppressive. Prostaglandin E2 (*PGE2*) expression by MSC induces interleukin-10 (*IL*-10) expression by macrophages (*M*), while reducing their expression of TNF-α. Differentiation of immature DC (*imDC*) to mature DC can be impeded by the production of IL-6 by MSC, which in turn can no longer activate T cells. MSC can also revert mature DC into immature regulatory DC which express high level of IL-10 and TGF-β and lower levels of IL-12 and TNF-α. These DC also express receptor Jagged-2 which induce CD4+ T cells to adopt a Th2 helper phenotype. *mouse only, not simians, **simians only, not mouse

surrounded by macrophages. These macrophages were shown to produce increased levels of the anti-inflammatory cytokine IL-10, both *in vivo* and *ex vivo* in response to bacterial LPS. *In vitro* assays suggested that the suppressive effect of MSC on the macrophage inflammatory response to LPS was dependent on the expression of the LPS receptor TLR4 by both cell types. These results suggested that LPS or

TNF-α-mediated activation of MSC and macrophages up-regulated the production of NO in both cell types, inducing the expression of cyclooxygenase 2 and prostaglandin E2 by MSC, which in turn binds to the EP2 and EP4 receptors on macrophages stimulating the production of IL-10 [53].

Via DC

Human and mouse MSC were shown to prevent monocyte-derived DC differentiation and maturation through a contact-dependent mechanism. Specifically, co-culture of immature DC with MSC downregulates LPS-induced upregulation of MHC class II molecules and CD40, CD80, and CD86 costimulatory molecules on DC [67, 68]. Other studies have reported that conditioned media from MSC could inhibit maturation and production of the pro-Th1 cytokine IL-12 in DC exposed to IFN-γ and LPS [68]. This effect was possibly mediated by a high quantity of IL-6 secreted by MSC [69]. In addition, it was suggested that MSC may alter DC migration by downregulating the expression of the chemokine receptor CCR7 and by preventing the downregulation of the anchoring protein E-cadherin, which maintains attachments between cells. [67]. A recent publication of Zhang *et al.* also established that MSC could induce the differentiation of mature DC into a nonreversible regulatory DC phenotype that adopts a more immature state characterized by low expression levels of CD40, CD80, CD86, and CD11c and elevated production of TGF-β and IL-10 as opposed to lower production of IL-12. Overall the new regulatory DC were shown to suppress T-cell proliferation through a contact-dependent mechanism. The authors identified Jagged-2, a Notch receptor ligand known to induce CD4 T-cell differentiation into Th2 helper cells, as the mediator responsible for the suppressive effect of the regulatory DC on T cells [70].

B Cells

The effect of MSC on B cells is unclear as both immunosuppressive and immunostimulatory responses have been reported. Rasmusson *et al.* demonstrated that cocultures of human activated B cells with human MSC at a ratio 10:1 increased the proliferation of IgG-secreting B cells through a mechanism mediated by both cell-cell contact and soluble factors [71]; IL-6 was identified as a potential mechanism because it was produced at high levels by MSC and is known to favor B-cell differentiation and antibody secretion [72]. On the other hand, Corcione *et al.* reported that human B cells activated with a combination of stimuli (CpG 2006, rCD40L, anti-immunoglobulin antibodies, IL-2, and IL-4) and cocultured with MSC at a ratio of 1:1 displayed reduced proliferation, differentiation, and chemotaxis in response to CXCL12 and CXCL13. This reduced chemotaxis was caused by decreased expression of the respective chemokine receptors, CXCR4 and CXCR5 [73]. This immunosuppressive effect was maintained when cells were separated in a trans-well system, thereby further suggesting that suppression was mediated by a soluble factor [73]. Research from our laboratory demonstrated that the production of an MMP-processed antagonistic form of CCL2 by MSC plays a key role in preventing B-cell maturation from plasmablasts to IgG-secreting plasma cells by promoting the proliferation of plasmablasts rather than their differentiation into plasma cells [66]. We demonstrated that the MMP-processed CCL2 blocked plasma cells from producing immunoglobulins by inducing expression of the transcription factor PAX5, which is known to repress immunoglobulin production. We further showed *in vivo* that MSC injection into ovalbumin-immunized mice reduced the ovalbumin-specific antibody response, thereby confirming the immunosuppressive effect of MSC on B cells [66].

Immunosuppressive Properties of MSC for Immunotherapy

Numerous *in vivo* results from animal disease models and early phase clinical trials in humans have supported the notion that MSC can be used as an immunosuppressive agent. Striking results obtained in phase II clinical trials using MSC for the treatment of steroid-resistant graft-versus-host disease (GVHD) have prompted the use of MSC in other immune-mediated diseases, such as organ transplant rejection, chronic inflammatory diseases, and auto-immune diseases. To date, 87 ongoing clinical trials around the world are testing the potency of MSC for the treatment of these and related conditions (www.clinicaltrials.gov) (Table 6.1).

MSC for the Prevention and Treatment of Steroid Refractory Acute GVHD

Allogeneic hematopoietic stem cell transplant can cure selected hematologic and lymphoid malignancies through two mechanisms. First, HSC can engraft and repopulate the hematopoietic niche and thus replace the malignant hematopoietic stem cells. Second, donor T cells from the graft can eliminate any remaining host malignant cells that may have survived the conditioning regimen through a process known as a graft-versus-leukemia (GVL) effect. Unfortunately, these donor T cells can also attack healthy host tissues and induce lethality through acute or chronic GVHD. Patients who develop GVHD following allogeneic HSC transplantation can be treated with immunosuppressive agents; however, these treatments render patients more susceptible to infections and cannot always alleviate the symptoms nor control GVHD [74]. Therefore, the immunosuppressive properties of MSC have made them an interesting biological agent for the treatment of GVHD.

MSC are currently used in several clinical trials and have generated promising results for the prevention and treatment of GVHD. Results from a study published in 2008 by Le Blanc *et al.*, involving 55 patients with severe steroid-resistant acute GVHD revealed the potential utility of MSC for the treatment of GVHD. Not only did none of the patients experience immediate toxicity due to the injection of the MSC,

Table 6.1 UCB: umbilical cord blood

Category	Conditions	Clinical phase	MSC origin	Number of studies	Country
Heart diseases	Myocardial infarct and heart failure	Phase I and II	Autologous and allogeneic	14	Denmark, United States, Finland, India, and France
Immune diseases	Multiple sclerosis	Phase I and II	Autologous and allogeneic	5	United Kingdom, Israel, United States, China
	Graf-versus-host disease	Phase I, II and III	Autologous, allogeneic and UCB-derived	16	Belgium, India, Spain, United States, Israel, Netherland, China, and Korea
	Crohn's disease	Phase II and III	Autologous and allogeneic	5	United States
	Organ transplantation and organ failure	Phase I and II	Autologous, allogeneic and UCB-derived	9	Netherland, Iran, China, Italy, Korea, and United States
	Diabetes mellitus	Phase I and II	Autologous and a llogeneic	3	China and United States
	Lupus	Phase I, II and III	Autologous, allogeneic and UCB-derived	2	China
Bone and cartilage defects	Bone and cartilage fracture, osteogenesis imperfecta and degenerative diseases	Phase I, II and III	Autologous, allogeneic and UCB-derived	18	United States, Belgium, Iran, China, Israel, France, Japan, Egypt, and Norway
Others	Miscellaneous	Phase I, II and III	Autologous and allogeneic	15	China, Korea, Japan, India, and Sweden

but 30 patients had a complete response (i.e., complete loss of all symptoms of acute GVHD). Of those 30 patients, 27 had received only one dose of MSC from a HLA-matched sibling, a haploidentical related donor, or an HLA-mismatched donor. Overall, the survival rate after 6 years in patients went from 2% with standard treatment against GVHD (steroid) to 52% after MSC treatment [75]. Although similar results were obtained in another study by Ringden *et al.* in which six out of eight patients show complete remission after MSC treatment [76], some failed to reach a positive outcome. In a clinical study by Bonin *et al.* only two patients (15%) out of 13 treated for steroid-refractory, acute GVHD showed complete remission following HLA-mismatch MSC transplantation [77]. Discrepancies between the methodology used by Le Blanc and Bonin to culture the MSC might explain the variation seen. Le Blanc's group cultured MSC in media supplemented with fetal bovine serum (FBS), while Bonin's group used platelet lysate instead of FBS. To date, no study has been conducted comparing MSC culture methods and the impact of the culture conditions on the immuno-modulatory properties of MSC. This analysis would greatly improve to use of MSC in immunotherapy has it could serve to optimize and standardize MSC culture.

Though numerous clinical studies using MSC for the treatment of GVHD have been conducted, the mechanism by which MSC mediate their immunosuppressive effect against GVHD is still unknown. It was described that acute GVHD is accompanied by a burst in cytokine production by activated donor immune cells, including IFN-γ and TNF-α [74]. In a mouse model of GVHD using MSC from iNOS-KO mice, Ren *et al.* demonstrated that NO produced by MSC in response to IFN-γ was principally responsible for the immunosuppressive effect observed [64]. Another study by Polchert *et al.* also identified IFN-γ stimulation of MSC as the key element promoting the immunosuppressive effect. The latter study suggested that the effectiveness of MSC infusions to treat mice undergoing allogeneic BMT varied with the phase of the disease and the level of circulating IFN-γ. Co-transplantation of MSC with HSC did not prevent the appearance of GVHD, whereas MSC infusion 20 days post-HSC transplantation, when the levels of IFN-γ are high, or implantation of MSC prestimulated with IFN-γ resulted in enhanced survival [78].

Additional studies hinted that the timing of MSC administration relative to HSC transplant may dramatically influence the outcome of both the GVL response and GVHD. Ning *et al.* reported results from a clinical trial in which patients with hematological malignancies underwent haploidentical PBSC transplantation, in combination with MSC or not. They observed a reduction of GVHD in MSC-injected patients but also a 60% tumor relapse, compared to a 20% relapse in the non-MSC treated group [79]. A study conducted on 199 patients with ALL who underwent HSC transplantation also reached the conclusion that a lower cancer relapse incidence was associated with GVHD, whereas a more effective GVHD prophylaxis or an absence of GVHD was associated with higher risk of cancer relapse [80]. Overall, these results demonstrate that controlled GVHD is likely beneficial to prevent cancer relapse and that the phase of the disease in which MSC are administered can drastically change the outcome. Future research should focus on optimizing the timeframe of allogeneic PBSC transplantation and MSC infusion in order to promote tumor rejection mediated by GVL, while diminishing GVHD-related symptoms and deaths.

MSC for the Treatment of Arthritic Diseases

The dual immune modulation capacity and mesenchymal plasticity of MSC has been explored for the treatment of inflammatory diseases such as rheumatoid arthritis (RA). Although results from ongoing clinical trials are not yet available, positive results were obtained in a mouse model of collagen-induced arthritis. For example, a single intraperitoneal injection of allogeneic MSC reduced damage to articular joints by downregulating the immune response specific to the collagen II (CII) self antigen, as seen by the reduced plasma level of TNF-α and by the higher level of CII-specific Treg cells in MSC-treated mice [58]. In addition, an *in vitro* study using T cells from RA patients demonstrated that MSC and chondrocyte-differentiated MSC could significantly inhibit the proliferation of CII-activated T cells when present at a ratio of 1:1. Moreover, in the presence of MSC, the production of pro-inflammatory cytokines IFN-γ and TNF-α by CII-reactive T cells was drastically reduced while the production of IL-4 and IL-10 was up-regulated [81].

MSC for the Treatment of Multiple Sclerosis

The immunosuppressive properties of MSC for the treatment of multiple sclerosis (MS) is to be examined in several planned clinical trials based on convincing results obtained from the experimental auto-immune encephalomyelitis (EAE) animal model of MS (www.clinicaltrials.gov). Research from Zappia *et al.* using a mouse model of EAE demonstrated that a single injection of MSC before and during the early phase of the disease reduced its severity. This was accompanied by a significant reduction in T cell and macrophage infiltration in the brain, as well as decreased demyelination of the brain and spinal cord [82]. However, MSC administration during the chronic phase of the disease had no effect in that study. In contrast, other results from a mouse model of chronic EAE showed a net reduction of inflammation in mice treated with MSC. MSC injected i.v. or intraventricularly were also found to localize to the lymph nodes and to down-regulate proliferation of lymphocytes in response to myelin antigens and mitogens [83]. In addition, recent results from our laboratory have identified MMP-processed CCL2 derived from MSC as a key factor ameliorating EAE in treated mice. Using CCL2 knockout MSC, we demonstrated that MMP-processed CCL2 secreted by MSC acts as an antagonist by inhibiting $CD4^+$ T-cell activation through the suppression of STAT3 phosphorylation. Furthermore, the MMP-processed CCL2 not only induced the upregulation of B7.H1 on $CD4^+$ T cells, which can inhibit activated T cells through PD-1 receptors, but also reduced plasma levels of pro-inflammatory cytokines IL-17 and TNF-α and decreased $CD4^+$ T-cell infiltration into the spinal cord of

MSC treated mice, therefore alleviating EAE symptoms cause by T-cells infiltration of the central nervous system [65].

MSC and Organ Transplantation

Rat models of allogeneic organ transplantation coupled with MSC infusion have shown successful engraftment and reduced rejection of transplanted organs such as pancreatic β-islets [60] and heart [59]. Both studies indicate that MSC co-transplantation prolonged graft survival, inhibited Th1 cell activation, and induced the production of either IL-10-secreting $CD4^+$ T cells or $CD4^+CD25^+FoxP3^+$ Treg, respectively [59, 60]. Ongoing clinical trials are currently testing the use of MSC for renal transplant and pancreatic β-islet cell transplant.

Immune Activation by MSC

Immune Recognition

Based on the extensive results demonstrating the immunosuppressive effect of MSC, it was suggested that MSC could use their immunosuppressive mechanisms to evade the immune system and therefore be used as an "off-the-shelf" universal donor product for clinical applications. However, results from our group and others showed that mouse MSC could induce an immune response in an allogeneic setting and be rejected [84, 85]. On the other hand MSC can also participate in immune recognition of invading pathogens. As mentioned earlier, MSC were shown to express Toll-like receptors that are essential for the initiation of the innate and subsequent adaptive immune responses. One study found that exposure of human MSC to TLR3 and TLR4 decreased their ability to suppress allogeneic T-cell proliferation through the down-regulation of Notch ligand Jagged-1 expression on the MSC [86]. A recent study from our group demonstrated that IFN-α or IFN-γ priming combined with TLR3 or TLR4 activation of human or mouse MSC induced the production of proinflammatory cytokines and chemokines, as well as *iNOS*/*NOS2A* gene expression. In addition, retrieval of subcutaneous-injected, matrigel-embedded mouse MSC for infiltration analyses suggested that TLR ligand-pretreated MSC were able to attract innate immune cells *in vivo*, such as granulocytes and NK cells [49].

MSC are Conditional APC

IFN-γ stimulation of MSC enables them to acquire APC-like features. Among other things, IFN-γ induces the surface expression of MHC class I and class II molecules [50, 51, 87, 88]. MSC are also able, following IFN-γ priming, to take up, process, and present exogenous antigens through their MHC class II molecules, leading to the activation of naïve $CD4^+$ T cells *in vitro* and *in vivo* [50, 51]. In addition, we recently reported that MSC, especially after IFN-γ priming, could present extracellular soluble proteins to $CD8^+$ T cells through their MHC class I molecules [87]. This function, a hallmark of professional APC such as DC, is known as cross-presentation and is critical for the formation of a T-cell immune response targeted at extracellular pathogens or tumor antigens. Although MSC do not express the classical B7 costimulatory molecules CD80 and CD86, they express other surface molecules (ICAM and LFA-3) and secrete cytokines (IL-6 and IL-7) that can deliver costimulatory signals necessary to T-cell activation [19]. We have developed the concept of using MSCs as a cell-based cancer vaccine by demonstrating complete rejection of ovalbumin-expressing EG7 lymphoma cells in naïve C57Bl/6 mice immunized with ovalbumin-pulsed IFN-γ-stimulated MSC [50]. Consequently, considering their ability to induce an antigen-specific T-cell activation conjugated with the ease with which MSCs can be isolated and expanded to desired numbers, they offer an interesting alternative to DC for cancer immunotherapy. However, although MSCs are easier to obtained than DC, their immunostimulatory potential in comparison to DC is lower since they do not express standard B7 costimulatory molecules. Further analysis comparing head-to-head MSCs versus DCs in a cancer vaccine model is needed.

Future Objectives Regarding MSC and Immunotherapy

Many aspects pertaining to the immune functions of MSC still need further investigation in order to optimize and fully exploit the immune regulatory properties of MSCs. For instance, although some mechanisms behind the immune mediated effects of MSCs have been confirmed *in vivo* (NO production, IFN-γ responsiveness, and MMP-cleaved-CCL2 production), gaps in our understanding remain. The field would greatly benefit from additional *in vivo* studies using MSC with knocked-out or knocked-down genes. In addition, it appears some discrepancies exist between the immunosuppressive mechanisms of mouse and human MSC. Indeed, a recent paper by Ren *et al.* demonstrated using MLR, that mouse-mediated T-cell immunosuppression is primarily due to the production of NO, while human and nonhuman primate MSC mediate T-cell immunosuppression by expressing IDO which depletes the micro-environment of tryptophan necessary to T-cell proliferation [89]. Nonetheless, MSC from distinct mammalian species studied can equivalently suppress T-cell activation and proliferation, although through different mechanisms. Finally, the dual ability of MSC to suppress or activate the immune system depending on the micro-environment is problematic. IFN-γ stimulation was shown to induce and/or up-regulate the antigen presentation potential of MSC in one hand, and in the other, to induce and/or up-regulate the expression of immune suppressive molecules (Fig. 6.2). How can we best skew MSC to a defined immunological phenotype in order to prevent unwanted behavior (i.e., immunosuppression in a cancer vaccine setting or T-cell priming in an immunosuppressive

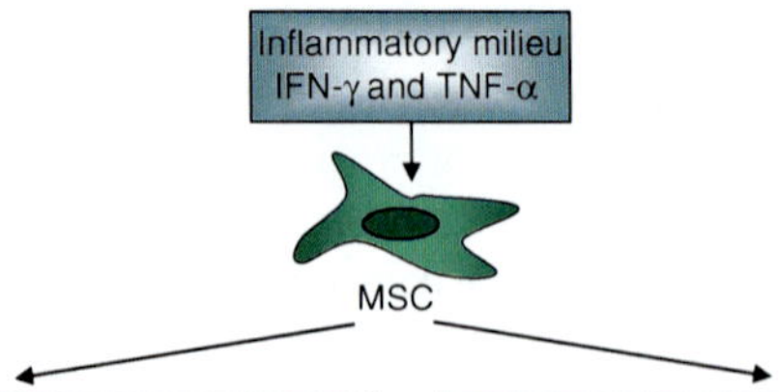

Immunostimulation	
Effects	**Outcomes**
Upregulate of MHC I surface expression	Increase CD8$^+$ T cells antigen priming
Surface expression of MHC II	CD4$^+$ T cells antigen priming
Conversion of proteasome into immunoproteasome	Increase antigenic peptides spectrum
Upregulate inflammatory cytokine expression (IL-6, IL7)	Induce costimulatory signaling in T cells

Immunosuppression	
Effects	**Outcomes**
Induction of iNOS production	Inhibit T cells proliferation
Induction of PGE2 production	Inhibit T cells proliferation
	Activate macrophages to produce IL-10
Upregulate B7.H1 surface expression	Inhibit activated T cells

Fig. 6.2 Dual immunomodulatory properties of MSC upon cytokines activation. Upon activation with inflammatory cytokines, *MSC*) have been demonstrated both *in vitro* and *in vivo* to display immunosuppressive properties and APC-like features depending on the context. IFN-γ priming of MSC induces the upregulation of surface expression of MHC I and induce MHC II surface expression therefore promoting CD8$^+$ and CD4$^+$ T cells antigen priming, respectively. In addition, IFN-γ induces the expression of immunoproteasome subunits which increases the pool of antigenic peptides produced by MSC. IFN-γ and TNF-α also upregulate IL-6 expression and induce IL-7 secretion which can both act as costimulatory signals toward T cells. On the other hand, IFN-γ activation of MSC also promotes the immunosuppressive potential of MSC. iNOS and PGE2 can both directly inhibit T cell proliferation. MSC-derived PGE2 can also activate macrophages to produce IL-10; indirectly suppressing T cell proliferation. Finally, B7.H1 expression is upregulated on INF-γ-activated MSC and inhibits activated T cells through PD.1

setting)? One solution would be to use *in vitro* methods of manipulation of the MSC, such as IFN-γ priming, in order to promote and boost their immune suppressive or APC functions.

MSC and Cancer

The possible implication of MSC in cancer development raised several concerns in the scientific and medical communities as this could be a serious drawback to the clinical applications of MSC, especially in cancer therapy [90]. MSCs were shown to be drawn to tumor sites by chemokines and to incorporate into the tumor stroma, thereby supporting tumor development through a series of mechanisms [91].

MSC produce growth factors, MMP and pro-angiogenic factors that can promote tumor growth and favor tumor metastasis [92]. Furthermore, the immunosuppressive functions of MSC can promote tumor cells proliferation by blocking the host immune response [93]. It is however impossible to determine if these cells will promote growth of preexistent tumors, which would impede their use in cancer patients.

Concluding Remarks

Since their discovery in 1976 by Friedenstein, the scientific interest in MSC has recently undergone a renaissance and they are now the object of intense preclinical and clinical scrutiny for their immune modulatory and regenerative properties. In addition to their physiological implications in tissue regeneration and HSC development, MSC have also been shown to modulate the immune response by either suppressing immune cells or behaving as conditional APC. Not only have their immunosuppressive properties been successfully applied to several immune disease animal models, but clinical researchers have also proven their beneficial effect for acquired disorders such as GVHD. MSC cell-based therapy is now a component of the field of immunotherapy and is currently being translated to the clinic, in some cases even without preclinical investigations in animal models. We, as researchers and clinicians, should focus our attention, however, to identifying the mechanistic underpinnings of MSC as a cell-based pharmaceutical, not only to prevent adverse effects but also to enhance their immuno-modulatory properties and optimize their safety and beneficial impact.

References

1. Friedenstein, A. J., J. F. Gorskaja, and N. N. Kulagina. (1976). Fibroblast precursors in normal and irradiated mouse hematopoietic organs. *Exp. Hematol.* 4:267.
2. Caplan, A. I. (2007). Adult mesenchymal stem cells for tissue engineering versus regenerative medicine. *J. Cell Physiol.* 213:341.
3. Dennis, J. E., A. Merriam, A. Awadallah, J. U. Yoo, B. Johnstone, and A. I. Caplan. (1999). A quadripotential mesenchymal progenitor cell isolated from the marrow of an adult mouse. *J. Bone Miner. Res.* 14:700.
4. Pittenger, M. F., A. M. Mackay, S. C. Beck, R. K. Jaiswal, R. Douglas, J. D. Mosca, M. A. Moorman, D. W. Simonetti, S. Craig, and D. R. Marshak. (1999). Multilineage potential of adult human mesenchymal stem cells. *Science* 284:143.
5. Sanchez-Ramos, J., S. Song, F. Cardozo-Pelaez, C. Hazzi, T. Stedeford, A. Willing, T. B. Freeman, S. Saporta, W. Janssen, N. Patel, D. R. Cooper, and P. R. Sanberg. (2000). Adult bone marrow stromal cells differentiate into neural cells *in vitro*. *Exp. Neurol.* 164:247.
6. Toma, C., M. F. Pittenger, K. S. Cahill, B. J. Byrne, and P. D. Kessler. (2002). Human mesenchymal stem cells differentiate to a cardiomyocyte phenotype in the adult murine heart. *Circulation* 105:93.
7. Caplan, A. I. (1991). Mesenchymal stem cells. *J. Orthop. Res.* 9:641.
8. Colter, D. C., R. Class, C. M. DiGirolamo, and D. J. Prockop. (2000). Rapid expansion of recycling stem cells in cultures of plastic-adherent cells from human bone marrow. *Proc. Natl. Acad. Sci. USA* 97:3213.

9. Colter, D. C., I. Sekiya, and D. J. Prockop. (2001). Identification of a subpopulation of rapidly self-renewing and multipotential adult stem cells in colonies of human marrow stromal cells. *Proc. Natl. Acad. Sci. USA* 98:7841.
10. Horwitz, E. M., B. K. Le, M. Dominici, I. Mueller, I. Slaper-Cortenbach, F. C. Marini, R. J. Deans, D. S. Krause, and A. Keating. (2005). Clarification of the nomenclature for MSC: the International Society for Cellular Therapy position statement. *Cytotherapy* 7:393.
11. Zuk, P. A., M. Zhu, P. Ashjian, D. A. De Ugarte, J. I. Huang, H. Mizuno, Z. C. Alfonso, J. K. Fraser, P. Benhaim, and M. H. Hedrick. (2002). Human adipose tissue is a source of multipotent stem cells. *Mol. Biol. Cell* 13:4279.
12. Sarugaser, R., D. Lickorish, D. Baksh, M. M. Hosseini, and J. E. Davies. (2005). Human umbilical cord perivascular (HUCPV) cells: a source of mesenchymal progenitors. *Stem Cells* 23:220.
13. In 't Anker, P. S., S. A. Scherjon, K. C. Kleijburg-van der, G. M. de Groot-Swings, F. H. Claas, W. E. Fibbe, and H. H. Kanhai. (2004). Isolation of mesenchymal stem cells of fetal or maternal origin from human placenta. *Stem Cells* 22:1338.
14. Jarocha, D., E. Lukasiewicz, and M. Majka. (2008). Advantage of mesenchymal stem cells (MSC) expansion directly from purified bone marrow $CD105^+$ and $CD271^+$ cells. *Folia Histochem. Cytobiol.* 46:307.
15. Simmons, P. J., and B. Torok-Storb. (1991). Identification of stromal cell precursors in human bone marrow by a novel monoclonal antibody, STRO-1. *Blood* 78:55.
16. Sorrentino, A., M. Ferracin, G. Castelli, M. Biffoni, G. Tomaselli, M. Baiocchi, A. Fatica, M. Negrini, C. Peschle, and M. Valtieri. (2008). Isolation and characterization of $CD146^+$ multipotent mesenchymal stromal cells. *Exp. Hematol.* 36:1035.
17. Dominici, M., B. K. Le, I. Mueller, I. Slaper-Cortenbach, F. Marini, D. Krause, R. Deans, A. Keating, D. Prockop, and E. Horwitz. (2006). Minimal criteria for defining multipotent mesenchymal stromal cells. The International Society for Cellular Therapy position statement. *Cytotherapy* 8:315.
18. Muguruma, Y., T. Yahata, H. Miyatake, T. Sato, T. Uno, J. Itoh, S. Kato, M. Ito, T. Hotta, and K. Ando. (2006). Reconstitution of the functional human hematopoietic microenvironment derived from human mesenchymal stem cells in the murine bone marrow compartment. *Blood* 107:1878.
19. Stagg, J. (2007). Immune regulation by mesenchymal stem cells: two sides to the coin. *Tissue Antigens* 69:1.
20. Dazzi, F., R. Ramasamy, S. Glennie, S. P. Jones, and I. Roberts. (2006). The role of mesenchymal stem cells in haemopoiesis. *Blood Rev.* 20:161.
21. Huang, G. P., Z. J. Pan, B. B. Jia, Q. Zheng, C. G. Xie, J. H. Gu, I. K. McNiece, and J. F. Wang. (2007). *Ex vivo* expansion and transplantation of hematopoietic stem/progenitor cells supported by mesenchymal stem cells from human umbilical cord blood. *Cell Transplant.* 16:579.
22. Mauch, P., L. Constine, J. Greenberger, W. Knospe, J. Sullivan, J. L. Liesveld, and H. J. Deeg. (1995). Hematopoietic stem cell compartment: acute and late effects of radiation therapy and chemotherapy. *Int. J. Radiat. Oncol. Biol. Phys.* 31:1319.
23. Awaya, N., K. Rupert, E. Bryant, and B. Torok-Storb. (2002). Failure of adult marrow-derived stem cells to generate marrow stroma after successful hematopoietic stem cell transplantation. *Exp. Hematol.* 30:937.
24. Bartsch, K., H. Al-Ali, A. Reinhardt, C. Franke, M. Hudecek, M. Kamprad, S. Tschiedel, M. Cross, D. Niederwieser, and C. Gentilini. (2009). Mesenchymal stem cells remain host-derived independent of the source of the stem-cell graft and conditioning regimen used. *Transplantation* 87:217.
25. Koc, O. N., S. L. Gerson, B. W. Cooper, S. M. Dyhouse, S. E. Haynesworth, A. I. Caplan, and H. M. Lazarus. (2000). Rapid hematopoietic recovery after coinfusion of autologous-blood stem cells and culture-expanded marrow mesenchymal stem cells in advanced breast cancer patients receiving high-dose chemotherapy. *J. Clin. Oncol.* 18:307.
26. Koc, O. N., C. Peters, P. Aubourg, S. Raghavan, S. Dyhouse, R. DeGasperi, E. H. Kolodny, Y. B. Yoseph, S. L. Gerson, H. M. Lazarus, A. I. Caplan, P. A. Watkins, and W. Krivit. (1999). Bone marrow-derived mesenchymal stem cells remain host-derived despite successful hematopoietic engraftment after allogeneic transplantation in patients with lysosomal and peroxisomal storage diseases. *Exp. Hematol.* 27:1675.

27. Gao, J., J. E. Dennis, R. F. Muzic, M. Lundberg, and A. I. Caplan. (2001). The dynamic *in vivo* distribution of bone marrow-derived mesenchymal stem cells after infusion. *Cells Tissues Organs* 169:12.
28. Devine, S. M., C. Cobbs, M. Jennings, A. Bartholomew, and R. Hoffman. (2003). Mesenchymal stem cells distribute to a wide range of tissues following systemic infusion into nonhuman primates. *Blood* 101:2999.
29. Ruster, B., S. Gottig, R. J. Ludwig, R. Bistrian, S. Muller, E. Seifried, J. Gille, and R. Henschler. (2006). Mesenchymal stem cells display coordinated rolling and adhesion behavior on endothelial cells. *Blood* 108:3938.
30. Honczarenko, M., Y. Le, M. Swierkowski, I. Ghiran, A. M. Glodek, and L. E. Silberstein. (2006). Human bone marrow stromal cells express a distinct set of biologically functional chemokine receptors. *Stem Cells* 24:1030.
31. Wynn, R. F., C. A. Hart, C. Corradi-Perini, L. O'Neill, C. A. Evans, J. E. Wraith, L. J. Fairbairn, and I. Bellantuono. (2004). A small proportion of mesenchymal stem cells strongly expresses functionally active CXCR4 receptor capable of promoting migration to bone marrow. *Blood* 104:2643.
32. Askari, A. T., S. Unzek, Z. B. Popovic, C. K. Goldman, F. Forudi, M. Kiedrowski, A. Rovner, S. G. Ellis, J. D. Thomas, P. E. DiCorleto, E. J. Topol, and M. S. Penn. (2003). Effect of stromal-cell-derived factor 1 on stem-cell homing and tissue regeneration in ischaemic cardiomyopathy. *Lancet* 362:697.
33. Spaeth, E., A. Klopp, J. Dembinski, M. Andreeff, and F. Marini. (2008). Inflammation and tumor microenvironments: defining the migratory itinerary of mesenchymal stem cells. *Gene Ther.* 15:730.
34. Horwitz, E. M., D. J. Prockop, L. A. Fitzpatrick, W. W. Koo, P. L. Gordon, M. Neel, M. Sussman, P. Orchard, J. C. Marx, R. E. Pyeritz, and M. K. Brenner. (1999). Transplantability and therapeutic effects of bone marrow-derived mesenchymal cells in children with osteogenesis imperfecta. *Nat. Med.* 5:309.
35. Horwitz, E. M., P. L. Gordon, W. K. Koo, J. C. Marx, M. D. Neel, R. Y. McNall, L. Muul, and T. Hofmann. (2002). Isolated allogeneic bone marrow-derived mesenchymal cells engraft and stimulate growth in children with osteogenesis imperfecta: implications for cell therapy of bone. *Proc. Natl. Acad. Sci. USA* 99:8932.
36. Makino, S., K. Fukuda, S. Miyoshi, F. Konishi, H. Kodama, J. Pan, M. Sano, T. Takahashi, S. Hori, H. Abe, J. Hata, A. Umezawa, and S. Ogawa. (1999). Cardiomyocytes can be generated from marrow stromal cells *in vitro*. *J. Clin. Invest* 103:697.
37. Nagaya, N., T. Fujii, T. Iwase, H. Ohgushi, T. Itoh, M. Uematsu, M. Yamagishi, H. Mori, K. Kangawa, and S. Kitamura. (2004). Intravenous administration of mesenchymal stem cells improves cardiac function in rats with acute myocardial infarction through angiogenesis and myogenesis. *Am. J. Physiol Heart Circ. Physiol.* 287:H2670–H2676.
38. Shake, J. G., P. J. Gruber, W. A. Baumgartner, G. Senechal, J. Meyers, J. M. Redmond, M. F. Pittenger, and B. J. Martin. (2002). Mesenchymal stem cell implantation in a swine myocardial infarct model: engraftment and functional effects. *Ann. Thorac. Surg.* 73:1919.
39. Hou, M., K. M. Yang, H. Zhang, W. Q. Zhu, F. J. Duan, H. Wang, Y. H. Song, Y. J. Wei, and S. S. Hu. (2007). Transplantation of mesenchymal stem cells from human bone marrow improves damaged heart function in rats. *Int. J. Cardiol.* 115:220.
40. Ohnishi, S., H. Sumiyoshi, S. Kitamura, and N. Nagaya. (2007). Mesenchymal stem cells attenuate cardiac fibroblast proliferation and collagen synthesis through paracrine actions. *FEBS Lett.* 581:3961.
41. Kinnaird, T., E. Stabile, M. S. Burnett, C. W. Lee, S. Barr, S. Fuchs, and S. E. Epstein. (2004). Marrow-derived stromal cells express genes encoding a broad spectrum of arteriogenic cytokines and promote *in vitro* and *in vivo* arteriogenesis through paracrine mechanisms. *Circ. Res.* 94:678.
42. Nakanishi, C., M. Yamagishi, K. Yamahara, I. Hagino, H. Mori, Y. Sawa, T. Yagihara, S. Kitamura, and N. Nagaya. (2008). Activation of cardiac progenitor cells through paracrine effects of mesenchymal stem cells. *Biochem. Biophys. Res. Commun.* 374:11.
43. Hung, S. C., R. R. Pochampally, S. C. Chen, S. C. Hsu, and D. J. Prockop. (2007). Angiogenic effects of human multipotent stromal cell conditioned medium activate the PI3K-Akt pathway

in hypoxic endothelial cells to inhibit apoptosis, increase survival, and stimulate angiogenesis. *Stem Cells* 25:2363.

44. Dai, Y., M. Xu, Y. Wang, Z. Pasha, T. Li, and M. Ashraf. (2007). HIF-1alpha induced-VEGF overexpression in bone marrow stem cells protects cardiomyocytes against ischemia. *J. Mol. Cell Cardiol.* 42:1036.
45. Nagaya, N., K. Kangawa, T. Itoh, T. Iwase, S. Murakami, Y. Miyahara, T. Fujii, M. Uematsu, H. Ohgushi, M. Yamagishi, T. Tokudome, H. Mori, K. Miyatake, and S. Kitamura. (2005). Transplantation of mesenchymal stem cells improves cardiac function in a rat model of dilated cardiomyopathy. *Circulation* 112:1128.
46. Ono, K., A. Matsumori, T. Shioi, Y. Furukawa, and S. Sasayama. (1998). Cytokine gene expression after myocardial infarction in rat hearts: possible implication in left ventricular remodeling. *Circulation* 98:149.
47. Guo, J., G. S. Lin, C. Y. Bao, Z. M. Hu, and M. Y. Hu. (2007). Anti-inflammation role for mesenchymal stem cells transplantation in myocardial infarction. *Inflammation* 30:97.
48. Pevsner-Fischer, M., V. Morad, M. Cohen-Sfady, L. Rousso-Noori, A. Zanin-Zhorov, S. Cohen, I. R. Cohen, and D. Zipori. (2007). Toll-like receptors and their ligands control mesenchymal stem cell functions. *Blood* 109:1422.
49. Romieu-Mourez, R., M. Francois, M. N. Boivin, M. Bouchentouf, D. E. Spaner, and J. Galipeau. (2009). Cytokine modulation of TLR expression and activation in mesenchymal stromal cells leads to a proinflammatory phenotype. *J. Immunol.* 182:7963.
50. Stagg, J., S. Pommey, N. Eliopoulos, and J. Galipeau. (2006). Interferon-gamma-stimulated marrow stromal cells: a new type of nonhematopoietic antigen-presenting cell. *Blood* 107:2570.
51. Chan, J. L., K. C. Tang, A. P. Patel, L. M. Bonilla, N. Pierobon, N. M. Ponzio, and P. Rameshwar. (2006). Antigen-presenting property of mesenchymal stem cells occurs during a narrow window at low levels of interferon-gamma. *Blood* 107:4817.
52. Trinchieri, G. (2003). Interleukin-12 and the regulation of innate resistance and adaptive immunity. *Nat. Rev. Immunol.* 3:133.
53. Nemeth, K., A. Leelahavanichkul, P. S. Yuen, B. Mayer, A. Parmelee, K. Doi, P. G. Robey, K. Leelahavanichkul, B. H. Koller, J. M. Brown, X. Hu, I. Jelinek, R. A. Star, and E. Mezey. (2009). Bone marrow stromal cells attenuate sepsis via prostaglandin E(2)-dependent reprogramming of host macrophages to increase their interleukin-10 production. *Nat. Med.* 15:42.
54. Nasef, A., A. Chapel, C. Mazurier, S. Bouchet, M. Lopez, N. Mathieu, L. Sensebe, Y. Zhang, N. C. Gorin, D. Thierry, and L. Fouillard. (2007). Identification of IL-10 and TGF-beta transcripts involved in the inhibition of T-lymphocyte proliferation during cell contact with human mesenchymal stem cells. *Gene Expr.* 13:217.
55. Tomchuck, S. L., K. J. Zwezdaryk, S. B. Coffelt, R. S. Waterman, E. S. Danka, and A. B. Scandurro. (2008). Toll-like receptors on human mesenchymal stem cells drive their migration and immunomodulating responses. *Stem Cells* 26:99.
56. Nasef, A., N. Ashammakhi, and L. Fouillard. (2008). Immunomodulatory effect of mesenchymal stromal cells: possible mechanisms. *Regen. Med.* 3:531.
57. Uccelli, A., L. Moretta, and V. Pistoia. (2008). Mesenchymal stem cells in health and disease. *Nat. Rev. Immunol.*
58. Augello, A., R. Tasso, S. M. Negrini, R. Cancedda, and G. Pennesi. (2007). Cell therapy using allogeneic bone marrow mesenchymal stem cells prevents tissue damage in collagen-induced arthritis. *Arthritis Rheum.* 56:1175.
59. Casiraghi, F., N. Azzollini, P. Cassis, B. Imberti, M. Morigi, D. Cugini, R. A. Cavinato, M. Todeschini, S. Solini, A. Sonzogni, N. Perico, G. Remuzzi, and M. Noris. (2008). Pretransplant infusion of mesenchymal stem cells prolongs the survival of a semiallogeneic heart transplant through the generation of regulatory T cells. *J. Immunol.* 181:3933.
60. Solari, M. G., S. Srinivasan, I. Boumaza, J. Unadkat, G. Harb, A. Garcia-Ocana, and M. Feili-Hariri. (2009). Marginal mass islet transplantation with autologous mesenchymal stem cells promotes long-term islet allograft survival and sustained normoglycemia. *J. Autoimmun.* 32:116.
61. Bingisser, R. M., P. A. Tilbrook, P. G. Holt, and U. R. Kees. (1998). Macrophage-derived nitric oxide regulates T cell activation via reversible disruption of the Jak3/STAT5 signaling pathway. *J. Immunol.* 160:5729.

62. Oh, I., K. Ozaki, K. Sato, A. Meguro, R. Tatara, K. Hatanaka, T. Nagai, K. Muroi, and K. Ozawa. (2007). Interferon-gamma and NF-kappaB mediate nitric oxide production by mesenchymal stromal cells. *Biochem. Biophys. Res. Commun.* 355:956.
63. Sato, K., K. Ozaki, I. Oh, A. Meguro, K. Hatanaka, T. Nagai, K. Muroi, and K. Ozawa. (2007). Nitric oxide plays a critical role in suppression of T-cell proliferation by mesenchymal stem cells. *Blood* 109:228.
64. Ren, G., L. Zhang, X. Zhao, G. Xu, Y. Zhang, A. I. Roberts, R. C. Zhao, and Y. Shi. (2008). Mesenchymal stem cell-mediated immunosuppression occurs via concerted action of chemokines and nitric oxide. *Cell Stem Cell* 2:141.
65. Rafei, M., P. M. Campeau, A. guilar-Mahecha, M. Buchanan, P. Williams, E. Birman, S. Yuan, Y. K. Young, M. N. Boivin, K. Forner, M. Basik, and J. Galipeau. (2009). Mesenchymal stromal cells ameliorate experimental autoimmune encephalomyelitis by inhibiting CD4 Th17 T cells in a CC chemokine ligand 2-dependent manner. *J. Immunol.* 182:5994.
66. Rafei, M., J. Hsieh, S. Fortier, M. Li, S. Yuan, E. Birman, K. Forner, M. N. Boivin, K. Doody, M. Tremblay, B. Annabi, and J. Galipeau. (2008). Mesenchymal stromal cell-derived CCL2 suppresses plasma cell immunoglobulin production via STAT3 inactivation and PAX5 induction. *Blood* 112:4991.
67. English, K., F. P. Barry, and B. P. Mahon. (2008). Murine mesenchymal stem cells suppress dendritic cell migration, maturation and antigen presentation. *Immunol. Lett.* 115:50.
68. Zhang, W., W. Ge, C. Li, S. You, L. Liao, Q. Han, W. Deng, and R. C. Zhao. (2004). Effects of mesenchymal stem cells on differentiation, maturation, and function of human monocyte-derived dendritic cells. *Stem Cells Dev.* 13:263.
69. Djouad, F., L. M. Charbonnier, C. Bouffi, P. Louis-Plence, C. Bony, F. Apparailly, C. Cantos, C. Jorgensen, and D. Noel. (2007). Mesenchymal stem cells inhibit the differentiation of dendritic cells through an interleukin-6-dependent mechanism. *Stem Cells* 25:2025.
70. Zhang, B., R. Liu, D. Shi, X. Liu, Y. Chen, X. Dou, X. Zhu, C. Lu, W. Liang, L. Liao, M. Zenke, and R. C. Zhao. (2009). Mesenchymal stem cells induce mature dendritic cells into a novel Jagged-2-dependent regulatory dendritic cell population. *Blood* 113:46.
71. Rasmusson, I., B. K. Le, B. Sundberg, and O. Ringden. (2007). Mesenchymal stem cells stimulate antibody secretion in human B cells. *Scand. J. Immunol.* 65:336.
72. Kishimoto, T. (2006). Interleukin-6: discovery of a pleiotropic cytokine. *Arthritis Res. Ther.* 8 Suppl 2:S2.
73. Corcione, A., F. Benvenuto, E. Ferretti, D. Giunti, V. Cappiello, F. Cazzanti, M. Risso, F. Gualandi, G. L. Mancardi, V. Pistoia, and A. Uccelli. (2006). Human mesenchymal stem cells modulate B-cell functions. *Blood* 107:367.
74. Shlomchik, W. D. (2007). Graft-versus-host disease. *Nat. Rev. Immunol.* 7:340.
75. Le, B. K., F. Frassoni, L. Ball, F. Locatelli, H. Roelofs, I. Lewis, E. Lanino, B. Sundberg, M. E. Bernardo, M. Remberger, G. Dini, R. M. Egeler, A. Bacigalupo, W. Fibbe, and O. Ringden. (2008). Mesenchymal stem cells for treatment of steroid-resistant, severe, acute graft-versus-host disease: a phase II study. *Lancet* 371:1579.
76. Ringden, O., M. Uzunel, I. Rasmusson, M. Remberger, B. Sundberg, H. Lonnies, H. U. Marschall, A. Dlugosz, A. Szakos, Z. Hassan, B. Omazic, J. Aschan, L. Barkholt, and B. K. Le. (2006). Mesenchymal stem cells for treatment of therapy-resistant graft-versus-host disease. *Transplantation* 81:1390.
77. von, B. M., F. Stolzel, A. Goedecke, K. Richter, N. Wuschek, K. Holig, U. Platzbecker, T. Illmer, M. Schaich, J. Schetelig, A. Kiani, R. Ordemann, G. Ehninger, M. Schmitz, and M. Bornhauser. (2009). Treatment of refractory acute GVHD with third-party MSC expanded in platelet lysate-containing medium. *Bone Marrow Transplant.* 43:245.
78. Polchert, D., J. Sobinsky, G. Douglas, M. Kidd, A. Moadsiri, E. Reina, K. Genrich, S. Mehrotra, S. Setty, B. Smith, and A. Bartholomew. (2008). IFN-gamma activation of mesenchymal stem cells for treatment and prevention of graft versus host disease. *Eur. J. Immunol.* 38:1745.
79. Ning, H., F. Yang, M. Jiang, L. Hu, K. Feng, J. Zhang, Z. Yu, B. Li, C. Xu, Y. Li, J. Wang, J. Hu, X. Lou, and H. Chen. (2008). The correlation between cotransplantation of mesenchymal stem cells and higher recurrence rate in hematologic malignancy patients: outcome of a pilot clinical study. *Leukemia* 22:593.

80. Nordlander, A., J. Mattsson, O. Ringden, K. Leblanc, B. Gustafsson, P. Ljungman, P. Svenberg, J. Svennilson, and M. Remberger. (2004). Graft-versus-host disease is associated with a lower relapse incidence after hematopoietic stem cell transplantation in patients with acute lymphoblastic leukemia. *Biol. Blood Marrow Transplant.* 10:195.
81. Zheng, Z. H., X. Y. Li, J. Ding, J. F. Jia, and P. Zhu. (2008). Allogeneic mesenchymal stem cell and mesenchymal stem cell-differentiated chondrocyte suppress the responses of type II collagen-reactive T cells in rheumatoid arthritis. *Rheumatology. (Oxford)* 47:22.
82. Zappia, E., S. Casazza, E. Pedemonte, F. Benvenuto, I. Bonanni, E. Gerdoni, D. Giunti, A. Ceravolo, F. Cazzanti, F. Frassoni, G. Mancardi, and A. Uccelli. (2005). Mesenchymal stem cells ameliorate experimental autoimmune encephalomyelitis inducing T-cell anergy. *Blood* 106:1755.
83. Kassis, I., N. Grigoriadis, B. Gowda-Kurkalli, R. Mizrachi-Kol, T. Ben-Hur, S. Slavin, O. Abramsky, and D. Karussis. (2008). Neuroprotection and immunomodulation with mesenchymal stem cells in chronic experimental autoimmune encephalomyelitis. *Arch. Neurol.* 65:753.
84. Eliopoulos, N., J. Stagg, L. Lejeune, S. Pommey, and J. Galipeau. (2005). Allogeneic marrow stromal cells are immune rejected by MHC class I- and class II-mismatched recipient mice. *Blood* 106:4057.
85. Nauta, A. J., G. Westerhuis, A. B. Kruisselbrink, E. G. Lurvink, R. Willemze, and W. E. Fibbe. (2006). Donor-derived mesenchymal stem cells are immunogenic in an allogeneic host and stimulate donor graft rejection in a nonmyeloablative setting. *Blood* 108:2114.
86. Liotta, F., R. Angeli, L. Cosmi, L. Fili, C. Manuelli, F. Frosali, B. Mazzinghi, L. Maggi, A. Pasini, V. Lisi, V. Santarlasci, L. Consoloni, M. L. Angelotti, P. Romagnani, P. Parronchi, M. Krampera, E. Maggi, S. Romagnani, and F. Annunziato. (2008). Toll-like receptors 3 and 4 are expressed by human bone marrow-derived mesenchymal stem cells and can inhibit their T-cell modulatory activity by impairing Notch signaling. *Stem Cells* 26:279.
87. Francois, M., R. Romieu-Mourez, S. Stock-Martineau, M. N. Boivin, J. L. Bramson, and J. Galipeau. (2009). Mesenchymal stromal cells cross-present soluble exogenous antigens as part of their antigen-presenting cell properties. *Blood* 114:2632.
88. Romieu-Mourez, R., M. Francois, M. N. Boivin, J. Stagg, and J. Galipeau. (2007). Regulation of MHC class II expression and antigen processing in murine and human mesenchymal stromal cells by IFN-gamma, TGF-beta, and cell density. *J. Immunol.* 179:1549.
89. Ren, G., J. Su, L. Zhang, X. Zhao, W. Ling, A. L'huillie, J. Zhang, Y. Lu, A. I. Roberts, W. Ji, H. Zhang, A. B. Rabson, and Y. Shi. (2009). Species variation in the mechanisms of mesenchymal stem cell-mediated immunosuppression. *Stem Cells* 27:1954.
90. Kidd, S., E. Spaeth, A. Klopp, M. Andreeff, B. Hall, and F. C. Marini. (2008). The (in) auspicious role of mesenchymal stromal cells in cancer: be it friend or foe. *Cytotherapy.* 10:657.
91. Dwyer, R. M., S. M. Potter-Beirne, K. A. Harrington, A. J. Lowery, E. Hennessy, J. M. Murphy, F. P. Barry, T. O'Brien, and M. J. Kerin. (2007). Monocyte chemotactic protein-1 secreted by primary breast tumors stimulates migration of mesenchymal stem cells. *Clin. Cancer Res.* 13:5020.
92. Karnoub, A. E., A. B. Dash, A. P. Vo, A. Sullivan, M. W. Brooks, G. W. Bell, A. L. Richardson, K. Polyak, R. Tubo, and R. A. Weinberg. (2007). Mesenchymal stem cells within tumour stroma promote breast cancer metastasis. *Nature* 449:557.
93. Djouad, F., P. Plence, C. Bony, P. Tropel, F. Apparailly, J. Sany, D. Noel, and C. Jorgensen. (2003). Immunosuppressive effect of mesenchymal stem cells favors tumor growth in allogeneic animals. *Blood* 102:3837.

Part III
T Cell Therapeutic Approaches

Chapter 7
Tumor-Specific Mutations as Targets for Cancer Immunotherapy

Brad H. Nelson and John R. Webb

Abstract The fundamental job of the immune system is to discriminate self from nonself. To achieve this, the immune system is actively tolerized against self proteins. When pathogens enter a host, they introduce foreign proteins to which the host is not tolerant, and an immune response ensues. In contrast, tumors represent a special case, as the vast majority of tumor proteins are "self" and hence do not trigger immune activation. However, mutation of genes important for regulation of cell growth is the underlying cause of cancer and any point mutation, insertion, reading frame-shift or protein fusion that generates a new protein sequence could theoretically be recognized as foreign by the immune system. With the advent of high-throughput sequencing technologies, we have entered an era where the tumor and germline genomes of individual patients can be sequenced, such that the entire repertoire of tumor-specific mutations can be known. To date, more than 78,000 somatic mutations have been reported in human cancer. While the prospect of targeting this huge diversity of mutations via pharmacological approaches appears daunting, T-cell-based treatments may offer a practical alternative owing to the enormous repertoire of antigen receptors expressed by the human T-cell compartment. To what extent are cancer mutations recognized by the immune system? To what extent can they be targeted by immunotherapy? Here, we review the work to date on these questions with a focus on tumor-specific mutations that have transitioned from basic laboratory investigations through to clinical trials in humans. Our goal is to identify the major issues that need to be resolved to enable advances in DNA sequencing to be translated to effective T-cell therapies in the clinic.

Keywords Somatic mutation • Cancer genomics • Tumor suppressor genes • High-throughput sequencing • Tumor-specific mutation

---

B.H. Nelson (✉)
Deeley Research Centre, BC Cancer Agency, Victoria, BC, Canada
e-mail: bnelson@bccancer.bc.ca

J. Medin and D. Fowler (eds.), *Experimental and Applied Immunotherapy*,
DOI 10.1007/978-1-60761-980-2_7,

A Brief Overview of Cancer Genomics

Tumors have unstable genomes and accumulate numerous genetic abnormalities ranging from point mutations to large-scale chromosomal aberrations. Until recently, genomic changes in cancer were detected by low-throughput techniques such as cytogenetics, transformation assays, and genetic manipulation of model organisms such as *Drosophila*. These efforts have been enormously successful, as they have led to the identification of many clinically-relevant oncogenes and tumor suppressor genes, including RAS, TP53, BCR-ABL, and HER-2/neu. With the advent of so-called "next-generation sequencing" it is increasingly feasible to sequence the entire transcriptome and/or genome of a tumor sample, which theoretically can identify and quantify all mutations at the DNA/RNA level. Indeed, there is optimism that the cost of sequencing the human genome will soon be as low as $1,000, which makes it reasonable to consider treatment strategies that exploit this information on an individual patient basis.

Several high-throughput sequencing studies over the past 6 years have yielded unprecedented insights into the frequency of somatic mutations in the cancer genome. Early studies focused on protein kinases, as this protein superfamily is intimately involved in cell growth regulation. Bardelli and colleagues [1] sequenced a large panel of tyrosine kinase domains in 35 colorectal cancer cell lines with validation in 147 primary colorectal cancers. They found that at least 30% of colorectal cancers contain one or more mutations in a tyrosine kinase. In a larger study, Greenman and colleagues [2] sequenced the coding exons of 518 protein kinase genes in 210 diverse human cancers and found more than 1,000 somatic mutations. Approximately 120 mutations showed evidence of playing a "driver" role; that is, such mutations would predictably confer a growth or survival advantage on cancer cells and would have been positively selected during the evolution of a cancer. The remaining mutations appeared to be "passengers"; that is, such mutations would have arose inadvertently during the random process of mutagenesis and would predictably confer no selective advantage to the tumor. Moving beyond the kinase superfamily, Ding and colleagues sequenced 623 genes with known or potential relationships to cancer in a panel of 188 human lung adenocarcinomas [3]. They found more than 1,000 somatic mutations across the samples and evidence for 26 driver mutations.

While the preceding studies focused on specific candidate genes, others have attempted to obtain a more genome-wide view of cancer mutations. Wood and colleagues sequenced 20,857 transcripts from a collection of 11 breast and 11 colorectal tumors [4] and found 1,718 genes (9.4% of the genes analyzed) with at least one nonsilent mutation. The large majority (93%) of alterations were single-base substitutions, resulting in missense changes (82%), stop codons (7%), or alterations of splice sites near start and stop codons (4%). The remaining somatic mutations were insertions, deletions, or duplications (7%). Overall, colorectal and breast tumors were each found to contain an average of 76 and 84 nonsilent mutations, respectively, of which 15 and 14 were putative driver mutations. The number of mutations per tumor was similar amongst colorectal tumors (ranging from 49 to 111) but was more variable in breast cancers (varying from 38 to 193). The breast cancer estimates compare favorably with a recent study in which the entire transcriptome and genome of a single lobular breast tumor was sequenced, which led to the identification of 32 somatic nonsynonymous coding mutations [5]. Based on their analysis, Wood and colleagues described the mutation landscapes of breast and colorectal cancers as being comprised of a small number of commonly mutated gene "mountains" amongst a backdrop of a much larger number of gene "hills" that are mutated at much lower frequency. In a similar analysis of 24 pancreatic cancers, this same group found that tumors contained an average of 63 genetic alterations, the majority of which were point mutations [6]. These mutations defined a core set of 12 cellular pathways and processes that were each genetically altered in 67–100% of the tumors. Finally, the group studied 22 cases of glioblastoma, where they found 685 genes (3.3% of the 20,661 genes analyzed) with at least one nonsilent somatic mutation [7].

Many cancers appear to have a "mutator phenotype," a term that refers to an increased somatic mutation rate [8]. For example, colorectal and endometrial cancers with defective DNA mismatch repair due to abnormalities in genes such as MLH1 and MSH2 exhibit increased rates of single nucleotide changes and small insertions/deletions at polynucleotide tracts [9]. In lung cancer, mutations in several genes implicated in DNA repair, including TP53, PRKDC, and SMG1, were positively correlated with higher mutation rates [3]. Intriguingly, increased numbers of tumor-infiltrating T cells are often found in cancers with defective DNA repair, including colorectal tumors with microsatellite instability, BRCA-mutant breast cancer, and BRCA1-mutant ovarian cancer [10–13]. These observations indicate that impaired DNA repair leads to a greater accumulation of mutations, which represent neo-antigens that could be recognized by the immune system.

The studies described above represent a mere sampling of the rapidly evolving field of cancer genomics. To date, more than 78,000 somatic mutations have been identified in human cancer (see the Catalogue of Somatic Mutations in Cancer or COSMIC) [14]. Adult epithelial cancers harbor on the order of 30–100 mutations, of which 5–20 are driver mutations [15, 16]. Looking to the future, the International Cancer Genome Consortium (ICGC) has the goal of comprehensively characterizing somatically acquired genetic events in at least 50 different types of cancer, which will involve high-coverage sequencing of at least 20,000 cancer genomes. This data will be integrated with expression and epigenetic profiles, as well as clinical annotation [17]. Similarly, the US National Institutes of Health's "The Cancer Genome Atlas" (TCGA) network will be generating sequence data for more than 20 tumor types and thousands of clinical samples over the next 5 years. Such efforts are predicted to increase the number of known somatic mutations in cancer from the current level of approximately 100,000 to over 100,000,000, an increase of three orders of magnitude. Based upon such staggering numbers it is easy to anticipate that there will be wealth of new potential targets for tumor immunologists to interrogate. However, before a tumor-specific mutation can be considered as a target for immunotherapy, it must meet a certain number of immunological criteria, as described below.

Basics of Cellular Immunology and the Potential for Immune Recognition of Tumor-Specific Mutations

Effective immune responses against tumors typically involve both $CD4^+$ and $CD8^+$ T cells. $CD4^+$ T cells recognize antigen in the form of peptides or "epitopes" presented on the cell surface by MHC class II molecules. These peptides are, in general, taken up by cells from the extracellular milieu via the phagolysosomal machinery. By contrast, $CD8^+$ T cells recognize peptides presented on the cell surface in the context of MHC class I molecules; these peptides, in general, are derived from the cytosolic compartment of cells (for a review of MHC class I and class II processing see [18]). MHC class I is expressed on all cell types except red blood cells. In contrast, MHC class II is expressed only by specific immune cells such as macrophages, B cells and dendritic cells, although there are reports of ectopic MHC class II expression by some epithelial tumors [19]. Given the more widespread expression of MHC class I compared to class II, $CD8^+$ T cells are usually considered the major arm of immunity that needs to be engaged for effective immunotherapy against solid tissue cancers.

How do peptides derived from cytosolic proteins end up on the cell surface, presented by MHC class I molecules? Both self and nonself proteins undergo natural turnover in cells. Protein degradation is mediated by the proteasome, a complex of at least 20 different protein subunits. The proteasome exists in at least two alternate forms: a constitutive form that is responsible for day-to-day housekeeping functions, and an inducible form (known as the "immunoproteasome") in which two components are replaced by the IFN-γ inducible LMP2 and LMP-7 subunits. Like all enzymatic processes, protein degradation by the proteasome and immunoproteasome shows substrate specificity, meaning that not all possible peptides from the cellular proteome are generated. However, the amino acid patterns underlying proteasome cleavage specificity are not completely understood and predictive algorithms are not yet very accurate [20].

Peptides produced by the proteasome and the immunoproteasome are subsequently shuttled into the lumen of the endoplasmic reticulum by a pair of membrane transporter molecules known as TAP-1 and TAP-2, which also display a certain level of substrate specificity. Once inside the ER, another molecule called TAPASIN facilitates the loading of proteasome-derived peptides into empty MHC class I molecules. But the complexity inherent in the process does not end here, as MHC class I molecules have a restricted peptide binding specificity. Indeed, the MHC class I locus is the most highly polymorphic in the human genome [21], and the greatest density of polymorphism is in the peptide-binding cleft, which is the site that binds proteasome-derived peptides. Owing to these polymorphisms, different MHC alleles bind a different spectrum of peptides, with selectivity based upon the primary sequence of peptides and, to a lesser extent, upon their length, which can vary from eight to ten amino acids. It is thought that such polymorphism confers an evolutionary advantage at the population level, as viral agents cannot readily escape immunity across an MHC diverse population by simply mutating residues within a single MHC class I-binding peptide.

In contrast to MHC class I processing, antigen processing and presentation by MHC class II molecules occurs by a distinct process. In general, extracellular antigens are picked up by antigen presenting cells (APCs) via the formation of phagosomes, which ultimately fuse with lysosomes to form phagolysosomes. Phagolysosomes contain nascent MHC class II molecules that become loaded with exogenous peptide and then migrate to the cell surface. Much like MHC class I, MHC class II molecules are also highly polymorphic and can bind a large array of peptides, depending upon the set of class II alleles expressed by a given individual. One important caveat to this scheme is a process known as cross-presentation, which is unique to dendritic cells (DCs). During cross-presentation, exogenous antigen is taken up via the phagosomal process normally reserved for MHC class II antigens, but ultimately ends up in an MHC class I presentation pathway [22]. Cross-presentation provides a mechanism whereby $CD8^+$ T-cell responses can be elicited against tumor-derived antigens that are released from tumor cells and taken up by phagocytic DCs.

Whereas the foregoing discussion concerns the generation of T-cell epitopes, the other essential requirement for cellular immunity is, of course, the presence of T cells that are capable of recognizing these epitopes. T cells express a cell surface receptor complex known as the T-cell Receptor (TCR), which binds peptides in the context of MHC class I (in the case of $CD8^+$ T cells) or class II (in the case of $CD4^+$ T cells). The TCR is highly selective in its ability to recognize peptide epitopes – even a single amino acid change in a peptide epitope can make the difference between recognition or nonrecognition. The average human T-cell compartment is estimated to contain a minimum of 2.5×10^7 unique clones [23], which are generated by random gene rearrangement. Nascent T cells are tested in the thymus for potential recognition of self proteins or "auto-reactivity." Most potentially auto-reactive T cells are deleted in the thymus by a process known as "central tolerance"; however, some auto-reactive T cells escape to the periphery where they are held in check by various mechanisms known collectively as "peripheral tolerance." The end result is a T-cell compartment with a highly diverse capacity for recognizing foreign antigens, but with limited ability to respond to self antigens. Central and peripheral tolerance represents unique challenges to the field of tumor immunotherapy since the vast majority of tumor proteins are "self," and hence are not expected to trigger immune activation. However, even a single point mutation in a peptide, as can occur in cancer, can theoretically generate a novel T-cell epitope that is no longer subject to central or peripheral tolerance.

Based on the above immunological considerations, there are a number of requirements that must be met before a tumor-specific mutation should be considered as a potential target for T-cell-based immunotherapy (summarized in Fig. 7.1):

1. The parent protein harbouring the tumor-specific mutation must be expressed at a sufficiently high level by tumor cells to allow for killing by armed effector T cells. Epitopes expressed at very low levels often do not surpass the minimum threshold of TCR triggering needed for appropriate T-cell activation.
2. The tumor-specific mutation must occur within a region of the protein that can be appropriately cleaved into suitable peptides by the proteasome. Although the proteasome is capable of cleaving proteins at a variety of positions, there are

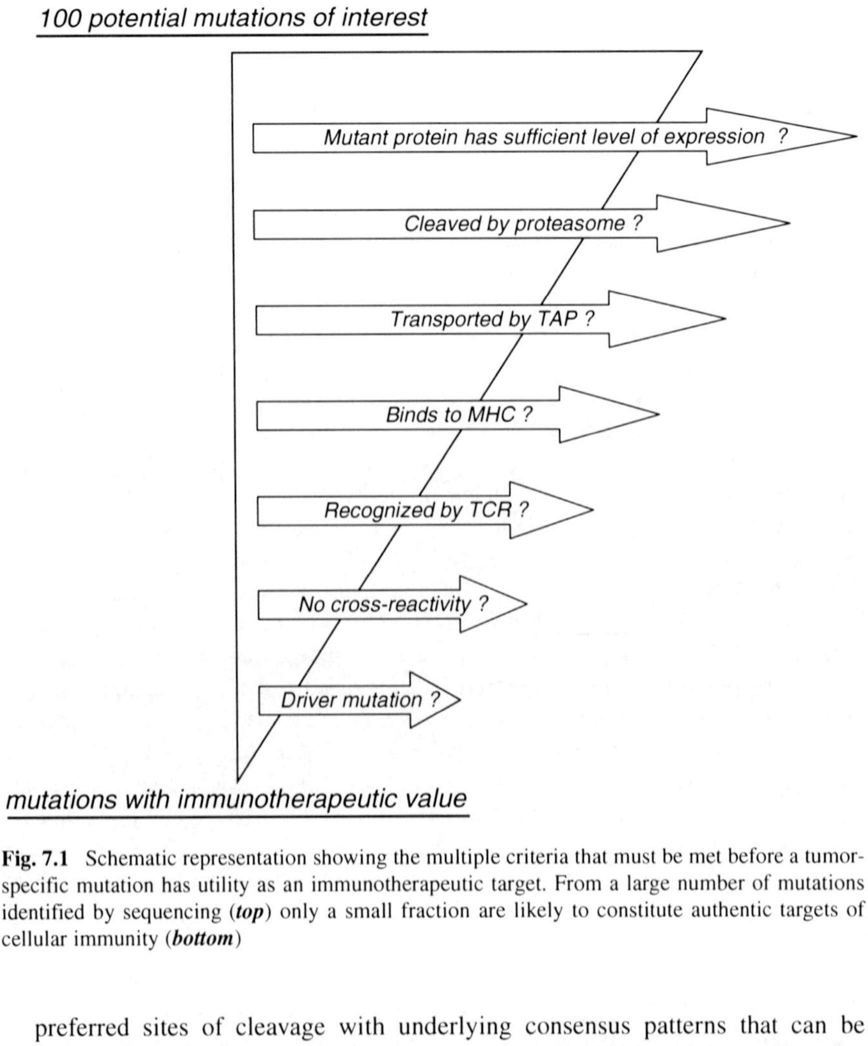

Fig. 7.1 Schematic representation showing the multiple criteria that must be met before a tumor-specific mutation has utility as an immunotherapeutic target. From a large number of mutations identified by sequencing (***top***) only a small fraction are likely to constitute authentic targets of cellular immunity (***bottom***)

preferred sites of cleavage with underlying consensus patterns that can be impacted by the introduction of point mutations. As a result, point mutations can create novel patterns of proteasome cleavage which indirectly result in the formation of new epitopes.

3. Substrate specificity exists with respect to which peptides are transported into the ER by the TAP transporter complex. Tumor-specific mutations that impact upon consensus sequences important for TAP transport could potentially impact the repertoire of peptides transported into the ER.
4. A significant amount of epitope selection occurs at the level of MHC binding. MHC alleles differ from one another with regards to the repertoire of peptides that can be accommodated within their peptide binding cleft. This specificity is predicated primarily by the presence of so-called hydrophobic "anchor" residues at appropriate positions in the binding peptide. Therefore, in order for a tumor-specific point mutation to be immunologically relevant, it must occur at a position

that does not disrupt MHC/peptide binding. At the same time, tumor-specific mutations must occur at a position that renders them visible to the TCR.

5. Even though all four of the above criteria may be met, it is essential that the T-cell compartment contain T cells that are able to recognize the mutation with sufficient avidity to trigger a functional immune response. As with MHC binding, recognition of peptides by the TCR requires a complex molecular interaction between the "contact residues" at the face of the TCR and specific amino acids within the peptide epitope. Not all positions of the peptide contribute to this process; therefore, in order for a tumor-specific mutation to be immunogenic, it must make a significant contribution to TCR recognition or cause a conformational shift in a neighboring amino acid that is critical for TCR binding.
6. T-cell responses against tumor-specific mutations should ideally demonstrate a minimal amount of "off-target" reactivity (cross-reactivity); otherwise, there may be unacceptable levels of collateral damage. One need look no further that the devastating effects of many autoimmune diseases to realize that the consequences of autoimmunity need to be minimized.
7. Lastly, targeting of a single epitope carries the risk that an already genetically unstable tumor might simply escape immune attack via down-regulation of the target protein. For these reasons, the optimal targets for immunotherapy are likely driver mutations. That said, passenger mutations that occur in essential genes such as α-actinin [24] might still represent effective targets because tumors would not be able to stop expressing such proteins without penalty; further alteration of the mutated epitope would represent the only option for antigen escape.

Evidence of Naturally Occurring Cellular Immunity Against Tumor-Specific Mutations

The first papers describing the recognition of mutated tumor proteins by T cells appeared in the mid-1990s, and a flood of reports has followed (for reviews see [25, 26]). There is a regularly updated online database of immunogenic, mutated cancer peptides on the Cancer Immunity website (http://www.cancerimmunity.org/peptidedatabase/Tcellepitopes.htm), which serves as an excellent resource for those in the field. Many studies have focused on identifying the antigens recognized by tumor-infiltrating lymphocytes (TILs), which serve as a convenient source of tumor-reactive T cells. Thus, $CD4^+$ and $CD8^+$ TILs have been shown to recognize point mutations in a wide variety of genes, including fibronectin [27], HSP70 [28], α-actinin-4 [24], trisphosphate isomerase [29], an RNA helicase [30], MHC class I [31], N-RAS [32], β-catenin [33], receptor-like protein tyrosine phosphatase kappa [34], MART-2 [35], and p14ARF [31]. All of these examples were seen in individual patients, highlighting the personalized nature of most tumor mutations. Intriguingly, these responses were often associated with better-than-expected survival, thus suggesting that the TIL response may be clinically beneficial [24, 36–38]. That said, tumor mutations can also be recognized by regulatory T cells, which may hamper anti-tumor immunity [39].

Note that some of the above-mentioned gene mutations appear to represent drivers (for example, N-RAS), whereas others appear to be passengers (for example, α-actinin-4). Point mutations need not reside in exons to be immunogenic, as intronic sequences can give rise to neo epitopes that are recognized by TILs [40]. Moreover, some mutations lead to the generation of neo epitopes that are remote from the mutation itself, thus suggesting that the mutation may alter the stability and/or processing of the parent protein to reveal cryptic epitopes [41, 42]. Whereas the above-mentioned studies also all concerned TILs, other studies have shown that T cells specific for tumor mutations can be generated by repeated *in vitro* stimulation of peripheral blood lymphocytes from cancer patients or normal donors. Examples include common oncogenic mutations in RAS [43–45] and B-RAF [46]; frameshift mutations in coding microsatellites of genes such as Caspase 5 [47], O-linked N-acetylglucosamine transferase [48], and TGFβIIR [49]; and fusion genes arising from chromosomal translocations, such as SYT-SSX in sarcoma [50], and ETV6-AML1 [51] along with BCR-ABL [52–54] in leukemia. Thus, even in cases where T-cell responses to a tumor mutation do not arise naturally, they can nonetheless be induced *in vitro* and, as discussed in the next section, through immunization *in vivo*.

Clinical Immunotherapy Trials That Have Targeted Tumor-Specific Mutations

As next-generation sequencing becomes more practical and affordable, an attractive immunotherapy strategy may be to generate peptide-based vaccines that target tumor-specific mutations, even on an individual patient basis. What is the likelihood of success of such strategies? We will address this critical question by reviewing the results of previous clinical trials that have targeted mutant versions of RAS and BCR-ABL (Table 7.1). These examples were selected because they represent well-described driver mutations that have been targeted with mutation-specific peptide-based vaccines in humans. Thus, they provide an excellent example of the type of therapeutic targets that next-generation sequencing can provide now, and in the future.

Ras

HRAS was the first human oncogene discovered [55, 56], and RAS family members (HRAS, KRAS, and NRAS) are frequently mutated in a variety of human cancers. Pre-clinical studies have shown that cancer patients and normal controls harbor $CD4^+$ and $CD8^+$ T cells that are able to specifically recognize mutant forms of RAS [43–45]. In general, these responses are undetectable directly *ex vivo*, however, they can be revealed through repeated *in vitro* stimulation with mutant peptides.

Table 7.1 Summary of immunological and clinical responses after vaccination of cancer patients against mutant RAS or BCR-ABL

Target	Patient cohort	Mode of immunotherapy	Immunological response	Clinical response	Reference
Mutant KRAS	Pancreatic cancer ($n=5$)	Peptide-pulsed PBMC, intravenous	Transient KRAS-specific T-cell proliferation in 2/5 patients	No major therapeutic responses seen.	Gjertsen 1995 [60]
Mutant KRAS	Pancreatic cancer ($n=48$)	Peptide vaccine, intradermal	Positive DTH and T-cell proliferation in 58% of patients	Responding patients showed prolonged survival.	Gjertsen 2001 [61]
Mutant KRAS	Various advanced cancers ($n=10$)	Peptide vaccine, subcutaneous	$CD4^+$ and $CD8^+$ T-cell responses in 3/10 patients	No major therapeutic responses seen.	Khleif 1999 [62]
Mutant KRAS	Pancreatic ($n=5$) and colorectal cancer ($n=7$)	Peptide-based vaccine, subcutaneous	T-cell responses (IFN-γ) in 5/11 patients	No major therapeutic responses reported.	Toubaji 2008 [64]
Mutant KRAS or p53	Various advanced cancers ($n=38$)	Peptide-loaded PBMC, intravenous	IFN-γ response (17/38) and CTL response (10/38)	IFN-γ response was positively associated with survival	Carbone 2005 [66]
Mutant N-RAS	Melanoma ($n=10$)	Peptide vaccine, intradermal	DTH (8/10)	Not reported	Hunger 2001 [67]
BCR-ABL	Chronic phase CML($n=12$)	Peptide vaccine, subcutaneous	DTH, T-cell proliferation and/or antibodies in 3/12 patients	Not assessable, as patients were on other treatments.	Pinilla-Ibarz 2000 [77]
BCR-ABL	Chronic phase CML ($n=14$)	Peptide vaccine, subcutaneous	DTH and/or T-cell proliferation in 14/14 patients	4/14 decreased Ph, 3/14 transiently PCR negative, 5/14 complete cytogenic remission	Cathcart 2004 [78]

(continued)

Table 7.1 (continued)

Target	Patient cohort	Mode of immunotherapy	Immunological response	Clinical response	Reference
BCR-ABL	CML ($n=16$)	Peptide vaccine, subcutaneous	DTH (11/16), $CD4^+$ T-cell proliferation (13/14), IFN-γ ELISPOT (5/5)	15/16 improved cytogenic response, 7/16 complete cytogenic remission	Bocchia 2005 [79]
BCR-ABL	Chronic phase CML ($n=19$)	Peptide vaccine, PADRE helper epitope, intradermal	Transient IFN-γ ELISPOT (14/19)	≥1-log fall in BCR-ABL transcript (13/19)	Rojas 2007 [80]
BCR-ABL	CML ($n=13$)	Heteroclitic and wild type peptide vaccine	IFN-γ ELISPOT (7/13), DTH (9/13), T-cell proliferation (11/13)	Negative BCR-ABL by FISH (2/13), other molecular tests inconclusive	Maslak 2008 [81]

Mutation-specific T-cell lines and clones were shown to kill tumor cells carrying the corresponding RAS mutation in some [57, 58], but not all [43, 59] studies, indicating that mutant RAS peptides can be naturally processed and presented, at least by some tumor cells. These preclinical studies provided a strong rationale for developing RAS-specific vaccines for cancer.

KRAS mutations are found in 90% of pancreatic cancers; therefore, it is fitting that the first clinical trial of a KRAS vaccine involved this tumor type. Five patients with pancreatic cancer were vaccinated with a synthetic peptide corresponding to the KRAS mutation found in each patient's tumor, all of which were variations at codon 12 [60]. The immunizing peptides were 17 amino acids in length, a size that is normally expected to primarily induce $CD4^+$ T-cell responses but may also trigger $CD8^+$ T-cell responses. The MHC status of patients was not considered; rather, the intent was that the 17-mers would by chance contain appropriate epitopes for presentation by diverse MHC class I and/or II molecules. Mutation-containing peptides were pulsed onto PBMCs, which were then delivered intravenously. After two to three rounds of vaccination, a transient KRAS-specific T-cell response was elicited in 2/5 patients, as assessed by thymidine incorporation of PBMCs after 7 days of *in vitro* culture. In one patient, the T-cell response was specific to the mutant peptide used for immunization, whereas in the other patient, T-cells responded to both wild-type and mutant peptides. The responses disappeared within a few weeks, and no major therapeutic responses were observed. However, tumor tissue obtained on autopsy showed dense T-cell infiltrates in the two responding patients.

Based on these results, this same group performed a larger Phase I/II trial in which 48 pancreatic cancer patients were vaccinated intradermally with KRAS peptides together with the adjuvant GM-CSF [61]. Ten of the patients were surgically resected whereas 38 had advanced, nonresectable disease. Resected patients received a single 17-mer peptide corresponding to the mutation present in their tumor. By contrast, nonresected patients received a pool of four different mutant peptides, as the mutational status of their tumor was not known. Vaccinations were generally well tolerated. Peptide-specific immunity was induced in 58% of evaluable patients, including a positive DTH reaction in 49% of patients and an *in vitro* proliferative response in 40% of vaccinated patients. Patients vaccinated with the four-peptide cocktail showed proliferative response to anywhere between one and four of the mutated peptides, but not to wild-type KRAS. In four of the responding patients with advanced disease, TILs were harvested from tumor biopsies, expanded in IL-2, and tested for reactivity to the four peptides from the vaccine. Negative results were obtained for three of the four patients. In the remaining patient, the expanded $CD4^+$ TIL reacted to KRAS peptide, indicating that vaccine-induced $CD4^+$ T cells had successfully homed to the tumor site. In addition, there was evidence of immunological memory against KRAS mutant peptides for up to 8 months, but repeated vaccinations were required to maintain a detectable response. Most intriguingly, in patients with advanced cancer, responders showed prolonged survival compared to nonresponders (148 vs 61 days, $p=0.0002$).

Khleif and colleagues conducted a similar phase I trial of a KRAS peptide vaccine in patients with a variety of advanced cancers [62]. The vaccine consisted of 13-mer peptides corresponding to the KRAS mutation found in each patient's tumor, all of which were variations at codon 12. Patients were vaccinated subcutaneously with peptide together with the adjuvant DETOX. Vaccinations were well tolerated. Three out of ten evaluable patients generated $CD4^+$ and/or $CD8^+$ T-cell responses specific to mutant KRAS [63]. Moreover, $CD8^+$ T cells were able to lyse an HLA-A2-matched tumor cell line carrying the corresponding mutant (but not wild-type) KRAS gene. Subsequently, these investigators conducted a Phase II clinical trial in which 12 patients (five pancreatic and seven colorectal cancer patients) with no evidence of active disease (NED) were vaccinated subcutaneously with 13-mer peptide corresponding to the KRAS codon 12 mutation present in their tumor [64]. DETOX was again used as the adjuvant. T-cell responses to the relevant mutant KRAS peptide were detected in 5/11 patients by quantitative real-time PCR measurement of IFN-γ expression in peripheral blood. Due to the small number of patients in the study and the disease status at time of treatment (NED), it was difficult to determine the effect of vaccination on disease outcome. Notably, however, patients who developed an immune response to mutant KRAS peptide showed increased overall survival compared to nonresponding patients ($p=0.043$).

In a small lung cancer study, patients with KRAS mutant tumors were immunized intradermally with a mixture of seven peptides representing the most common KRAS mutations, with GM-CSF administered as an adjuvant [65]. Three patients, all with stage III disease, received the full vaccine course. While the vaccine was well tolerated, KRAS-specific T-cell responses were not detected *ex vivo*. Nonetheless, one patient developed a positive DTH reaction. Carbone and colleagues also vaccinated 38 patients with 17-mer peptides corresponding to patient-specific mutations in the KRAS or TP53 genes [66]. As in other studies, there was no attempt to determine the patients' HLA type, or to predict epitopes for the different mutations. Peptides were loaded onto irradiated autologous PBMC, which served as APCs. Patients had a variety of solid tumors, and either had evident disease or were classified as NED with >50% chance of recurrence. Eligible mutations included nonsilent point mutations, frame-shift mutations, or insertions/deletions internal to the coding sequence. In general, the vaccines were well tolerated. Overall, 17/38 patients demonstrated a positive IFN-γ response and 10/38 patients had a positive CTL response against autologous peptide-pulsed B cells. Patients in the NED group were more likely to show a response than those with evident disease: 4/9 versus 6/28 by the CTL assay, and 8/9 versus 8/28 by the IFN-γ assay. By contrast, responses against influenza challenge were equivalent between the two groups. In multivariate analysis, CTL and IFN-γ responses were associated with each other, and an IFN-γ response was positively associated with survival.

Finally, a second member of the RAS family, N-RAS, has also been targeted by vaccination. In a Phase I study, ten melanoma patients were immunized intradermally with a pool of four 25-mer N-RAS peptides (all with codon 61 mutations), using GM-CSF as adjuvant [67]. Patients were not typed for HLA haplotype nor examined for expression of the four mutations in tumor tissue. Nonetheless, 8/10 patients showed strong DTH reactions, and an *in vitro* T-cell response was detected in 2/10 patients. The specificity of these responses was confirmed by cloning peptide-specific CD4$^+$ T cells from peripheral blood: in these studies, it was demonstrated that T cells recognized mutant but not wild-type peptides. Collectively, these studies show that RAS vaccines can be used safely in the clinic and can induce CD4$^+$ and CD8$^+$ T-cell responses to driver mutations involving single amino acid changes.

BCR-ABL

BCR-ABL is a chimeric gene formed by the translocation of the c-ABL proto-oncogene on chromosome 9 to the breakpoint cluster region within the BCR gene on chromosome 22. The resulting 210 kD BCR-ABL protein shows tyrosine kinase activity and is necessary and sufficient for transformation of leukemic cells in patients with CML, thus meeting the criteria of a driver mutation. In 95% of patients, there are common breakpoints in the BCR gene with two alternative junctions [68]. The BCR-ABL fusion protein is an ideal tumor-specific antigen, as the junction contains an amino acid sequence that is not expressed in normal cells. Moreover, the fusion region can be presented by MHC class I and II to CD8$^+$ and CD4$^+$ T cells, respectively [52–54]. In particular, four breakpoint peptides have been identified that bind with high/intermediate affinity to HLA-A3, A11, B8, and A2.1, and elicit MHC-restricted cytotoxicity *in vitro* [52, 69–72]. One of these peptides has been shown by mass spectrometry to be naturally processed and presented on leukemic cells [73]. Breakpoint peptides have also been identified that bind MHC class II and elicit CD4$^+$ T-cell responses *in vitro* [54, 70, 74–76].

Based on these encouraging pre-clinical studies, Pinilla-Ibarz and colleagues conducted the first clinical trial of a vaccine targeting BCR-ABL [77]. Twelve patients with chronic phase CML received a cocktail of five peptides (four that bound MHC class I, and one that bound MHC class II) corresponding to a BCR-ABL breakpoint sequence, together with QS-21 as an adjuvant. All patients had the appropriate BCR-ABL breakpoint; 11/12 had at least one MHC allele relevant to one of the peptides; and 7/12 expressed both MHC class I and II alleles relevant to the peptides. Vaccinations were well tolerated. Collectively, some form of response was seen in 3/12 patients. Specifically, in 3/6 patients treated at the two highest dose levels of vaccine, peptide-specific T-cell proliferative responses ($n=3$) and/or DTH responses ($n=2$) were generated and lasted up to 5 months after vaccination. Two of these patients also developed antibody responses to BCR-ABL. CTL responses were not seen, although the authors commented that more sensitive assays such as ELISPOT or MHC class I tetramers might reveal such responses. Thus, a BCR-ABL peptide vaccine can elicit specific immune responses in patients with chronic phase CML even in the presence of active disease. These same investigators subsequently conducted a Phase 2 trial in which 14 patients with chronic phase CML were vaccinated with 6 BCR-ABL fusion peptides (five that bound MHC class I, and one that bound MHC class II), again with QS-21 as an adjuvant [78]. Patients received other concurrent treatments, including IFN-α, imatinib, and allogeneic donor lymphocyte infusions (DLI). No significant toxicities were observed. In 14/14 patients,

DTH and/or $CD4^+$ T-cell proliferative responses developed. Moreover, positive IFN-γ ELISPOT results were seen in $CD4^+CD45RO^+$ T cells (11/14 patients) and $CD8^+$ T cells (4/14 patients). Encouraging clinical responses were also seen: (a) four patients in hematologic remission had a decrease in their Philadelphia chromosome (Ph) percentages; (b) three patients in molecular relapse after allogenic transplantation became transiently PCR negative after vaccination; and (c) five patients reached complete cytogenetic remission. While encouraging, none of these responses could be directly attributed to vaccination due to the concurrent treatments patients received. Nonetheless, the authors concluded that a BCR-ABL peptide vaccine can safely and reliably elicit peptide-specific $CD4^+$ T-cell responses in CML patients when administered in conjunction with standard treatments.

In a similar study [79], 16 CML patients were given a pooled peptide vaccine corresponding to the same BCR-ABL breakpoint used above (four peptides bound MHC class I, and one peptide bound MHC class II). Molgramostim, QS-21, and GM-CSF were used as adjuvants. Of the 16 patients, 14 expressed MHC class II alleles appropriate for these peptides, and 8 expressed appropriate MHC class I alleles. Patients developed peptide-specific DTH (11/16 patients), $CD4^+$ T-cell proliferative responses (13/14 patients), and IFN-γ ELISPOT responses (5/5 patients). Again, promising clinical responses were seen. Of ten patients concurrently on imatinib, all showed improved cytogenetic responses after six vaccinations, with five patients reaching complete cytogenic remission (of which three achieved undetectable amounts of BCR-ABL transcript). Of six patients on concurrent IFN-α treatment, all but one had improved cytogenetic responses, and two patients reached complete cytogenic remission after vaccination. Based on these results, the authors suggested that the addition of BCR-ABL peptide vaccination to conventional treatments in CML patients might reduce residual disease and increase the number of patients reaching a molecular response [77].

In a study by Rojas and colleagues [80], 19 imatinib-treated CML patients in first chronic phase were vaccinated with BCR-ABL breakpoint peptides, some of which were linked to a pan-MHC class II (DR) epitope called PADRE, which was intended to augment $CD4^+$ T-cell help. Patients were vaccinated with a cocktail of three peptides together with GM-CSF. T-cell responses to PADRE were seen in all patients, and 14/19 patients developed T-cell responses to BCR-ABL peptides as assessed by IFN-γ ELISPOT. T-cell responses were transient, disappearing by day 148 in all but one case. Nonetheless, the development of anti-BCR-ABL T-cell responses correlated with a subsequent fall in BCR-ABL transcripts. Specifically, of 14 patients who had experienced a major cytogenetic response at baseline, 13 showed at least a 1-log fall in BCR-ABL transcripts. This occurred several months after completing vaccination, which is consistent with an effect at a primitive CML stem cell level. The authors conclude that BCR-ABL peptide vaccination may improve control of CML, especially in patients responding well to imatinib.

Finally, two groups have explored the potential value of using heteroclitic peptides to induce $CD8^+$ T-cell responses to otherwise weak epitopes from the breakpoint region of BCR-ABL [81, 82]. Heteroclitic peptides are designed to have increased HLA binding affinity due to selective substitutions in the HLA binding region (typically 1–3 amino acids). Vaccination of CML patients with peptide pools containing heteroclitic HLA-A2-binding peptides was shown to trigger $CD8^+$ T-cell responses against the heterclitic peptide in 6/6 patients and to the corresponding wild-type epitope in 4/6 patients [81]. Unfortunately, at the end of study, all patients remained positive for BCR-ABL transcript in either the blood or bone marrow, suggesting that despite the presence of a vaccine-induced immune response, underlying disease was still present. In the second study patients were immunized with a mixture of wild-type and heterclitic MHC class I-binding peptides combined with native MHC class II-binding peptides corresponding to BCR-ABL breakpoint sequences [82]. Peptides were combined with GM-CSF as adjuvant and were delivered a total of 15 times over a period of 12 months. Of ten patients treated, three achieved a 1-log reduction in BCR-ABL transcript levels, and three other patients achieved a major molecular response. Taken together, these data suggest that heteroclitic peptides can potentially be used to induce T-cell responses to mutated epitopes with otherwise weak HLA binding properties.

Roadmap for the Field

The clinical trials described above with RAS and BCR-ABL vaccines (summarized in Table 7.1) provide proof-of-principle that tumor-specific driver mutations can be targeted immunologically. While clinical responses have only been documented in some patients, one can imagine that continued improvements in vaccine formulations and delivery will lead to even more potent immunological responses followed by increasingly impressive clinical responses. Thus, the concept of creating personalized peptide vaccines based on the mutations identified in individual tumors remains attractive and feasible. Going forward, how can we efficiently prioritize tumor mutations for immunotherapeutic targeting?

Fortunately, the tools associated with epitope prediction are improving rapidly as more mass spectrometry data becomes available describing the repertoire of natural MHC class I-associated peptides. Initially, programs such as the Parker algorithm [83] and SYFPEITHI [84] were designed to predict MHC binding affinity. More advanced algorithms predict not only MHC binding but also proteasomal and immunoproteasomal cleavage and TAP transport efficiency. Many of these epitope prediction programs are available online at sites such as NetCTL [85, 86], NetMHC [87], EpiJen [88] and MAPPP [89]. One of the most comprehensive online resources is the Immune Epitope Database Resource [90], which includes algorithms for each step of MHC class I presentation pathway. The MHC class I binding activity of epitopes that are predicted by the such algorithms can be subsequently validated using synthetic peptides or arrays of peptides. At least two commercial enterprises (ProImmune's Reveal and Prove [91] and BeckmanCoulter's iTopia [92]) offer such analysis on a fee-for-service basis, allowing potential epitopes to be synthesized and validated across a panel of different MHC class I alleles in a matter of weeks.

Once MHC binding has been demonstrated, the next step is to determine whether T-cell responses can be generated against the mutated epitope. Three complementary approaches can be followed: (a) T-cell responses against the mutated peptide of interest can be elicited *in vitro*. Using this approach, peptide-pulsed or antigen-expressing DCs are used to stimulate autologous T cells *in vitro*. This approach is often able to prime tumor antigen-specific T-cell responses, even using PBMC from antigen-naïve, healthy donors [93, 94]; these observations imply that tumor antigen-specific responses have been expanded out of the naïve T-cell repertoire. (b) An alternative to this approach is to use TILs as a source of tumor antigen-reactive T cells. In this latter scenario, a positive response against a mutated tumor antigen would be particularly informative, as it would infer that the mutated epitope is presented at the tumor cell surface by MHC molecules and is recognized by a relevant T cell. Conversely, the inability of TILs to recognize a given tumor-specific mutation should not be construed as a lack of immunogenicity; that is, it is possible that T-cell responses may not have been appropriately primed *in vivo*. (c) As an alternative to *in vitro* T-cell priming, mice that are transgenic for human MHC molecules can serve as a convenient surrogate system to study the potential immunogenicity of tumor-specific mutations. HLA transgenic mice can be immunized with either single peptides or pools of peptides spanning the point mutation of interest, and the responses elicited against each peptide can be measured by IFN-γ ELISPOT or other methods. Although the number of human MHC molecules that can be assessed this way is currently limited, the number of HLA transgenic mouse strains is growing quickly [95].

Once T-cell reactivity has been validated, the final step is to assess whether the mutant epitope is expressed on the tumor cell surface at sufficient levels to sensitize target cells for recognition and destruction. Unfortunately, sufficient quantities of viable tumor cells are often not available for immunological assays. As an alternative, one can attempt to use mass spectrometry to assess whether an epitope is presented by MHC class I or II on the surface of tumor cells. However, current mass spectrometry approaches also require large numbers of cells for epitope analysis; thus, refinements are necessary before routine use of this methodology can be envisioned. Notably, none of the RAS or BCR-ABL clinical trials discussed above incorporated a validation step to ensure that the target epitope was expressed at sufficient levels by the tumor; this lack of validation may in part explain the low frequency of clinical responses. Thus, technological advances at this step are greatly needed to move both pre-clinical and clinical studies forward.

As an alternative to the potentially lengthy series of *in vitro* validation steps described above, a more direct approach would be to immunize patients with a panel of peptides encoding tumor-specific mutations identified through genomic efforts. The advantages of this approach are that peptides can be readily synthesized at GMP-grade for relatively low cost, making them useable as clinical reagents. In addition, direct immunization with candidate peptides would engage the patient's own immune system to directly respond to any immunogenic peptides, thus removing all biases and dramatically shortening the timeline for clinical intervention. The major disadvantage to the direct immunization approach is the potential for off-target autoreactivity. However, even with the extensive *in vitro* validation approaches described above, the potential for autoreactivity in response to vaccination is still relatively unknown until the point of immunization. Thus, for patients with advanced disease, it may be worth considering the risk-benefit ratios of moving directly toward vaccination with peptides containing tumor-specific mutations.

Summary

The vast repertoire of T-cell receptors in the human immune system offers great promise as a means to therapeutically target tumor-specific mutations in cancer. Although the number of estimated mutations per tumor (30–100) may initially seem daunting from an immunological perspective, it is important to remember that not all mutations will meet the necessary criteria for immunogenicity. Moreover, the presence of multiple mutations may actually improve the chance of conferring clinical benefit when targeting a single specific mutation, since tumor killing may cause the release of additional mutated proteins and priming of endogenous immunity against the same. In this regard, there is abundant evidence of natural T-cell recognition of tumor-specific mutations in cancer. Importantly, tumor mutations appear to increase in number as patients undergo standard treatments, owing not only to ongoing tumor evolutionary processes but also to the direct mutagenic effects of radiation and chemotherapy, as indicated by genomic sequencing studies in glioma [7, 96, 97] and breast cancer [5]. Even highly targeted treatments such as the BCR-ABL inhibitor imatinib [98] and the EGF-R inhibitor erlotinib [99] promote the outgrowth of tumor cells harboring drug-resistance mutations. While the acquisition of new mutations can lead to resistance to conventional treatments, such mutations provide a rich source of potential antigens for immunotherapy. Thus, with appropriate therapeutic enhancement, the immune response to cancer can potentially evolve in step with tumors, offering a personalized approach to cancer treatment that far surpasses what can be imagined with current pharmaceutical approaches.

References

1. Bardelli A, Parsons DW et al (2003). Mutational analysis of the tyrosine kinome in colorectal cancers. Science 300:949.
2. Greenman C, Stephens P et al (2007). Patterns of somatic mutation in human cancer genomes. Nature 446:153–158.
3. Ding L, Getz G et al (2008). Somatic mutations affect key pathways in lung adenocarcinoma. Nature 455:1069–1075.
4. Wood LD, Parsons DW et al (2007). The genomic landscapes of human breast and colorectal cancers. Science 318:1108–1113.
5. Shah SP, Morin RD et al (2009). Mutational evolution in a lobular breast tumour profiled at single nucleotide resolution. Nature 461:809–813.
6. Jones S, Zhang X et al (2008). Core signaling pathways in human pancreatic cancers revealed by global genomic analyses. Science 321:1801–1806.

7. Parsons DW, Jones S et al (2008). An integrated genomic analysis of human glioblastoma multiforme. Science 321:1807–1812.
8. Loeb LA, Bielas JH et al (2008). Cancers exhibit a mutator phenotype: clinical implications. Cancer Res 68:3551–3557; discussion 3557.
9. Lengauer C, Kinzler KW et al (1998). Genetic instabilities in human cancers. Nature 396:643–649.
10. Buckowitz A, Knaebel HP et al (2005). Microsatellite instability in colorectal cancer is associated with local lymphocyte infiltration and low frequency of distant metastases. Br J Cancer 92:1746–1753.
11. Lakhani SR (1999). The pathology of familial breast cancer: morphological aspects. Breast Cancer Res 1:31–35.
12. Kuroda H, Tamaru J et al (2005). Immunophenotype of lymphocytic infiltration in medullary carcinoma of the breast. Virchows Arch 446:10–14.
13. Clarke B, Tinker AV et al (2009). Intraepithelial T cells and prognosis in ovarian carcinoma: novel associations with stage, tumor type, and BRCA1 loss. Mod Pathol 22:393–402.
14. Forbes SA, Bhamra G et al (2008). The Catalogue of Somatic Mutations in Cancer (COSMIC). Curr Protoc Hum Genet Chapter 10: Unit 10 11.
15. Schinzel AC, and Hahn WC (2008). Oncogenic transformation and experimental models of human cancer. Front Biosci 13:71–84.
16. Beerenwinkel N, Antal T et al (2007). Genetic progression and the waiting time to cancer. PLoS Comput Biol 3:e225.
17. Stratton MR, Campbell PJ et al (2009). The cancer genome. Nature 458:719–724.
18. Vyas JM, Van der Veen AG et al (2008). The known unknowns of antigen processing and presentation. Nat Rev Immunol 8:607–618.
19. Rangel LB, Agarwal R et al (2004). Anomalous expression of the HLA-DR alpha and beta chains in ovarian and other cancers. Cancer Biol Ther 3:1021–1027.
20. Liu T, Liu W et al (2009). Computational prediction of the specificities of proteasome interaction with antigen protein. Cell Mol Immunol 6:135–142.
21. Horton R, Wilming L et al (2004). Gene map of the extended human MHC. Nat Rev Genet 5:889–899.
22. Bevan MJ (2006). Cross-priming. Nat Immunol 7:363–365.
23. Arstila TP, Casrouge A et al (1999). A direct estimate of the human alphabeta T cell receptor diversity. Science 286:958–961.
24. Echchakir H, Mami-Chouaib F et al (2001). A point mutation in the alpha-actinin-4 gene generates an antigenic peptide recognized by autologous cytolytic T lymphocytes on a human lung carcinoma. Cancer Res 61:4078–4083.
25. Sensi M, and Anichini A (2006). Unique tumor antigens: evidence for immune control of genome integrity and immunogenic targets for T cell-mediated patient-specific immunotherapy. Clin Cancer Res 12:5023–5032.
26. Jia J, Cui J et al (2009). Genome-scale search of tumor-specific antigens by collective analysis of mutations, expressions and T-cell recognition. Mol Immunol 46:1824–1829.
27. Wang HY, Zhou J et al (2002). Identification of a mutated fibronectin as a tumor antigen recognized by $CD4^{+}$ T cells: its role in extracellular matrix formation and tumor metastasis. J Exp Med 195:1397–1406.
28. Gaudin C, Kremer F et al (1999). A hsp70-2 mutation recognized by CTL on a human renal cell carcinoma. J Immunol 162:1730–1738.
29. Pieper R, Christian RE et al (1999). Biochemical identification of a mutated human melanoma antigen recognized by $CD4^{(+)}$ T cells. J Exp Med 189:757–766.
30. Baurain JF, Colau D et al (2000). High frequency of autologous anti-melanoma CTL directed against an antigen generated by a point mutation in a new helicase gene. J Immunol 164:6057–6066.
31. Huang J, El-Gamil M et al (2004). T cells associated with tumor regression recognize frameshifted products of the CDKN2A tumor suppressor gene locus and a mutated HLA class I gene product. J Immunol 172:6057–6064.

32. Linard B, Bezieau S et al (2002). A RAS-mutated peptide targeted by CTL infiltrating a human melanoma lesion. J Immunol 168:4802–4808.
33. Robbins PF, El-Gamil M et al (1996). A mutated beta-catenin gene encodes a melanoma-specific antigen recognized by tumor infiltrating lymphocytes. J Exp Med 183:1185–1192.
34. Novellino L, Renkvist N et al (2003). Identification of a mutated receptor-like protein tyrosine phosphatase kappa as a novel, class II HLA-restricted melanoma antigen. J Immunol 170:6363–6370.
35. Kawakami Y, Wang X et al (2001). Isolation of a new melanoma antigen, MART-2, containing a mutated epitope recognized by autologous tumor-infiltrating T lymphocytes. J Immunol 166:2871–2877.
36. Karanikas V, Colau D et al (2001). High frequency of cytolytic T lymphocytes directed against a tumor-specific mutated antigen detectable with HLA tetramers in the blood of a lung carcinoma patient with long survival. Cancer Res 61:3718–3724.
37. Mami-Chouaib F, Echchakir H et al (2002). Antitumor cytotoxic T-lymphocyte response in human lung carcinoma: identification of a tumor-associated antigen. Immunol Rev 188:114–121.
38. Zorn E, and Hercend T (1999). A natural cytotoxic T cell response in a spontaneously regressing human melanoma targets a neoantigen resulting from a somatic point mutation. Eur J Immunol 29:592–601.
39. Wang HY, Peng G et al (2005). Recognition of a new ARTC1 peptide ligand uniquely expressed in tumor cells by antigen-specific $CD4^+$ regulatory T cells. J Immunol 174:2661–2670.
40. Coulie PG, Lehmann F et al (1995). A mutated intron sequence codes for an antigenic peptide recognized by cytolytic T lymphocytes on a human melanoma. Proc Natl Acad Sci USA 92:7976–7980.
41. Topalian SL, Gonzales MI et al (2002). Revelation of a cryptic major histocompatibility complex class II-restricted tumor epitope in a novel RNA-processing enzyme. Cancer Res 62:5505–5509.
42. Maccalli C, Li YF et al (2003). Identification of a colorectal tumor-associated antigen (COA-1) recognized by $CD4^{(+)}$ T lymphocytes. Cancer Res 63:6735–6743.
43. Van Elsas A, Nijman HW et al (1995). Induction and characterization of cytotoxic T-lymphocytes recognizing a mutated p21RAS peptide presented by HLA-A*0201. Int J Cancer 61:389–396.
44. Qin H, Chen W et al (1995). $CD4^+$ T-cell immunity to mutated RAS protein in pancreatic and colon cancer patients. Cancer Res 55:2984–2987.
45. Abrams SI, Hand PH et al (1996). Mutant RAS epitopes as targets for cancer vaccines. Semin Oncol 23:118–134.
46. Sharkey MS, Lizee G et al (2004). $CD4^{(+)}$ T-cell recognition of mutated B-RAF in melanoma patients harboring the V599E mutation. Cancer Res 64:1595–1599.
47. Schwitalle Y, Linnebacher M et al (2004). Immunogenic peptides generated by frameshift mutations in DNA mismatch repair-deficient cancer cells. Cancer Immun 4:14.
48. Ripberger E, Linnebacher M et al (2003). Identification of an HLA-A0201-restricted CTL epitope generated by a tumor-specific frameshift mutation in a coding microsatellite of the OGT gene. J Clin Immunol 23:415–423.
49. Linnebacher M, Gebert J et al (2001). Frameshift peptide-derived T-cell epitopes: a source of novel tumor-specific antigens. Int J Cancer 93:6–11.
50. Worley BS, van den Broeke LT et al (2001). Antigenicity of fusion proteins from sarcoma-associated chromosomal translocations. Cancer Res 61:6868–6875.
51. Yotnda P, Garcia F et al (1998). Cytotoxic T cell response against the chimeric ETV6-AML1 protein in childhood acute lymphoblastic leukemia. J Clin Invest 102:455–462.
52. Yotnda P, Firat H et al (1998). Cytotoxic T cell response against the chimeric p210 BCR-ABL protein in patients with chronic myelogenous leukemia. J Clin Invest 101:2290–2296.
53. Nieda M, Nicol A et al (1998). Dendritic cells stimulate the expansion of bcr-abl specific $CD8^+$ T cells with cytotoxic activity against leukemic cells from patients with chronic myeloid leukemia. Blood 91:977–983.

54. Bosch GJ, Joosten AM et al (1996). Recognition of BCR-ABL positive leukemic blasts by human CD4+ T cells elicited by primary *in vitro* immunization with a BCR-ABL breakpoint peptide. Blood 88:3522–3527.
55. Reddy EP, Reynolds RK et al (1982). A point mutation is responsible for the acquisition of transforming properties by the T24 human bladder carcinoma oncogene. Nature 300:149–152.
56. Tabin CJ, Bradley SM et al (1982). Mechanism of activation of a human oncogene. Nature 300:143–149.
57. Fossum B, Olsen AC et al (1995). CD8+ T cells from a patient with colon carcinoma, specific for a mutant p21-RAS-derived peptide (Gly13-->Asp), are cytotoxic towards a carcinoma cell line harbouring the same mutation. Cancer Immunol Immunother 40:165–172.
58. Gjertsen MK, Bjorheim J et al (1997). Cytotoxic CD4+ and CD8+ T lymphocytes, generated by mutant p21-RAS (12Val) peptide vaccination of a patient, recognize 12Val-dependent nested epitopes present within the vaccine peptide and kill autologous tumour cells carrying this mutation. Int J Cancer 72:784–790.
59. Juretic A, Jurgens-Gobel J et al (1996). Cytotoxic T-lymphocyte responses against mutated p21 RAS peptides: an analysis of specific T-cell-receptor gene usage. Int J Cancer 68:471–478.
60. Gjertsen MK, Bakka A et al (1995). Vaccination with mutant RAS peptides and induction of T-cell responsiveness in pancreatic carcinoma patients carrying the corresponding RAS mutation. Lancet 346:1399–1400.
61. Gjertsen MK, Buanes T et al (2001). Intradermal RAS peptide vaccination with granulocyte-macrophage colony-stimulating factor as adjuvant: clinical and immunological responses in patients with pancreatic adenocarcinoma. Int J Cancer 92:441–450.
62. Khleif SN, Abrams SI et al (1999). A phase I vaccine trial with peptides reflecting RAS oncogene mutations of solid tumors. J Immunother 22:155–165.
63. Abrams SI, Khleif SN et al (1997). Generation of stable CD4+ and CD8+ T cell lines from patients immunized with RAS oncogene-derived peptides reflecting codon 12 mutations. Cell Immunol 182:137–151.
64. Toubaji A, Achtar M et al (2008). Pilot study of mutant RAS peptide-based vaccine as an adjuvant treatment in pancreatic and colorectal cancers. Cancer Immunol Immunother 57:1413–1420.
65. Meyer RG, Korn S et al (2007). An open-label, prospective phase I/II study evaluating the immunogenicity and safety of a RAS peptide vaccine plus GM-CSF in patients with non-small cell lung cancer. Lung Cancer 58:88–94.
66. Carbone DP, Ciernik IF et al (2005). Immunization with mutant p53- and K-RAS-derived peptides in cancer patients: immune response and clinical outcome. J Clin Oncol 23:5099–5107.
67. Hunger RE, Brand CU et al (2001). Successful induction of immune responses against mutant RAS in melanoma patients using intradermal injection of peptides and GM-CSF as adjuvant. Exp Dermatol 10:161–167.
68. Shtivelman E, Lifshitz B et al (1986). Alternative splicing of RNAs transcribed from the human ABL gene and from the BCR-ABL fused gene. Cell 47:277–284.
69. Bocchia M, Wentworth PA et al (1995). Specific binding of leukemia oncogene fusion protein peptides to HLA class I molecules. Blood 85:2680–2684.
70. Bocchia M, Korontsvit T et al (1996). Specific human cellular immunity to BCR-ABL oncogene-derived peptides. Blood 87:3587–3592.
71. Greco G, Fruci D et al (1996). Two BRC-ABL junction peptides bind HLA-A3 molecules and allow specific induction of human cytotoxic T lymphocytes. Leukemia 10:693–699.
72. Buzyn A, Ostankovitch M et al (1997). Peptides derived from the whole sequence of BCR-ABL bind to several class I molecules allowing specific induction of human cytotoxic T lymphocytes. Eur J Immunol 27:2066–2072.
73. Clark RE, Dodi IA et al (2001). Direct evidence that leukemic cells present HLA-associated immunogenic peptides derived from the BCR-ABL b3a2 fusion protein. Blood 98:2887–2893.
74. ten Bosch GJ, Toornvliet AC et al (1995). Recognition of peptides corresponding to the joining region of p210BCR-ABL protein by human T cells. Leukemia 9:1344–1348.

75. Pawelec G, Max H et al (1996). BCR/ABL leukemia oncogene fusion peptides selectively bind to certain HLA-DR alleles and can be recognized by T cells found at low frequency in the repertoire of normal donors. Blood 88:2118–2124.
76. Mannering SI, McKenzie JL et al (1997). HLA-DR1-restricted bcr-abl (b3a2)-specific CD4+ T lymphocytes respond to dendritic cells pulsed with b3a2 peptide and antigen-presenting cells exposed to b3a2 containing cell lysates. Blood 90:290–297.
77. Pinilla-Ibarz J, Cathcart K et al (2000). Vaccination of patients with chronic myelogenous leukemia with BCR-ABL oncogene breakpoint fusion peptides generates specific immune responses. Blood 95:1781–1787.
78. Cathcart K, Pinilla-Ibarz J et al (2004). A multivalent BCR-ABL fusion peptide vaccination trial in patients with chronic myeloid leukemia. Blood 103:1037–1042.
79. Bocchia M, Gentili S et al (2005). Effect of a p210 multipeptide vaccine associated with imatinib or interferon in patients with chronic myeloid leukaemia and persistent residual disease: a multicentre observational trial. Lancet 365:657–662.
80. Rojas JM, Knight K et al (2007). Clinical evaluation of BCR-ABL peptide immunisation in chronic myeloid leukaemia: results of the EPIC study. Leukemia 21:2287–2295.
81. Maslak PG, Dao T et al (2008). A pilot vaccination trial of synthetic analog peptides derived from the BCR-ABL breakpoints in CML patients with minimal disease. Leukemia 22:1613–1616.
82. Jain N, Reuben JM et al (2009). Synthetic tumor-specific breakpoint peptide vaccine in patients with chronic myeloid leukemia and minimal residual disease: a phase 2 trial. Cancer 115:3924–3934.
83. Parker KC, Bednarek MA et al (1994). Scheme for ranking potential HLA-A2 binding peptides based on independent binding of individual peptide side-chains. J Immunol 152:163–175.
84. Rammensee H, Bachmann J et al (1999). SYFPEITHI: database for MHC ligands and peptide motifs. Immunogenetics 50:213–219.
85. Larsen MV, Lundegaard C et al (2007). Large-scale validation of methods for cytotoxic T-lymphocyte epitope prediction. BMC Bioinformatics 8:424.
86. Larsen MV, Lundegaard C et al (2005). An integrative approach to CTL epitope prediction: a combined algorithm integrating MHC class I binding, TAP transport efficiency, and proteasomal cleavage predictions. Eur J Immunol 35:2295–2303.
87. Lundegaard C, Lamberth K et al (2008). NetMHC-3.0: accurate web accessible predictions of human, mouse and monkey MHC class I affinities for peptides of length 8–11. Nucleic Acids Res 36:W509–512.
88. Doytchinova IA, Guan P et al (2006). EpiJen: a server for multistep T cell epitope prediction. BMC Bioinformatics 7:131.
89. Hakenberg J, Nussbaum AK et al (2003). MAPPP: MHC class I antigenic peptide processing prediction. Appl Bioinformatics 2:155–158.
90. Zhang Q, Wang P et al (2008). Immune epitope database analysis resource (IEDB-AR). Nucleic Acids Res 36:W513–518.
91. Westrop SJ, Grageda N et al (2009). Novel approach to recognition of predicted HIV-1 Gag B3501-restricted CD8 T-cell epitopes by HLA-B3501(+) patients: confirmation by quantitative ELISpot analyses and characterisation using multimers. J Immunol Meth 341:76–85.
92. Wulf M, Hoehn P et al (2009). Identification of human MHC class I binding peptides using the iTOPIA- epitope discovery system. Meth Mol Biol 524:361–367.
93. Wilson CC, Olson WC et al (1999). HIV-1-specific CTL responses primed *in vitro* by blood-derived dendritic cells and Th1-biasing cytokines. J Immunol 162:3070–3078.
94. Tuting T, Wilson CC et al (1998). Autologous human monocyte-derived dendritic cells genetically modified to express melanoma antigens elicit primary cytotoxic T cell responses *in vitro*: enhancement by cotransfection of genes encoding the Th1-biasing cytokines IL-12 and IFN-alpha. J Immunol 160:1139–1147.
95. Pascolo S (2005). HLA class I transgenic mice: development, utilisation and improvement. Expert Opin Biol Ther 5:919–938.

96. Hunter C, Smith R et al (2006). A hypermutation phenotype and somatic MSH6 mutations in recurrent human malignant gliomas after alkylator chemotherapy. Cancer Res 66:3987–3991.
97. The Cancer Genome Atlas (2008). Comprehensive genomic characterization defines human glioblastoma genes and core pathways. Nature 455:1061–1068.
98. O'Hare T, Eide CA et al (2008). New BCR-ABL inhibitors in chronic myeloid leukemia: keeping resistance in check. Expert Opin Investig Drugs 17:865–878.
99. Yamamoto H, Toyooka S et al (2009). Impact of EGFR mutation analysis in non-small cell lung cancer. Lung Cancer 63:315–321.

Chapter 8
Counteracting Subversion of MHC Class II Antigen Presentation by Tumors

Jacques Thibodeau, Marie-Claude Bourgeois-Daigneault, and Réjean Lapointe

Abstract The success of immunotherapy against human cancers relies on somewhat ill-defined correlates. While a role for $CD4^+$ T lymphocytes in the control of tumor growth is well established, we are still looking for ways to harness the MHC class II antigen presentation pathway for the development of an efficient immune response. Learning the mechanisms by which tumors circumvent the immune responses is the first step towards the development of cell-based vaccines. Here, we discuss the variability of MHC II expression by tumor cells and the impact on the immune response. Also, we address how tumor cells or dendritic cells can be modified *ex vivo* to activate circulating tumor-specific T cells in the fight against cancers.

Keywords Dendritic cell • HLA-DM • MHC • Tumor • Vaccine

Tumors and the Immune System

For decades the idea of curing cancers through immunotherapy has been a motivation to immunologists [1]. The capacity of the immune system to mount an effective antitumor response has been established in a number of experimental systems. The immunogenic nature of tumors was clearly demonstrated by successfully preventing growth of a transplanted, chemically-induced tumor following vaccination of syngeneic mice with killed cancer cells. Accordingly, solid tumors, stroma cells, and the neighboring tissues are generally infiltrated by a panoply of immune cells, including members of both the adaptive and innate immunity arms. In humans, the immune system has been harnessed in the fight against cancers. However, the results of the first-generation of immunotherapy in clinical trials have not met early expectations [2]. Impressive credible tumor regressions have been reported in some patients;

J. Thibodeau (✉)
Laboratoire d'Immunologie Moléculaire, Département de Microbiologie et Immunologie, Université de Montréal, CP 6128 Succ Centre ville, Montréal, QC, Canada
e-mail: jacques.thibodeau@umontreal.ca

J. Medin and D. Fowler (eds.), *Experimental and Applied Immunotherapy*,
DOI 10.1007/978-1-60761-980-2_8, © Springer Science+Business Media, LLC 2011

however, future advances will require an improved understanding of the intermediate immunological surrogates associated with such responses. Understanding the reasons why the immune system has in general proven to be insufficient holds the key to the development of more efficacious anticancer vaccines, be they therapeutic or prophylactic. The frequent inertia of our defense system may be due to the existence of numerous pathways leading to the development, establishment, and dissemination of cancers, as well as to the genetic variability in the human population. Also, we have only begun to fully appreciate the complexity of the interplay between tumor cells, their immediate environment (stroma), and the immune system [3]. It is clear that some tumors are not only passively invisible to an otherwise therapeutic immune system but also actively neutralize the body's anticancer artillery.

Why only few tumors become immunogenic and targets for the immune system? The answer to this question depends, at least in part, on the fact that the life-cycle of cancer cells often depends on the aberrant expression of molecules that become recognized as foreign antigens (Ags) by T cells. Impairment in the presentation or recognition of peptides derived from these tumor-associated antigens (TAAs) in the context of either MHC class I or class II molecules favors tumor cell evasion from the immune system.

In this day and age of proteomic, genomic, and other "omic" approaches, detailed characteristics of tumor cells and the extent to which they differ from normal cells is increasingly apparent. Such differences are particularly important given the fact that the adaptive immune system is sensitive to the "foreign" nature of the tumor [4]. Coupled with the capacity of discriminating dangerous from innocuous new encounters, the "self" versus "nonself" recognition of tissue components is the cornerstone of the immune system's evolution. Our defense mechanisms culminate with the destruction of available material into smaller pieces to scan proteins and lipids that have never been seen, thereby identifying foreign material that could activate an alarmed immune system. The viral origin of cancer has been established in only a minority of cases, and as such, tumor presentation to the immune system is undoubtedly more typical of self identification rather than microbial identification. Mechanisms developed by viruses to subvert the immune system have been the subject of many excellent reviews in recent years (for example, see [5]); this aspect of tumor biology will not be addressed in this chapter. By analogy to the antiviral response, it is precisely the subtle changes that distinguish tumor cells from normal cells that we must identify in order not only to better understand carcinogenesis, but also to design effective immune therapies. Indeed, like self antigens, tumors do not induce a significant danger signal to the immune system; the mechanisms underlying this lack of danger signals are multiple and complex, yet must be overcome because efficient therapies will undoubtedly rely on a potent adaptive immune response.

Many tumors do not express MHC molecules but most cell types, including those of non-bone marrow origin, upregulate the MHC II antigen presentation machinery in the presence of IFN-γ [6]. Depending on the cell type, MHC expression can be further modulated positively or negatively by TNF-α [7]. The presence of such inflammatory cytokines, which are typically present in the tumor microenvironment, is primarily dictated either directly or indirectly by the adaptive T-cell response.

The current generation of researchers is certainly aware that the fight against cancer will be most likely to succeed if the attack comes from multiple directions simultaneously. CD4⁺ T helper (Th) cells, which are frequently identified in the battlefield that constitutes the tumor microenvironment, represent a cell population of particular interest in terms of attempts to modulate tumor biology. This review will address the importance of Th cells in immunotherapy and summarize current knowledge of the MHC II antigen processing and presentation machinery in tumors.

Role of Adaptive $CD4^+$ T-Cell Responses in Tumor Eradication

The evidence for a role of $CD4^+$ T cells in the antitumor response in mouse and human is compelling (see recent review, [8]). Tumor immunity following immunization with tumor cells or specific peptides relies on a functional $CD4^+$ T-cell effector compartment, even in the case of MHC class II-negative tumors [9]. Indeed, early experiments suggested that the need for Th cells in the antitumor response could be bypassed when tumor cells were engineered to secrete IL-2 [10], which is the prototypical cytokine produced by Th cells. Accordingly, tumor-infiltrating $CD4^+$ T lymphocytes (TILs) from a variety of human tumors such as melanoma have been shown to secrete arrays of cytokines when cocultivated with autologous cancer cells. Such TAA-specific lymphocytes, which can secrete either Th1, Th2 or a mixed pattern of cytokines, provide help to distal immune effectors, including DCs, eosinophils, macrophages, NK cells, and cytotoxic T cells [11]. Very recently, a role for Th cells in the direct mobilization of effector CTLs to some virus-infected tissues has been demonstrated and such interplay may also prove to be critical in some cancers [12]. However, an early study found no differences in the nature of the inflammatory infiltrate between HLA-DR positive and negative breast tumors, suggesting that activation of T cells by TAAs occurred on professional APCs [13]. Although the role of tumor-specific antibodies in controlling tumor growth can be debated, it should be noted that Th cells are critical for effective antibody production using vaccines that contain a suitable Th epitope (for example, see [14]). In general, it appears that a Th1-type cytokine profile, which is characterized by IFN-γ and IL-2 secretion, is the preferred Th cytokine phenotype for antitumor immunity [10]. The exact role of Th17 cells, which represent a more recently described Th subset that also secretes an inflammatory pattern of cytokines, remains to be established. These cells have been identified in the mice tumor microenvironment and studies using IL-17-deficient mice suggest that Th17 cells may either promote or prevent tumor growth [15, 16].

In addition to the release of soluble mediators that can act in a paracrine fashion, Th cells mediate some biological functions through cell-cell contacts. Such important effector functions are exemplified by the priming of CTLs by APC activation through the CD40 pathway [17]. In addition, it is well known that $CD4^+$ Th cells can directly mediate cytotoxicity against tumor cells [18]; the role of Th cytotoxicity in determining antitumor immunity certainly deserves further exploration. From a negative regulatory standpoint, it is also clear now that Th cell activation that results in the generation of regulatory T cells (Tregs) can represent a significant drawback. Indeed, Tregs specific for the immunogen could arise and inhibit the antitumor response. While experimental conditions for the development of such cells can be avoided *in vitro*, pre-existing or therapy-generated Tregs can pose limit to the success of immunotherapy [19].

Th cells therefore help dictate the efficacy of the antitumor response, and as such, this information should be informative for the design of therapeutic vaccines that augment the Th arm of the immune response. The search for TAAs and their encoded T-cell epitopes has intensified in recent years; epitopes in TAAs such as tyrosinase, MAGE, NY-ESO-1, and gp100 have each been implemented [20]. Using this knowledge, many therapeutic cancer vaccines aimed at stimulating T cell help are being developed. In lung and melanoma cancer patients, there are clear indications that vaccination with the MAGE-3 tumor antigen induces $CD4^+$ T-cell responses [21]. Despite such evidence in support of the role of helper T-cell responses, clinical trial results in melanoma patients injected with both class I- and class II-restricted peptides yielded discordant results as to the impact of the class II epitope [22, 23]. These last studies highlight the importance of carefully monitoring the CD4 response and continuing the search for optimal antigens and vaccine delivery methods.

Tumor Cells as APCs

The debate as to the importance of tumor immunosurveillance still continues [24]. The immunogenicity of cancer cells, albeit typically weak, has certainly been demonstrated. Despite the fact that MHC class I and II negative cancer cells can be eliminated in some experimental systems [25], tumor cell loss of MHC molecules results in a growth advantage, thereby illustrating that the adaptive immune system exerts a pressure against tumor progression. Thus, tumors are able at some point in their natural history to present antigens and act as APCs. However, optimal activation of naïve T cells also requires the capture of tumor antigens by surrounding APCs; such APC can then home to regional lymph nodes and cross-present tumor Ags for the subsequent activation of $CD8^+$ T cells [26]. Then, once effector cells return to the tissue, MHC class I-positive tumor cells are capable of being recognized and attacked.

The same concepts as above apply to the role of MHC II molecules in immunosurveillance. However, the tissue distribution of MHC II molecules is restricted relative to the more ubiquitous expression of MHC I; that is, many solid tumors do not express MHC II molecules. Thus, involvement of $CD4^+$ T cells is mainly dependent on infiltrating APCs that pick up available antigens or that phagocytose tumor cells. For example, adoptive transfer of experienced $CD4^+$ T cells can induce regression of an established MHC class II-negative tumor. This observation suggested that professional APCs were able to process and present tumor antigens [27]. The IL-2 and IFN-γ-producing T cells present in the vicinity of the tumor help create an inflammatory DTH-type of environment, thereby facilitating tumor clearance. Interestingly, many tumors such as melanoma or glioma will express MHC II molecules in the presence of IFN-γ or as a result of tumorigenesis; such tumor cells would therefore be capable of presenting native TAAs to MHC II-restricted Th1 cells [28]. As such, these tumor cells may directly present endogenous antigens and become targets of the antitumor response. A tumor expressing MHC II molecules could amplify the immune response and even present new T-cell epitopes [29, 30]. The regulation of the MHC II antigen presentation machinery in cancer cells will be discussed below.

Why Does the Antitumor T-Cell Response Often Prove Defective?

It is unlikely that a cancer cell will be totally devoid of TAA expression during tumorigenesis. Activation of αβ T cells specific for TAAs requires the processing of proteins and the display of immunogenic peptides on MHC molecules. In this review, we start from the premise that tumors indeed harbor TAAs, which include $CD4^+$ T-cell epitopes amenable to immunotherapy. Considering the diversity of defense mechanisms that contribute to antitumor immunity, it is surprising that spontaneously arising cancer cells can proliferate to an extent that is lethal to the host. Many review articles have addressed the issue of escape from antitumor immunity in-depth (see [31, 32], for example). The same immune evasion mechanisms are likely to blunt immunotherapy efforts. Such mechanisms include: the presence of an increased number of regulatory T cells; reduced tumor cell expression of adhesion or co-stimulatory molecules; increased tumor cell expression of FasL; the presence of inhibitory factors or regulatory cytokines such as IDO, TGF-β, and IL-10; and altered signal transduction pathways in tumor-infiltrating T cells, leading to T-cell unresponsiveness [33, 34]. The interplay between tumor, stroma, and immune cells also needs to be dissected. However, features of the abortive immune responses mentioned above are covered in other chapters of this text, and as such, we have focused our review on MHC class II antigen processing and peptide display.

Subversion of MHC II Antigen Presentation in Tumors

Overview of the MHC-II Antigen Presentation Pathway

MHC class II molecules are heterodimers composed of two glycosylated transmembrane chains (a and b) [35]. As opposed to MHC class I molecules, classical MHC class II molecules (HLA-DR, -DP, and -DQ) do not associate with peptides in the ER [35]. Although peptide binding is possible in this compartment, it is prevented by the presence of the invariant chain (Ii). This chaperone is expressed at high levels and associates with folding MHC II molecules, occupying the peptide binding groove and preventing aggregation [36]. MHC-II molecules subsequently exit the ER, and in the endosomes, Ii is degraded by a panoply of proteases that ultimately leave only a short class II-associated invariant chain peptide (CLIP) inside the groove of the MHC-II molecules.

The peptide binding groove of most MHC II alleles must be freed by the action of the nonclassical MHC-II molecule called HLA-DM [37] (Fig. 8.1). The latter

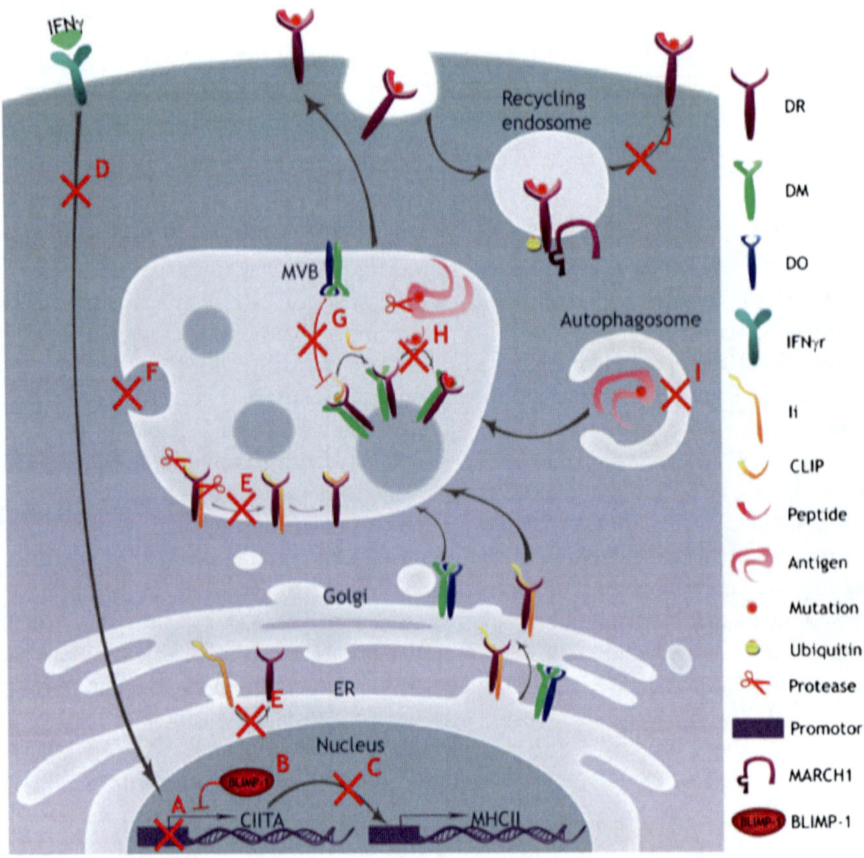

Fig. 8.1 Subversion of the MHC II antigen presentation pathway in tumor cells. Some tumor cells do not express MHC II molecules due to epigenetic events affecting the gene promoter (*A*). Some tumors see their expression of MHC II molecules shut down due to the interplay between Blimp-1 and CIITA in the nucleus (*B*, *C*). Most cells up-regulate MHC II expression in response to IFN-g but this pathway is blunted in many tumors (*D*). The absence of Ii, or the presence of nonphosphorylated Iip35 could prohibit MHC II egress from the endoplasmic reticulum. Lack of Ii cleavage would result in retention of MHC II molecules in the endocytic pathway (*E*). Mutations in the machinery responsible for the formation of multivesicular bodies (*MVBs*) may inhibit the transfer of HLA-DR and HLA-DM to the internal vesicles where peptide loading occurs (*F*). Overexpression of HLA-DO would also inhibit the sorting of HLA-DM to internal membranes (*G*). Lack of available peptides from TAAs (sometimes bearing a mutation creating a new T-cell epitope) could be the result of increased proteolytic activity in endocytic compartments (*H*). However, some TAAs may never gain access to the MHC II-rich compartments due to inefficient autophagy (*I*). Finally, MARCH1 expression would lead to the internalization and degradation of mature MHC II-peptide complexes

molecule is nonpolymorphic and encodes a cytoplasmic lysosomal sorting signal [37, 38]. HLA-DM also assists peptide loading by stabilizing empty MHC-II molecules [39]. In most resting APCs, the function of HLA-DM is regulated by HLA-DO [40]. The purpose of HLA-DO is not clearly established but has been shown to negatively regulate the function of HLA-DM in early compartments of the endocytic pathway. Thus, in B cells, HLA-DO favors the presentation of antigens specifically endocytosed through the B-cell receptor and which are degraded in late acidic vesicles [40, 41].

Antigens, whether self or foreign, are degraded indiscernibly in the endocytic pathway. Most MHC II molecules are loaded with self antigens at any given point in time. Endogenous TAAs will gain access to MHC-II loading compartments by many different means. For example, transmembrane proteins from the plasma membrane will be endocytosed and sent to lysosomes for degradation. Cytoplasmic and nuclear antigens can be engulfed by autophagy and find themselves in the presence of classical MHC II molecules and HLA-DM [42]. The MHC class II antigen processing pathway can undergo significant modification as part of tumorigenesis, thereby precluding efficient presentation of T-cell epitopes. The next sections will highlight some of these aberrations reported in cancer cells.

Patterns of MHC Class II Expression in Tumor Cells

Unusual HLA expression has been reported in many different cancers [43]. Total or partial loss of MHC class I protein levels is a common trait among human neoplasms and has been associated with rapid growth, tumor evasion, and metastasis in various tumors. The reasons for this are numerous and include: deficiencies in the key players of antigen processing pathway, and the occurrence of epigenetic events [44]. Importantly, such phenomenon can be progressive and exacerbated as a result of selective pressure by the immune system or by immunotherapy [45].

On the other hand, most normal tissues are usually devoid of HLA class II antigens, except in pathological conditions such as inflammation and autoimmunity. In the past 30 years or so, research has intensified in order to describe and characterize mechanistically the patterns of MHC class II expression on human and mouse tumor cell lines or primary samples of various origins. Results varied greatly between tumors of a given origin for different patients; as such, the prognostic value of MHC II expression is certainly not universal. Because of the large body of literature on this matter, only a few studies are described in detail below to exemplify important concepts underlying regulation of the MHC II pathway in tumors.

Tumors such as the ones derived from colorectal or breast tissue often express MHC II molecules; however, correlation of such expression with clinical outcome has not been readily apparent [46]. The breast epithelium does not typically express MHC class II molecules, and as such, the MHC expression phenotype is thought to arise in response to hormones or cytokines [47]. On the other hand,

MHC II⁺ cells often lose expression of essential components of this pathway. For example, B-cell lymphomas of high-grade malignancy are sometimes negative for MHC II [48]. Also, differential constitutive or inducible expression of MHC II isotypes, principally DR and DQ, is common and has been described in many tumor types [49]. In other cell types such as larynx, rectal, and breast carcinomas, however, some studies did not correlate the expression of MHC II with a better prognosis [50, 51]. Functional studies have addressed the capacity of MHC II$^+$ tumor cells to present antigens. For example, despite high levels of surface MHC II molecules, peripheral blood B cells from B-CLL patients were shown to be poor stimulators in mixed lymphocyte reactions (MLR) and have reduced capacity to present a model soluble antigen [52]. Altogether, these results suggest that the impact of MHC II molecules on the final outcome for the patient will be the result of a delicate balance between intrinsic tumor factors and the capacity to generate either an efficient immune response or tolerance.

The causes for the appearance of different phenotypes across tumors or individuals with similar malignancies remain nebulous but likely involve both transcriptional and posttranscriptional mechanisms. Lung tumor cell lines were shown to vary in the constitutive and inducible expression of MHC II [53]. As a general rule, genes involved in MHC II antigen presentation are coregulated by CIITA [54]. Aberrant overtranscription of MHC class II antigen presentation genes in B-CLL has been reported and correlated with enhanced expression of CIITA [55]. This transcription factor binds to shared promoter elements involved in constitutive MHC II expression as well as in IFN-γ-mediated induction. Some tumors do not up-regulate MHC II molecules in response to IFN-γ; this functional deficit may arise from problems at various levels, including the CIITA gene transcription, mRNA translation, or protein stability [56].

The CIITA gene is itself down-regulated by Blimp-1, a transcription regulator expressed in plasma cells. Tumors of the B-cell compartment usually display MHC II molecules at the cell surface. However, situations occur where MHC II expression is reduced, such as in cases of diffuse large B-cell lymphoma [57]. It is currently not clear if this is due to observed over-expression of Blimp-1 [58]. Clearly, expression of Blimp-1 does not predict IFN-γ response as CIITA expression is up-regulated in multiple myeloma cells by IFN-γ [59].

Finally, the display of MHC II molecules may be regulated indirectly by modifications in the endocytic pathway of tumors or directly by the interaction with chaperones such as Ii or MARCH ubiquitin ligases. MARCH1 and 8 have been shown to add ubiquitin to the cytoplasmic tail of MHC-II molecules, causing their intracellular sequestration and degradation. While MARCH8 is ubiquitously expressed, MARCH1 is inducible by IL-10 in monocytes and down-regulated by TLR4 stimulation in DCs. [60, 61] MARCH1 is also expressed in resting murine [62] and human B lymphocytes (Lapointe R., Steimle V., and Thibodeau J., unpublished data). Future studies will establish if a link exists in some tumors between the poor MHC-II display and the presence of MARCH proteins. Clearly however, IL-10 produced locally by some tumor cells and T-regulatory cells may affect MARCH1 expression and MHC II display.

Patterns of Ii Expression in Tumor Cells

Early studies revealed the existence of an "Ia-associated invariant chain" (CD74 or Ii) that co-precipitated with MHC-II molecules [35]. Overall, in normal and neoplastic cells, the pattern of Ii expression correlates with the one of MHC II molecules, with no expression at the final stage of B-cell maturation [63]. However, further analysis revealed numerous examples of discoordinate expression between the two molecules (see [64], for example). Although the Ii gene shares common, CIITA-dependent, regulatory elements with MHC II genes, the human and mouse Ii promoters also contain two functional NF-kB/Rel binding sites that either activate or inhibit expression depending on the cell type [65].

The level of expression, the proportion of isoforms, and the presence of cleavage products are some of the variables associated with Ii expression in various tumor types and patients. In humans, Ii exists in four forms that originate from alternative splicing and alternative initiation of translation [36]. The Iip35 isoform is translated from the most 5′ AUG triplet and encodes an RxR (Arg-x-Arg) ER retention motif that is masked upon MHC II binding and Ii phosphorylation by PKC [36]. Intriguingly, in hairy cell leukemia (HCL) and some B-CLL, high levels of Ii, especially Iip35, are found [66]. This correlates with an increase in the proportion of SDS-stable, compact MHC II molecules containing Iip35. The significance of this finding is not known but it was postulated that formation of such a complex would prevent binding of endogenous tumor antigens [67]. Hairy leukemic cells showed alterations in the expression of various cleavage products or posttranslationally modified forms of Ii [68].

Furthermore, expression of Ii on renal cell cancers correlated with the degree of lymphocyte infiltration [69]. Paradoxically, high levels of Ii correlated with less lymphocytic infiltration and poor prognosis in high-grade tumors of the colon as well as in gastric carcinoma [70]. Also, recent data has shown that patients with pancreatic ductal adenocarcinoma displaying lower expression of Ii had a favorable survival rate [71].

The impact of Ii on endogenous antigen presentation by MHC class II molecules has been principally addressed in the context of tumor vaccines. It was shown in 2008 that tumor cells genetically engineered to express MHC II molecules are very efficient in activating the immune system, provided that they do not express Ii [72]. It is assumed that in the absence of Ii, the palette of antigens, including TAAs, capable of binding MHC II molecules increases over a wider range of compartments [73]. Also, expression of Ii may alter the presentation of antigens by MHC I [74].

Patterns of HLA-DM and -DO Expression in Tumors

The action of HLA-DM and HLA-DO will affect the level of CLIP at the cell surface [75]. CLIP located in the groove of classical MHC-II molecules can be detected by flow cytometry using specific mAbs [76]. Given that Ii is normally expressed, there is either an inverse or direct correlation between CLIP levels and that of HLA-DM or HLA-DO, respectively. As CLIP prevents the binding of antigenic peptides, these nonclassical chaperones will have a definitive impact on the immune response [75]. Still, the importance of HLA-DM is debated as it was shown in a mouse model that tumor cells transfected with MHC II molecules without Ii, or with Ii and HLA-DM are highly immunogenic [29]. Thus, it is likely that HLA-DM is critical only in the context of Ii expression.

HLA-DM is coregulated with HLA-DR [77] and low CLIP occupancy of MHC II molecules has been reported in a number of malignancies. For example, some Burkitt's lymphoma and MHC class II$^+$ breast carcinoma cells were shown to display little CLIP in association with HLA-DR or HLA-DQ [78, 79]. Recently, tumor cell expression of HLA-DM was associated with a Th1 cytokine profile and was shown to predict improved survival in breast carcinoma patients [80]. It was suggested that HLA-DM reduces CLIP at the cell surface, thereby avoiding Th2 polarization, which is promoted with high levels of CLIP expression [81]. More recently, microarray analysis of ovarian cancer cells revealed that high HLA-DMβ expression correlated with improved survival [82]. Accordingly, some pre-B ALL (ETV6-AML1) show only little CLIP; it was postulated that such cells would induce a favorable immune response, thereby explaining the delayed relapse of malignancy in these patients [83]. On the other hand, other cells such as Reed-Sternberg cells in malignant Hodgkin's disease or myeloid leukemic blasts present high levels of CLIP, which in the latter case predicts a poor clinical outcome [84, 85]. Interestingly, it has been reported that MHC-II-CLIP complexes become the target of autoreactive T cells in cyclosporine-treated animals receiving an autologous bone marrow transplant [86]. The capacity to mount a autologous graft-versus-leukemia (GVL) response using cyclosporine has been tested clinically; however, no definitive conclusion could be drawn regarding decreased relapse rates and improved disease-free survival after autologous bone marrow transplantation and subsequent cyclosporine therapy in patients with leukemia or lymphoma [87].

Little is known on the possible implication of HLA-DO in the antitumor response. Interestingly, an amino acid change in HLA-DOα was found in a patient suffering from CML [88]. However, this mutation does not appear to affect the function of HLA-DO. In B-CLL, the HLA-DOα mRNA expression was increased and, although it did not translate into more HLA-DO protein, it was established that it correlated with poor survival [55]. More comprehensive studies looking at all the components of the MHC II antigen presentation pathway will be needed to better understand the impact of CLIP and peptide loading on various clinical parameters of tumor immunology.

Modulation of MHC II Accessory Molecules in Tumors

Presentation of peptides in normal cells is a function of efficient synthesis, sorting, and processing of the antigens as well as proper trafficking and maturation of MHC II molecules. Intrinsic modifications by tumor cells of a cellular compartment and its components such as lipids and enzymes are likely to influence, directly or indirectly, the processing, loading, and presentation of antigens to T cells. A few examples of such potentially clinically-relevant perturbations are given below.

Defects in autophagy have been associated with cellular transformation [89]. Because autophagy has been intimately linked to antigen processing by MHC II in a variety of systems [42], it is likely that some tumors will exhibit defects in the processing of certain antigens [90]. On the other hand, through mechanisms ranging from gene amplification to posttranscriptional modification, tumors often upregulate cathepsins or reduce their inhibitors, cystatins [91]. Such endosomal/lysosomal protease regulation has been shown to have a tremendous negative impact on the generation of T-cell epitopes and on Ii degradation [92].

The endocytic pathway may also be the target of subtle changes. Dysregulation of "endosomal sorting complex required for transport" (ESCRT) proteins is involved in the development of various cancers [93]. Components of the ESCRT complexes, such as TSG-101, have been described as tumor suppressors since their inactivation prevents proper targeting and degradation of activated receptors. ESCRT protein complexes are part of an elaborate machinery responsible, among other things, for the inward budding of vesicles from the outer membrane of vesicular bodies. Knowing that HLA-DM and -DR must interact on the internal membranes of MVBs to efficiently achieve peptide loading [94], it will be interesting to determine if mutations affecting the ESCRT machinery will impact antigen presentation.

As mentioned above, many murine tumor cell lines do not express or upregulate the MHC II Ag presentation machinery in response to IFN-γ [56]. For example, absence of gamma-interferon-inducible lysosomal thiol reductase (GILT) in melanomas disrupts T-cell recognition of select immunodominant epitopes [95]. As a second example, in the setting of head and neck cancer cells, CIITA does not induce cathepsin S, which is a cysteine protease involved in the late stage of Ii processing [96]. Numerous new alterations are likely to be described in tumor cells and their repercussions on the adaptive response in the context of immune evasion will undoubtedly uncover some surprises.

Counteracting Subversion of Antigen Presentation

Tumors show heterogeneous expression of antigen presentation molecules and the impact on local leukocyte infiltration, cytokine production and, ultimately, prognosis will remain variable. Moreover, subversion mechanisms may differ depending on the cancer types. Researchers must continue to decipher the mechanisms by which tumors evade the immune response and also must define the correlates behind recent successes of immunotherapy or vaccination [97]. Improving antigen processing and presentation is the first step in the development of an efficient adaptive response.

Many methods have been envisaged to maximize antigen presentation. Tumor cells expressing MHC molecules are being exploited as vaccines. However, a more commonly utilized approach is to transfer *in vitro*-manipulated natural or artificial APCs displaying defined antigens loaded in controlled conditions. Other *in vivo* approaches are being developed in order to limit manipulations of host cells and avoid cumbersome patient-specific immunotherapy. In this last section, we will address the need to discover more TAAs, improve cellular vaccines, and define alternative methods in the quest to effectively stimulate $CD4^+$ T cells.

Discovery of Novel TAAs and T-Cell Epitopes

Tumors of various origins will express different antigens, and as such, no universal immunogen can be derived for immunotherapy purposes. Coupled with the fact that immune pressure can select for cancer cells that have lost the expression of a given antigen, this situation demonstrates the need to define a library of antigens for each type of tumor and probably also for different recognizable stages of a given cancer. Some TAAs may be encoded for by alternative ORFs [98], following chromosomal rearrangements [99], or through mutations to generate tumor-specific antigens (TSAs), which represent a subclass of TAAs [100].

In recent years, TAAs recognized by $CD4^+$ or $CD8^+$ T cells have been defined at a regular pace and for many cancers. As TAAs are often found in normal tissues, breaking tolerance to these antigens through vaccination may result in tumor recognition but also in autoimmunity. The classical example is the skin depigmentation (vitiligo) observed in melanoma patients immunized against gp100 and other melanoma antigens [101]. We also have to be aware that other treatments such as chemotherapy might modify the proteome of cancer cells and provide new targets for immunotherapy [102].

Another hurdle to the development of effective immunotherapy is the genetic diversity at the MHC I and II locus. Currently, antigens are usually identified for the most common alleles such as HLA-A2 and HLA-DP4 [22]. The identification of new epitopes recognized in the context of a series of isotypes and alleles will open the door to universal use of immunotherapy. In this context, defining the immunopeptidome for different cancer-patient combinations is likely to produce valuable information in the future [103]. The mass spectrometry approach to the mapping of MHC class I or II binding antigens is constantly improving in terms of sensitivity and efficacy. Once peptides of very low abundance could be identified, it will be possible to define some new TSAs that may arise through processes such as protein splicing [104]. The immunological response to antigens is usually directed against a narrow set of immunodominant peptides derived from complex antigens. Other epitopes are hidden because of inadequate processing or low affinity for MHC molecules. The possibility of using cryptic epitopes is attractive because one is usually tolerant only to the dominant determinants of self-proteins and the T-cell repertoire against cryptic determinants remains inert in the host [105].

Cellular Vaccines

Tumor Vaccines

First-generation tumor vaccines have failed to deliver significant clinical success [2]. Using different vaccination strategies, measurable immunizations were achieved, but with only few clinical responses of documented tumor regression. This therapeutic limitation exemplifies the complexity of breaking tolerance to self targets; as such, we need to define more efficient immunization platforms and combine different approaches to reach a threshold that ultimately leads to tumor regression. Here, we review selected examples involving modulation of MHC class II antigenic presentation to enhance immunization.

The idea behind these vaccines is that tumor cells, although poorly immunogenic themselves, express a full complement of endogenous TAAs. Trials in breast cancer patients have utilized autologous tumor cells as well as allogeneic cell lines [106]. Tumor vaccines can be genetically modified with the MHC II machinery of processing and presentation in order to increase T-cell activation [106]. These vaccines rely on endogenously synthesized proteins, which represent a source of antigens qualitatively limited in the context of MHC class II molecules. Understanding why some cells preferentially present endogenous as opposed to exogenous or transmembrane antigens in the context of MHC class II would help in the design of cancer vaccines [79]. Treatment of cells with cytokines that promote the processing of endogenous (even nuclear) antigens through autophagy might increase the variety of T-cell epitopes generated in tumor cell vaccines [107]. Interestingly, many groups reported that Ii expression is detrimental to the presentation of endogenous antigens by mouse and human tumor cells. For example, knocking down Ii expression by various means increased presentation of some antigens and improved the immunotherapy [108]. Despite the fact that TAAs are endogenously expressed, tumor cells have also been transfected with MHC II molecules covalently linked to antigenic peptides to increase the response [109].

Some tumor cells, especially those originating from the B lymphocyte lineage, already express MHC II molecules. These cells may contain high levels of HLA-DO, which is a proven inhibitor of HLA-DM and peptide presentation. As HLA-DM was shown to dictate the cryptic and immunodominant fate of epitopes [110], knocking down expression of HLA-DO by shRNA will diminish the display of CLIP and potentially increase the presentation of important TAAs. On the other hand, increasing the expression of HLA-DO may reveal cryptic T-cell epitopes for which no tolerance has been established. Importantly, HLA-DM and HLA-DO are generally monomorphic, thereby making their overexpression easily amenable to the clinic.

As mentioned above, many tumors do not express classical or nonclassical MHC class II molecules and need to be further manipulated *in vitro*. IFN-γ will up-regulate the MHC II antigen presentation machinery as well as more than 200 other genes [6]. Some tumors were shown to gain full antigen presentation capabilities in these conditions (see above). For those tumors not responding to this cytokine, genetic modification can be envisaged. Introduction of CIITA has been achieved in cellular vaccines but some tumors do not fully respond to the transactivator and some genes have been reported to remain silent [111]. Careful monitoring of the gene expression profile is needed in these conditions to ensure that the whole antigen presentation machinery is up-regulated.

DC Vaccines

The most promising therapeutic cancer vaccines are based on DCs [112]. Although costly and cumbersome, the adoptive transfer of *ex vivo*-modified, monocyte-derived autologous DCs offers the advantage of controlling the display of peptides and the cellular context in which the antigens will be presented. DCs have the unique capacity to activate naïve T cells and overcome T-cell nonresponsiveness *in vivo*. While it is desirable to use mature DCs for the adoptive transfer, the endocytic capacity of immature DCs is immense and may favor antigen capture [113]. Following DC maturation, the MARCH1 ubiquitin ligase and CIITA are down-regulated and the peptide-loaded MHC II molecules become stable at the plasma membrane [60]. However, in plasmacytoid DCs (pDCs), MARCH1 and CIITA levels remain high, allowing continuous presentation of endogenous antigens [114]. These observations suggested to Villadangos and collaborators that activated pDCs may represent better cellular vaccines than monocyte-derived DCs when using DNA-based delivery of antigens (see below) [114]. Accordingly, presentation of endogenous antigens by the tumor vaccines described above would benefit from overexpressing CIITA and knocking down MARCH1.

A multitude of methods have been used to display desired T-cell epitopes. DCs can phagocytose apoptotic and necrotic tumor cells; alternatively, hybrids can be made by electrofusion [115]. Still yet, tumor cell lysates have been pulsed onto DCs [116]. However, there are certain TAAs such as GA733-2 expressed in colon, breast, lung and some nonepithelial tumors, which inhibit antigen processing upon uptake by APCs [117]. The use of recombinant Ags has also been described [118] and these can be used as immune complexes with adjuvants such as ISCOM or coupled to monoclonal antibodies directed to surface markers such as DEC-205 [119].

Synthetic peptides corresponding to carefully selected epitopes represent the handiest source of antigens. Their formulation has evolved in recent years. For example, multi-epitope Trojan antigen peptide vaccines or peptides with overlapping CD4 and CD8 epitopes induce both CTL and Th immune responses [120, 121]. However, although DCs express empty MHC II molecules at their surface, the loading is rather inefficient. Recently, chemicals capable of breaking hydrogen bonds between low-affinity peptides and HLA-DR were discovered [122]. Other small molecules capable of enhancing the catalytic activity of HLA-DM and/or peptide binding have recently been identified by high-throughput screening [123, 124]. Such compounds may be very useful for the loading of exogenous synthetic peptides on DCs [125]. In order to maximize peptide loading of synthetic peptides to DCs, we have recently genetically modified DCs to express HLA-DM at the plasma membrane. We found that the loading of exogenous peptides, including a DR7-restricted T-cell epitope of gp100, was increased (Pezeshki, M. and Thibodeau, J., submitted). Along the same line, the group of Watts suggested to use agents known to stabilize proteins in their native conformation, such as DMSO and glycerol [126]. Also, just as for tumor cell vaccines, overexpression of CIITA in DCs increased MHC II expression and improved immunostimulatory activity [127]. Additional genetic approaches aimed at delivering antigens to DCs for the induction of a $CD4^{+}$ T-cell response have been described [128]. Such approaches include the use of various viral vectors such as oncoretroviruses [129], mRNA electroporation, and the gene gun [130, 131]. Impairment in the trafficking and processing of the recombinant antigens has been ameliorated by merging the luminal sequences of

TAAs to endosomal targeting signal sequences from melanosome-associated proteins [132]. Alternatively, Ii-based DNA vaccines have been used to deliver antigens to peptide loading compartments. In this strategy, defined T-cell epitopes or polypeptides are cloned in place of CLIP-coding region or to the 3′ region of the Ii cDNA to generate a fusion protein [128]. For example, DCs expressing a pancreas carcinoma antigen fused to invariant chain inhibited tumor growth in mice [133]. Finally, small fragments such as cell-penetrating peptides (CPPs) or "Ii-key" have been fused to T-cell epitopes to increase their potency [134, 135].

B-Cell Vaccines

Alternative sources of APCs have also been evaluated *in vitro*. For instance, B lymphocytes stimulated by CD40L have been shown to proliferate in high numbers [136], to display a wider array of MHC class II epitopes due to a downmodulation of HLA-DM/DO ratio [41], and to be suitable as APCs [137]. B cells have been shown to serve as efficient APCs for expansion of TAA-specific $CD8^+$ [138] and $CD4^+$ [41] T cells. Antigens can be loaded onto B cells by pulsing with tumor lysates [41] or peptides [139], by retroviral transduction [138], and by gene electroporation [132]. Interestingly, in addition to possessing the capacity of presenting MHC class II epitopes independently of the specificity of the BCR when pulsed exogenously, B cells have the capacity to promote MHC class I cross-presentation [140]. Although B cells therefore represent a potentially suitable source of APCs, their clinical applicability is currently limited by the poor availability of a clinical-grade source of CD40L.

Surrogate APCs

To overcome the need for live autologous hematopoietic cells in immunotherapy, alternatives have been developed where the minimum essential antigen presentation requirements are expressed on "artificial" supports. Such alternatives include cellular-based systems such as fibroblasts and *Drosophila* cells [141] or acellular artificial APCs that contain microbeads, liposomes, or exosomes [142–144]. These methods also bring their share of technical challenges but may ultimately allow standardization of cancer vaccination protocols.

Conclusion

Although surgery, radiotherapy, and chemotherapy have demonstrated efficacy in some settings, alternative strategies such as immunotherapy are needed to recognize and kill tumors. Understanding the role of $CD4^+$ T cells in the antitumor response, especially in the context of the apparent counter-productive influence of $CD4^+$ Tregs, will require that studies in humans decipher the paths leading to the generation of the various helper subsets, and presentation of immunogenic as opposed to suppressive MHC class II epitopes. Finally, to make a difference in the clinic, we must develop cancer immunotherapy methods that increase antigen presentation via *in vivo* targeting of immunogens. TAAs or peptides can be targeted to DCs after coupling to antibodies specific for surface markers combined and combined approaches. As an example, electrodes have been designed for *in vivo* DNA delivery by direct electroporation in humans; such an approach may seem extreme, but may be necessary to improve the delivery of genes coding for TAAs and other adjuvant genes [145]. Of course, we also need ingenious ideas to spark the field of cancer vaccination and as usual, only imagination will be a limit to innovation.

Acknowledgements MCBD is supported by a studentship from the Cole Foundation. This work was supported by a grant from the Canadian Cancer Society (# 17230) to JT and grants from the Canadian Institutes of Health Research and the Cancer Research Society to RL.

References

1. Old LJ. (1992) Tumor immunology: the first century. Curr Opin Immunol 4:603–607.
2. Rosenberg SA, Yang JC, Restifo NP. (2004) Cancer immunotherapy: moving beyond current vaccines. Nat Med 10:909–915.
3. Dunn GP, Bruce AT, Ikeda H, et al. (2002) Cancer immunoediting: from immunosurveillance to tumor escape. Nat Immunol 3:991–998.
4. Pardoll D. (2003) Does the immune system see tumors as foreign or self? Annu Rev Immunol 21:807–839.
5. Coscoy L. (2007) Immune evasion by Kaposi's sarcoma-associated herpesvirus. Nat Rev Immunol 7:391–401.
6. Boehm U, Klamp T, Groot M, et al. (1997) Cellular responses to interferon-gamma. Annu Rev Immunol 15:749–795.
7. Watanabe Y, Jacob CO. (1991) Regulation of MHC class II antigen expression. Opposing effects of tumor necrosis factor-alpha on IFN-gamma-induced HLA-DR and Ia expression depends on the maturation and differentiation stage of the cell. J Immunol 146:899–905.
8. Knutson KL, Disis ML. (2005) Tumor antigen-specific T helper cells in cancer immunity and immunotherapy. Cancer Immunol Immunother 54:721–728.
9. Corthay A, Skovseth DK, Lundin KU, et al. (2005) Primary antitumor immune response mediated by $CD4^+$ T cells. Immunity 22:371–383.
10. Fearon ER, Pardoll DM, Itaya T, et al. (1990) Interleukin-2 production by tumor cells bypasses T helper function in the generation of an antitumor response. Cell 60:397–403.
11. Pardoll DM, Topalian SL. (1998) The role of $CD4^+$ T cell responses in antitumor immunity. Curr Opin Immunol 10:588–594.
12. Nakanishi Y, Lu B, Gerard C, et al. (2009) $CD8^{(+)}$ T lymphocyte mobilization to virus-infected tissue requires $CD4^{(+)}$ T-cell help. Nature 462:510–513.
13. Concha A, Ruiz-Cabello F, Cabrera T, et al. (1995) Different patterns of HLA-DR antigen expression in normal epithelium, hyperplastic and neoplastic malignant lesions of the breast. Eur J Immunogenet 22:299–310.
14. Disis ML, Goodell V, Schiffman K, et al. (2004) Humoral epitope-spreading following immunization with a HER-2/neu peptide based vaccine in cancer patients. J Clin Immunol 24:571–578.
15. Wang L, Yi T, Kortylewski M, et al. (2009) IL-17 can promote tumor growth through an IL-6-Stat3 signaling pathway. J Exp Med 206:1457–1464.

16. Martin-Orozco N, Muranski P, Chung Y, et al. (2009) T helper 17 cells promote cytotoxic T cell activation in tumor immunity. Immunity 31:787–798.
17. Ridge JP, Di Rosa F, Matzinger P. (1998) A conditioned dendritic cell can be a temporal bridge between a CD4+ T- helper and a T-killer cell. Nature 393:474–478.
18. Knutson KL, Disis ML. (2005) Augmenting T helper cell immunity in cancer. Curr Drug Targets Immune Endocr Metab Disord 5:365–371.
19. Nizar S, Copier J, Meyer B, et al. (2009) T-regulatory cell modulation: the future of cancer immunotherapy? Br J Cancer 100:1697–1703.
20. Wang RF. (2002) Enhancing antitumor immune responses: intracellular peptide delivery and identification of MHC class II-restricted tumor antigens. Immunol Rev 188:65–80.
21. Atanackovic D, Altorki NK, Stockert E, et al. (2004) Vaccine-induced CD4+ T cell responses to MAGE-3 protein in lung cancer patients. J Immunol 172:3289–3296.
22. Khong HT, Yang JC, Topalian SL, et al. (2004) Immunization of HLA-A*0201 and/or HLA-DPbeta1*04 patients with metastatic melanoma using epitopes from the NY-ESO-1 antigen. J Immunother 27:472–477.
23. Phan GQ, Touloukian CE, Yang JC, et al. (2003) Immunization of patients with metastatic melanoma using both class I- and class II-restricted peptides from melanoma-associated antigens. J Immunother 26:349–356.
24. Ostrand-Rosenberg S. (2008) Immune surveillance: a balance between protumor and antitumor immunity. Curr Opin Genet Dev 18:11–18.
25. Egilmez NK, Hess SD, Chen FA, et al. (2002) Human CD4+ effector T cells mediate indirect interleukin-12- and interferon-gamma-dependent suppression of autologous HLA-negative lung tumor xenografts in severe combined immunodeficient mice. Cancer Res 62:2611–2617.
26. Huang AY, Golumbek P, Ahmadzadeh M, et al. (1994) Role of bone marrow-derived cells in presenting MHC class I-restricted tumor antigens. Science 264:961–965.
27. Greenberg PD, Kern DE, Cheever MA. (1985) Therapy of disseminated murine leukemia with cyclophosphamide and immune Lyt-1+,2- T cells. Tumor eradication does not require participation of cytotoxic T cells. J Exp Med 161:1122–1134.
28. Soos JM, Krieger JI, Stuve O, et al. (2001) Malignant glioma cells use MHC class II transactivator (CIITA) promoters III and IV to direct IFN-gamma-inducible CIITA expression and can function as nonprofessional antigen presenting cells in endocytic processing and CD4(+) T-cell activation. Glia 36:391–405.
29. Armstrong TD, Clements VK, Martin BK, et al. (1997) Major histocompatibility complex class II-transfected tumor cells present endogenous antigen and are potent inducers of tumor-specific immunity. Proc Natl Acad Sci U S A 94:6886–6891.
30. Mortara L, Frangione V, Castellani P, et al. (2009) Irradiated CIITA-positive mammary adenocarcinoma cells act as a potent anti-tumor-preventive vaccine by inducing tumor-specific CD4+ T cell priming and CD8+ T cell effector functions. Int Immunol 21:655–665.
31. Willimsky G, Blankenstein T. (2007) The adaptive immune response to sporadic cancer. Immunol Rev 220:102–112.
32. Rabinovich GA, Gabrilovich D, Sotomayor EM. (2007) Immunosuppressive strategies that are mediated by tumor cells. Annu Rev Immunol 25:267–296.
33. Zea AH, Curti BD, Longo DL, et al. (1995) Alterations in T cell receptor and signal transduction molecules in melanoma patients. Clin Cancer Res 1:1327–1335.
34. Fischer WH, thor SP, Terheyden P, et al. (1999) Function and dysfunction of CD4(+) T cells in the immune response to melanoma. Cancer Immunol Immunother 48:363–370.
35. Cresswell P. (1994) Assembly, transport, and function of MHC class II molecules. Annu Rev Immunol 12:259–293.
36. Stumptner-Cuvelette P, Benaroch P. (2002) Multiple roles of the invariant chain in MHC class II function. Biochim Biophys Acta 1542:1–13.
37. Kelly AP, Monaco JJ, Cho S, et al. (1991) A new human HLA class II-related locus, DM. Nature 353:571–573.
38. Marks MS, Roche PA, van Donselaar E, et al. (1995) A lysosomal targeting signal in the cytoplasmic tail of the b chain directs HLA-DM to MHC class II compartments. J Cell Biol 131:351–369.

39. Kropshofer H, Hammerling GJ, Vogt AB. (1997) How HLA-DM edits the MHC class II peptide repertoire: Survival of the fittest? Immunol Today 18:77–82.
40. Alfonso C, Liljedahl M, Winqvist O, et al. (1999) The role of H2-O and HLA-DO in major histocompatibility complex class II-restricted antigen processing and presentation. Immunol Rev 172:255–266.
41. Lapointe R, Bellemare-Pelletier A, Housseau F, et al. (2003) CD40-stimulated B lymphocytes pulsed with tumor antigens are effective antigen-presenting cells that can generate specific T cells. Cancer Res 63:2836–2843.
42. Crotzer VL, Blum JS. (2009) Autophagy and its role in MHC-mediated antigen presentation. J Immunol 182:3335–3341.
43. Garrido F, Cabrera T, Concha A, et al. (1993) Natural history of HLA expression during tumour development. Immunol Today 14:491–499.
44. Wright KL, Ting JP. (2006) Epigenetic regulation of MHC-II and CIITA genes. Trends Immunol 27:405–412.
45. Restifo NP, Marincola FM, Kawakami Y, et al. (1996) Loss of functional beta 2-microglobulin in metastatic melanomas from five patients receiving immunotherapy. J Natl Cancer Inst 88:100–108.
46. Durrant LG, Ballantyne KC, Armitage NC, et al. (1987) Quantitation of MHC antigen expression on colorectal tumours and its association with tumour progression. Br J Cancer 56:425–432.
47. Tabibzadeh SS, Sivarajah A, Carpenter D, et al. (1990) Modulation of HLA-DR expression in epithelial cells by interleukin 1 and estradiol-17 beta. J Clin Endocrinol Metab 71:740–747.
48. Momburg F, Herrmann B, Moldenhauer G, et al. (1987) B-cell lymphomas of high-grade malignancy frequently lack HLA-DR, -DP and -DQ antigens and associated invariant chain. Int J Cancer 40:598–603.
49. Oldford SA, Robb JD, Watson PH, et al. (2004) HLA-DRB alleles are differentially expressed by tumor cells in breast carcinoma. Int J Cancer 112:399–406.
50. de Bruin EC, van de Velde CJ, van Krieken JH, et al. (2008) Epithelial human leukocyte antigen-DR expression predicts reduced recurrence rates and prolonged survival in rectal cancer patients. Clin Cancer Res 14:1073–1079.
51. Moller P, Mattfeldt T, Gross C, et al. (1989) Expression of HLA-A, -B, -C, -DR, -DP, -DQ, and of HLA-D-associated invariant chain (Ii) in non-neoplastic mammary epithelium, fibroadenoma, adenoma, and carcinoma of the breast. Am J Pathol 135:73–83.
52. Dazzi F, D'Andrea E, Biasi G, et al. (1995) Failure of B cells of chronic lymphocytic leukemia in presenting soluble and alloantigens. Clin Immunol Immunopathol 75:26–32.
53. Redondo M, Ruiz-Cabello F, Concha A, et al. (1998) Differential expression of MHC class II genes in lung tumour cell lines. Eur J Immunogenet 25:385–391.
54. Ting JP, Trowsdale J. (2002) Genetic control of MHC class II expression. Cell 109 Suppl:S21–S33.
55. Souwer Y, Chamuleau ME, van de Loosdrecht AA, et al. (2009) Detection of aberrant transcription of major histocompatibility complex class II antigen presentation genes in chronic lymphocytic leukaemia identifies HLA-DOA mRNA as a prognostic factor for survival. Br J Haematol 145:334–343.
56. Walter W, Lingnau K, Schmitt E, et al. (2000) MHC class II antigen presentation pathway in murine tumours: tumour evasion from immunosurveillance? Br J Cancer 83:1192–1201.
57. Jordanova ES, Philippo K, Giphart MJ, et al. (2003) Mutations in the HLA class II genes leading to loss of expression of HLA-DR and HLA-DQ in diffuse large B-cell lymphoma. Immunogenetics 55:203–209.
58. Garcia JF, Roncador G, Garcia JF, et al. (2006) PRDM1/BLIMP-1 expression in multiple B and T-cell lymphoma. Haematologica 91:467–474.
59. Zhao M, Flynt FL, Hong M, et al. (2007) MHC class II transactivator (CIITA) expression is upregulated in multiple myeloma cells by IFN-gamma. Mol Immunol 44:2923–2932.
60. De Gassart A., Camosseto V, Thibodeau J, et al. (2008) MHC class II stabilization at the surface of human dendritic cells is the result of maturation-dependent MARCH I down-regulation. Proc Natl Acad Sci U S A 105(9):3491–3496.

61. Thibodeau J, Bourgeois-Daigneault MC, Huppe G, et al. (2008) Interleukin-10-induced MARCH1 mediates intracellular sequestration of MHC class II in monocytes. Eur J Immunol 38:1225–1230.
62. Ishido S, Goto E, Matsuki Y, et al. (2009) E3 ubiquitin ligases for MHC molecules. Curr Opin Immunol 21:78–83.
63. Paulnock-King D, Sizer KC, Freund YR, et al. (1985) Coordinate induction of Ia alpha, beta, and Ii mRNA in a macrophage cell line. J Immunol 135:632–636.
64. Degener T, Momburg F, Moller P. (1988) Differential expression of HLA-DR, HLA-DP, HLA-DQ and associated invariant chain (Ii) in normal colorectal mucosa, adenoma and carcinoma. Virchows Arch A Pathol Anat Histopathol 412:315–322.
65. Brown AM, Linhoff MW, Stein B, et al. (1994) Function of NF-kappa B/Rel binding sites in the major histocompatibility complex class II invariant chain promoter is dependent on cell-specific binding of different NF-kappa B/Rel subunits. Mol Cell Biol 14:2926–2935.
66. Spiro RC, Sairenji T, Humphreys RE. (1984) Identification of hairy cell leukemia subset defining p35 as the human homologue of Ii. Leuk Res 8:55–62.
67. Narni F, Kudo J, Mars W, et al. (1986) HLA-DR-associated invariant chain is highly expressed in chronic lymphocytic leukemia. Blood 68:372–377.
68. Yao XR, Humphreys RE. (1988) Decreased expression of Ii-derived p25 and of P67/69 in hairy leukemic cells. Leuk Res 12:787–794.
69. Saito T, Kimura M, Kawasaki T, et al. (1997) MHC class II antigen-associated invariant chain on renal cell cancer may contribute to the anti-tumor immune response of the host. Cancer Lett 115:121–127.
70. Jiang Z, Xu M, Savas L, et al. (1999) Invariant chain expression in colon neoplasms. Virchows Arch 435:32–36.
71. Nagata S, Jin YF, Yoshizato K, et al. (2009) CD74 Is a Novel Prognostic Factor for Patients with Pancreatic Cancer Receiving Multimodal Therapy. Ann Surg Oncol 16(9):2531–2538.
72. Thompson JA, Srivastava MK, Bosch JJ, et al. (2008) The absence of invariant chain in MHC II cancer vaccines enhances the activation of tumor-reactive type 1 $CD4^+$ T lymphocytes. Cancer Immunol Immunother 57:389–398.
73. Ostrand-Rosenberg S. (2001) Mouse SaI sarcoma tumor model. Curr Protoc Immunol Chapter 20:Unit 20.3.
74. Lin X, Wang X, Capek HL, et al. (2009) Effect of invariant chain on major histocompatibility complex class I molecule expression and stability on human breast tumor cell lines. Cancer Immunol Immunother 58:729–736.
75. van Ham M, van Lith M, Griekspoor A, et al. (2000) What to do with HLA-DO? Immunogenetics 51:765–770.
76. Denzin LK, Cresswell P. (1995) HLA-DM induces CLIP dissociation from MHC class II ab dimers and facilitates peptide loading. Cell 82:155–165.
77. Westerheide SD, Louis-Plence P, Ping DS, et al. (1997) HLA-DMA and HLA-DMB gene expression functions through the conserved S-X-Y region. J Immunol 158:4812–4821.
78. Khanna R, Burrows SR, Thomson SA, et al. (1997) Class I processing-defective Burkitt's lymphoma cells are recognized efficiently by $CD4^+$ EBV-specific CTLs. J Immunol 158:3619–3625.
79. Harris PE, Maffei A, Colovai AI, et al. (1996) Predominant HLA-class II bound self-peptides of a hematopoietic progenitor cell line are derived from intracellular proteins. Blood 87:5104–5112.
80. Oldford SA, Robb JD, Codner D, et al. (2006) Tumor cell expression of HLA-DM associates with a Th1 profile and predicts improved survival in breast carcinoma patients. Int Immunol 18:1591–1602.
81. Chaturvedi P, Hengeveld R, Zechel MA, et al. (2000) The functional role of class II-associated invariant chain peptide (CLIP) in its ability to variably modulate immune responses. Int Immunol 12:757–765.
82. Callahan MJ, Nagymanyoki Z, Bonome T, et al. (2008) Increased HLA-DMB expression in the tumor epithelium is associated with increased CTL infiltration and improved prognosis in advanced-stage serous ovarian cancer. Clin Cancer Res 14:7667–7673.

83. Jastaniah WA, Alessandri AJ, Reid GS, et al. (2006) HLA-DM expression is elevated in ETV6-AML1 translocation-positive pediatric acute lymphoblastic leukemia. Leuk Res 30:487–489.
84. Bosshart H, Jarrett RF. (1998) Deficient major histocompatibility complex class II antigen presentation in a subset of Hodgkin's disease tumor cells. Blood 92:2252–2259.
85. Chamuleau ME, Souwer Y, Van Ham SM, et al. (2004) Class II-associated invariant chain peptide expression on myeloid leukemic blasts predicts poor clinical outcome. Cancer Res 64:5546–5550.
86. Hess AD, Bright EC, Thoburn C, et al. (1997) Specificity of effector T lymphocytes in autologous graft-versus-host disease: role of major histocompatibility complex class II invariant chain peptide. Blood 89:2203–2209.
87. Yeager AM, Vogelsang GB, Jones RJ, et al. (1993) Cyclosporine-induced graft-versus-host disease after autologous bone marrow transplantation for acute myeloid leukemia. Leuk Lymphoma 11:215–220.
88. van Lith M, van Ham M, Neefjes J. (2002) Novel polymorphisms in HLA-DOA and HLA-DOB in B-cell malignancies. Immunogenetics 54:591–595.
89. Chen N, Karantza-Wadsworth V. (2009) Role and regulation of autophagy in cancer. Biochim Biophys Acta 1793(9):1516–1523.
90. Mathew R, Karp CM, Beaudoin B, et al. (2009) Autophagy suppresses tumorigenesis through elimination of p62. Cell 137:1062–1075.
91. Cox JL. (2009) Cystatins and cancer. Front Biosci 14:463–474.
92. Watts C, Moss CX, Mazzeo D, et al. (2003) Creation versus destruction of T cell epitopes in the class II MHC pathway. Ann N Y Acad Sci 987:9–14.
93. Tanaka N, Kyuuma M, Sugamura K. (2008) Endosomal sorting complex required for transport proteins in cancer pathogenesis, vesicular transport, and non-endosomal functions. Cancer Sci 99:1293–1303.
94. Zwart W, Griekspoor A, Kuijl C, et al. (2005) Spatial separation of HLA-DM/HLA-DR interactions within MIIC and phagosome-induced immune escape. Immunity 22:221–233.
95. Goldstein OG, Hajiaghamohseni LM, Amria S, et al. (2008) Gamma-IFN-inducible-lysosomal thiol reductase modulates acidic proteases and HLA class II antigen processing in melanoma. Cancer Immunol Immunother 57:1461–1470.
96. Meissner M, Whiteside TL, Kaufmann R, et al. (2009) CIITA versus IFN-gamma induced MHC class II expression in head and neck cancer cells. Arch Dermatol Res 301:189–193.
97. Appay V, Douek DC, Price DA. (2008) $CD8^+$ T cell efficacy in vaccination and disease. Nat Med 14:623–628.
98. Wang RF, Johnston SL, Zeng G, et al. (1998) A breast and melanoma-shared tumor antigen: T cell responses to antigenic peptides translated from different open reading frames. J Immunol 161:3598–3606.
99. Wang RF, Wang X, Rosenberg SA. (1999) Identification of a novel major histocompatibility complex class II-restricted tumor antigen resulting from a chromosomal rearrangement recognized by $CD4^{(+)}$ T cells. J Exp Med 189:1659–1668.
100. Pieper R, Christian RE, Gonzales MI, et al. (1999) Biochemical identification of a mutated human melanoma antigen recognized by $CD4^{(+)}$ T cells. J Exp Med 189:757–766.
101. Kawakami Y, Rosenberg SA. (1997) Immunobiology of human melanoma antigens MART-1 and gp100 and their use for immuno-gene therapy. Int Rev Immunol 14:173–192.
102. Zhang JT, Liu Y. (2007) Use of comparative proteomics to identify potential resistance mechanisms in cancer treatment. Cancer Treat Rev 33:741–756.
103. Fortier MH, Caron E, Hardy MP, et al. (2008) The MHC class I peptide repertoire is molded by the transcriptome. J Exp Med 205:595–610.
104. Berkers CR, de JA, Ovaa H, et al. (2009) Transpeptidation and reverse proteolysis and their consequences for immunity. Int J Biochem Cell Biol 41:66–71.
105. Cibotti R, Kanellopoulos JM, Cabaniols JP, et al. (1992) Tolerance to a self-protein involves its immunodominant but does not involve its subdominant determinants. Proc Natl Acad Sci USA 89:416–420.

106. Anderson KS. (2009) Tumor vaccines for breast cancer. Cancer Invest 27:361–368.
107. English L, Chemali M, Desjardins M. (2009) Nuclear membrane-derived autophagy, a novel process that participates in the presentation of endogenous viral antigens during HSV-1 infection. Autophagy 5(7):1026–1029.
108. Qiu G, Goodchild J, Humphreys RE, et al. (1999) Cancer immunotherapy by antisense suppression of Ii protein in MHC-class-II-positive tumor cells. Cancer Immunol Immunother 48:499–506.
109. Ladanyi A, Nishimura MI, Rosenberg SA, et al. (2000) Tumorigenicity and immunogenicity of murine tumor cells expressing an MHC class II molecule with a covalently bound antigenic peptide. J Immunother 23:36–47.
110. Blum JS, Ma C, Kovats S. (1997) Antigen-presenting cells and the selection of immunodominant epitopes. Crit Rev Immunol 17:411–417.
111. Baton F, Deruyffelaere C, Chapin M, et al. (2004) Class II transactivator (CIITA) isoform expression and activity in melanoma. Melanoma Res 14:453–461.
112. Tuyaerts S, Aerts JL, Corthals J, et al. (2007) Current approaches in dendritic cell generation and future implications for cancer immunotherapy. Cancer Immunol Immunother 56:1513–1537.
113. Villadangos JA, Schnorrer P, Wilson NS. (2005) Control of MHC class II antigen presentation in dendritic cells: a balance between creative and destructive forces. Immunol Rev 207:191–205.
114. Young LJ, Wilson NS, Schnorrer P, et al. (2008) Differential MHC class II synthesis and ubiquitination confers distinct antigen-presenting properties on conventional and plasmacytoid dendritic cells. Nat Immunol 9:1244–1252.
115. Parkhurst MR, DePan C, Riley JP, et al. (2003) Hybrids of dendritic cells and tumor cells generated by electrofusion simultaneously present immunodominant epitopes from multiple human tumor-associated antigens in the context of MHC class I and class II molecules. J Immunol 170:5317–5325.
116. Nestle FO, Alijagic S, Gilliet M, et al. (1998) Vaccination of melanoma patients with peptide- or tumor lysate-pulsed dendritic cells. Nat Med 4:328–332.
117. Gutzmer R, Li W, Sutterwala S, et al. (2004) A tumor-associated glycoprotein that blocks MHC class II-dependent antigen presentation by dendritic cells. J Immunol 173:1023–1032.
118. Parkhurst MR, Riley JP, Robbins PF, et al. (2004) Induction of $CD4^+$ Th1 lymphocytes that recognize known and novel class II MHC restricted epitopes from the melanoma antigen gp100 by stimulation with recombinant protein. J Immunother 27:79–91.
119. Steinman RM. (1996) Dendritic cells and immune-based therapies. Exp Hematol 24:859–862.
120. Zeng G, Li Y, El-Gamil M, et al. (2002) Generation of NY-ESO-1-specific $CD4^+$ and $CD8^+$ T cells by a single peptide with dual MHC class I and class II specificities: a new strategy for vaccine design. Cancer Res 62:3630–3635.
121. Lu J, Higashimoto Y, Appella E, et al. (2004) Multiepitope Trojan antigen peptide vaccines for the induction of antitumor CTL and Th immune responses. J Immunol 172:4575–4582.
122. Falk K, Lau JM, Santambrogio L, et al. (2002) Ligand exchange of major histocompatibility complex class II proteins is triggered by H-bond donor groups of small molecules. J Biol Chem 277:2709–2715.
123. Hopner S, Dickhaut K, Hofstatter M, et al. (2006) Small organic compounds enhance antigen loading of class II major histocompatibility complex proteins by targeting the polymorphic P1 pocket. J Biol Chem 281:38535–38542.
124. Nicholson MJ, Moradi B, Seth NP, et al. (2006) Small molecules that enhance the catalytic efficiency of HLA-DM. J Immunol 176:4208–4220.
125. Marin-Esteban V, Falk K, Rotzschke O. (2004) "Chemical analogues" of HLA-DM can induce a peptide-receptive state in HLA-DR molecules. J Biol Chem 279:50684–50690.
126. Ghumman B, Bertram EM, Watts TH. (1998) Chemical chaperones enhance superantigen and conventional antigen presentation by HLA-DM-deficient as well as HLA-DM-sufficient antigen-presenting cells and enhance IgG2a production *in vivo*. J Immunol 161:3262–3270.
127. Marten A, Ziske C, Schottker B, et al. (2001) Transfection of dendritic cells (DCs) with the CIITA gene: increase in immunostimulatory activity of DCs. Cancer Gene Ther 8:211–219.
128. Bonehill A, Heirman C, Thielemans K. (2005) Genetic approaches for the induction of a $CD4^+$ T cell response in cancer immunotherapy. J Gene Med 7:686–695.

129. Lapointe R, Royal RE, Reeves ME, et al. (2001) Retrovirally transduced human dendritic cells can generate T cells recognizing multiple MHC class I and class II epitopes from the melanoma antigen glycoprotein 100. J Immunol 167:4758–4764.
130. Kim D, Hoory T, Monie A, et al. (2008) Enhancement of DNA vaccine potency through coadministration of CIITA DNA with DNA vaccines via gene gun. J Immunol 180:7019–7027.
131. Bonehill A, Van Nuffel AM, Corthals J, et al. (2009) Single-step antigen loading and activation of dendritic cells by mRNA electroporation for the purpose of therapeutic vaccination in melanoma patients. Clin Cancer Res 15:3366–3375.
132. Lepage S, Lapointe R. (2006) Melanosomal targeting sequences from gp100 are essential for MHC class II-restricted endogenous epitope presentation and mobilization to endosomal compartments. Cancer Res 66:2423–2432.
133. Nagaraj S, Neumann J, Winzen B, et al. (2008) Pancreas carcinoma antigen fused to invariant chain elicits T-cell response and tumor growth inhibition. Pancreas 37:321–327.
134. Humphreys RE, Adams S, Koldzic G, et al. (2000) Increasing the potency of MHC class II-presented epitopes by linkage to Ii-Key peptide. Vaccine 18:2693–2697.
135. Voutsas IF, Gritzapis AD, Mahaira LG, et al. (2007) Induction of potent $CD4^+$ T cell-mediated antitumor responses by a helper HER-2/neu peptide linked to the Ii-Key moiety of the invariant chain. Int J Cancer 121:2031–2041.
136. Banchereau J, de PP, Valle A, et al. (1991) Long-term human B cell lines dependent on interleukin-4 and antibody to CD40. Science 251:70–72.
137. Schultze JL, Grabbe S, von Bergwelt-Baildon MS. (2004) DCs and CD40-activated B cells: current and future avenues to cellular cancer immunotherapy. Trends Immunol 25:659–664.
138. Kondo E, Topp MS, Kiem HP, et al. (2002) Efficient generation of antigen-specific cytotoxic T cells using retrovirally transduced CD40-activated B cells. J Immunol 169:2164–2171.
139. Schultze JL, Michalak S, Seamon MJ, et al. (1997) CD40-activated human B cells: an alternative source of highly efficient antigen presenting cells to generate autologous antigen-specific T cells for adoptive immunotherapy. J Clin Invest 100:2757–2765.
140. Leclerc D, Beauseigle D, Denis J, et al. (2007) Proteasome-independent major histocompatibility complex class I cross-presentation mediated by papaya mosaic virus-like particles leads to expansion of specific human T cells. J Virol 81:1319–1326.
141. Sun S, Cai Z, Langlade-Demoyen P, et al. (1996) Dual function of *Drosophila* cells as APCs for naive $CD8^+$ T cells: implications for tumor immunotherapy. Immunity 4:555–564.
142. Kim JV, Latouche JB, Riviere I, et al. (2004) The ABCs of artificial antigen presentation. Nat Biotechnol 22:403–410.
143. Steenblock ER, Wrzesinski SH, Flavell RA, et al. (2009) Antigen presentation on artificial acellular substrates: modular systems for flexible, adaptable immunotherapy. Expert Opin Biol Ther 9:451–464.
144. Thery C, Ostrowski M, Segura E. (2009) Membrane vesicles as conveyors of immune responses. Nat Rev Immunol 9:581–593.
145. Low L, Mander A, McCann KJ, et al. (2009) DNA vaccination with electroporation induces increased antibody responses in patients with prostate cancer. Hum Gene Ther 20(11): 1269–1278.

Chapter 9
Mechanisms and Implications of Immunodominance in CD8⁺ T-Cell Responses

Claude Perreault

Abstract Immunodominant Major Histocompatibility Complex class I (MHC I)-associated epitopes suppress T-cell responses against cryptic epitopes. That is, when confronted with complex antigens, $CD8^+$ T cells respond to only a few "immunodominant" epitopes and neglect other "cryptic" peptides that would otherwise be immunogenic when presented alone. Immunodominant epitopes are better immunogens than cryptic epitopes. Compared with T cells specific for cryptic epitopes, $CD8^+$ T cells that recognize immunodominant epitopes interact with their antigen with higher avidity, are primed after a shorter duration of antigen presentation, expand more swiftly and extensively, and generate more potent effector function. Furthermore, by curtailing the duration of Ag presentation [through deletion or exhaustion of antigen presenting cells (APCs)], immunodominant $CD8^+$ effector T cells selectively impair priming against cryptic epitopes. Immunodominance results in one major advantage and one potential drawback: that is, immunodominance favors expansion of the fittest effector T cells but may enhance the risk of immune escape by antigen-loss variants. Targeting immunodominant epitopes is probably crucial not only for success of immune responses against pathogens but also in cancer immunotherapy. Indeed, $CD8^+$ T cells targeted to immunodominant but not cryptic minor histocompatibility antigens can eradicate leukemia and melanoma in mice. In this chapter, I will review the current state-of-the-art regarding T-cell immunodominance and discuss key elements of ongoing and future research in this area.

Keywords Cell differentiation • Cytotoxic T cells • Major histocompatibility complex • Peptide • T-cell receptor

C. Perreault (✉)
Department of Medicine, Université de Montréal, Maisonneuve-Rosemont Hospital, Montréal, QC, Canada
e-mail: c.perreault@videotron.ca

J. Medin and D. Fowler (eds.), *Experimental and Applied Immunotherapy*,
DOI 10.1007/978-1-60761-980-2_9,

Introduction: Definition of Immunodominance

The term "immunodominance" is widely used but loosely defined [1]. In general, immunodominant T-cell responses or epitopes are those that induce the immune response of a greatest magnitude (strength of response) within a single subject; such responses are most frequently detected in a group of individuals (frequency of recognition) or are thought to confer the best protection against pathogens or tumor cells [1–4]. In this chapter, we shall define immunodominant epitopes as those that elicit the immune response of greatest magnitude under real-life conditions, that is, when the immune system is confronted simultaneously with numerous epitopes. In this chapter, the terms epitope and MHC-associated peptide are used interchangeably.

The immunogenicity of an epitope in experimental models is commonly assessed by immunizing subjects solely against the epitope of interest. In this situation, immunogenicity depends on two factors: (1) whether the epitope is adequately presented by MHC molecules on competent APCs; and (2) whether epitope-reactive T cells are present in the T-cell repertoire. However, under most circumstances, such as with infection and transplantation, immune responses are triggered by APCs that present a multitude of nonself epitopes. In this case, T cells respond to only a few "immunodominant" epitopes and neglect other peptides that are otherwise immunogenic when presented alone. Thus, immunodominant epitopes are always immunogenic whereas immunorecessive (or cryptic) epitopes are immunogenic only when presented alone. "Immunodomination" refers to the process whereby immunodominant epitopes suppress T-cell responses against immunorecessive epitopes [5]. My main objective is to review the mechanisms responsible for immunodomination and to discuss their biological relevance. I will focus on $CD8^+$ T-cell responses because the dominance hierarchy of $CD4^+$ T-cell responses has been studied less extensively and may not be contrived by immunodomination (Table 9.1) [6, 7]. Features such as the stability of pepMHC II complexes and TCR:epitope on-rate do influence the amplitude and diversity of CD4 T cell responses [8, 9]. However, there is no substantive evidence that immunodominant pepMHC II epitopes suppress CD4 T-cell responses to other epitopes.

Table 9.1 Important issues to be addressed regarding immunodominance

- Is immunodomination strictly a $CD8^+$ T cell phenomenon or is it operative in $CD4^+$ T cells as well?
- Observation: $CD8^+$ T-cell competition for APC resources is responsible for immunodomination and exerts its effect in the first 5 h after immunization. Unresolved biology: (1) what are the underlying mechanisms? (2) and specifically, do such mechanisms involve production of TNF-α?
- Does immunodominance significantly increase the risk of immune escape by Ag-loss variants?
- What might be the best set of markers to predict which epitopes are immunodominant in humans?
- In humans, do all cancer cells express immunodominant MiHAs and/or TAAs?

H7^a and HY: Two Model Epitopes That Lie at Opposite Ends of the Immunodominance Hierarchy in H2^b Mice

Implicit in the definition of immunodomination is the notion that whether or not an epitope will elicit a response depends on its own characteristics as well as on those of co-presented epitopes. Therefore, because epitope A may dominate B but be dominated by C, its designation as dominant or not is relative and potentially ambiguous. Comparison of $H7^a$ (AKA $B6^{dom1}$) and HY minor histocompatibility antigens (MiHA) has therefore been an instructive paradigm to study immunodomination [3, 10–15]. Indeed, these two epitopes lie at opposite ends on the dominance scale in $H2^b$ mice. HY elicits $CD8^+$ cytotoxic responses when presented alone but not when presented with a single or numerous autosomal MiHAs [15, 16]. On the contrary, $H7^a$ is almost always dominant and induces $CD8^+$ cytotoxic effectors when presented with a multitude of MiHAs [14, 15]. Both $H7^a$ and HY are $H2D^b$-associated nonapeptides. HY (WMHHNMDLI) is encoded by the *Uty* gene whereas $H7^a$ (KAPDNRETL) is encoded by the *Stt3b* gene at the telomeric end of chromosome 9 [17–19].

Immunodomination Results from Competition for APC Resources

We conducted a series of experiments where $H7^{a-b+}$ female mice were immunized with APCs expressing $H7^a$ and/or HY. These mice generated $CD8^+$ T-cell responses against $H7^a$ and/or HY presented alone, but responded only against $H7^a$ when the two MiHAs were presented on the same APC [12, 15]. Notably, immunodomination disappeared in two circumstances: (1) when $H7^a$ and HY were presented simultaneously but on separate APCs; and (2) when huge numbers of APCs co-expressing the two MiHAs were used for priming [12, 15]. These observations demonstrate that immunodomination results from competition for APC resources. The generality of this concept has been validated with other MiHAs, ovalbumin peptides, and viral peptides [20–24].

The mechanisms of competition for APC resources that lead to immunodomination remain ill-defined. Two studies reported that injection of very large numbers of memory $CD8^+$ T cells specific for cryptic Ags did not enable these T cells to compete more successfully against T cells that recognized dominant epitopes [25, 26]. Thus, except perhaps in some extreme situations [27], immunodominant T cells do not win the competition at the T cell/APC interface because they are more abundant than other T-cell clonotypes in preimmune animals [11]. Some models suggest that immunodomination is mediated via APC killing [13, 28]. Thus, we used a model in which APCs were injected into the foreleg footpads of naive recipient mice; APC numbers in the draining (axillary and brachial) lymph nodes were assessed at various times after injection [13]. Rapid elimination of APCs occurred following interactions with MHC I-restricted, but not MHC II-restricted T cells and was observed when APCs presented an immunodominant ($H7^a$), but not a cryptic (HY), epitope. However, at least under certain conditions, immunodomination can be driven by other means than merely APC destruction [29].

Therefore, an alternative possibility would be that early interactions with CTL specific for immunodominant epitopes lead to functional exhaustion or inactivation of the APCs themselves. Willis *et al.* addressed this question in 2006 using a system based on adoptive transfer of naïve $CD8^+$ T cells expressing transgenic OT-I or P14 TCR, followed by immunization with DCs loaded with the peptides recognized by the two TCRs [24]. Transfer of OT-I cells simultaneously with P14 cells resulted in competition between the two T-cell populations. However, when OT-I cells were transferred as little as 5 h after immunization, competition was no longer observed. These results show that $CD8^+$ T cell competition for DCs is an early event that exerts its effect in the first 5 h after immunization [24]. At 5 h after immunization, T cells have not divided yet. The question remains as to what T-cell effector mechanism(s) might be operative after such a short time interval. The prime candidate might be production of TNF-α. Indeed, naive virus-specific TCR-transgenic $CD8^+$ T cells stimulated with either their cognate peptide ligand or virus-infected cells produced soluble and membrane-bound TNF-α as early as 2 h post-stimulation, with production peaking by 4 h [30]. Evidence suggests that soluble TNF-α can interfere with APC maturation during T-cell activation and reduce the viability of the APCs [30]. Furthermore, mice deficient in TNFRI and TNFRII (p55R and p75R, respectively) were able to control an infection with LCMV but generated significantly higher frequencies of virus-specific $CD8^+$ T cells compared with wild-type mice during the acute phase of infection and in memory [31]. These findings suggest that TNF-α production by immunodominant $CD8^+$ T cells may cause immunodomination by rapidly impairing the survival or function of APCs, thereby preventing activation of immunorecessive T cells.

The Transcriptome of Anti-HY and Anti-$H7^a$ $CD8^+$ T Cells

In order to decipher the mechanisms and the ultimate role of immunodominance, we sought to compare the differentiation program of T cells specific for dominant and cryptic Ags. We therefore analyzed global patterns of gene expression in effector $CD8^+$ T cells specific for $H7^a$ and HY MiHAs. Our experimental protocol led to expansion of anti-HY and anti-$H7^a$ $CD8^+$ T cells that were primed concomitantly in the same host and received similar $CD4^+$ T cell help. Thus, B10.$H7^b$ female mice were primed by i.p. injection of a cell mixture containing B10 and B10.$H7^b$ male splenocytes. Because of the immunodomination phenomenon, $H7^a$ abrogates recognition of HY presented on the same APC (B10 male splenocytes), but not of HY presented on separate APCs (B10.$H7^b$ male splenocytes) [3]. Thus, with this immunization schema, each population of APC triggered $CD8^+$ T cells specific for a single MHC I-associated epitope. Recognition of B10.$H7^b$ male splenocytes led to expansion of CD8⁺ T cells specific for the H2D^b-restricted H7^a Ag, whereas male B10 APCs engendered selective expansion of CD8⁺ T cells specific for the H2D^b-restricted H7^a Ag [12, 14, 15, 18, 19]. Of note, both populations of Ag-specific CD8⁺ T cells received CD4⁺ T cell help solely from CD4⁺ T cells specific for the MHC II-restricted HY Ag [32]. After depletion of B220⁺ and CD4⁺ cells, splenocytes were stained with anti-CD8 Ab as well as H7^a and HY tetramers (Tet) (H7^a and HY Tet were labeled with different fluorochromes). Then, three populations of CD8⁺ splenocytes were purified using FACS cell sorting: HYTet⁺, H7^{a}Tet⁺, and Tet⁻. RNA of sorted T cells was extracted and linearly amplified, cRNA was prepared, and Affymetrix Mouse Genome 430 2.0 oligonucleotide arrays were used to analyze gene expression [10].

We first assessed the gene expression profile of effector CD8⁺ T cells primed against the immunodominant H7^a MiHA. Differentially expressed genes were defined according to two criteria: a $\geq$2.5-fold difference in transcript levels between H7^{a}Tet⁺ and Tet⁻ cells, and a p-value$\leq$0.02. Based on these criteria, 222 genes were induced and 86 were repressed in anti-H7^a T cells relative to Tet⁻ CD8⁺ T cell controls [10]. Strikingly, taking a$>$1.5-fold difference and a p-value$<$0.05 as selection criteria, we found that only 15 of 308 genes were differentially expressed in H7^a-versus HY-specific T cells. Thus, about the time of maximal T cell expansion (day 14), Ag specificity had a relatively modest influence on the repertoire of genes that are transcriptionally modulated by the CD8⁺ T-cell differentiation program. Nevertheless, at least some of the differentially expressed genes are key determinants of T-cell fitness and effector function: *Il7r* (IL-7 receptor), *Klrg1* (killer cell lectin-like receptor subfamily G, member 1), *Sell* (L-selectin) and *Gzma* (granzyme A) [33–37]. Expression of *Il7r* and *Sell* characterizes central memory T cells with substantial proliferative potential whereas *Klrg1* is up-regulated on exhausted T cells. Granzyme A contributes to CD8 T-cell cytotoxicity. Those studies performed at the zenith of T-cell expansion characterize specific features of effector T cells that win or lose the immunodomination competition. It must be pointed out, however, that differentially expressed genes at that time (day 14) are not necessarily responsible for immunodomination. Because immunodomination appears to result from some very early event following the T cell/APC interaction, the discovery of such causative genes remains an unmet challenge because it would necessitate performance of time sequential high-throughput expression profiling experiments during the most relevant time frame, that is, the first 5 h after immunization [24]. Furthermore, assignment of a causative role for a gene would require performance of specific loss-of-function and gain-of-function studies.

Asynchronous Differentiation of CD8⁺ T Cells That Recognize Dominant and Cryptic Antigens

Time is an essential dimension of complex systems [38]. We therefore analyzed how differences in gene expression profile of anti-H7^a and anti-HY T cells would be modulated during various phases of the immune response. From day 7 to day 100, we monitored the expansion of anti-$H7^a$ and anti-HY T cells and their expression of selected factors linked to effector function and survival fitness: Sell, Il7R, Klrg1 and the activation-induced glycoform of CD43 (recognized by the 1B11 Ab) [10].

At all time points, the frequencies of $H7^a Tet^+$ T cells were greater than those of $HYTet^+$ T cells. Because the frequencies of anti-$H7^a$ and anti-HY T cell precursors in the preimmune repertoire are similar [11], this means that anti-$H7^a$ T cells underwent more extensive expansion than anti-HY T cells. The salient finding though, was the lack of synchronicity between expansion of the two T-cell populations. Anti-$H7^a$ T cells reached maximal expansion (100%) on day 10 when anti-HY T cells had only reached 20% of their peak response. Expansion of anti-HY T cells reached its maximum between day 15 and 20; at that point, anti-$H7^a$ T cells were well into their contraction phase.

IL-7 receptor (Il7R) is down-regulated following Ag stimulation and is subsequently re-expressed on memory T cells [35]. The proportion of $IL7R^+$ T cells decreased to attain a nadir of about 12% in both anti-$H7^a$ and anti-HY T-cell populations; thereafter, expression increased in the memory T-cell compartment. The lowest proportions of $IL7R^+$ T cells were observed at about the same time as the expansion peaks. Thus, IL7R modulation lingered 5 days behind in anti-HY cells relative to anti-$H7^a$ T cells. However, the picture was different with Sell expression. The overall evolution of Sell as a function of time was similar for the two T-cell populations; however, during the effector phase (day 7–30), the proportion of $Sell^+$ elements was consistently lower in anti-$H7^a$ T cells compared to anti-HY T cells. In the memory T-cell pool (day 100), the proportions of $IL7R^+$ and $Sell^+$ elements were similar for anti-HY and anti-$H7^a$ T cells.

Expression of 1B11 epitopes (activation-induced CD43 glycoform) on $CD8^+$ T cells correlates with perforin/granzyme mediated cytotoxicity [39, 40]. From day 10 to 20, the proportion of $1B11^+$ cells, as well as the MFI of $1B11^+$ cells, were slightly lower for HY-specific T cells relative to $H7^a$-specific T cells. Here, the notable point is that for both T-cell populations, expression of 1B11 epitopes reached a peak between day 7 and 10 and decreased rapidly thereafter. On day 10, the numbers of anti-$H7^a$ T cells were at their zenith whereas anti-HY T cells had reached only 20% of their maximal expansion.

Klrg1 is an inhibitory receptor whose up-regulation correlates with replicative senescence in $CD8^+$ T cells [41, 42]. Klrg1 ligation by cadherins hampers $CD8^+$ T-cell proliferation and cytotoxic activity [43, 44]. We found that accumulation of $Klrg1^+$ cells in the two T-cell populations proceeded at a similar pace for the first 10 days after immunization, but diverged by day 15. From day 15–100, the proportion of $Klrg1^+$ cells was consistently higher for HY-specific T cells relative to $H7^a$-specific T cells. Overall, two main conclusions may be drawn from these studies. First, expansion and development of effector function are well synchronized in anti-$H7^a$ T cells but are disconnected in anti-HY T cells. And second, preferential up-regulation of Klrg1 on HY-specific T cells suggests that upon rechallenge, these T cells would exhibit poor expansion potential and would therefore have minimal "protective" value [45].

T-Cell Avidity and TCR Affinity

The question remains as to why immunodominant T cells expand more swiftly and extensively than other T cells. The strength of T-cell responses is thought to be determined by three factors: epitope density, the frequency of epitope-specific T cells in the preimmune repertoire, and TCR:epitope affinity. High epitope density and high frequencies of epitope-specific T cells both have a positive impact on immunogenicity but are neither necessary nor sufficient for immunodominance [11, 14, 25, 46–49]. However, as exemplified by $H7^a$, all immunodominant T cells nevertheless appear to share one cardinal feature: they interact with APCs expressing their cognate antigen with high avidity [1, 50, 51]. T-cell avidity is mainly determined by the affinity of TCR:epitope interactions and can be assessed by staining with MHC:peptide multimers [52–56]. The results of tetramer association (on-rate) and tetramer decay (off-rate) assays closely correspond to the K_{on} and K_{off} rates, respectively, of the interaction between the soluble TCR and immobilized pepMHC complexes [55, 57]. $CD8^+$ T cells specific for $H7^a$ and HY express similar levels of CD8 and TCR and display similar TCR:pepMHC off-rates [11]. However, tetramer concentrations required to reach 50% and 75% normalized fluorescence were 3.0 nM and 7.5 nM for anti-$H7^a$ T cells versus 30.0 nM and 130.0 nM for anti-HY T cells ($p<0.001$, Student's t test) [11]. Thus, the TCR:pepMHC on-rate is much more rapid for $H7^a$-specific T cells relative to HY-specific T cells. High affinity interactions maximize CD8 T-cell expansion by enhancing their proliferation and survival [58, 59].

Conclusion: The Role of Immunodominance

In summary, immunodominant $CD8^+$ T cells display four main features: (1) they interact with high avidity with their antigen; (2) their priming requires presentation of antigen for only a short duration; (3) they expand more swiftly and extensively and generate more potent effectors than T cells specific for cryptic epitopes; and (4) they suppress priming of $CD8^+$ T cells recognizing cryptic epitopes through competition for APC resources. Since cryptic epitopes are poor immunogens, they require longer duration of Ag presentation than dominant epitopes [60, 61]. By curtailing the duration of Ag presentation (through APC deletion or exhaustion, possibly mediated by TNF-α), immunodominant $CD8^+$ effector T cells selectively impair priming against cryptic epitopes. This model explains why cryptic epitopes such as HY are immunogenic when presented alone (unlimited duration of Ag presentation), but silent when presented with $H7^a$ on the same APC (limited duration of Ag presentation). As a corollary, given the key role of TCR affinity in immunodominance, tetramer association and tetramer decay assays may represent the best surrogate markers to predict immunodominance.

During the course of infection, a 24–48 h delay in generation of effector $CD8^+$ T cells can make the difference between life and death [62, 63]. By leading to selective expansion of the fittest CD8+ effector T cells, immunodomination may be beneficial to the host. Indeed, during the course of immune responses, T cells compete for Ag and other APC resources as well as for prosurvival cytokines [64–66]. We propose that inhibition of T-cell responses against cryptic Ags ensures that host resources for which T cells compete are devoted to T cells that have optimal effector potential. This biology might favor rapid clearance of pathogens not only at the time of primary antigenic exposure but also upon subsequent encounters because immunodomination has a major (though not exclusive) role in shaping the repertoire of the memory T cell compartment [26, 67, 68]. Immunodominance may have one drawback: restricting the diversity of CD8+ T-cell responses to cancer-related antigens may increase the risk of immune escape by Ag-loss variants [69–71]. However, we postulate that, in general, favoring expansion of the fittest effector T cells may have more importance than increasing the diversity of the T cell repertoire. Nevertheless, in vaccine development, it may be advantageous to expand T cells that recognize both dominant and nondominant epitope by presenting the various epitopes on different APCs (e.g., by injecting antigens at different sites) [15, 23].

Two types of epitopes can be targeted for cancer immunotherapy: MiHAs and tumor-associated antigens (TAAs) [72–74]. The success of adoptive immunotherapy using T cells targeted against immunodominant but not against cryptic mouse MiHAs suggests that the prime importance of immunodominance can be extended to cancer immunotherapy [40, 75, 76]. However, translation of this concept into the clinic will require in-depth studies to evaluate fundamental questions and perform a critical evaluation of candidate target antigens. Thus, the question remains: do all human tumors express immunodominant epitopes (MiHAs and/or TAAs)? And, given the importance of TCR-Ag affinity in immunodominance, can TAA-specific T cells interact with their cognate epitope with the same level of avidity as MiHA-specific T cells? Evidence suggests that because of constraints imposed by self-tolerance mechanisms, TAAs are recognized essentially by low avidity T cells [74]. Therefore an emerging strategy is to redirect peripheral blood T cells to TAAs by the genetic transfer of TCRs or chimeric antigen receptors [77]. Some of the most pressing unanswered questions regarding the mechanisms of immunodominance and targeting of immunodominant epitopes for cancer immunotherapy are listed in Table 9.1.

References

1. Bihl, F., N. Frahm, L. Di Giammarino, J. Sidney, M. John, K. Yusim, T. Woodberry, K. Sango, H. S. Hewitt, L. Henry, C. H. Linde, J. V. Chisholm, T. M. Zaman, E. Pae, S. Mallal, B. D. Walker, A. Sette, B. T. Korber, D. Heckerman, and C. Brander. 2006. Impact of HLA-B alleles, epitope binding affinity, functional avidity, and viral coinfection on the immunodominance of virus-specific CTL responses. *J. Immunol.* 176: 4094–4101.
2. Gallimore, A., T. Dumrese, H. Hengartner, R. M. Zinkernagel, and H. G. Rammensee. 1998. Protective immunity does not correlate with the hierarchy of virus- specific cytotoxic T cell responses to naturally processed peptides. *J. Exp. Med.* 187: 1647–1657.

3. Perreault, C., D. C. Roy, and C. Fortin. 1998. Immunodominant minor histocompatibility antigens: the major ones. *Immunol. Today* 19: 69–74.
4. Yewdell, J. W. 2006. Confronting complexity: real-world immunodominance in antiviral CD8+ T cell responses. *Immunity* 25: 533–543.
5. Yewdell, J. W., and J. R. Bennink. 1999. Immunodominance in major histocompatibility complex class I-restricted T lymphocyte responses. *Annu. Rev. Immunol.* 17: 51–88.
6. Smith, A. L., M. E. Wikstrom, and B. Fazekas de St. Groth. 2000. Visualizing T cell competition for peptide/MHC complexes: a specific mechanism to minimize the effect of precursor frequency. *Immunity* 13: 783–794.
7. Creusot, R. J., L. L. Thomsen, J. P. Tite, and B. M. Chain. 2003. Local cooperation dominates over competition between CD4+ T cells of different antigen/MHC specificity. *J. Immunol.* 171: 240–246.
8. Baumgartner, C. K., A. Ferrante, M. Nagaoka, J. Gorski, and L. P. Malherbe. 2010. Peptide-MHC class II complex stability governs CD4 T cell clonal selection. *J. Immunol.* 184: 573–581.
9. Huppa, J. B., M. Axmann, M. A. Mortelmaier, B. F. Lillemeier, E. W. Newell, M. Brameshuber, L. O. Klein, G. J. Schutz, and M. M. Davis. 2010. TCR-peptide-MHC interactions in situ show accelerated kinetics and increased affinity. *Nature* 463: 963–967.
10. Baron, C., M. C. Meunier, E. Caron, C. Côté, M. J. Cameron, D. J. Kelvin, R. LeBlanc, V. Rineau, and C. Perreault. 2006. Asynchronous differentiation of CD8 T cells that recognize dominant and cryptic antigens. *J. Immunol.* 177: 8466–8475.
11. Roy-Proulx, G., C. Baron, and C. Perreault. 2005. CD8 T-cell ability to exert immunodomination correlates with T-cell receptor:epitope association rate. *Biol. Blood Marrow Transplant.* 11: 260–271.
12. Roy-Proulx, G., M. C. Meunier, A. M. Lanteigne, S. Brochu, and C. Perreault. 2001. Immunodomination results from functional differences between competing CTL. *Eur. J. Immunol.* 31: 2284–2292.
13. Loyer, V., P. Fontaine, S. Pion, F. Hétu, D. C. Roy, and C. Perreault. 1999. The *in vivo* fate of APCs displaying minor H antigens and/or MHC differences is regulated by CTLs specific for immunodominant class I-associated epitopes. *J. Immunol.* 163: 6462–6467.
14. Pion, S., G. J. Christianson, P. Fontaine, D. C. Roopenian, and C. Perreault. 1999. Shaping the repertoire of cytotoxic T-lymphocyte responses: explanation for the immunodominance effect whereby cytotoxic T lymphocytes specific for immunodominant antigens prevent recognition of nondominant antigens. *Blood* 93: 952–962.
15. Pion, S., P. Fontaine, M. Desaulniers, J. Jutras, J. G. Filep, and C. Perreault. 1997. On the mechanisms of immunodominance in cytotoxic T lymphocyte responses to minor histocompatibility antigens. *Eur. J. Immunol.* 27: 421–430.
16. Wettstein, P. J. 1986. Immunodominance in the T cell response to multiple non-H-2 histocompatibility antigens. III. Single histocompatibility antigens dominate the male antigen. *J. Immunol.* 137: 2073–2079.
17. Eden, P. A., G. J. Christianson, P. Fontaine, P. J. Wettstein, C. Perreault, and D. C. Roopenian. 1999. Biochemical and immunogenetic analysis of an immunodominant peptide ($B6^{dom1}$) encoded by the classical *H7* minor histocompatibility locus. *J. Immunol.* 162: 4502–4510.
18. Greenfield, A., D. Scott, D. Pennisi, I. Ehrmann, P. Ellis, L. Cooper, E. Simpson, and P. Koopman. 1996. An H-YDb epitope is encoded by a novel mouse Y chromosome gene. *Nat. Genet.* 14: 474–478.
19. McBride, K., C. Baron, S. Picard, S. Martin, D. Boismenu, A. Bell, J. Bergeron, and C. Perreault. 2002. The model $B6^{dom1}$ minor histocompatibility antigen is encoded by a mouse homolog of the yeast *STT3* gene. *Immunogenetics* 54: 562–569.
20. Wolpert, E. Z., P. Grufman, J. K. Sandberg, A. Tegnesjö, and K. Kärre. 1998. Immunodominance in the CTL response against minor histocompatibility antigens: interference between responding T cells, rather than with presentation of epitopes. *J. Immunol.* 161: 4499–4505.
21. Grufman, P., E. Z. Wolpert, J. K. Sandberg, and K. Kärre. 1999. T cell competition for the antigen-presenting cell as a model for immunodominance in the cytotoxic T lymphocyte response against minor histocompatibility antigens. *Eur. J. Immunol.* 29: 2197–2204.

22. Kedl, R. M., W. A. Rees, D. A. Hildeman, B. Schaefer, T. Mitchell, J. Kappler, and P. Marrack. 2000. T cells compete for access to antigen-bearing antigen-presenting cells. *J. Exp. Med.* 192: 1105–1114.
23. Rodriguez, F., S. Harkins, M. K. Slifka, and J. L. Whitton. 2002. Immunodominance in virus-induced CD8+ T-cell responses is dramatically modified by DNA immunization and is regulated by gamma interferon. *J. Virol.* 76: 4251–4259.
24. Willis, R. A., J. W. Kappler, and P. C. Marrack. 2006. CD8 T cell competition for dendritic cells *in vivo* is an early event in activation. *Proc. Natl. Acad. Sci. USA* 103: 12063–12068.
25. Chen, W., L. C. Anton, J. R. Bennink, and J. W. Yewdell. 2000. Dissecting the multifactorial causes of immunodominance in class I-restricted T cell responses to viruses. *Immunity* 12: 83–93.
26. Choi, E. Y., Y. Yoshimura, G. J. Christianson, T. J. Sproule, S. Malarkannan, N. Shastri, S. Joyce, and D. C. Roopenian. 2001. Quantitative analysis of the immune response to mouse non-MHC transplantation antigens *in vivo*: the H60 histocompatibility antigen dominates over all others. *J. Immunol.* 166: 4370–4379.
27. Kotturi, M. F., I. Scott, T. Wolfe, B. Peters, J. Sidney, H. Cheroutre, M. G. von Herrath, M. J. Buchmeier, H. Grey, and A. Sette. 2008. Naive precursor frequencies and MHC binding rather than the degree of epitope diversity shape CD8+ T cell immunodominance. *J. Immunol.* 181: 2124–2133.
28. Wong, P., and E. G. Pamer. 2003. Feedback regulation of pathogen-specific T cell priming. *Immunity* 18: 499–511.
29. Chen, W., K. Pang, K.-A. Masterman, G. Kennedy, S. Basta, N. Dimopoulos, F. Hornung, M. Smyth, J. R. Bennink, and J. W. Yewdell. 2004. Reversal in the immunodominance hierarchy in secondary CD8+ T cell responses to influenza A virus: roles for cross-presentation and lysis-independent immunodomination. *J. Immunol.* 173: 5021–4805.
30. Brehm, M. A., K. A. Daniels, and R. M. Welsh. 2005. Rapid production of TNF-a following TCR engagement of naive CD8 T cells. *J. Immunol.* 175: 5043–5049.
31. Suresh, M., A. Singh, and C. Fischer. 2005. Role of tumor necrosis factor receptors in regulating CD8 T-cell responses during acute lymphocytic choriomeningitis virus infection. *J. Virol.* 79: 202–213.
32. Scott, D., C. Addey, P. Ellis, E. James, M. J. Mitchell, N. Saut, S. Jurcevic, and E. Simpson. 2000. Dendritic cells permit identification of genes encoding MHC class II restricted epitopes of transplantation antigens. *Immunity* 12: 711–720.
33. von Andrian, U. H., and C. R. Mackay. 2000. T-cell function and migration. Two sides of the same coin. *N. Engl. J. Med.* 343: 1020–1034.
34. Barry, M., and R. C. Bleackley. 2002. Cytotoxic T lymphocytes: all roads lead to death. *Nat. Rev. Immunol.* 2: 401–409.
35. Kaech, S. M., J. T. Tan, E. J. Wherry, B. T. Konieczny, C. D. Surh, and R. Ahmed. 2003. Selective expression of the interleukin 7 receptor identifies effector CD8 T cells that give rise to long-lived memory cells. *Nat. Immunol.* 4: 1191–1198.
36. Li, J., G. Huston, and S. L. Swain. 2003. IL-7 promotes the transition of CD4 effectors to persistent memory cells. *J. Exp. Med.* 198: 1807–1815.
37. D'Souza, W. N., and S. M. Hedrick. 2006. Cutting edge: latecomer CD8 T cells are imprinted with a unique differentiation program. *J. Immunol.* 177: 777–781.
38. Ahn, A. C., M. Tewari, C. S. Poon, and R. S. Phillips. 2006. The limits of reductionism in medicine: could systems biology offer an alternative? *PLoS. Med.* 3: e208.
39. Harrington, L. E., M. Galvan, L. G. Baum, J. D. Altman, and R. Ahmed. 2000. Differentiating between memory and effector CD8 T cells by altered expression of cell surface O-glycans. *J. Exp. Med.* 191: 1241–1246.
40. Meunier, M. C., J. S. Delisle, J. Bergeron, V. Rineau, C. Baron, and C. Perreault. 2005. T cells targeted against a single minor histocompatibility antigen can cure solid tumors. *Nat. Med.* 11: 1222–1229.
41. Hand, T. W., M. Morre, and S. M. Kaech. 2007. Expression of IL-7 receptor a is necessary but not sufficient for the formation of memory CD8 T cells during viral infection. *Proc. Natl. Acad. Sci. USA* 104: 11730–11735.

42. Joshi, N. S., W. Cui, A. Chandele, H. K. Lee, D. R. Urso, J. Hagman, L. Gapin, and S. M. Kaech. 2007. Inflammation directs memory precursor and short-lived effector CD8⁺ T cell fates via the graded expression of T-bet transcription factor. *Immunity* 27: 281–295.
43. Grundemann, C., M. Bauer, O. Schweier, N. von Oppen, U. Lassing, P. Saudan, K. F. Becker, K. Karp, T. Hanke, M. F. Bachmann, and H. Pircher. 2006. Cutting edge: identification of E-cadherin as a ligand for the murine killer cell lectin-like receptor G1. *J. Immunol.* 176: 1311–1315.
44. Ito, M., T. Maruyama, N. Saito, S. Koganei, K. Yamamoto, and N. Matsumoto. 2006. Killer cell lectin-like receptor G1 binds three members of the classical cadherin family to inhibit NK cell cytotoxicity. *J. Exp. Med.* 203: 289–295.
45. Voehringer, D., M. Koschella, and H. Pircher. 2002. Lack of proliferative capacity of human effector and memory T cells expressing killer cell lectinlike receptor G1 (KLRG1). *Blood* 100: 3698–3702.
46. Choi, E. Y., G. J. Christianson, Y. Yoshimura, T. J. Sproule, N. Jung, S. Joyce, and D. C. Roopenian. 2002. Immunodominance of H60 is caused by an abnormally high precursor T cell pool directed against its unique minor histocompatibility antigen peptide. *Immunity* 17: 593–603.
47. Pamer, E. G., A. J. Sijts, M. S. Villanueva, D. H. Busch, and S. Vijh. 1997. MHC class I antigen processing of Listeria monocytogenes proteins: implications for dominant and subdominant CTL responses. *Immunol. Rev.* 158: 129–136.
48. Busch, D. H., and E. G. Pamer. 1998. MHC class I/peptide stability: implications for immunodominance, *in vitro* proliferation, and diversity of responding CTL. *J. Immunol.* 160: 4441–4448.
49. Choi, E. Y., G. J. Christianson, Y. Yoshimura, N. Jung, T. J. Sproule, S. Malarkannan, S. Joyce, and D. C. Roopenian. 2002. Real-time T-cell profiling identifies H60 as a major minor histocompatibility antigen in murine graft-versus-host disease. *Blood* 100: 4259–4264.
50. Price, D. A., J. M. Brenchley, L. E. Ruff, M. R. Betts, B. J. Hill, M. Roederer, R. A. Koup, S. A. Migueles, E. Gostick, L. Wooldridge, A. K. Sewell, M. Connors, and D. C. Douek. 2005. Avidity for antigen shapes clonal dominance in CD8⁺ T cell populations specific for persistent DNA viruses. *J. Exp. Med.* 202: 1349–1361.
51. Trautmann, L., M. Rimbert, K. Echasserieau, X. Saulquin, B. Neveu, J. Dechanet, V. Cerundolo, and M. Bonneville. 2005. Selection of T cell clones expressing high-affinity public TCRs within human cytomegalovirus-specific CD8 T cell responses. *J. Immunol.* 175: 6123–6132.
52. Kersh, G. J., E. N. Kersh, D. H. Fremont, and P. M. Allen. 1998. High- and low-potency ligands with similar affinities for the TCR: the importance of kinetics in TCR signaling. *Immunity* 9: 817–826.
53. Kersh, E. N., A. S. Shaw, and P. M. Allen. 1998. Fidelity of T cell activation through multistep T cell receptor z phosphorylation. *Science* 281: 572–575.
54. Monks, C. R., B. A. Freiberg, H. Kupfer, N. Sciaky, and A. Kupfer. 1998. Three-dimensional segregation of supramolecular activation clusters in T cells. *Nature* 395: 82–86.
55. Savage, P. A., J. J. Boniface, and M. M. Davis. 1999. A kinetic basis for T cell receptor repertoire selection during an immune response. *Immunity* 10: 485–492.
56. Messaoudi, I., J. A. G. Patino, R. Dyall, J. LeMaoult, and J. Nikolich-Zugich. 2002. Direct link between *mhc* polymorphism, T cell avidity, and diversity in immune defense. *Science* 298: 1797–1800.
57. Busch, D. H., and E. G. Pamer. 1999. T cell affinity maturation by selective expansion during infection. *J. Exp. Med.* 189: 701–710.
58. Hommel, M., and P. D. Hodgkin. 2007. TCR affinity promotes CD8⁺ T cell expansion by regulating survival. *J. Immunol.* 179: 2250–2260.
59. Zehn, D., S. Y. Lee, and M. J. Bevan. 2009. Complete but curtailed T-cell response to very low-affinity antigen. *Nature* 458: 211–214.
60. Gett, A. V., F. Sallusto, A. Lanzavecchia, and J. Geginat. 2003. T cell fitness determined by signal strength. *Nat. Immunol.* 4: 355–360.

61. Prlic, M., G. Hernandez-Hoyos, and M. J. Bevan. 2006. Duration of the initial TCR stimulus controls the magnitude but not functionality of the CD8⁺ T cell response. *J. Exp. Med.* 203: 2135–2143.
62. Ehl, S., P. Klenerman, P. Aichele, H. Hengartner, and R. M. Zinkernagel. 1997. A functional and kinetic comparison of antiviral effector and memory cytotoxic T lymphocyte populations *in vivo* and *in vitro*. *Eur. J. Immunol.* 27: 3404–3413.
63. Li, Q., P. J. Skinner, S. J. Ha, L. Duan, T. L. Mattila, A. Hage, C. White, D. L. Barber, L. O'Mara, P. J. Southern, C. S. Reilly, J. V. Carlis, C. J. Miller, R. Ahmed, and A. T. Haase. 2009. Visualizing antigen-specific and infected cells in situ predicts outcomes in early viral infection. *Science* 323: 1726–1729.
64. Blattman, J. N., J. M. Grayson, E. J. Wherry, S. M. Kaech, K. A. Smith, and R. Ahmed. 2003. Therapeutic use of IL-2 to enhance antiviral T-cell responses *in vivo*. *Nat. Med.* 9: 540–547.
65. Seder, R. A., and R. Ahmed. 2003. Similarities and differences in CD4⁺ and CD8⁺ effector and memory T cell generation. *Nat. Immunol.* 4: 835–842.
66. Melchionda, F., T. J. Fry, M. J. Milliron, M. A. McKirdy, Y. Tagaya, and C. L. Mackall. 2005. Adjuvant IL-7 or IL-15 overcomes immunodominance and improves survival of the CD8⁺ memory cell pool. *J. Clin. Invest.* 115: 1177–1187.
67. Blattman, J. N., D. J. Sourdive, K. Murali-Krishna, R. Ahmed, and J. D. Altman. 2000. Evolution of the T cell repertoire during primary, memory, and recall responses to viral infection. *J. Immunol.* 165: 6081–6090.
68. Crowe, S. R., S. J. Turner, S. C. Miller, A. D. Roberts, R. A. Rappolo, P. C. Doherty, K. H. Ely, and D. L. Woodland. 2003. Differential antigen presentation regulates the changing patterns of CD8⁺ T cell immunodominance in primary and secondary influenza virus infections. *J. Exp. Med.* 198: 399–410.
69. Allen, T. M., D. H. O'Connor, P. Jing, J. L. Dzuris, B. R. Mothe, T. U. Vogel, E. Dunphy, M. E. Liebl, C. Emerson, N. Wilson, K. J. Kunstman, X. Wang, D. B. Allison, A. L. Hughes, R. C. Desrosiers, J. D. Altman, S. M. Wolinsky, A. Sette, and D. I. Watkins. 2000. Tat-specific cytotoxic T lymphocytes select for SIV escape variants during resolution of primary viraemia. *Nature* 407: 386–390.
70. Erickson, A. L., Y. Kimura, S. Igarashi, J. Eichelberger, M. Houghton, J. Sidney, D. McKinney, A. Sette, A. L. Hughes, and C. M. Walker. 2001. The outcome of hepatitis C virus infection is predicted by escape mutations in epitopes targeted by cytotoxic T lymphocytes. *Immunity* 15: 883–895.
71. Barouch, D. H., J. Kunstman, M. J. Kuroda, J. E. Schmitz, S. Santra, F. W. Peyerl, G. R. Krivulka, K. Beaudry, M. A. Lifton, D. A. Gorgone, D. C. Montefiori, M. G. Lewis, S. M. Wolinsky, and N. L. Letvin. 2002. Eventual AIDS vaccine failure in a rhesus monkey by viral escape from cytotoxic T lymphocytes. *Nature* 415: 335–339.
72. Perreault, C., and S. Brochu. 2002. Adoptive cancer immunotherapy: discovering the best targets. *J. Mol. Med.* 80: 212–218.
73. Bleakley, M., and S. R. Riddell. 2004. Molecules and mechanisms of the graft-versus-leukaemia effect. *Nat. Rev. Cancer* 4: 371–380.
74. Offringa, R. 2009. Antigen choice in adoptive T-cell therapy of cancer. *Curr. Opin. Immunol.* 21: 190–199.
75. Fontaine, P., G. Roy-Proulx, L. Knafo, C. Baron, D. C. Roy, and C. Perreault. 2001. Adoptive transfer of T lymphocytes targeted to a single immunodominant minor histocompatibility antigen eradicates leukemia cells without causing graft-versus-host disease. *Nat. Med.* 7: 789–794.
76. Meunier, M. C., C. Baron, and C. Perreault. 2009. Two host factors regulate persistence of $H7^a$-specific T cells injected in tumor bearing mice. *PLoS ONE* 4: e4116.
77. Sadelain, M., R. Brentjens, and I. Riviere. 2009. The promise and potential pitfalls of chimeric antigen receptors. *Curr. Opin. Immunol.* 21: 215–223.

Chapter 10
T Regulatory Cells and Cancer Immunotherapy

Adele Y. Wang and Megan K. Levings

Abstract The immune system is comprised of a network of regulatory and effector mechanisms that is balanced to eliminate threats to the body, while preventing uncontrolled inflammatory and auto-immune responses that would otherwise harm self. T regulatory (Treg) cells are central to this process and are required to maintain self-tolerance by actively inhibiting self-reactive T cells by a variety of direct and indirect mechanisms. Numerous studies have shown that Treg cells also exist in tumors, promote a suppressive environment that interferes with antitumor immunity, and are a major barrier to successful immunotherapy. In this review we will discuss evidence for how different types of $CD4^+$ Treg cells are involved in antitumor immunity in both mice and humans. We also highlight strategies that have successfully inhibited Treg cells in animal models of cancer and in human clinical trials. In addition, the known effects of Treg cells on current immunotherapy approaches will be discussed. A better understanding of how Treg cells hinder antitumor immunity is required to optimize current therapeutic regimens and to make effective cancer immunotherapy a reality for an increased number of patients.

Keywords T regulatory cells • Immunotherapy • FOXP3 • Cancer • Tolerance

Introduction

Immune homeostasis is maintained by a balance of regulatory and effector mechanisms that elicit protection against extracellular pathogens while preventing autoimmunity. It is now well established that T regulatory (Treg) cells are necessary for normal immune homeostasis through their capacity to control the activation and expansion of self-reactive T cells and thereby maintain self-tolerance. For example,

M.K. Levings (✉)
Department of Surgery, University of British Columbia,
and Child and Family Research Institute, Vancouver, BC, Canada
e-mail: megan.levings@ubc.ca

J. Medin and D. Fowler (eds.), *Experimental and Applied Immunotherapy*,

DOI 10.1007/978-1-60761-980-2_10,

depletion of Treg cells results in the spontaneous development of autoimmune diseases including type 1 diabetes, inflammatory bowel disease and thyroiditis [1]. On the other hand, a deficit in Treg cells triggers antitumor immunity in tumor-bearing animals and boosts immune responses to pathogens [2]. Indeed, increasing evidence suggests that Treg-cell-mediated suppression of responses to tumor antigens is a major barrier to successful cancer immunotherapy [3–5]. In this review we will discuss how different types of CD4[+] Treg cells impact antitumor immunity in mice and humans. We will also illustrate strategies that have been successfully employed to deplete and inhibit Treg cells in mouse tumor models and discuss how these findings are being translated to human clinical trials. Finally, we will review how Treg cells may impact the effectiveness of a variety of approaches to boost the antitumor immune response.

Subsets of CD4$^+$ Treg Cells

Naturally Occurring Treg Cells in Mice and Humans

Naturally occurring (n) FOXP3$^+$ Treg cells were originally defined on the basis of constitutive expression of CD25 (also known as interleukin (IL)-2 receptor α chain) [6], and represent 5–10% of the total CD4$^+$ T-cell population in humans and mice [7, 8]. These nTreg cells exit the thymus as a distinct and functionally mature T-cell subpopulation with a diverse T-cell receptor (TCR) repertoire that is poised to suppress responses to self-antigens [9]. As discussed in more detail below, since most tumor antigens are self-antigens, nTreg cells represent a major mechanism by which effective antitumor immunity is suppressed. In humans, other proteins that are characteristically expressed by nTreg cells include LAP, IL-1R, CD39, and CD73, but not CD127, CD49d, IL-2, or IFN-γ [1, 10, 11]. The development, maintenance, and function of nTreg cells depends on the expression of the forkhead/winged-helix family transcription regulator, FOXP3 [12]. A genetic deficiency of *FOXP3* in mice results in hyper-activation of CD4$^+$ and CD8$^+$ T cells that causes severe organ-specific autoimmunity. Evidence that adoptive transfer of nTreg cells can reverse auto-immunity indicates that the pathology is a direct sequelae of a deficiency in Treg cells [12].

It is now clear that although there are many parallels between the phenotype and function of Treg cells in mice and humans, there are several important differences. As in mice, mutation of *FOXP3* in humans also causes auto-immunity, and is the underlying genetic defect in patients with immune dysregulation, polyendocrinopathy, enteropathy, and X-linked syndrome (IPEX). IPEX is a rare X-linked immunodeficiency disease that typically causes type 1 diabetes, inflammatory bowel disease, allergies, and hyper IgE production [13, 14]. The cellular changes in IPEX patients, however, are diverse and do not parallel those in *FOXP3*$^{-/-}$ mice. Specifically, the magnitude of the defect in Treg cells in IPEX patients depends on the type of *FOXP3* mutation: Only patients with null mutations appear to have a complete block of the development and function of Treg cells. In addition, there is a parallel defect in cytokine production by conventional T (Tconv) cells, suggesting that in humans, FOXP3 may have a function outside the Treg subset [13]. We have also shown that humans express two different splice variants of FOXP3, with the smaller FOXP3b form lacking exon 2 [15]. It was recently shown that FOXP3b lacks part of the transcriptional repression domain and cannot interact with ROR-α or ROR-γt, two transcription factors associated with T helper 17 (Th17)-cell development [16, 17]. Thus, because human T cells coexpress both isoforms of FOXP3, cooperation with other transcription factors and transcriptional regulation is likely distinct from that in mice.

Another major difference between mouse and human cells is that FOXP3 is transiently expressed in human Tconv cells, but at levels that are insufficient to suppress cytokine production or confer suppressive function [2, 18, 19]. This finding has significant implications for studies that have simply used analysis of FOXP3 expression (with or without CD25 assessment) to track and enumerate Treg cells. In the absence of assays to determine whether the putative FOXP3$^+$ Treg cells have repressed cytokines and/or suppressive capacity, accurate conclusions about changes in Treg numbers cannot be made. The fact that transient expression of FOXP3 is not sufficient to confer a Treg phenotype suggests that coexpression of other transcription factors and stable epigenetic changes are required for Treg development. This notion is supported by our finding that only upon stable and long-term expression of FOXP3 can the Treg phenotype be recapitulated in naive or memory human CD4$^+$ T cells [20]. Recent evidence indicating that a subset of FOXP3$^+$ cells secretes IL-17 and may or may not be suppressive underscores the need for better markers and standardized assays to identify Treg subsets [21–23].

Inducible Treg Cells

In addition to thymus-derived Treg cells, there are also subsets of Treg cells that develop in the periphery when they encounter antigen in a tolerogenic context. The best-characterized types of induced (i) CD4$^+$ Treg cells are those that are phenotypically similar to the nTreg cells described above (i.e., FOXP3$^+$CD25$^+$) or the IL-10 secreting type 1 T regulatory (Tr1) cells [24]. Through their capacity to produce both IL-10 and TGF-β, many tumor cells influence the development of iTreg cells via direct and indirect mechanisms [10]. The main indirect mechanism is mediated by effects on antigen-presenting cells (APCs), which when exposed to IL-10 and/or TGF-β become tolerogenic and contribute to the development of different types of iTreg cells [5].

In mice, the development of FOXP3$^+$ iTreg cells (which were previously known as Th3 cells) can be stimulated by many different mechanisms. For example, exposure to TGF-β in the absence or presence of retinoic acid, a vitamin A metabolite produced by CD103$^+$ dendritic cells (DCs) in the gut-associated lymphoid tissue, leads to the development of FOXP3$^+$ Treg cells that can suppress auto-immunity in *FOXP3*-deficient mice [25]. Activation of naive T cells by $\alpha_V\beta_8$-integrin-expressing DCs [26] or DCs lacking in SOCS3 [27] can contribute to the generation of peripheral Treg cells. Blockade of the phosphatidylinositol 3-kinase (PI3K)/Akt/mammalian target of rapamycin (mTOR) pathway also stimulates iTreg development [28]. This finding has had broad clinical implications since rapamycin (Sirolomus), a commonly used immunosuppressive drug that inhibits the mTOR pathway, has the potential to specifically block Tconv cells and spare/promote nTreg and iTreg cells [28–30]. FOXP3$^+$ T cells can also be induced by TGF-β in humans, but compared to their mouse counterparts, the resulting T cells do not have detectable suppressive capacity *in vitro* [19]. This indicates that more research is required to understand if the process of iTreg development is fundamentally different in mice and humans, or if current methods to detect suppression *in vitro* are inadequate to reveal their true suppressive potential.

In contrast to FOXP3$^+$ T cells, Tr1 cell development occurs exclusively in the periphery when naïve CD4$^+$ T cells are primed by tolerogenic APCs in the presence of IL-10 [24]. IL-27 has recently been shown to induce IL-10-producing Tr1 cells in mice; however, its role in human Tr1 cells has yet to be defined [31, 32]. Tr1 cells characteristically produce high levels of IL-10 and TGF-β, but do not constitutively express FOXP3 [24]. In humans, Tr1 cells also secrete low levels of IFN-γ, which contributes to their suppressive mechanism through the inhibition of T-cell proliferation and the induction of apoptosis [33]. Presently, there are no known specific cell-surface markers to isolate and track Tr1 cells *ex vivo*; thus, their biological characterization and correlation with human disease remains challenging.

Suppressive Mechanisms of Treg Cells

Treg cells have a remarkable ability to suppress the proliferation and effector function of many different types of immune cells including CD4$^+$ T cells, CD8$^+$ T cells, NK cells, NK-T cells, DCs, monocytes, and B cells [1, 10]. Knowledge of how Treg cells interact with other immune cells to prevent antitumor immunity is fundamental for the discovery of therapeutic targets to enhance the antitumor response. In the text below, we review the major suppressive mechanisms that are thought to be employed by CD4$^+$ Treg cells. Although most of this knowledge regarding mechanisms of action has not been generated in the setting of cancer specifically, findings in other disease contexts can likely be translated to tumor-specific Treg cells.

Inhibitory Cytokines

IL-10 and TGF-β are immunosuppressive cytokines that are central mechanisms of suppression. Although FOXP3$^+$ Treg cells were originally thought to operate via a cytokine-independent, cell-contact-dependent mechanism, it is now clear that this paradigm needs to be revised based on recent data that, at least in mice, FOXP3$^+$ Treg cells express cell surface-bound TGF-β and mediate cytokine-dependent suppression and infectious tolerance [11, 34]. TGF-β affects many aspects of T-cell function, most notably suppression of proliferation and cytokine production [35]. IL-10 also has similar antiinflammatory and inhibitory effects on most hematopoietic cells [24] and is required for normal immune homeostasis. Similarly to TGF-β, IL-10 acts via a positive feedback pathway to promote its own expression, thereby reinforcing immune suppression and expanding the immune-regulatory network [24, 34].

Another inhibitory cytokine known as IL-35 is preferentially expressed by mouse FOXP3$^+$ Treg cells and contributes to their suppressive activity [36]. IL-35, which is a member of the IL-12 family and a heterodimer composed of Ebi3 and p35 subunits, is secreted by Treg cells but only in the context of a suppression assay in the presence of Tconv cells. Treg cells that are deficient in IL-35 cannot suppress T-cell proliferation and fail to rescue inflammatory bowel disease *in vivo*. A recent study found that some of the effects of IL-35 rely on IL-10 production, and indeed both IL-10 and IL-35 are required for maximal Treg-cell-mediated suppression *in vitro* [37]. In contrast, human Treg cells isolated *ex vivo* and FOXP3-transduced T cells do not express detectable amount of IL-35 [20, 38]; however, more studies are required to address the possibility that this cytokine may only be induced in the presence of non-Treg cells or other types of accessory cells. Whether or not IL-35 contributes to antitumor immunity remains to be defined.

Cytolytic Pathways

Classically, CD8$^+$ cytotoxic T lymphocytes (CTLs) and NK cells are thought to be the primary mediators of cytolysis via perforin and granzymes. More recently, it has become evident that Treg cells can also express perforin and/or granzymes and thereby directly kill their targets of suppression. For example, activated human Treg cells express granzyme A and perforin, and can eliminate T cells and APCs through this pathway [39]. In addition, the induction of apoptosis in Tconv cells by mouse Treg cells is dependent on granzyme B expression [40]; Treg cells from granzyme-B-deficient mice have decreased suppressive capacity *in vitro*. Treg cells can also suppress B cell function by killing them through a granzyme B- and perforin-dependent manner [41]. In addition to perforin and granzyme, Treg cells also express Fas and FasL [42], and FasL-induced apoptosis of target cells is another possible mechanism of suppression. Indeed, a recent study found that human Treg cells induce Fas-mediated apoptosis of autologous CD8$^+$ T cells [43], and that this process is amplified in patients with cancer. Taken together with evidence indicating that Treg cells interfere with tumor cell removal by granzyme B and perforin-mediated killing of NK cells and CTLs [44], these data suggest that Treg mediated cytolysis is relevant to antitumor immunity.

Metabolic Dysregulation

Treg cells constitutively express the high affinity IL-2 receptor, which is a heterotrimeric receptor composed of CD25 (α-chain), CD122 (β-chain), and CD132 (γ-chain). The trimeric complex has a 100-fold higher affinity for IL-2 than the dimeric (CD122 and CD132) form [17]. Therefore, it has been speculated that Treg cells might starve

Tconv cells by competitive consumption of IL-2 [45]. Although we have not found experimental evidence to support this mechanism in humans [8], a recent report showed that mouse Treg cells consume IL-2 and induce cytokine-deprivation apoptosis of Tconv cells *in vitro* [46]. On the other hand, another study suggests that IL-2 deprivation alone is not sufficient to suppress cytokine production from Tconv cells [47]. Interestingly, Leveque *et al.* [48] suggested that IL-2 could contribute to antitumor immunity by interfering with Treg suppression; upon culture with IL-2, ovarian cancer-associated Treg cells were converted into proinflammatory Th17 cells and lost their suppressive ability. Therefore, more work is required to better understand if IL-2 consumption is truly a significant mechanism of suppression.

Another potential way that Treg cells can inhibit the metabolic environment is by expression of CD39 and CD73, which are ectoenzymes that catalyze the conversion of cAMP to adenosine [49, 50]. Treg cells can also suppress Tconv cells by using membrane gap junctions to transfer cAMP, a potent inhibitor of proliferation and IL-2 synthesis in T cells [51]. Adenosine is an immunosuppressive molecule that has long been known to inhibit Tconv cell proliferation via activation of the adenosine receptor 2A. Notably adenosine can also enhance the development of iTreg cells by promoting the production of TGF-β [52]. Although the accumulation of extracellular adenosine in the tumor microenvironment has been established [53], it remains to be determined what proportion is derived from Treg cells [54]. Studies to investigate whether Treg cells utilize these novel metabolic dysregulation pathways to suppress antitumor immunity are required.

Interaction with APCs

Increasing evidence indicates that Treg cells communicate with APCs, in particular with DCs and monocytes to modulate their maturation and function [55–57]. One molecule that is critical to this process is CLTA-4. Although CTLA-4 has long been known to be associated with the immunosuppressive function of Treg cells [58], it has been difficult to establish the mechanistic basis for its effects. A recent study by Onishi *et al.*, published in 2008, provided evidence that the role of CTLA-4 in Treg cell-mediated suppression is mediated via DCs, not via direct effects on T cells [56]. Specifically, Treg cells were found to aggregate around DCs and cause down-regulation of CD80/86 expression by a mechanism that required both LFA-1 and CTLA-4. Indeed, a Treg-specific deficiency in CTLA-4 impairs the suppressive capacity of Treg cells *in vitro* and *in vivo,* and eliminates Treg-mediated down-regulation of CD80/86 on DCs [59]. CTLA4^{+} Treg cells can also stimulate the production of IDO production by DCs. IDO is an enzyme that degrades the essential amino acid tryptophan [60], representing yet another mechanism of Treg-mediated immunosuppression. Notably, mice with CTLA-4^{-} Treg cells have robust antitumor immunity [59], highlighting the importance of Treg-DC interactions in cancer. Since Treg cells found in human tumors express CTLA-4 [61], and IDO^{+} APCs are found both in tumors and their draining lymph nodes [60], CTLA-4-mediated suppression of APCs is likely a major mechanism that counteracts antitumor immunity.

Another mechanism by which Treg cells can suppress APCs is via induction of B7-H4, a member of the B7 family of T-cell costimulatory molecules [62, 63]. Ovarian-tumor-associated macrophages, but not normal macrophages, express B7-H4 and contribute to the suppression of tumor-associated antigen (TAA)-specific T-cell responses [64]. Since Treg cells can enhance expression of B7-H4 on APCs, targeting B7-H4 is another potential approach to enhance T-cell immunity [61].

Emerging evidence suggests that B7-H1 (PD-L1), which is broadly and constitutively expressed by B cells, DCs, macrophages, and T cells [65], is important for the maintenance of self-tolerance [66] and has a major role in regulation of tumor immunity [67]. Its ligand, program death-1 (PD-1) is up-regulated on activated T cells and on engagement reverses the activation pathway [66]. Mechanistically, stimulation of PD-1 may directly impact Treg cells since ligation by PD-L1 promotes sustained expression of FOXP3 [68]. Interestingly, PD-1 blockade in melanoma patients appeared to interfere with the suppressive capacity of Treg cells [67] and enhanced the generation of melanoma Ag-specific CTLs. Therefore, blockade of PD-1 is another promising therapeutic approach to improve tumor immunity.

Evidence for Treg Cell-Mediated Suppression in Tumor Immunity

As discussed above, tumors recruit a wide selection of defence mechanisms to suppress host immunity, interfere with the antitumor response by altering T cell and APC function, and establish an immunosuppressive cytokine environment. This topic has been extensively reviewed [69–71] and here we will briefly discuss data from mouse models and then focus on evidence that active suppression by Treg cells inhibits the antitumor response in humans.

Mouse Tumor Models

More than 20 years ago, Robert North and his colleagues found that suppressor T cells inhibited antitumor responses [72] and that they could be selectively eliminated by cyclophosphamide [73]. When suppressor T cells were redefined as $CD4^{+}CD25^{+}$ Treg cells in the 1990s, interest in the role of these cells in tumor immunology was reignited. Early studies in mouse models showed that tumor formation induced by adoptive transfer of tumor cell lines could be reduced by depleting $CD25^{+}$ cells [74]; more recently, it was shown that tumor reduction upon Treg depletion was associated with the development of tumor-specific $CD8^{+}$ T and NK-cell responses [75]. Furthermore, T cells stimulated in the absence of $CD25^{+}$ T cells have improved lytic ability and increased production of IFN-γ allowing for more efficient elimination of tumor cells [76, 77].

In addition to these studies with tumor cell lines, research on the development of *de novo* tumors in animal models also supports the notion that Treg cells interfere with tumor rejection. For example, tumors induced by the chemical carcinogen methylcholanthrene (MCA) develop more slowly and less frequently in mice depleted of $CD25^+$ T cells [78, 79]. Notably, the majority of T cells infiltrating the MCA-induced tumor are $FOXP3^+CD4^+$ Treg cells. Increased percentages of Treg cells are also found in benzopyrene-induced tumors; such Treg cells were shown to express high levels of FOXP3 and IL-10 [80]. Overall, these accumulated findings suggest that Treg cells are central in the prevention of tumor immunity through the suppression of Tconv cells and APCs.

Human Tumors

Upon identification of CD25, and subsequently FOXP3, as proteins that could be used to track and enumerate Treg cells in humans, many clinical studies have attempted to correlate changes in Treg cells with outcomes in various types of cancer. As in mice, a higher frequency of putative Treg cells is seen in patients with a wide variety of cancers (reviewed by Zou [5]). Due to the lack of specific markers, it has been difficult to define the relative contribution of nTreg cells versus different subsets of iTreg cells, but it is widely assumed that both types of cells coexist in the tumor microenvironment.

Unfortunately, the results of many of these studies are difficult to interpret since both CD25 and FOXP3 are also expressed on activated Tconv cells. The recent finding that a subset of $FOXP3^+$ cells produces IL-17 [23, 81] further underlines the need for more precise tools to monitor *bona fide* Treg cells in human diseases. One strategy is to couple analysis of FOXP3 expression with cytokine production; for example, a recent study enumerated $FOXP3^+CD4^+$ Treg cells in conjunction IL-2 and IFN-γ levels and found a significant increase in the proportion of Treg cells in intratumoral and peritumoral sections of metastatic melanoma tumors, but not in peripheral blood [82]. Another approach is to monitor the methylation status of the FOXP3 promoter because cells that only express FOXP3 transiently and at low levels do not have stable demethylation of defined regions [83]. Notably, Huehn and colleagues developed a quantitative real-time PCR-based methylation assay to enable more precise identification of Treg numbers by measuring the degree of demethylation at the Treg-specific demethylated region of the *FOXP3* promoter. Using this method, they found that Treg numbers were increased in the blood of patients with IL-2-treated melanoma and in tissue from patients with lung and colon carcinomas [84].

More direct evidence for the role of Treg cells in human cancers has come from elegant studies on patients with ovarian cancer. $FOXP3^+$ Treg cells were found to be abundant in the ovarian-tumor microenvironment, inhibit TAA-specific $CD8^+$ T-cell cytotoxicity, and predict patient survival [61]. Evidence that the Treg cells isolated from peripheral blood, ascites, or solid tumors of ovarian cancer patients were equivalent in their capacity to suppress T-cell activation *in vitro* [61] suggests that tumor-infiltrating Treg cells are not more suppressive than those in peripheral blood and indicates that the decrease in antitumor immunity is related to their increased numbers rather than altered function at the tumor site [5]. Preferential trafficking of

Treg cells to tumors may be mediated by CC-chemokine ligand (CCL) 22, which is the ligand for CCR4 that is highly expressed on Treg cells [61].

Further questions are: what is the antigen-specificity of Treg cells that infiltrate tumors; and, does their repertoire overlap with that of the desired TAA-specific Tconv cells? Theoretically, TAA-specific Treg cells may pre-exist as part of the normal nTreg repertoire and/or be actively induced in the periphery via one of the mechanisms discussed above. There is strong evidence that TAA-specific Treg cells do indeed exist and share common targets with Tconv cells. For example, among a panel of $CD4^+$ T-cell clones from human melanoma-infiltrating lymphocytes, $FOXP3^+$ Treg cell-clones specific for LAGE-1 were identified [85]. Notably, these Treg-cell clones inhibited LAGE-1-stimulated T-cell proliferation. TAA-specific Treg cells can also prevent the priming of TAA-specific T effector cells: Only upon removal of $CD4^+CD25^+$ T cells were NY-ESO-1-specific $CD4^+$ T cells generated from naïve populations [86].

Evidence That Treg Cells Inhibit Cancer Immunotherapy

DCs are potent APCs that have superior ability to take up, process, and present antigens. DC-based vaccines have been widely employed as an immunotherapy strategy to enhance TAA-specific immunity in cancer patients. Various methods of loading antigen onto DCs include pulsing with soluble peptides [87], RNA transfection [88], and virus-based gene transfer [89]. Unfortunately, there has not been consistent tumor regression in patients who have participated in DC-based regimens, including one randomized phase III clinical trial [90, 91]. It is hypothesized that the efficiency of DC vaccines is often impeded by the presence of Treg cells. For example, co-culture of DCs with human lung carcinoma cells in order to load them with TAAs results in increased TGF-β production that in turn generates $FOXP3^+$ Treg cells [92]. Similarly, although DCs fused with breast carcinomas stimulate the expansion of Tconv cells, the Treg cells that expand in parallel ultimately suppress T-cell responses [93]. The effects of adoptively transferred TAA-specific T cells is also inhibited by Treg cells. For example, in a phase I trial of melanoma patients, although transfer of TAA-specific T cells together with fludarabine resulted in a 2.9-fold improvement in the life-span of T cells, there was a parallel increase in endogenous $FOXP3^+$ Treg cells [94]. Thus, in order to amplify the antitumor response elicited by cancer immunotherapy, it is necessary to control Treg cell proliferation while allowing the expansion of T effector cells.

Strategies to Deplete/Inhibit Treg Cells to Enhance Antitumor Immunity in Mice

There is extensive evidence from mouse tumor models that the removal of Treg cells augments antitumor responses. As many of these studies have been reviewed by others [5, 95], here we will focus on strategies to deplete Treg cells and promote tumor immunity that have been published within the last 5 years (Fig. 10.1).

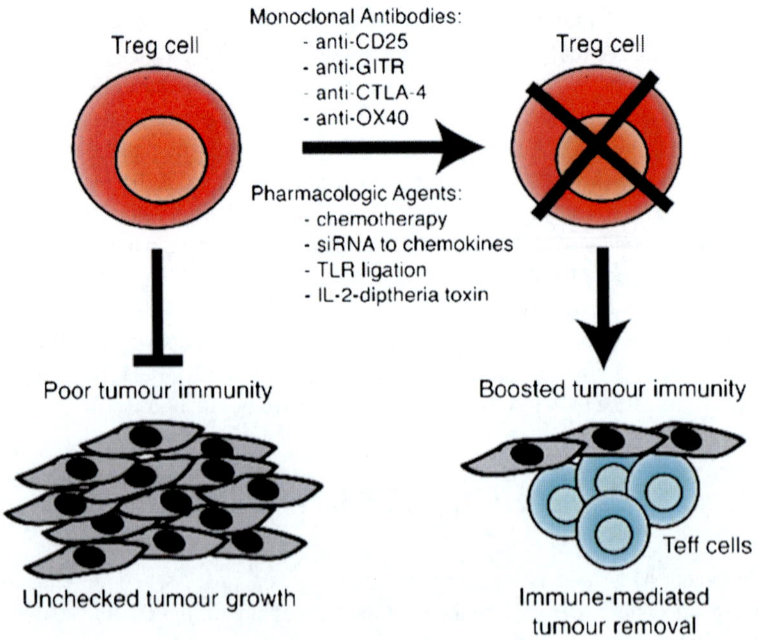

Fig. 10.1 Strategies to remove Treg cells and/or inhibit their function. The presence of Treg cells interferes with the antitumor responses allowing tumor cells to evade host immunity. Animal models and clinical trials have effectively employed various strategies to render Treg cells inactive and boost antitumor immunity

Anti-CD25 mAbs

In mice, the most widely used strategy to remove Treg cells and boost antitumor immunity is to treat with an anti-CD25 monoclonal antibody (mAb), usually PC61 [96]. Although there is some debate about whether this mAb truly depletes Treg cells [97, 98], there are many reports indicating that *in vivo* administration of PC61 causes tumor regression and is associated with a reduction in circulating Treg cells [99, 100]. On the other hand, administration of PC61 appears to be less efficient when given to mice with established tumors; this observation is likely due to the simultaneous depletion of activated $CD25^+$ T effector cells in this setting [78]. More recent studies have combined the use of anti-CD25 mAbs with other therapies, such as DC-based vaccination, to achieve synergistic antitumor response [101, 102]. A drawback to this approach as mentioned above is that, since T effector cells also express CD25, the anti-CD25 mAbs have the potential to inadvertently remove desirable TAA-specific T effector cells as well as Tregs.

Anti-GITR mAbs

In addition to CD25, GITR-specific mAbs have also been tested in mice since Treg cells are known to express this protein. Anti-GITR mAbs protect mice from developing melanoma upon challenge with the B16 tumor cell line [103] and allow the generation of antitumor immunity in Meth A-induced sarcomas [104]. Some anti-GITR mAbs are agonistic; that is, GITR ligation on T effector cells is thought to limit their capacity to be suppressed by Treg cells [105]. Supporting this concept, it was recently reported that injection of tumor-primed $CD4^+$ T cells in combination with GITR ligation *in vivo* resulted in effective $CD8^+$ T-cell-dependent immunity in a breast tumor model [106]. Notably, like CD25, GITR is also expressed by activated Tconv cells and on some APCs [105]; as such, even though GITR-specific Abs can impact Treg cells, it is unclear whether this could ever be used as a Treg-specific therapy.

Anti-CTLA-4 mAbs

Treg cells constitutively express CTLA-4 [58]; interestingly, FOXP3 directly controls CTLA-4 expression [59]. Therefore, treatment with anti-CTLA-4 mAbs represents another possible approach to inhibit Treg cell function. Indeed, *in vivo* injection of anti-CTLA-4 mAbs can result in tumor rejection and immunity upon re-exposure to tumor cells [107]. It should be noted that since CTLA-4 is also expressed by Tconv cells, the mechanistic basis for the effects of CTLA-4 blockade is not necessarily due to altered Treg cell function. Indeed, it has been reported that CTLA-4 blockade increased the number of TAA-specific Tconv cells with no apparent impact on the number of peripheral Treg cells [5]. Furthermore, a recent study showed that CTLA-4^+ self Ag-specific Tconv cells were responsible for tissue inflammation, further supporting the notion that the effects of CTLA-4 block-ade may not be exclusively due to effects on Treg cells [108].

Chemotherapy

Cyclophosphamide is an alkylating agent that mediates DNA cross-linking and can specifically deplete Treg cells in mice bearing B16 melanomas [103] or neu-expressing tumors [109]. A combination of cyclophosphamide with TLR agonists also leads to regression of established colon tumors by depleting Treg cells [110]. Combination therapy with cyclophosphamide and stimulation of the OX40 costimulatory molecule has been tested and results in strong tumor immunity [111]. Interestingly, this combined treatment specifically reduces Treg cell numbers in the tumor but not in the periphery and is associated with an influx of $CD8^+$ T cells into the tumor. However, cyclophosphamide-based treatments are certainly not exclusively Treg cell specific as they can also deplete $CD4^+$ T-effector cells [112]. The long-term efficiency of cyclophosphamide on Treg cells remains to be explored.

TLRs

Another potential strategy to abrogate Treg cells in cancer therapy is through stimulation of Toll-like receptors (TLRs). This strategy may either work indirectly via stimulating DC maturation or directly on the Treg cells themselves. For example, direct stimulation of TLR 8 reverses the suppressive capacity of Treg cells [113]. Experiments involving adoptive transfer of a human tumor cell line into immunodeficient mice elegantly demonstrated that treatment of Treg cells with poli-G10, a TLR8 ligand, blocks their ability to dampen the *in vivo* $CD8^+$ T-cell response to the tumor [113]. TLR9 signalling has also been implicated in the prevention of Treg-cell activity [114]. That is, TLR9 is highly expressed by IL-10-producing Treg cells derived in the presence of vitamin D3; stimulation of these Treg cells with TLR9 agonist, CpG oligonucleotides, leads to loss of their regulatory function. These promising results indicate that targeted stimulation of TLRs may be an effective approach to restraining Treg cell activity.

Clinical Trials to Deplete/Inhibit Treg Cells to Enhance Antitumor Immunity

Based on the extensive data from mouse models indicating that removal of Treg cells and/or inhibition of Treg-cell-mediated suppressive activity contributes to successful tumor clearance, many clinical trials have been initiated to test whether similar strategies will work in humans (Table 10.1). Below we discuss evidence from human clinical trials that demonstrate Treg-cell depletion and removal is beneficial for cancer immunotherapy.

Agents Targeting IL-2 or CD25

Anti-CD25 mAbs such as Daclizumab and Basiliximab are available for clinical use and can block the interaction between IL-2 and its receptor [115, 116]. The ability of Daclizumab to deplete Treg cells is controversial. In early results from an ongoing clinical trial, the administration of the drug in patients with metastatic breast cancer is associated with a significant and sustained elimination of $CD25^+$ $FOXP3^+$ Treg cells in peripheral blood [117]. On the other hand, this mAb has actually been shown to maintain Treg cell function in recipients of heart transplants leading to lower incidence of acute rejection [118]. Additionally, exposure of peripheral blood mononuclear cells from cancer patients to low doses of Basiliximab *in vitro* significantly reduced the frequency of $CD25^{high}$ T cells and augmented IFN-γ production [119]. Moreover, *in vivo* administration of Basiliximab to 9 metastatic cancer patients demonstrated no adverse effects; future studies to determine whether pretreatment with this mAb could transiently block Treg cells and augment the efficacy of cancer immunotherapy strategies are warranted [119]. The primary manufacturer of Daclizumab has recently discontinued this product, thus the effects of CD25 blockade will have to be further defined using Basiliximab or other therapeutics.

There are many clinical trials that have employed denileukin diftitox (also known as ONTAK), which is a fusion protein between the active domain of diphtheria toxin

Table 10.1 Summary of clinical trials aimed at depleting Treg cells in humans

Tumor type	Treatment	Treg-cell effect	Antitumor response	References
Carcinoembryonic antigen (CEA)–expressing malignancies	Denileukin diftitox and DCs	Depletion of Treg cells	Increase of CEA-specific $CD4^+$ and $CD8^+$ T-cell immunity	[122]
Melanoma	Denileukin diftitox	Depletion of Treg cells	Increase of tumor antigen-specific $CD8^+$ T cells which can lyse target cells and secrete IFN-γ	[132]
Renal cell carcinoma	Denileukin diftitox and DCs	Reduced number of Treg cells and abrogated Treg-mediated immunosuppressive activity *in vivo*	Increase of tumor-specific $CD4^+$ and $CD8^+$ T-cell immunity	[120]
Ovary, breast and lung cancers	Denileukin diftitox	Depletion of Treg cells and impaired $CD25^+$ suppressive capacity	Increase of IFN-γ production from $CD8^+$ T effector cells	[133]
Melanoma	CTLA4-specific antibody	No decrease in suppression, number of Treg cells, or in *FOXP3* mRNA	Increase of tumor-infiltrating and tumor-specific $CD8^+$ T cells	[125, 126]
Bladder cancer	CTLA4-specific antibody	Increased ratio of T effector cells to Treg cells	Increase of $CD4^+ICOS^+$ T effectors that are tumor-specific and produce IFN-γ	[127]
Melanoma	CTLA4-specific antibody	No reduction in Treg numbers or function but an increase in their absolute number	Transient resistance of peripheral T cell proliferation to the inhibitory effects of Treg cells	[134]

and IL-2 [120]. The complex binds to cells expressing high levels of CD25, resulting in internalization, blockade of protein synthesis, and cell death [121]. Since the half-life of denileukin diftitox is short (70–80 min), it is speculated that the impact on subsequently activated T-effector cells is limited [122]. In various human cancers treated with denileukin diftitox, a significant reduction of Treg cell numbers has been reported, and this appears to correlate with increased antitumor immunity [120, 122] and impaired suppressive function *in vivo* [120]. Moreover, antitumor vaccination with RNA-transfected DCs in combination with Treg-cell depletion by denileukin diftitox improved stimulation of tumor-specific T cells [120]. Evidence that ONTAK also abrogated DC-mediated activation of T cells indicates that this reagent should only be used in a prevaccination setting [120].

Anti-CTLA-4 mAbs

In addition to depleting Treg cells, anti-CTLA-4 mAbs can also prevent their suppressive function. Fully humanized anti-CTLA-4 mAbs have been tested in several clinical trials with promising results. For example, treatment of patients with metastatic melanoma resulted in increased TAA-specific B cell and T-cell immune responses, and several patients had durable objective clinical responses [123]. Evidence that periodic infusions of anti-CTLA-4 mAbs enhances antitumor immunity in melanoma patients vaccinated with irradiated, autologous tumor cells engineered to secrete GM-CSF suggests that CTLA-4 blockade may also be a useful combination strategy [124]. There are conflicting data on whether the clinical benefit of anti-CTLA-4 mAbs is associated with changes in Treg cells, however. In melanoma patients receiving the mAb, the suppressive activity of Treg cells *in vitro* and *in vivo* was not inhibited [125, 126]. In fact, there was no decrease in the number of circulating $CD4^{+}CD25^{+}$ T cells, nor a decrease in *FOXP3* gene expression [125, 126]. On the other hand, a recent report demonstrated that, in the tumor tissue of bladder cancer patients, FOXP3 expression was lowered in $CD4^{+}$ cells after CTLA-4 blockade and there was a clear increase in the ratio of T effector to Treg cells [127]. Notably, anti-CTLA-4 therapy is associated with an increase in Th17 cells, providing a likely mechanism for the possible inflammatory and auto-immune side-effects of this treatment [128]. More studies are needed to establish a clear relationship between CTLA-4 blockade and Treg cells, and to evaluate whether the induction of undesirable auto-immunity could preclude the routine use of anti-CTLA-4 in cancer treatment.

Other Strategies to Decrease Treg Cells

Beyond the use of mAbs to specifically target Treg cells, several other indirect strategies have also proven to be effective. For example, autologous DCs pulsed with lysate derived from three melanoma cell lines resulted in a reduced number of $CD4^{+}TGF\text{-}\beta^{+}$ Treg cells and an enhanced anti-melanoma immune response [129].

Another method is to use siRNA to knock-down expression of CCL22 and CCL7, chemokines which attract Treg cells, on monocyte-derived DCs [130]. This strategy diminishes Treg-cell numbers and increases infiltrating $CD8^+$ T cells in human tumor xenografts in athymic nude mice. Alternatively, radiofrequency thermal ablation in lung cancer patients can also reduce $FOXP3^+$ Treg cells and enhance IFN-γ production by $CD4^+$ T-effector cells [131]. Overall, these studies demonstrate that there are many different ways to effectively deplete Treg cells in humans. In the next 5 years, the most feasible and effective regimens will likely become routine as more strategies shown to work in animal models are translated into the clinic.

Conclusions

Tumors have a variety of defence mechanisms to actively escape immune responses. A significant amount of evidence demonstrates that the ability of tumor cells to enhance Treg cell development, expansion, and/or function plays a major role in dampening anticancer immunotherapy. Possible strategies to overcome the effects of Treg cells and increase tumor immunity include: cellular depletion, neutralization of their suppressive capacity, blocking their differentiation, and/or inhibiting trafficking into tumors. Many outstanding concerns and questions must be addressed as therapeutic strategies to block Treg cells are brought to the clinic. For example, what is the best way to achieve the balance between inhibition of Treg-cell activity and the possibility of co-incident autoimmunity? Since Treg cells might be involved to various degrees in different stages and kinds of cancers, what is the optimal time point to target these cells? Further, since the majority of current strategies target $CD25^+FOXP3^+$ Treg cells, will other types of Treg cells simply take over the job of suppressing antitumor responses? Despite effective depletion strategies, will tumor cells rapidly induce the de novo induction of Treg cells and effectively negate this approach? Combination therapies consisting of reagents that target multiple mechanisms involved in suppression of antitumor immunity need to be explored. Despite these gaps in knowledge, encouraging results from early clinical trials suggest that manipulation of Treg cells is a promising approach to enhance cancer immunotherapy and should inspire future research to optimize this approach.

References

1. Vignali DA, Collison LW, Workman CJ (2008) How regulatory T cells work. Nat Rev Immunol 8: 523–532
2. Allan SE, Crome SQ, Crellin NK et al (2007) Activation-induced FOXP3 in human T effector cells does not suppress proliferation or cytokine production. Int Immunol 19: 345–354
3. Dunn GP, Old LJ, Schreiber RD (2004) The immunobiology of cancer immunosurveillance and immunoediting. Immunity 21: 137–148
4. Sakaguchi S (2005) Naturally arising Foxp3-expressing CD25+CD4+ regulatory T cells in immunological tolerance to self and non-self. Nat Immunol 6: 345–352

5. Zou W (2006) Regulatory T cells, tumour immunity and immunotherapy. Nat Rev Immunol 6: 295–30710.1038/nri1806
6. Sakaguchi S, Sakaguchi N, Asano M et al (1995) Immunologic self-tolerance maintained by activated T cells expressing IL-2 receptor alpha-chains (CD25). Breakdown of a single mechanism of self-tolerance causes various autoimmune diseases. J Immunol 155: 1151–1164
7. Hori S, Nomura T, Sakaguchi S (2003) Control of regulatory T cell development by the transcription factor Foxp3. Science 299: 1057–1061
8. Levings MK, Sangregorio R, Roncarolo MG (2001) Human cd25(+)cd4(+) t regulatory cells suppress naive and memory T cell proliferation and can be expanded *in vitro* without loss of function. J Exp Med 193: 1295–1302
9. Fontenot JD, Dooley JL, Farr AG et al (2005) Developmental regulation of Foxp3 expression during ontogeny. J Exp Med 202: 901–906
10. Allan SE, Broady R, Gregori S et al (2008) CD4+ T-regulatory cells: toward therapy for human diseases. Immunol Rev 223: 391–421
11. Tran DQ, Andersson J, Hardwick D et al (2009) Selective expression of latency-associated peptide (LAP) and IL-1 receptor type I/II (CD121a/CD121b) on activated human FOXP3+ regulatory T cells allows for their purification from expansion cultures. Blood 113: 5125–5133
12. Josefowicz SZ, Rudensky A (2009) Control of regulatory T cell lineage commitment and maintenance. Immunity 30: 616–625
13. Bacchetta R, Passerini L, Gambineri E et al (2006) Defective regulatory and effector T cell functions in patients with FOXP3 mutations. J Clin Invest 116: 1713–1722
14. Gambineri E, Perroni L, Passerini L et al (2008) Clinical and molecular profile of a new series of patients with immune dysregulation, polyendocrinopathy, enteropathy, X-linked syndrome: inconsistent correlation between forkhead box protein 3 expression and disease severity. J Allergy Clin Immunol 122: 1105–1112 e1101
15. Allan SE, Passerini L, Bacchetta R et al (2005) The role of 2 FOXP3 isoforms in the generation of human CD4+ Tregs. J Clin Invest 115: 3276–3284
16. Lopes JE, Torgerson TR, Schubert LA et al (2006) Analysis of FOXP3 reveals multiple domains required for its function as a transcriptional repressor. J Immunol 177: 3133–3142
17. Zhou L, Lopes JE, Chong MM et al (2008) TGF-beta-induced Foxp3 inhibits T(H)17 cell differentiation by antagonizing RORgammat function. Nature 453: 236–240
18. Gavin MA, Torgerson TR, Houston E et al (2006) Single-cell analysis of normal and FOXP3-mutant human T cells: FOXP3 expression without regulatory T cell development. Proc Natl Acad Sci USA 103: 6659–6664
19. Tran DQ, Ramsey H, Shevach EM (2007) Induction of FOXP3 expression in naive human CD4+FOXP3 T cells by T-cell receptor stimulation is transforming growth factor-beta dependent but does not confer a regulatory phenotype. Blood 110: 2983–2990
20. Allan SE, Alstad AN, Merindol N et al (2008) Generation of potent and stable human CD4+ T regulatory cells by activation-independent expression of FOXP3. Mol Ther 16: 194–202
21. Elyaman W, Bradshaw EM, Uyttenhove C et al (2009) IL-9 induces differentiation of TH17 cells and enhances function of FoxP3+ natural regulatory T cells. Proc Natl Acad Sci USA. doi:10.1073/pnas.0812530106
22. Miyara M, Yoshioka Y, Kitoh A et al (2009) Functional Delineation and Differentiation Dynamics of Human CD4(+) T Cells Expressing the FoxP3 Transcription Factor. Immunity. doi:10.1016/j.immuni.2009.03.019
23. Voo KS, Wang YH, Santori FR et al (2009) Identification of IL-17-producing FOXP3+ regulatory T cells in humans. Proc Natl Acad Sci USA 106: 4793–4798
24. Roncarolo MG, Gregori S, Battaglia M et al (2006) Interleukin-10-secreting type 1 regulatory T cells in rodents and humans. Immunol Rev 212: 28–50
25. Mucida D, Park Y, Kim G et al (2007) Reciprocal TH17 and regulatory T cell differentiation mediated by retinoic acid. Science 317: 256–260
26. Travis MA, Reizis B, Melton AC et al (2007) Loss of integrin alpha(v)beta8 on dendritic cells causes autoimmunity and colitis in mice. Nature 449: 361–365

27. Matsumura Y, Kobayashi T, Ichiyama K et al (2007) Selective expansion of foxp3-positive regulatory T cells and immunosuppression by suppressors of cytokine signaling 3-deficient dendritic cells. J Immunol 179: 2170–2179
28. Haxhinasto S, Mathis D, Benoist C (2008) The AKT-mTOR axis regulates de novo differentiation of CD4+Foxp3+ cells. J Exp Med 205: 565–574
29. Battaglia M, Stabilini A, Migliavacca B et al (2006) Rapamycin promotes expansion of functional CD4+CD25+FOXP3+ regulatory T cells of both healthy subjects and type 1 diabetic patients. J Immunol 177: 8338–8347
30. Sauer S, Bruno L, Hertweck A et al (2008) T cell receptor signaling controls Foxp3 expression via PI3K, Akt, and mTOR. Proc Natl Acad Sci U S A 105: 7797–7802
31. Awasthi A, Carrier Y, Peron JP, Bettelli E et al (2007) A dominant function for interleukin 27 in generating interleukin 10-producing anti-inflammatory T cells. Nat Immunol 8: 1380–1389
32. Fitzgerald DC, Zhang GX, El-Behi M et al (2007) Suppression of autoimmune inflammation of the central nervous system by interleukin 10 secreted by interleukin 27-stimulated T cells. Nat Immunol 8: 1372–1379
33. Wood KJ, Sawitzki B (2006) Interferon gamma: a crucial role in the function of induced regulatory T cells *in vivo*. Trends Immunol 27: 183–187
34. Shevach EM, Davidson TS, Huter EN et al (2008) Role of TGF-Beta in the induction of Foxp3 expression and T regulatory cell function. J Clin Immunol 28: 640–646
35. Li MO, Wan YY, Sanjabi S et al (2006) Transforming growth factor-beta regulation of immune responses. Annu Rev Immunol 24: 99–146
36. Collison LW, Workman CJ, Kuo TT, Boyd K, Wang Y, Vignali KM, Cross R, Sehy D, Blumberg RS, Vignali DA (2007) The inhibitory cytokine IL-35 contributes to regulatory T-cell function. Nature 450: 566–569
37. Collison LW, Pillai MR, Chaturvedi V et al (2009) Regulatory T cell suppression is potentiated by target T cells in a cell contact, IL-35- and IL-10-dependent manner. J Immunol 182: 6121–6128
38. Bardel E, Larousserie F, Charlot-Rabiega P et al (2008) Human CD4+ CD25+ Foxp3+ regulatory T cells do not constitutively express IL-35. J Immunol 181: 6898–6905
39. Grossman WJ, Verbsky JW, Barchet W et al (2004) Human T regulatory cells can use the perforin pathway to cause autologous target cell death. Immunity 21: 589–601
40. Gondek DC, Lu LF, Quezada SA et al (2005) Cutting edge: contact-mediated suppression by CD4+CD25+ regulatory cells involves a granzyme B-dependent, perforin-independent mechanism. J Immunol 174: 1783–1786
41. Zhao DM, Thornton AM, DiPaolo RJ et al (2006) Activated CD4+CD25+ T cells selectively kill B lymphocytes. Blood 107: 3925–3932
42. Strauss L, Bergmann C, Szczepanski M et al (2007) A unique subset of CD4+CD25highFoxp3+ T cells secreting interleukin-10 and transforming growth factor-beta1 mediates suppression in the tumor microenvironment. Clin Cancer Res 13: 4345–4354
43. Strauss L, Bergmann C, Whiteside TL (2009) Human circulating CD4+CD25highFoxp3+ regulatory T cells kill autologous CD8+ but not CD4+ responder cells by Fas-mediated apoptosis. J Immunol 182: 1469–1480
44. Cao X, Cai SF, Fehniger TA et al (2007) Granzyme B and perforin are important for regulatory T cell-mediated suppression of tumor clearance. Immunity 27: 635–646
45. Thornton AM, Shevach EM (1998) CD4+CD25+ immunoregulatory T cells suppress polyclonal T cell activation *in vitro* by inhibiting interleukin 2 production. J Exp Med 188: 287–296
46. Pandiyan P, Zheng L, Ishihara S et al (2007) CD4+CD25+Foxp3+ regulatory T cells induce cytokine deprivation-mediated apoptosis of effector CD4+ T cells. Nat Immunol 8: 1353–1362
47. Oberle N, Eberhardt N, Falk CS et al (2007) Rapid suppression of cytokine transcription in human CD4+CD25 T cells by CD4+Foxp3+ regulatory T cells: independence of IL-2 consumption, TGF-beta, and various inhibitors of TCR signaling. J Immunol 179: 3578–3587
48. Leveque L, Deknuydt F, Bioley G et al (2009) Interleukin 2-mediated conversion of ovarian cancer-associated CD4+ regulatory T cells into proinflammatory interleukin 17-producing helper T cells. J Immunother 32: 101–108

49. Borsellino G, Kleinewietfeld M, Di Mitri D et al (2007) Expression of ectonucleotidase CD39 by Foxp3+ Treg cells: hydrolysis of extracellular ATP and immune suppression. Blood 110: 1225–1232
50. Deaglio S, Dwyer KM, Gao W et al (2007) Adenosine generation catalyzed by CD39 and CD73 expressed on regulatory T cells mediates immune suppression. J Exp Med 204: 1257–1265
51. Bopp T, Becker C, Klein M et al (2007) Cyclic adenosine monophosphate is a key component of regulatory T cell-mediated suppression. J Exp Med 204: 1303–1310
52. Zarek PE, Huang CT, Lutz ER et al (2008) A2A receptor signaling promotes peripheral tolerance by inducing T-cell anergy and the generation of adaptive regulatory T cells. Blood 111: 251–259
53. Ohta A, Gorelik E, Prasad SJ et al (2006) A2A adenosine receptor protects tumors from antitumor T cells. Proc Natl Acad Sci U S A 103: 13132–13137
54. Sitkovsky MV, Kjaergaard J, Lukashev D et al (2008) Hypoxia-adenosinergic immunosuppression: tumor protection by T regulatory cells and cancerous tissue hypoxia. Clin Cancer Res 14: 5947–5952
55. Misra N, Bayry J, Lacroix-Desmazes S et al (2004) Cutting edge: human CD4+CD25+ T cells restrain the maturation and antigen-presenting function of dendritic cells. J Immunol 172: 4676–4680
56. Onishi Y, Fehervari Z, Yamaguchi T et al (2008) Foxp3+ natural regulatory T cells preferentially form aggregates on dendritic cells *in vitro* and actively inhibit their maturation. Proc Natl Acad Sci USA 105: 10113–10118
57. Veldhoen M, Moncrieffe H, Hocking RJ et al (2006) Modulation of dendritic cell function by naive and regulatory CD4+ T cells. J Immunol 176: 6202–6210
58. Takahashi T, Tagami T, Yamazaki S et al (2000) Immunologic self-tolerance maintained by CD25(+)CD4(+) regulatory T cells constitutively expressing cytotoxic T lymphocyte-associated antigen 4. J Exp Med 192: 303–310
59. Wing K, Onishi Y, Prieto-Martin P et al (2008) CTLA-4 control over Foxp3+ regulatory T cell function. Science 322: 271–275
60. Mellor AL, Munn DH (2004) IDO expression by dendritic cells: tolerance and tryptophan catabolism. Nat Rev Immunol 4: 762–774
61. Curiel TJ, Coukos G, Zou L et al (2004) Specific recruitment of regulatory T cells in ovarian carcinoma fosters immune privilege and predicts reduced survival. Nat Med 10: 942–949
62. Kryczek I, Wei S, Zhu G et al (2007) Relationship between B7-H4, regulatory T cells, and patient outcome in human ovarian carcinoma. Cancer Res 67: 8900–8905
63. Zang X, Loke P, Kim J et al (2003) B7x: a widely expressed B7 family member that inhibits T cell activation. Proc Natl Acad Sci USA 100: 10388–10392
64. Kryczek I, Zou L, Rodriguez P et al (2006) B7-H4 expression identifies a novel suppressive macrophage population in human ovarian carcinoma. J Exp Med 203: 871–881
65. Chen L (2004) Co-inhibitory molecules of the B7-CD28 family in the control of T-cell immunity. Nat Rev Immunol 4: 336–347
66. Keir ME, Francisco LM, Sharpe AH (2007) PD-1 and its ligands in T-cell immunity. Curr Opin Immunol 19: 309–314
67. Wang W, Lau R, Yu D, Zhu W et al (2009) PD1 blockade reverses the suppression of melanoma antigen-specific CTL by CD4+ CD25(Hi) regulatory T cells. Int Immunol 21: 1065–1077
68. Francisco LM, Salinas VH, Brown KE et al (2009) PD-L1 regulates the development, maintenance, and function of induced regulatory T cells. J Exp Med 206: 3015–3029
69. Albini A, Sporn MB (2007) The tumour microenvironment as a target for chemoprevention. Nat Rev Cancer 7: 139–147
70. Curiel TJ (2007) Tregs and rethinking cancer immunotherapy. J Clin Invest 117: 1167–1174
71. Whiteside TL (2008) The tumor microenvironment and its role in promoting tumor growth. Oncogene 27: 5904–5912
72. North RJ, Bursuker I (1984) Generation and decay of the immune response to a progressive fibrosarcoma. I. Ly-1+2- suppressor T cells down-regulate the generation of Ly-1-2+ effector T cells. J Exp Med 159: 1295–1311

73. North RJ (1982) Cyclophosphamide-facilitated adoptive immunotherapy of an established tumor depends on elimination of tumor-induced suppressor T cells. J Exp Med 155: 1063–1074
74. Shimizu J, Yamazaki S, Sakaguchi S (1999) Induction of tumor immunity by removing CD25+CD4+ T cells: a common basis between tumor immunity and autoimmunity. J Immunol 163: 5211–5218
75. Shimizu K, Fujii S (2009) DC therapy induces long-term NK reactivity to tumors via host DC. Eur J Immunol 39: 457–468
76. Casares N, Arribillaga L, Sarobe P et al (2003) CD4+/CD25+ regulatory cells inhibit activation of tumor-primed CD4+ T cells with IFN-gamma-dependent antiangiogenic activity, as well as long-lasting tumor immunity elicited by peptide vaccination. J Immunol 171: 5931–5939
77. Chen ML, Pittet MJ, Gorelik L et al (2005) Regulatory T cells suppress tumor-specific CD8 T cell cytotoxicity through TGF-beta signals *in vivo*. Proc Natl Acad Sci USA 102: 419–424
78. Betts G, Twohig J, Van den Broek M et al (2007) The impact of regulatory T cells on carcinogen-induced sarcogenesis. Br J Cancer 96: 1849–1854
79. Tawara I, Take Y, Uenaka A et al (2002) Sequential involvement of two distinct CD4+ regulatory T cells during the course of transplantable tumor growth and protection from 3-methylcholanthrene-induced tumorigenesis by CD25-depletion. Jpn J Cancer Res 93: 911–916
80. Chen YL, Fang JH, Lai MD et al (2008) Depletion of CD4(+)CD25(+) regulatory T cells can promote local immunity to suppress tumor growth in benzo[a]pyrene-induced forestomach carcinoma. World J Gastroenterol 14: 5797–5809
81. Beriou G, Costantino CM, Ashley CW et al (2009) IL-17-producing human peripheral regulatory T cells retain suppressive function. Blood 113: 4240–4249
82. Ahmadzadeh M, Felipe-Silva A, Heemskerk B et al (2008) FOXP3 expression accurately defines the population of intratumoral regulatory T cells that selectively accumulate in metastatic melanoma lesions. Blood 112: 4953–4960
83. Huehn J, Polansky JK, Hamann A (2009) Epigenetic control of FOXP3 expression: the key to a stable regulatory T-cell lineage? Nat Rev Immunol 9: 83–89
84. Wieczorek G, Asemissen A, Model F et al (2009) Quantitative DNA methylation analysis of FOXP3 as a new method for counting regulatory T cells in peripheral blood and solid tissue. Cancer Res 69: 599–608
85. Wang HY, Lee DA, Peng G et al (2004) Tumor-specific human CD4+ regulatory T cells and their ligands: implications for immunotherapy. Immunity 20: 107–118
86. Nishikawa H, Jager E, Ritter G et al (2005) CD4+ CD25+ regulatory T cells control the induction of antigen-specific CD4+ helper T cell responses in cancer patients. Blood 106: 1008–1011
87. Mincheff M, Zoubak S, Altankova I et al (2003) Human dendritic cells genetically engineered to express cytosolically retained fragment of prostate-specific membrane antigen prime cytotoxic T-cell responses to multiple epitopes. Cancer Gene Ther 10: 907–917
88. Su Z, Dannull J, Yang BK et al (2005) Telomerase mRNA-transfected dendritic cells stimulate antigen-specific CD8+ and CD4+ T cell responses in patients with metastatic prostate cancer. J Immunol 174: 3798–3807
89. Tuettenberg A, Jonuleit H, Tuting T et al (2004) Early adenoviral gene expression mediates immunosuppression by transduced dendritic cell (DC): implications for immunotherapy using genetically modified DC. J Immunol 172: 1524–1530
90. Rosenberg SA, Yang JC, Schwartzentruber D et al (1998) Immunologic and therapeutic evaluation of a synthetic peptide vaccine for the treatment of patients with metastatic melanoma. Nat Med 4: 321–327
91. Schadendorf D, Ugurel S, Schuler-Thurner B et al (2006) Dacarbazine (DTIC) versus vaccination with autologous peptide-pulsed dendritic cells (DC) in first-line treatment of patients with metastatic melanoma: a randomized phase III trial of the DC study group of the DeCOG. Ann Oncol 17: 563–570
92. Dumitriu IE, Dunbar DR, Howie SE et al (2009) Human dendritic cells produce TGF-beta 1 under the influence of lung carcinoma cells and prime the differentiation of CD4+CD25+Foxp3+ regulatory T cells. J Immunol 182: 2795–2807

93. Vasir B, Wu Z, Crawford K et al (2008) Fusions of dendritic cells with breast carcinoma stimulate the expansion of regulatory T cells while concomitant exposure to IL-12, CpG oligodeoxynucleotides, and anti-CD3/CD28 promotes the expansion of activated tumor reactive cells. J Immunol 181: 808–821
94. Wallen H, Thompson JA, Reilly JZ et al (2009) Fludarabine modulates immune response and extends *in vivo* survival of adoptively transferred CD8 T cells in patients with metastatic melanoma. PLoS One 4: e4749
95. Gallimore A, Godkin A (2008) Regulatory T cells and tumour immunity – observations in mice and men. Immunology 123: 157–163
96. Colombo MP, Piconese S (2007) Regulatory-T-cell inhibition versus depletion: the right choice in cancer immunotherapy. Nat Rev Cancer 7: 880–887
97. Kohm AP, McMahon JS, Podojil JR et al (2006) Cutting Edge: Anti-CD25 monoclonal antibody injection results in the functional inactivation, not depletion, of CD4+CD25+ T regulatory cells. J Immunol 176: 3301–3305
98. Zelenay S, Demengeot J (2006) Comment on "Cutting edge: anti-CD25 monoclonal antibody injection results in the functional inactivation, not depletion, of CD4+CD25+ T regulatory cells." J Immunol 177: 2036–2037
99. Imai H, Saio M, Nonaka K et al (2007) Depletion of CD4+CD25+ regulatory T cells enhances interleukin-2-induced antitumor immunity in a mouse model of colon adenocarcinoma. Cancer Sci 98: 416–423
100. Johnson BD, Jing W, Orentas RJ (2007) CD25+ regulatory T cell inhibition enhances vaccine-induced immunity to neuroblastoma. J Immunother 30: 203–214
101. Maes W, Rosas GG, Verbinnen B et al (2009) DC vaccination with anti-CD25 treatment leads to long-term immunity against experimental glioma. Neuro Oncol. doi:10.1215/15228517-2009-004
102. Prasad SJ, Farrand KJ, Matthews SA et al (2005) Dendritic cells loaded with stressed tumor cells elicit long-lasting protective tumor immunity in mice depleted of CD4+CD25+ regulatory T cells. J Immunol 174: 90–98
103. Turk MJ, Guevara-Patino JA, Rizzuto GA et al (2004) Concomitant tumor immunity to a poorly immunogenic melanoma is prevented by regulatory T cells. J Exp Med 200: 771–782
104. Ko K, Yamazaki S, Nakamura K et al (2005) Treatment of advanced tumors with agonistic anti-GITR mAb and its effects on tumor-infiltrating Foxp3+CD25+CD4+ regulatory T cells. J Exp Med 202: 885–891
105. Shimizu J, Yamazaki S, Takahashi T et al (2002) Stimulation of CD25(+)CD4(+) regulatory T cells through GITR breaks immunological self-tolerance. Nat Immunol 3: 135–142
106. Liu Z, Tian S, Falo LD Jr et al (2009) Therapeutic Immunity by Adoptive Tumor-primed CD4(+) T-cell Transfer in Combination With *In vivo* GITR Ligation. Mol Ther. doi:10.1038/mt.2009.100
107. Leach DR, Krummel MF, Allison JP (1996) Enhancement of antitumor immunity by CTLA-4 blockade. Science 271: 1734–1736
108. Ise W, Kohyama M, Nutsch KM et al (2010) CTLA-4 suppresses the pathogenicity of self antigen-specific T cells by cell-intrinsic and cell-extrinsic mechanisms. Nat Immunol 11: 129–135
109. Ercolini AM, Ladle BH, Manning EA et al (2005) Recruitment of latent pools of high-avidity CD8(+) T cells to the antitumor immune response. J Exp Med 201: 1591–1602
110. Roux S, Apetoh L, Chalmin F et al (2008) CD4+CD25+ Tregs control the TRAIL-dependent cytotoxicity of tumor-infiltrating DCs in rodent models of colon cancer. J Clin Invest 118: 3751–3761
111. Hirschhorn-Cymerman D, Rizzuto GA, Merghoub T et al (2009) OX40 engagement and chemotherapy combination provides potent antitumor immunity with concomitant regulatory T cell apoptosis. J Exp Med 206: 1103–1116
112. Zhang H, Chua KS, Guimond M et al (2005) Lymphopenia and interleukin-2 therapy alter homeostasis of CD4+CD25+ regulatory T cells. Nat Med 11: 1238–1243
113. Peng G, Guo Z, Kiniwa Y et al (2005) Toll-like receptor 8-mediated reversal of CD4+ regulatory T cell function. Science 309: 1380–1384
114. Urry Z, Xystrakis E, Richards DF et al (2009) Ligation of TLR9 induced on human IL-10-secreting Tregs by 1alpha,25-dihydroxyvitamin D3 abrogates regulatory function. J Clin Invest 119: 387–398

115. Ramirez CB, Marino IR (2007) The role of basiliximab induction therapy in organ transplantation. Expert Opin Biol Ther 7: 137–148
116. Waldmann TA (2007) Daclizumab (anti-Tac, Zenapax) in the treatment of leukemia/lymphoma. Oncogene 26: 3699–3703
117. Rech AJ, Vonderheide RH (2009) Clinical use of anti-CD25 antibody daclizumab to enhance immune responses to tumor antigen vaccination by targeting regulatory T cells. Ann N Y Acad Sci 1174: 99–106
118. Vlad G, Ho EK, Vasilescu ER et al (2007) Anti-CD25 treatment and FOXP3-positive regulatory T cells in heart transplantation. Transpl Immunol 18: 13–21
119. Okita R, Yamaguchi Y, Ohara M et al (2009) Targeting of CD4+CD25high cells while preserving CD4+CD25low cells with low-dose chimeric anti–CD2 antibody in adoptive immunotherapy of cancer. Int J Oncol. 34(2): 563–572
120. Dannull J, Su Z, Rizzieri D et al (2005) Enhancement of vaccine–mediated antitumor immunity in cancer patients after depletion of regulatory T cells. J Clin Invest 115: 3623–3633
121. Waters CA, Schimke PA, Snider CE et al (1990) Interleukin 2 receptor-targeted cytotoxicity. Receptor binding requirements for entry of a diphtheria toxin-related interleukin 2 fusion protein into cells. Eur J Immunol 20: 785–791
122. Morse MA, Hobeika AC, Osada T et al (2008) Depletion of human regulatory T cells specifically enhances antigen-specific immune responses to cancer vaccines. Blood 112: 610–618
123. Yuan J, Gnjatic S, Li H et al (2008) CTLA-4 blockade enhances polyfunctional NY-ESO-1 specific T cell responses in metastatic melanoma patients with clinical benefit. Proc Natl Acad Sci USA 105: 20410–20415
124. Hodi FS, Butler M, Oble DA et al (2008) Immunologic and clinical effects of antibody blockade of cytotoxic T lymphocyte-associated antigen 4 in previously vaccinated cancer patients. Proc Natl Acad Sci USA 105: 3005–3010
125. Klein O, Ebert LM, Nicholaou T et al (2009) Melan-A-specific cytotoxic T cells are associated with tumor regression and autoimmunity following treatment with anti-CTLA-4. Clin Cancer Res 15: 2507–2513
126. Ribas A, Comin-Anduix B, Economou JS et al (2009) Intratumoral immune cell infiltrates, FoxP3, and indoleamine 2,3-dioxygenase in patients with melanoma undergoing CTLA4 blockade. Clin Cancer Res 15: 390–399
127. Liakou CI, Kamat A, Tang DN et al (2008) CTLA-4 blockade increases IFNgamma-producing CD4+ICOShi cells to shift the ratio of effector to regulatory T cells in cancer patients. Proc Natl Acad Sci USA 105: 14987–14992
128. von Euw E, Chodon T, Attar N et al (2009) CTLA4 blockade increases Th17 cells in patients with metastatic melanoma. J Transl Med 7: 35
129. Lopez MN, Pereda C, Segal G et al (2009) Prolonged survival of dendritic cell-vaccinated melanoma patients correlates with tumor-specific delayed type IV hypersensitivity response and reduction of tumor growth factor beta-expressing T cells. J Clin Oncol 27: 945–952
130. Kang S, Xie J, Ma S et al (2009) Targeted knock down of CCL22 and CCL17 by siRNA during DC differentiation and maturation affects the recruitment of T subsets. Immunobiology. doi:10.1016/j.imbio.2009.03.001
131. Fietta AM, Morosini M, Passadore I et al (2009) Systemic inflammatory response and down-modulation of peripheral CD25(+)Foxp3(+) T-regulatory cells in patients undergoing radiofrequency thermal ablation for lung cancer. Hum Immunol. doi:10.1016/j.humimm.2009.03.012
132. Mahnke K, Schonfeld K, Fondel S et al (2007) Depletion of CD4+CD25+ human regulatory T cells *in vivo*: kinetics of Treg depletion and alterations in immune functions *in vivo* and *in vitro*. Int J Cancer 120: 2723–2733
133. Barnett B, Kryczek I, Cheng P et al (2005) Regulatory T cells in ovarian cancer: biology and therapeutic potential. Am J Reprod Immunol 54: 369–377
134. Menard C, Ghiringhelli F, Roux S et al (2008) Ctla-4 blockade confers lymphocyte resistance to regulatory T-cells in advanced melanoma: surrogate marker of efficacy of tremelimumab? Clin Cancer Res 14: 5242–5249

Chapter 11
Negative Regulators in Cancer Immunology and Immunotherapy

Wolfgang Zimmermann and Robert Kammerer

Abstract It is now well established that the immune system is able to detect and destroy tumors in a process termed tumor immunosurveillance. However, the "dark side" of tumor immunity is immune evasion. That is, by the time a patient suffers from a clinically-detectable tumor, the tumor has already successfully evaded cancer immunosurveillance and often has established effective mechanisms to actively suppress the immune system, particularly in the tumor microenvironment. Therefore, cell contact-dependent and -independent immunosuppressive networks represent a significant barrier to effective immunity and immunotherapy. In this chapter, we describe some of these immunosuppressive mechanisms and components that are linked in complex networks. A better understanding of these mechanisms will eventually lead to improvements of cancer immunotherapies.

Keywords Drug targets • Immune evasion/escape • Immunosuppression • Suppressor cells • Tumor immunotherapy

Suppressor Cells

Regulatory Lymphocytes

CD4$^+$ as superscript in CD4$^+$ in analogy to FOXP3 + Treg

CD4$^+$ regulatory T cells (Treg) are characterized by the expression of the forkhead/winged-helix transcription factor FOXP3. FOXP3$^+$ T cells can be divided into natural thymus-derived Treg (nTreg) and peripherally induced adaptive Treg (iTreg or aTreg).

R. Kammerer (✉)
Institute of Immunology, Friedrich-Loeffler-Institute, Tuebingen, Germany
e-mail: Robert.Kammerer@fli.bund.de

J. Medin and D. Fowler (eds.), *Experimental and Applied Immunotherapy*,
DOI 10.1007/978-1-60761-980-2_11, © Springer Science+Business Media, LLC 2011

nTreg are self-reactive and mediate peripheral self-tolerance, whereas iTregs dampen inflammation. Both types of Treg cells can accumulate in tumor tissues [1]. Because the role of Treg in tumor immunity is discussed in detail elsewhere in this book (see Chapter 10), we will focus on the position of Treg within the suppressive networks. FOXP3$^+$ Treg cells in the tumor microenvironment are highly activated and have undergone extensive proliferation. Both the proliferation of nTreg and the conversion of iTreg are critically dependent on the presence of TGF-β, which is produced by many tumors in large amounts. In addition, several other metabolic pathways that are heightened in the tumor microenvironment can also stimulate Treg conversion, including IDO, arginase, and Cox-2 [2]. Furthermore, dysfunctional or tolerogenic APCs, induced by tumor-derived factors within the tumor microenvironment, favor conversion and expansion of Treg cells [2]. On the other hand, intratumoral Treg cells can secrete high amounts of IL-10 on ICOS-ICOSL engagement, thereby contributing to the immunosuppressive milieu in the tumor environment [3]. Taken together, emerging evidence indicates that Treg cells play a central role in tumor antigen-specific immune suppression, making the reduction of Treg a logical therapeutic strategy. Although depletion of Treg cells is most-often studied, it might be a more successful strategy to manage rather than eliminate Treg cells because depletion strategies might foster rapid Treg regeneration [4, 5] (Fig. 11.1).

Other T_{reg}

In addition to CD4$^+$ Treg, other T cells with suppressive functions have been described, including CD8$^+$ Treg and regulatory natural killer T cells (NKT). In contrast to CD4$^+$ Treg cells, CD8$^+$ Treg are not well defined. However, similar to naturally occurring CD4$^+$ Treg, a subset of CD8$^+$ Treg cells express FOXP3 and CTLA-4; furthermore, their suppressive phenotype was found to be dependent on cell contact [6]. In addition, several subsets of adaptive CD8$^+$ CD28$^-$ Treg cells have been reported in humans, such as CD8$^+$ Treg cells that are induced by allo-APCs and CD8$^+$ Treg that are induced by a concerted action of monocytes, IL-2, and GM-CSF [6]. The role of CD8$^+$ Treg in cancer is not well established; however, increasing evidence indicates that CD8$^+$ Treg cells also accumulate in the tumor microenvironment [7].

NKT cells are true antigen-specific T cells that also have innate properties, thereby forming a bridge between the innate and adaptive immune systems. In contrast to conventional T cells, NKT cells do not recognize peptides presented by classical class I or II molecules; rather, NKT cells recognize a lipid antigen presented by the nonclassical class I-like MHC molecule, CD1d [8]. Recently, a subset of NKT cells termed "type II NKT cells" was identified that mediates immune suppressive properties. Interestingly, suppressive NKT cells use an indirect mechanism to inhibit cytotoxic T cell-mediated tumor immunosurveillance; specifically, such cells secret IL-13, which then induces myeloid cells to make TGF-β, which ultimately suppresses T-cell-mediated immunity [9].

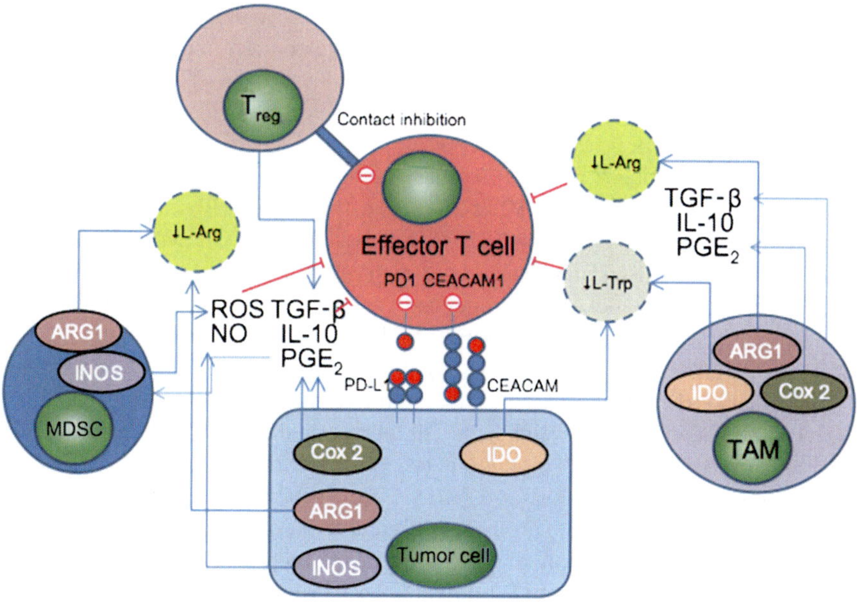

Fig. 11.1 Immunosuppressive tumor microenvironment for effector T cells. Immunosuppressive activity is exerted by tumor cells and tumor infiltrating normal cells, including T_{reg}, TAM, and MDSC. MDSC and TAM create, together with the tumor cells themselves, an immunosuppressive environment by depleting L-tryptophan and L-arginine and by secretion of TGF-β, IL-10 and PGE_2 as well as ROS and NO. T_{reg} cells inhibit T effector cells in an antigen-specific manner. In addition, tumor cells express ligands (PD-L1 and CEACAMs) for inhibitory receptors on effector T cells that suppress effector T-cell function on cell-cell contact with the tumor cell

Myeloid-Derived Suppressor Cells

Myeloid-derived suppressor cells (MDSC) comprise a heterogeneous group of immature hematopoietic cells of myeloid origin. Such cells, which include myeloid progenitor cells as well as immature macrophages, granulocytes, and DCs, expand during infection and inflammation to protect the host against overshooting immune reactions and immune-mediated damage. MDSC accumulate in tumors because of a partial block in differentiation. Activation of MDSC leads to the expression of immune suppressive factors such as arginase 1 and inducible nitric oxide (NO) synthase (iNOS or NOS2). Their catalytic activity results in formation of toxic reactive oxygen (ROS) and NO species (including peroxynitrite), which inhibit tumor-directed T cell and innate immune responses [10]. Direct cell-cell contact between MDSC and target cells facilitates the production and delivery of these highly reactive, short-lived compounds [11]. Granulocytic and monocytic subpopulations of MDSC exist that differ in their suppression mechanisms, relying on ROS and NO production, respectively; however, each subset is capable of suppressing T-cell responses in tumor-bearing mice [12].

Signals for expansion and activation of MDSC have been elucidated. Expansion is achieved through tumor-derived factors that foster myelopoiesis, inhibit differentiation and promote survival; conversely, factors produced by activated T cells and tumor stromal cells mediate MDSC activation, a prerequisite for their suppressive ability. Typically, PGE_2, IL-1β, IL-6, GM-CSF, and VEGF favor accumulation of MDSC. The signaling cascades induced by these factors primarily result in JAK/STAT3 activation. On the other hand, engagement of STAT1/STAT6 pathways as well as signaling through TLRs leads to NF-κB activation and results in activation of MDSC. Main extracellular mediators of STAT1/STAT6 pathway activation are IFN-γ, IL-4, IL-13, and TLR ligands, which are operational in bacterial and viral infections [10]. Some of these factors and signaling pathways might provide molecular targets for therapeutic interference provided that this intervention does not impair immune effector mechanisms.

A number of therapeutic strategies that aim to diminish the immunosuppressive effects of MDSC have been tested mostly in murine models. Such strategies include promotion of myeloid cell differentiation, inhibition of MDSC attraction, expansion, and function, as well as elimination of MDSC by chemotherapy [10]. For example, all-*trans* retinoic acid (ATRA) promotes differentiation of MDSC by up-regulation of glutathione synthesis and concomitant ROS reduction [13]. Use of ATRA therapy improved the antitumor effects of different cancer vaccines in mice and abrogated the T-cell inhibitory effects of MDSC from RCC patients [14, 15]. In over 30 clinical trials, patients with solid tumors are currently treated with the clinically-approved compound ATRA (Tretinoin) in combination with different chemotherapeutic drugs, cytokines, or tumor vaccines (*ClinicalTrials.gov*). MDSC expansion is controlled by tumor-derived factors. Recently it was shown that blockage of SCF, which signals via its receptor KIT on MDSC, reversed immune tolerance in mice with advanced tumors [16]. COX-2 is overexpressed in most cancers and is critical to the formation of PGE_2, which in turn up-regulates arginase 1 and NOS2 activity in MDSC [17]. COX-2 inhibitors such as celecoxib can enhance T-cell responses and improve the efficiency of tumor vaccines in mice [18]; such agents are now being tested, mostly in combination with chemotherapy, worldwide in nearly 200 clinical trials in patients with a wide spectrum of solid cancers (*ClinicalTrials.gov*). Phosphodiesterase 5 (PDE5) inhibitors, such as the approved drug sildenafil, inhibit the formation of GMP by hydrolysis of cyclic GMP, thereby leading to down-regulation of *ARG1* and *NOS2* expression in MDSC. In a recent preclinical study, PDE5 inhibition enhanced endogenous antitumor immunity [19]. In summary, because several substances that have been already approved for other therapeutic indications can favorably modulate MDSC biology, ongoing and newly initiated clinical trials with these substances will soon clarify the validity of MDSC inhibition as a tumor therapy concept.

Tumor-Associated Macrophages

During the last few years, the link between chronic inflammation and tumor progression has become increasingly clear. Macrophages play a major role in the establishment and maintenance of inflammation in the stroma of many tumors and, therefore, are thought to contribute significantly to tumor progression. Two types of macrophages can be discerned that are polarized toward the Th1 cytokine phenotype (M1 macrophages) or the Th2 phenotype (M2 macrophages). The classical M1-polarized macrophages represent a major group of APC, produce proinflammatory Th1 cytokines like IL-12, and can kill tumor cells and ingested intracellular microorganisms by release of NO, ROS, and TNF. Tumor-associated macrophages (TAM), on the other hand, exhibit the M2 phenotype and dampen inflammation as well as facilitate angiogenesis, lymphangiogenesis, and tissue remodeling, thereby enhancing tumor progression and invasion. Consequently, one would expect patients with large TAM tumor infiltrates to have a worse prognosis. Indeed, for the majority of a wide spectrum of tumor types, a positive correlation exists between the abundance of TAMs and a poor patient prognosis [20, 21]. Tumor cells assist TAM accumulation and survival in the tumor by secretion of IL-6, IL-10, and colony-stimulation factor1 (CSF-1), whereas TAM contribute to immune tolerance by production of tolerogenic factors such as IL-10, TGF-β, and PGE_2 [22, 23].

Because TAM are crucial for the maintenance of chronic inflammation in the tumor microenvironment, which is favorable for tumor progression, TAM appear to represent an attractive target for tumor therapies. Therapeutic strategies could address inhibition of recruitment of TAM and their survival in the tumor, inhibition of their angiogenic potential and prometastatic tissue remodeling capacity, and alleviation of their immunosuppressive properties. Chemokine and chemokine receptor inhibitors that disrupt the unwanted alliance between TAM and tumor cells by inhibiting TAM attraction to the tumor site, as well as inhibition of their tumor cell-mobilizing proteolytic activity, have shown promise for the suppression of tumor growth in preclinical models; as such, this approach might represent a valuable strategy for clinical trials in the future [24].

Suppressive Ligands and Receptors

Soluble Factors

Soluble factors secreted by tumor cells seem to have a dramatic impact on the tumor microenvironment with respect to tumor survival and progression. In particular, some of these factors have the ability to suppress immune effector functions. The best characterized immune suppressive factors are PGE_2, TGF-β, and IL-10.

Prostaglandins

Elevated concentrations of prostaglandins, in particular PGE_2, in the tumor microenvironment are caused by the up-regulation of COX-2 expression by tumor cells. COX enzymes, which consist of the constitutively expressed COX-1 and the inducible

COX-2, play key roles in the biosynthesis of prostaglandins from arachidonic acid via an intermediate PGH_2. Specific prostaglandin synthases generate individual prostaglandins from PGH_2, including PGI_2, PGD_2, $PGF_{2\alpha}$, and PGE_2. PGE_2 mediates its action by engagement of specific cell surface G protein-coupled receptors designated EP1-EP4 [25]. EP1-EP4 are coupled through their intracellular sequences to specific G protein with different second messenger signaling pathways [26]. PGE_2 has pleiotropic effects in the tumor microenvironment, including tumor maintenance and progression, metastatic spread, and perhaps even participation in tumor initiation (for review see Ref. [25]). Here, we want to focus on the effects of PGE_2 on antitumor immune responses. PGE_2 can influence the activity of T cells, B cells, and professional APC such as DCs and macrophages. T cells respond to PGE_2 by enhanced production of IL-4, IL-5, and IL-10 (Th2 cytokines that can have immune suppressive qualities) and reduced secretion of IL-2 and IFN-γ (Th1 cytokines that in general help mediate protective antitumor responses). PGE_2 induces isotype class switching in B cells to produce IgG1 and IgE. On encounter with PGE_2, APCs up-regulate production of IL-10 and reduce expression of IL-12 [27]. Overall, PGE_2 signaling leads to a bias of the immune reaction toward a Th2 response at the expense of a Th1 response, a shift that does not favor tumor destruction. As mentioned above, COX-2 inhibitors are being tested intensively in clinical trials, mostly in combination with chemotherapy in patients with a variety of cancers. However, prostaglandin synthases specifically involved in the generation of PGE_2 may be more suitable targets. In particular, the inducible microsomal prostaglandin E synthase 1 may be a new drug target for the immunotherapy of cancer [28].

Transforming Growth Factor-β

TGF-β is a pleiotropic cytokine with numerous effects on cell proliferation, differentiation, migration, and survival that affects multiple biological processes, including development, carcinogenesis, fibrosis, wound healing, and immune responses [29]. TGF-β is synthesized as an inactive precursor molecule. After proteolytic processing, active TGF-β exists either as a cell surface-bound or soluble cytokine. Three different TGF-β isoforms have been described that bind to a heterodimeric complex consisting of two transmembrane receptor serine/threonine kinases known as type I and II receptors. Intracellular signal transduction is mediated by several Smad transcription factors as well as by Smad-independent pathways. There are two main sources of TGF-β in the tumor microenvironment: the tumor cells themselves and myeloid-derived suppressor cells [29]. TGF-β exerts its effects on multiple immune cell types, including T cells, NK cells, and DCs within the tumor microenvironment. TGF-β inhibits T-cell proliferation through suppression of IL-2 secretion and IL-2-independent mechanisms, and also suppresses T-cell effector function by limiting the expression of effector molecules such as IFN-γ and perforin. Furthermore, the presence of TGF-β promotes the generation of Treg cells, which in turn suppress the effector function of T cells [30]. Furthermore, TGF-β suppresses NK cell function by down-regulating NK cell receptor expression, thereby resulting in decreased secretion of IFN-γ and diminished killing function. A third mechanism by which TGF-β can suppress tumor antigen-specific immune responses is via regulation of DC function. Both DC antigen presentation and cytokine production are impaired on administration of exogenous TGF-β [31]. Therefore, blockade of TGF-β signaling may improve cancer immunotherapies. Various TGF-β or TGF-β signaling inhibitors, including antibodies, small-molecule inhibitors, and oligonucleotides have been evaluated in preclinical studies; subsequently, some inhibitors have already entered clinical trials (summarized in ref. [30]).

Interleukin-10

IL-10 was originally termed cytokine synthesis inhibitory factor caused by its ability to suppress production of proinflammatory cytokines by various cell types, including Th1 cells, DCs, and macrophages. Now it is established that IL-10 is the "master cytokine" for the containment and suppression of inflammatory responses by the adaptive immune system. For example, IL-10 down-regulates expression of MHC molecules as well as costimulatory molecules such as CD80 and CD86 on APCs, thereby favoring the generation of Treg cells. Not surprisingly, several viruses encode closely-related homologues of IL-10 in their genome, which bind to the IL-10 receptor and inhibit the host immune response. This biology may represent a mechanism of immune evasion adapted by the virus, further emphasizing the important role of IL-10 in suppression of immune responses [32]. Elevated expression of IL-10 has also been demonstrated in a variety of malignant human tumors, including melanoma, basal cell carcinomas, squamous cell carcinomas, renal cell carcinomas, colorectal carcinomas, ovarian carcinoma, malignant gliomas, lung cancer, breast cancer, and B cell lymphomas. More importantly, high levels of IL-10 have been correlated with poor prognosis. The source of IL-10 can be the tumor cells themselves, TAM, myeloid suppressor cells, DC, Th2 and Treg cells [33]. In addition to the production of TGF-β, IL-10 release appears to be the main effector mechanism of various subtypes of Treg. Numerous preclinical experiments have identified that IL-10 has a strong impact on the suppression of an effective antitumor immune response [33]. Although in some clinical trials the administration of IL-10 was used as a therapeutic agent to treat autoimmune diseases, administration of blocking anti-IL-10 and anti-IL-10 receptor antibodies is still evaluated preclinically but may be envisaged to enhance the efficiency of antitumor immunotherapies.

Suppressive Cellular Receptors

Any immune response is initiated by key receptors such as the T-cell and B-cell receptors. However, the quality of an immune response is influenced by additional costimulatory and coinhibitory signals that are also communicated to immune cells during antigen presentation. Recently, the known number of these molecules has increased enormously because of their identification by large-scale genome sequencing projects. Coinhibitory receptors certainly represent interesting targets for tumor immunotherapies. Inhibitory receptors can be divided into two groups: receptors with ligands on other immune cells (particularly, on professional APCs) and receptors that encounter their ligands on tumor cells.

B7-1/B7-2 Ligands and CD28/CTLA-4 Receptors

CTLA-4 represents one of the major checkpoints of T-cell activation. When the critical role of CD28-dependent signaling in supporting T-cell receptor (TCR)-mediated T-cell activation was firmly established in the early 1990s, CTLA-4, which is up-regulated on T cells on activation, was recognized as a closely-related paralog of CD28 that interacts with the same ligands (that is, B7-1 and B7-2). Interestingly, CTLA-4 transduces negative signals into T cells [34, 35]. Today, it is well established that CTLA-4 blocks AKT signaling by activation of the phosphatase PP2A, thereby inhibiting TCR signaling. In addition, because of the higher affinity of CTLA-4 relative to CD28 for binding to ligands of the B7 family, up-regulation of CTLA-4 leads to displacement of CD28 from the CD28-B7 complex and termination of CD28-mediated costimulation in T cells [36]. CTLA-4 is especially important for T-cell priming and expansion and less important for T-cell effector function because the ligands for CTLA-4 (B7-1 and B7-2) are exclusively expressed by professional APC. The role of CTLA-4 in limiting lymphocyte expansion was convincingly demonstrated by the phenotype of CTLA-4 knockout mice, which succumb to polyclonal lymphoproliferation that is rapidly lethal within a few weeks of birth [37].

Blockade of CTLA-4 with a mAb results in a lower threshold for TCR-mediated T-cell activation and a more pronounced T-cell immune response. Because tumor antigens are in general weakly immunogenic and often even self-antigens, CTLA-4 is a promising candidate target molecule for improvement of cancer immunotherapy. Indeed, in a number of preclinical investigations, monotherapy with anti-CTLA-4 antibodies induced rejection of several types of transplantable tumors in mice, including lymphomas and colon, prostate, and renal cell carcinomas; importantly, the success of this anti-CTLA-4 therapy seems to critically depend on inherent tumor immunogenicity [38]. Based on these encouraging data, the strategy of CTLA-4 blockade was transferred into the clinic.

Two human mAb directed against CTLA-4 are currently in clinical development: ipilimumab and tremelimumab. Because CTLA-4 blockade targets the immune system and not the tumor directly, immunogenic tumors should be most amenable to anti-CTLA-4 mAb-based therapies. Not surprisingly, most clinical trials tested CTLA-4 blockade as a monotherapy in metastatic melanomas and renal cell carcinomas, although this strategy has also been tested in prostate, ovarian, breast, and colon carcinomas. Evidence for tumor regression with prolonged time to progression has been seen in melanoma patients and durable responses have been observed in patients with melanoma, ovarian, prostate, and renal cell carcinomas [38, 39]. Interestingly, antitumor responses showed a unique characteristic: that is, a short-term disease progression was often followed by delayed regression, leading often to a prolonged duration of clinical responses. However, inhibition of CTLA-4 in tumor patients also induces significant side-effects termed immune-related adverse events (IRAEs), which may be caused by breakage of tolerance to self-antigens. The most common presentations of IRAE are rash, colitis, or hepatitis; however, other types of inflammation such as hypophysitis (inflammation of the pituitary gland) have also been observed. Importantly, the occurrence of IRAE has been associated with tumor regression and prolonged time-to-relapse in patients with resected high-risk melanomas, thereby indicating that there is a link between induction of autoimmunity and antitumor responses [40]. Because preclinical studies indicated that anti-CTLA-4 monotherapy was insufficient in poorly immunogenic tumors, the poor immunogenicity of many human tumors may limit the clinical use of anti-CTLA-4 monotherapy.

Accordingly, combinatorial therapies may represent the next logical step forward. However, initial results of published studies indicated that coadministration of melanoma peptides and anti-CTLA-4 antibody was not superior to anti-CTLA-4 monotherapy with respect to response rates and remission duration [40]. Surprisingly, anti-CTLA-4 blockade did not result in a measurable increase of an antipeptide response in peripheral blood; the mechanism accounting for this lack of immune activation remains undefined. In addition, the link between IRAE and tumor response has not been elucidated at the molecular level. Furthermore, evidence for antigen specificity of the T-cell clones responsible for toxicity is lacking and it is therefore questionable whether the antitumor immune response is antigen-specific. Alternatively, in an effort to more fully support endogenous antitumor immunity, it may be reasonable to combine anti-CTLA-4 therapy with manipulation of molecules that impact additional immunological checkpoints. Possible candidate molecules are discussed below.

PD-1 Ligand and PD-1 Receptor

Programmed cell death1(PD-1, PDCD1, CD279) is another receptor of the CD28 family expressed by activated T and B cells as well as by myeloid-derived cells. Over the past 15 years it has become clear that the primary function of PD-1 is to attenuate the immune response. PD-1 binds PD-L1 (B7-H1, CD274) and PD-L2 (B7-DC, CD273). Although PD-L2 expression is limited to professional APC, PD-L1 is expressed by many normal tissues and also by many tumors, including glioma, melanoma, and cancers of the larynx, lung, stomach, colon, breast, cervix, ovary, kidney, bladder, and liver [41]. Interestingly, expression of PD-L1 in human tumor cells *in vitro* is up-regulated upon IFN-γ stimulation. More recently, it was found that expression of PD-L1 by tumors *in vivo* correlates with poor prognosis in hepatocellular carcinoma, ovarian and pancreatic cancer but not in esophageal cancer [42]. Although PD-1 may be involved in regulation of T-cell activation during T-cell priming, it is thought to be especially important for the maintenance of peripheral tolerance [36].

PD-1 has a signaling motif in its cytoplasmic tail that is comprised of an immunoreceptor tyrosine-based inhibitory motif (ITIM) followed by an immunoreceptor tyrosine-based switch motif (ITSM), which are only found in PD-1 and in the

B and T lymphocyte attenuator (BTLA) within the CD28 family. Such motifs are also present in several other signaling molecules, including various Siglecs and CEACAM1 (see below). The critical immune regulatory function of PD-1 was initially demonstrated in a lymphocytic choriomeningitis virus (LCMV) model. Chronic virus infections, which cannot be cleared by a proper T-cell response, may lead to high expression of PD-1 on virus-specific T cells after prolonged time of infection [43]. These virus-specific T cells become largely nonfunctional and are referred to as "exhausted" T cells [43, 44]. Mice lacking PD-L1 can die from immune-mediated pathology when infected by a certain LCMV strain, thereby indicating that PD-1-mediated T-cell exhaustion can protect the host from a nonresolving immune response. Importantly, blocking PD-1 and PD-L1 interaction can restore some of the function of exhausted T cells.

Taking into account the function of PD-1 on T cells, one would speculate that expression of PD-L1 may lead to exhaustion of T cells within the tumor microenvironment. However, a prerequisite for this scenario would be expression of PD-1 on tumor-infiltrating T cells. Indeed, Rosenberg and colleagues reported recently that the majority of TILs isolated from melanoma tumors expressed PD-1, including MART-1/Melan-A melanoma antigen-specific $CD8^+$ T cells; in contrast, T cells in normal tissues and peripheral blood did not have increased PD-1 expression [45]. Because it was shown that cosignaling via PD-1 in addition to TCR signaling can also inhibit direct cytotoxic activity of $CD8^+$ T cells, it is likely that PD-1/PD-L1 interaction is involved in T-cell suppression within the tumor microenvironment; these data provide a rationale for blocking PD-1 signaling as a therapeutic strategy for tumor immunotherapy.

Preclinical studies have demonstrated that blocking of the PD-1/PD-L1 pathway can induce a substantial antitumor effect in different murine tumor models [46, 47]. The first clinical trial of an anti-PD-1 approach, which was published in 2008, used a humanized mAb in a phase I study in patients with advanced hematologic malignancies. It was reported that a single dose of antibody was safe and well tolerated [48]. In addition, clinical benefit was observed in 6 out of 18 patients, including four with stable disease, one minimal response and one complete response. The patient with the complete remission had been diagnosed with stage III follicular lymphoma [48]. Presently, fully humanized mAbs against PD-1 are being tested in ongoing clinical phase I trials (*ClinicalTrials.gov*).

CEACAM1 Inhibitory Receptor

A third class of inhibitory receptors is represented by CEACAM1 (carcinoembryonic antigen [CEA]-related cell adhesion molecule), which is a member of the CEA family; the CEA family is itself a subfamily of the immunoglobulin superfamily. In contrast to PD-1, which is expressed primarily on B and T cells, CEACAM1 is expressed on B, T and NK cells. CEACAM1 appears to represent a broad modulator of the cellular effectors of antitumor immunity [49]. The signaling motifs in the cytoplasmic tail of CEACAM1 are very similar to that of PD-1; however, during evolution of primates, the ITSM was changed to a classical ITIM motif [93]. The inhibition of T cells by CEACAM1 is SHP1/PTPN6 phosphatase-dependent. CEACAM1 associates with and recruits SHP1 to the TCR/CD3 complex, thereby leading to decreased phosphorylation of CD3ζ and ZAP-70 and consequently decreased activation of the elements downstream of ZAP-70 [50]. Physiological ligands of CEACAM1 belong to the CEA family, including CEACAM1, CEACAM5 (CEA), and putatively CEACAM6 under certain circumstances. Because CEACAM1 is also expressed by DCs, the CEACAM1-CEACAM1 interaction may be involved in the regulation of T-cell priming [51]. These homophilic and heterophilic interactions have enormous consequences for the role of CEACAM1 in tumor immunology. It is well known that CEA is a tumor-associated antigen strongly expressed by a large number of cancers, including colon, breast, and lung carcinomas. In addition, CEACAM1 itself is expressed in numerous tumor entities, including melanomas. In some tumors such as renal cell carcinomas, CEACAM1 is up-regulated upon IFN-γ stimulation [52]. Therefore, immune effector cells that express CEACAM1 will most likely encounter one of its ligand at the tumor site, thereby leading to suppression of the antitumor effector function of cytotoxic TIL. Indeed, it was found by us and others that interaction of CEACAM1 with CEA or CEACAM1 on tumor cells can suppress the cytotoxic activity of effector cells [53–55]. Furthermore, in some tumors, including renal cell carcinoma and melanoma, infiltration of IFN-γ-secreting cells induces CEACAM1 up-regulation on the tumor cells, which leads to suppression of the infiltrating immune cells [52, 56]. Blockade of CEACAM1, therefore, may prevent negative signaling during lymphocyte effector function; such a therapeutic approach is of particular interest because this negative feedback mechanism works not only in T cells (as is the case for PD-1 and CTLA-4), but also in NK cells [55]. These observations indicate that additional blockade of CEACAM1-CEACAM1 interaction may be more beneficial for promotion of antitumor responses than disruption of CTLA-4 and PD-1 signaling alone. However, only a limited number of preclinical studies have been performed to analyze the effect of CEACAM1 blockade *in vivo*. In one such study, Blumberg and colleagues showed that blockade of CEACAM1 can reduce T helper cell-dependent murine colitis [57]. It has not yet been determined whether CEACAM1 blockade will exert a favorable effect on antitumor immunity.

Enzymes with Immunosuppressive Function

Indoleamine-2,3-Dioxygenase

Tryptophan-2,3-dioxygenase (TDO) and IDO catalyze the first, rate-limiting step in the catabolism of the essential amino acid L-tryptophan of the kynurenine pathway. Closely-linked paralogous IDO genes, IDO1 and the phylogenetically older IDO2, exist in mammals. In contrast to TDO, both IDO genes are inducible in a variety of cells by proinflammatory signals, in particular by IFN-γ. Their products, however, exhibit a 10-fold difference in substrate affinity ($K_{mIDO1} < K_{mIDO2}$) [58, 59].

IDO1 appears to be a key player in the innate immune system that confers inhibition of intracellular pathogens by tryptophan depletion and/or generation of toxic metabolites. In addition, IDO1 generates a tolerogenic state under various physiological and pathophysiological conditions such as pregnancy. The important role of IDO in promoting tolerance was initially revealed by experiments that identified the rejection of allogeneic concepti after treatment of female mice with the IDO inhibitor 1-methyl tryptophan (1-MT) [60]. APC-like subsets of plasmacytoid DC and macrophages can mediate T-cell tolerance via IDO through activation of suppressor function in T_{reg} cells or cell-cycle arrest in effector T cells, which are particularly sensitive to tryptophan deprivation and kynurenine catabolites [61]. Tryptophan withdrawal through increase of uncharged $tRNA^{Trp}$ leads to the activation of GCN2 kinase, which in turn initiates various T-cell responses [62]. Despite this detailed knowledge of the signaling cascade, the molecular basis for the differential sensitivity of T cells is not yet understood.

Furthermore, overexpression of IDO1 in immunogenic cancer cells has been shown to prevent tumor rejection by T and NK cells; in addition, treatment of host mice with IDO inhibitors or IDO1-specific siRNA postponed tumor formation [63–65]. Cancer cells of a number of human malignancies constitutively express *IDO1*, which is possibly mediated by the loss of activity of the tumor suppressor *Bin1* [63, 66]. Indeed, expression of *IDO1* in tumor cells in a number of solid tumors and leukemias (acute myelogenous leukemia, colon, endometrial, serous ovarian cancer, and hepatocellular carcinoma) has been correlated with a worse patient progress, thus supporting the notion that IDO contributes to the immunosuppressive milieu in the tumor microenvironment by inhibition of effector T cells through tryptophan depletion [67–71]. *IDO1* expression in tumor endothelial cells in renal cell carcinoma patients, which is associated with reduced malignant cell proliferation, or in TILs in HCC, however, appears to favor longer survival of patients [72, 73]. These results suggest that the role of IDO in the tumor microenvironment is probably more complex than anticipated; careful consideration should therefore be exerted when conducting clinical trials relying on IDO inhibition (see below). Elevated IDO levels are also found in a subset of plasmacytoid DC in tumor-draining lymph nodes; this effect may be induced by chronic inflammation, which is commonly thought to support tumor formation [74, 75]. Taken together, IDO1 both in tumor cells and DCs in tumor-draining lymph nodes appears to contribute to tumor immune escape. Experiments with *IDO1* knockout mice indeed support this notion [76].

Based on these observations, it has been suggested to use IDO inhibition as an adjunct to existing tumor therapies [23, 76]. Preclinical models have evaluated the effect of interference of IDO expression in both tumor cells and DCs. For example, ballistic delivery into the skin of an IDO-specific shRNA-encoding plasmid served as an immune adjuvant for tumor antigen vaccination (neu tumor antigen). Silencing of IDO mRNA in DC led to enhanced T-cell infiltration into the tumor and increased cytotoxic activity, which mediated tumor growth reduction [77]. In addition, a number of synthetic and natural small molecule IDO inhibitors have been tested (1-MT, MTH-Trp, brassinins). Most commonly the competitive IDO inhibitor 1-MT has been utilized. Antitumor effects of 1-MT were exerted, especially when used in combination with chemotherapy in preclinical mouse models, surprisingly by the *dextro* (D) stereoisomer, which preferentially inhibits IDO2 whereas the *levo* (L) form inhibits IDO1 [76, 78]. Paradoxically, neither in DCs nor in tumor cells does IDO2, although expressed, appear to be responsible for the immune suppressive effect mediated by tryptophan catabolism [79, 80]. Therefore, other tryptophan-interacting molecules and off-target effects of 1-MT should be considered [81]. A possible target of competitive IDO inhibitors could be the L-amino acid transport system (LAT1/CD98), by which also tryptophan is transported, or other tryptophan transport systems [82, 83]. Unfortunately, with respect to the further characterization of these transporters, only the racemate D,L-1-MT form was tested. Therefore, it remains unclear whether the stereoselective antitumor effect of 1-MT is mediated through a possible differential inhibition of amino acid transporters by the two 1-MT stereoisomers. Despite these unresolved issues, four clinical trials have been initiated in the USA; in each case, patients with relapsed or refractory solid tumors are being treated with D-1-MT (Table 11.1). Besides identification of a tolerable inhibitor dose, these studies might also shed some light on the role of tryptophan catabolism in cancer immune tolerance in humans.

Arginase 1 and Inducible Nitric Oxide Synthase

MDSC are the main source of the L-arginine-metabolizing enzymes arginase 1 (which converts L-arginine to urea and L-ornithine) and inducible nitric oxide synthase (iNOS/NOS2). However, some tumor types like prostate carcinoma can up-regulate expression of the corresponding genes [86]. There are several mechanisms that help explain immune suppression that occurs with up-regulation of *ARG1* and *NOS2*. First, local depletion of the nonessential amino acid L-arginine in MDSC, which occurs by enhanced catabolism caused by elevated arginase 1 activity, leads to a shortage of this amino acid with concomitant inhibition of T-cell proliferation [87]. Shortage of L-arginine can trigger down-regulation of the CD3ζ chain of the TCR complex, which is indispensable for T-cell activation and occurs in most tumor patients [88]. Interestingly, deprivation of another amino acid, L-tryptophan, has similar effects on T-cell function (see above). Second, ROS are instrumental for the suppressive activity of MDSC; blockade of ROS formation completely abolishes the immune suppressive effects of these cells from tumor-bearing mice and patients *in vitro* [10]. ROS are formed from L-arginine and oxygen (O_2) by NOS2 catalysis leading to NO and L-citrulline or, at low arginine levels, to the highly reactive superoxide anion (O_2^-), which combines with NO to form peroxynitrite. Perxoynitrite chemically modifies proteins at a number of amino acid residues, notably tyrosine, thereby yielding nitrotyrosine. Up-regulation of arginase and NOS2 in tumors, which is associated with tumor progression in many types of cancer, leads to increased tyrosine nitrosylation combined with unresponsiveness of infiltrating $CD8^+$ T lymphocytes that can be reversed by specific arginase and NOS inhibitors in prostate carcinoma explant cultures [86]. Nitrosylation of tyrosines in the TCR-CD8 complex has been shown to convey reversible T-cell unresponsiveness [89].

Table 11.1 Drug targets for reversal of tumor-mediated immunosuppression

Target	Ligand or receptor	Function	Cellular source	Inhibitor	Development status	References
COX-2	n.a.	Production of prostaglandins formation of Tregs	Tumor cells TAM	Celecoxib	Phase I-phase III	Clinical.trial.gov
mPGES-1	n.a.	Production of PGE_2	Tumor cells	n.a.	Preclinical	[28]
TGF-β	TGF-β RI	Down-regulation of immune response	Tumor cells-Treg	mAb TGF-β antisense Fc-TGF-β RII fusion sTGF-β RIII small molecule inhibitors	Phase I–phase III	clinical.trial.gov[91]
IL-10	IL-10 R	Down-regulation of immune responses	Tumor cells MDSC TAM	Anti-IL10 mAb Anti-IL10R mAb IL-10R-Fc fusion proteins	Preclinical	
Arginase 1	n.a.	Depletion of arginine	MDSC	n.a	Preclinical	
iNOS	n.a.	Generation of ROS Perxoynitrite	Tumor cells MDSC TAM	GW274150	Preclinical	
IDO/IDO2	n.a.	Suppression of T cells	Tumor cells DC	1-methyl-D-tryptophan	Phase I	clinical.trial.gov
CD25 (Treg)	IL-2	Treg marker	Treg activated T cells	Depleting anti-CD25 mAb denileukin diftitox (IL-2/diphteria toxin fusion protein)	Phase I-phase II approved by FDA for the treatment of CD25+ cutaneous T-cell leukemia and lymphoma	clinical.trial.gov

CTLA-4	B7-1 B7-2	Negative coregulator of T cells	Activated T cells	Tremelimumab Ipilimumab MDX-CTLA-4 (antibodies) MDX-01 (recombinant CTLA-4)	Phase I-phase III	[92]
PD-1	PD-L1 PD-L2	Negative coregulator of T cells	Activated T cells	mAb ONO-4538 CT-011	Phase I	[48]
CEACAM1	CEACAMs	Negative coregulator of T cells and NK cells	Various cell types	mAb	Preclinical	

n.a. not applicable

Therapeutic approaches to abrogate the suppressive impact of arginine metabolites have been discussed in the section above on MDSC.

Perspectives and Future Directions

In order to profoundly understand tumor immune reactions, the dynamics of immune cell/tumor cell interactions also have to be taken into consideration [90]. Recently, it was shown that the dynamics of tumor cell killing by tumor antigen-specific CTL in the tumor microenvironment is quite different from that in other compartments of the body such as the lymph nodes, and even more different from *in vitro* killing [91]. Surprisingly, it takes on average 6 h for one CTL to kill

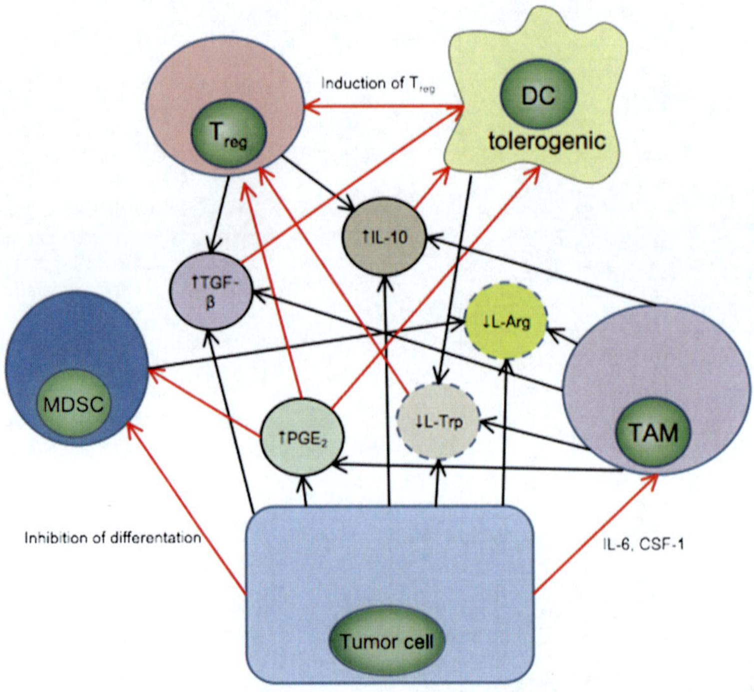

Fig. 11.2 Immunosuppressive network in the tumor microenvironment. Different cell types participate in the immune suppressive network by secretion of key regulatory factors, thereby driving each other into a suppressive state. Black arrows indicate the production of factors whereas red arrows indicate that this factor stimulates the indicated cell type to elicit full immunosuppression. The red arrows between the cells indicate that one cell either attracts the other or induces the suppressive nature of the other cell by partly unknown mechanisms. Key factors which are up-regulated are TGF-β, IL-10 and PGE_2, whereas L-tryptophan and L-arginine are depleted in the tumor microenvironment. Of note, each immunosuppressive cell population modifies the concentration of these key factors in the same way and the concentration of each factor is modified by at least two different cell types

a single tumor cell in the tumor microenvironment. In contrast, B cell killing in lymph nodes was previously estimated to require much less than one hour [92]. The molecular basis for the delayed tumor cell killing in the tumor microenvironment remains to be addressed. Clearly, the unique dynamics of immune cell responses within the tumor microenvironment may further complicate immunotherapeutic strategies for cancer treatment.

As discussed above, a complex network of mutual functional enhancement exists between the immunosuppressive cells that exist in the tumor (Fig. 11.2). However, the interplay between these cell populations is only starting to be understood. It seems likely that, once we better understand how these cells communicate with each other, we can identify new cellular and molecular targets for therapeutic efforts to disrupt the vicious circle of immunosuppression that exists within the tumor microenvironment.

References

1. Zhou G, Levitsky HI (2007). Natural regulatory T cells and de novo-induced regulatory T cells contribute independently to tumor-specific tolerance. J Immunol 178:2155–2162.
2. Qin FX (2009). Dynamic behavior and function of Foxp3+ regulatory T cells in tumor bearing host. Cell Mol Immunol 6:3–13.
3. Ito T, Hanabuchi S, Wang YH et al (2008). Two functional subsets of FOXP3+ regulatory T cells in human thymus and periphery. Immunity 28:870–880.
4. Colombo MP, Piconese S (2007). Regulatory-T-cell inhibition versus depletion: the right choice in cancer immunotherapy. Nat Rev Cancer 7:880–887.
5. Curiel TJ (2008). Regulatory T cells and treatment of cancer. Curr Opin Immunol 20:241–246.
6. Aandahl EM, Torgersen KM, Tasken K (2008). $CD8^+$ regulatory T cells-A distinct T-cell lineage or a transient T-cell phenotype? Hum Immunol 69:696–699.
7. Wang RF (2008). $CD8^+$ regulatory T cells, their suppressive mechanisms, and regulation in cancer. Hum Immunol 69:811–814.
8. Berzofsky JA, Terabe M (2008). NKT cells in tumor immunity: opposing subsets define a new immunoregulatory axis. J Immunol 180:3627–3635.
9. Park JM, Terabe M, van den Broeke LT et al (2005). Unmasking immunosurveillance against a syngeneic colon cancer by elimination of $CD4^+$ NKT regulatory cells and IL-13. Int J Cancer 114:80–87.
10. Gabrilovich DI, Nagaraj S (2009). Myeloid-derived suppressor cells as regulators of the immune system. Nat Rev Immunol 9:162–174.
11. Kusmartsev S, Nefedova Y, Yoder D et al (2004). Antigen-specific inhibition of $CD8^+$ T cell response by immature myeloid cells in cancer is mediated by reactive oxygen species. J Immunol 172:989–999.
12. Movahedi K, Guilliams M, Van den BJ et al (2008). Identification of discrete tumor-induced myeloid-derived suppressor cell subpopulations with distinct T cell-suppressive activity. Blood 111:4233–4244.
13. Nefedova Y, Fishman M, Sherman S et al (2007). Mechanism of all-trans retinoic acid effect on tumor-associated myeloid-derived suppressor cells. Cancer Res 67:11021–11028.
14. Kusmartsev S, Cheng F, Yu B et al (2003). All-trans-retinoic acid eliminates immature myeloid cells from tumor-bearing mice and improves the effect of vaccination. Cancer Res 63:4441–4449.
15. Kusmartsev S, Su Z, Heiser A et al (2008). Reversal of myeloid cell-mediated immunosuppression in patients with metastatic renal cell carcinoma. Clin Cancer Res 14:8270–8278.

16. Pan PY, Wang GX, Yin B et al (2008). Reversion of immune tolerance in advanced malignancy: modulation of myeloid-derived suppressor cell development by blockade of stem-cell factor function. Blood 111:219–228.
17. Rodriguez PC, Hernandez CP, Quiceno D et al (2005). Arginase I in myeloid suppressor cells is induced by COX-2 in lung carcinoma. J Exp Med 202:931–939.
18. Haas AR, Sun J, Vachani A et al (2006). Cycloxygenase-2 inhibition augments the efficacy of a cancer vaccine. Clin Cancer Res 12:214–222.
19. Serafini P, Meckel K, Kelso M et al (2006). Phosphodiesterase-5 inhibition augments endogenous antitumor immunity by reducing myeloid-derived suppressor cell function. J Exp Med 203:2691–2702.
20. Bingle L, Brown NJ, Lewis CE (2002). The role of tumour-associated macrophages in tumour progression: implications for new anticancer therapies. J Pathol 196:254–265.
21. Takayama H, Nishimura K, Tsujimura A et al (2009). Increased infiltration of tumor associated macrophages is associated with poor prognosis of bladder carcinoma in situ after intravesical bacillus Calmette-Guerin instillation. J Urol 181:1894–1900.
22. Allavena P, Sica A, Garlanda C et al (2008). The Yin-Yang of tumor-associated macrophages in neoplastic progression and immune surveillance. Immunol Rev 222:155–161.
23. Muller AJ, Scherle PA (2006). Targeting the mechanisms of tumoral immune tolerance with small-molecule inhibitors. Nat Rev Cancer 6:613–625.
24. Allavena P, Sica A, Solinas G et al (2008). The inflammatory micro-environment in tumor progression: the role of tumor-associated macrophages. Crit Rev Oncol Hematol 66:1–9.
25. Greenhough A, Smartt HJ, Moore AE et al (2009). The COX-2/PGE2 pathway: key roles in the hallmarks of cancer and adaptation to the tumour microenvironment. Carcinogenesis 30:377–386.
26. Breyer RM, Bagdassarian CK, Myers SA et al (2001). Prostanoid receptors: subtypes and signaling. Annu Rev Pharmacol Toxicol 41:661–690.
27. Harizi H, Gualde N (2005). The impact of eicosanoids on the crosstalk between innate and adaptive immunity: the key roles of dendritic cells. Tissue Antigens 65:507–514.
28. Samuelsson B, Morgenstern R, Jakobsson PJ (2007). Membrane prostaglandin E synthase-1: a novel therapeutic target. Pharmacol Rev 59:207–224.
29. Bierie B, Moses HL (2006). Tumour microenvironment: TGF-β: the molecular Jekyll and Hyde of cancer. Nat Rev Cancer 6:506–520.
30. Wrzesinski SH, Wan YY, Flavell RA (2007). Transforming growth factor-beta and the immune response: implications for anticancer therapy. Clin Cancer Res 13:5262–5270.
31. Geissmann F, Revy P, Regnault A et al (1999). TGF-β 1 prevents the noncognate maturation of human dendritic Langerhans cells. J Immunol 162:4567–4575.
32. Maynard CL, Weaver CT (2008). Diversity in the contribution of interleukin-10 to T-cell-mediated immune regulation. Immunol Rev 226:219–233.
33. O'Garra A, Barrat FJ, Castro AG et al (2008). Strategies for use of IL-10 or its antagonists in human disease. Immunol Rev 223:114–131.
34. Brunet JF, Denizot F, Luciani MF et al (1987). A new member of the immunoglobulin superfamily – CTLA-4. Nature 328:267–270.
35. Linsley PS, Greene JL, Tan P et al (1992). Coexpression and functional cooperation of CTLA-4 and CD28 on activated T lymphocytes. J Exp Med 176:1595–1604.
36. Riley JL (2009). PD-1 signaling in primary T cells. Immunol Rev 229:114–125.
37. Waterhouse P, Penninger JM, Timms E et al (1995). Lymphoproliferative disorders with early lethality in mice deficient in CTLA-4. Science 270:985–988.
38. Peggs KS, Quezada SA, Korman AJ et al (2006). Principles and use of anti-CTLA-4 antibody in human cancer immunotherapy. Curr Opin Immunol 18:206–213.
39. Fong L, Small EJ (2008). Anti-cytotoxic T-lymphocyte antigen-4 antibody: the first in an emerging class of immunomodulatory antibodies for cancer treatment. J Clin Oncol 26:5275–5283.
40. Attia P, Phan GQ, Maker AV et al (2005). Autoimmunity correlates with tumor regression in patients with metastatic melanoma treated with anti-cytotoxic T-lymphocyte antigen-4. J Clin Oncol 23:6043–6053.

41. Blank C, Mackensen A (2007). Contribution of the PD-L1/PD-1 pathway to T-cell exhaustion: an update on implications for chronic infections and tumor evasion. Cancer Immunol Immunother 56:739–745.
42. Gao Q, Wang XY, Qiu SJ et al (2009). Overexpression of PD-L1 significantly associates with tumor aggressiveness and postoperative recurrence in human hepatocellular carcinoma. Clin Cancer Res 15:971–979.
43. Barber DL, Wherry EJ, Masopust D et al (2006). Restoring function in exhausted CD8 T cells during chronic viral infection. Nature 439:682–687.
44. Wherry EJ, Ha SJ, Kaech SM et al (2007). Molecular signature of $CD8^+$ T cell exhaustion during chronic viral infection. Immunity 27:670–684.
45. Ahmadzadeh M, Johnson LA, Heemskerk B et al (2009). Tumor antigen-specific CD8 T cells infiltrating the tumor express high levels of PD-1 and are functionally impaired. Blood 114(8):1537–1544.
46. Nomi T, Sho M, Akahori T et al (2007). Clinical significance and therapeutic potential of the programmed death-1 ligand/programmed death-1 pathway in human pancreatic cancer. Clin Cancer Res 13:2151–2157.
47. Parekh VV, Lalani S, Kim S et al (2009). PD-1/PD-L blockade prevents anergy induction and enhances the anti-tumor activities of glycolipid-activated invariant NKT cells. J Immunol 182:2816–2826.
48. Berger R, Rotem-Yehudar R, Slama G et al (2008). Phase I safety and pharmacokinetic study of CT-011, a humanized antibody interacting with PD-1, in patients with advanced hematologic malignancies. Clin Cancer Res 14:3044–3051.
49. Kammerer R, Hahn S, Singer BB et al (1998). Biliary glycoprotein (CD66a), a cell adhesion molecule of the immunoglobulin superfamily, on human lymphocytes: structure, expression and involvement in T cell activation. Eur J Immunol 28:3664–3674.
50. Chen Z, Chen L, Qiao SW et al (2008). Carcinoembryonic antigen-related cell adhesion molecule 1 inhibits proximal TCR signaling by targeting ZAP-70. J Immunol 180: 6085–6093.
51. Kammerer R, Stober D, Singer BB et al (2001). Carcinoembryonic antigen-related cell adhesion molecule 1 on murine dendritic cells is a potent regulator of T cell stimulation. J Immunol 166:6537–6544.
52. Kammerer R, Riesenberg R, Weiler C et al (2004). The tumour suppressor gene CEACAM1 is completely but reversibly downregulated in renal cell carcinoma. J Pathol 204:258–267.
53. Kammerer R, von Kleist S (1994). CEA expression of colorectal adenocarcinomas is correlated with their resistance against LAK-cell lysis. Int J Cancer 57:341–347.
54. Stern N, Markel G, Arnon TI et al (2005). Carcinoembryonic antigen (CEA) inhibits NK killing via interaction with CEA-related cell adhesion molecule 1. J Immunol 174:6692–6701.
55. Markel G, Lieberman N, Katz G et al (2002). CD66a interactions between human melanoma and NK cells: a novel class I MHC-independent inhibitory mechanism of cytotoxicity. J Immunol 168:2803–2810.
56. Markel G, Seidman R, Cohen Y et al (2009). Dynamic expression of protective CEACAM1 on melanoma cells during specific immune attack. Immunology 126:186–200.
57. Iijima H, Neurath MF, Nagaishi T et al (2004). Specific regulation of T helper cell 1-mediated murine colitis by CEACAM1. J Exp Med 199:471–482.
58. Ball HJ, Sanchez-Perez A, Weiser S et al (2007). Characterization of an indoleamine 2,3-dioxygenase-like protein found in humans and mice. Gene 396:203–213.
59. Yuasa HJ, Takubo M, Takahashi A et al (2007). Evolution of vertebrate indoleamine 2,3-dioxygenases. J Mol Evol 65:705–714.
60. Munn DH, Zhou M, Attwood JT et al (1998). Prevention of allogeneic fetal rejection by tryptophan catabolism. Science 281:1191–1193.
61. Mellor AL, Munn DH (2004). IDO expression by dendritic cells: tolerance and tryptophan catabolism. Nat Rev Immunol 4:762–774.
62. Mellor AL, Munn DH (2008). Creating immune privilege: active local suppression that benefits friends, but protects foes. Nat Rev Immunol 8:74–80.

63. Uyttenhove C, Pilotte L, Theate I et al (2003). Evidence for a tumoral immune resistance mechanism based on tryptophan degradation by indoleamine 2,3-dioxygenase. Nat Med 9:1269–1274.
64. Yoshida N, Ino K, Ishida Y et al (2008). Overexpression of indoleamine 2,3-dioxygenase in human endometrial carcinoma cells induces rapid tumor growth in a mouse xenograft model. Clin Cancer Res 14:7251–7259.
65. Zheng X, Koropatnick J, Li M et al (2006). Reinstalling antitumor immunity by inhibiting tumor-derived immunosuppressive molecule IDO through RNA interference. J Immunol 177:5639–5646.
66. Prendergast GC (2008). Immune escape as a fundamental trait of cancer: focus on IDO. Oncogene 27:3889–3900.
67. Brandacher G, Perathoner A, Ladurner R et al (2006). Prognostic value of indoleamine 2,3-dioxygenase expression in colorectal cancer: effect on tumor-infiltrating T cells. Clin Cancer Res 12:1144–1151.
68. Chamuleau ME, van de Loosdrecht AA, Hess CJ et al (2008). High INDO (indoleamine 2,3-dioxygenase) mRNA level in blasts of acute myeloid leukemic patients predicts poor clinical outcome. Haematologica 93:1894–1898.
69. Ino K, Yamamoto E, Shibata K et al (2008). Inverse correlation between tumoral indoleamine 2,3-dioxygenase expression and tumor-infiltrating lymphocytes in endometrial cancer: its association with disease progression and survival. Clin Cancer Res 14:2310–2317.
70. Pan K, Wang H, Chen MS et al (2008). Expression and prognosis role of indoleamine 2,3-dioxygenase in hepatocellular carcinoma. J Cancer Res Clin Oncol 134:1247–1253.
71. Takao M, Okamoto A, Nikaido T et al (2007). Increased synthesis of indoleamine-2,3-dioxygenase protein is positively associated with impaired survival in patients with serous-type, but not with other types of, ovarian cancer. Oncol Rep 17:1333–1339.
72. Ishio T, Goto S, Tahara K et al (2004). Immunoactivative role of indoleamine 2,3-dioxygenase in human hepatocellular carcinoma. J Gastroenterol Hepatol 19:319–326.
73. Riesenberg R, Weiler C, Spring O et al (2007). Expression of indoleamine 2,3-dioxygenase in tumor endothelial cells correlates with long-term survival of patients with renal cell carcinoma. Clin Cancer Res 13:6993–7002.
74. Muller AJ, Sharma MD, Chandler PR et al (2008). Chronic inflammation that facilitates tumor progression creates local immune suppression by inducing indoleamine 2,3 dioxygenase. Proc Natl Acad Sci USA 105:17073–17078.
75. Munn DH, Sharma MD, Hou D et al (2004). Expression of indoleamine 2,3-dioxygenase by plasmacytoid dendritic cells in tumor-draining lymph nodes. J Clin Invest 114:280–290.
76. Hou DY, Muller AJ, Sharma MD et al (2007). Inhibition of indoleamine 2,3-dioxygenase in dendritic cells by stereoisomers of 1-methyl-tryptophan correlates with antitumor responses. Cancer Res 67:792–801.
77. Yen MC, Lin CC, Chen YL et al (2009). A novel cancer therapy by skin delivery of indoleamine 2,3-dioxygenase siRNA. Clin Cancer Res 15:641–649.
78. Metz R, Duhadaway JB, Kamasani U et al (2007). Novel tryptophan catabolic enzyme IDO2 is the preferred biochemical target of the antitumor indoleamine 2,3-dioxygenase inhibitory compound D-1-methyl-tryptophan. Cancer Res 67:7082–7087.
79. Lob S, Konigsrainer A, Schafer R et al (2008). Levo- but not dextro-1-methyl tryptophan abrogates the IDO activity of human dendritic cells. Blood 111:2152–2154.
80. Lob S, Konigsrainer A, Zieker D et al (2009). IDO1 and IDO2 are expressed in human tumors: levo- but not dextro-1-methyl tryptophan inhibits tryptophan catabolism. Cancer Immunol Immunother 58:153–157.
81. Lob S, Konigsrainer A, Rammensee HG et al (2009). Inhibitors of indoleamine-2,3-dioxygenase for cancer therapy: can we see the wood for the trees? Nat Rev Cancer 9: 445–452.
82. Kudo Y, Boyd CA (2001). Characterisation of L-tryptophan transporters in human placenta: a comparison of brush border and basal membrane vesicles. J Physiol 531:405–416.
83. Seymour RL, Ganapathy V, Mellor AL et al (2006). A high-affinity, tryptophan-selective amino acid transport system in human macrophages. J Leukoc Biol 80:1320–1327.

84. Schlingensiepen KH, Fischer-Blass B, Schmaus S et al (2008). Antisense therapeutics for tumor treatment: the TGF-β 2 inhibitor AP 12009 in clinical development against malignant tumors. Recent Results Cancer Res 177:137–150.
85. Ribas A (2008). Overcoming immunologic tolerance to melanoma: targeting CTLA-4 with tremelimumab (CP-675,206). Oncologist 13(Suppl 4):10–15.
86. Bronte V, Kasic T, Gri G et al (2005). Boosting antitumor responses of T lymphocytes infiltrating human prostate cancers. J Exp Med 201:1257–1268.
87. Rodriguez PC, Quiceno DG, Ochoa AC (2007). L-arginine availability regulates T-lymphocyte cell-cycle progression. Blood 109:1568–1573.
88. Baniyash M (2004). TCR zeta-chain downregulation: curtailing an excessive inflammatory immune response. Nat Rev Immunol 4:675–687.
89. Nagaraj S, Gupta K, Pisarev V et al (2007). Altered recognition of antigen is a mechanism of $CD8^+$ T cell tolerance in cancer. Nat Med 13:828–835.
90. Pittet MJ (2009). Behavior of immune players in the tumor microenvironment. Curr Opin Oncol 21:53–59.
91. Breart B, Lemaitre F, Celli S et al (2008). Two-photon imaging of intratumoral $CD8^+$ T cell cytotoxic activity during adoptive T cell therapy in mice. J Clin Invest 118:1390–1397.
92. Mempel TR, Pittet MJ, Khazaie K et al (2006). Regulatory T cells reversibly suppress cytotoxic T cell function independent of effector differentiation. Immunity 25:129–141.
93. Kammerer R, Zimmermann W (2010). Coevolution of activating and inhibitory receptors within mammalian carcinoembryonic antigen families. BMC Biol 4:8–12.

Chapter 12
Genetically Engineered Antigen Specificity in T Cells for Adoptive Immunotherapy

Daniel J. Powell, Jr. and Bruce L. Levine

Abstract Antigen specificity has been genetically conferred to human T cells for therapy through the transfer of sequences encoding antigen-specific T-cell receptors (TCRs) or through the antigen-specific antibody-based chimeric antigen receptors (CARs). By either method, the advantage of the adoptive transfer of modified T cells over immunotherapies developed to date such as vaccines or cytokines alone is that T cells have the ability to directly traffic to the site of tumor or infection, to multiply at that site, and to persist in a memory state for months to years.

Keywords T-Cell receptor • Chimerical antigen receptor • Gene therapy • Adoptive transfer • Immunotherapy

Introduction

Adoptive transfer, as originally described by Medawar [1], describes the passive transfer of immunocompetent cells for the treatment of cancer or infection disease. Adoptive immunotherapy with antigen-specific T cells has emerged as a powerful approach for the treatment of advanced refractory cancer and chronic viral infection. Autologous T cells may be selected for high avidity against antigen and then activated and expanded *ex vivo* to induce strong effector functions. Culture conditions can be customized for various T-cell subsets and modification of T-cell specificities through genetic engineering which allows for redirection of T cell specificity towards tumor or pathogens. Conditioning regimens are now known to play an important role in the efficiency of adoptively transferred T cells by providing a favorable environment for engraftment, proliferation, effector function and memory

D.J. Powell, Jr. (✉)
Pathology and Laboratory Medicine, Clinical Cell and Vaccine Production Facility, Ovarian Cancer Research Center, The University of Pennsylvania School of Medicine, Philadelphia, PA, USA
e-mail: Poda@mail.med.upenn.edu

J. Medin and D. Fowler (eds.), *Experimental and Applied Immunotherapy*,
DOI 10.1007/978-1-60761-980-2_12,

formation. At present, antigen-specific T cells for use in adoptive immunotherapy trials are broadly divided into two categories, naturally occurring and genetically modified T cells. Naturally occurring antigen-specific T cells have generally been harvested and expanded from the peripheral blood of patients following vaccination or following antigen-specific stimulation *in vitro*, or from autologous tumor deposits. Virus derived proteins recognized as nonself antigens naturally elicit vigorous T-cell responses and T-cell memory formation providing the opportunity for transfer of T cells with exquisite specificity for antigen. However, the ability to identify, isolate, and expand naturally occurring tumor antigen-specific T cells from patients with cancer has been a more daunting task since the majority of tumor antigens are in fact self tissue proteins that result in T-cell tolerization. Still, in metastatic melanoma where tumor-reactive T cells can be readily isolated from resected tumor and amplified *ex vivo*, administration of naturally occurring tumor-reactive T cells in combination with immune-depleting conditioning can elicit objective clinical responses even in patients with bulky refractory disease [2, 3]. These studies and others have demonstrated the power of the T-cell transfer approach when antigen-specific T cells are present and can be easily isolated and expanded.

Table 12.1 lists some points to consider in engineering effective T lymphocyte therapies in light of our current understanding of T-lymphocyte biology and homeostasis. Unfortunately, until recently many trials were carried out before the complexity of T-cell biology and costimulation was understood. As a result, cells were propagated in what are now understood to be suboptimal conditions that impair the essential functions and long-term engraftment of T cells. Similarly, *in vitro* activity observed in preclinical experiments may not translate to *in vivo* activity due to a lack of costimulation in the local tumor environment or any of several other reasons listed in Table 12.1. The deficiency of naturally occurring antigen-specific T cells in many tumors, and the inability to break tolerance to self/tumor antigens along with the inability to raise sufficient numbers of properly functional effector T cells are major obstacles facing the field of cancer immunotherapy. Genetically retargeting T cells circumvents these limitations and now provides an alternative approach toward the de novo generation of antigen-specific T cells for therapy.

Table 12.1 Points to consider in *ex vivo* engineering of an effective T lymphocyte therapy

- Infused T-cell population should retain properties that permit persistence and homing to tumor or lymph node
- Engraftment and persistence of adoptively transferred CTLs may well depend on adequate CD4 T-cell help or exogenous cytokine support
- Transduced gene product of transferred cells should not be immunogenic
- T-cell manufacturing process should avoid induction of replicative senescence
- Effector to target ratio: host tumor burden should not exceed killing capacity of adoptively transferred cells
- Tumor or HIV antigen-specific T cells may have been deleted or tolerized in the donor by previous chemotherapy or by the tumor itself
- Lack of costimulation may induce anergy or apoptosis of adoptively transferred cells

The genetically engineered T-cell approach in many cases eliminates the need for surgery to obtain antigen-specific T cells, has the potential to rescue nonreactive T cells, and can introduce enhanced specificities not found in many patients. Further, since the specificity of antigen recognition is known, the adoptive transfer of genetically retargeted T cells can be combined with cancer vaccination to enhance immune surveillance.

The T-cell receptor (TCR) is an ab heterodimeric structure that confers antigen specificity and functional avidity to T cells, in association with the CD3 signaling heterocomplex (Fig. 12.1). Through the TCR, T cells naturally recognize intracellular protein derivatives in an MHC restricted and antigen processing dependent manner. Since TCR ab gene sequences can be isolated from tumor-reactive T-cell clones, one possible approach to generate tumor antigen-specific T cells for therapy is to transfer these ab sequences to a polyclonal or nonreactive T-cell population to redirect their specificity. An alternative method of T-cell redirection builds upon the ability of antibodies to recognize intact cell surface antigens independent of MHC and antigen processing restrictions. Antigen-specific antibody-based chimeric antigen receptors (CARs), alternatively labeled chimeric immune receptors (CIRs) or "T-bodies," are comprised of an extracellular antibody antigen binding domain combined with intracellular T-cell-signaling receptor domains (Fig. 12.1). Accordingly, T cell redirected with CARs can be endowed with the capacity to

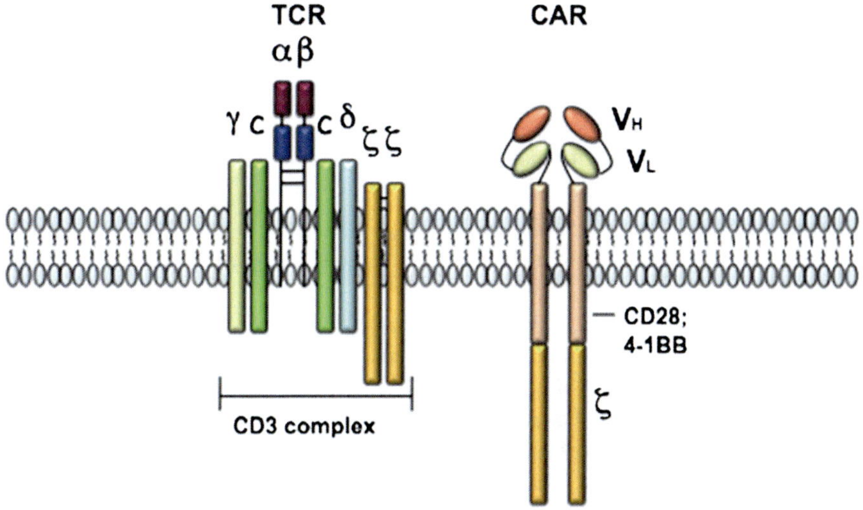

Fig. 12.1 Endogenous T cells express a single heterodimeric T-cell receptor (*TCR*). Bispecific retargeted T cells are created by the introduction of genes that encode proteins that recognize target antigens. *Left panel*: These genes can encode natural TCRs or TCRs with enhanced affinity that function in the same MHC-restricted manner as endogenous TCRs. *Right panel*: Alternatively, these genes can encode chimeric antigen-specific receptors (*CAR*) that target surface antigens in an MHC independent fashion. CARs express an extracellular ligand generally derived from an antibody and intracellular signaling modules derived from T-cell–signaling proteins

recognize and respond to specific target surface antigens, independent of MHC haplotype.

In this chapter, we will discuss how antigen specificity has been genetically conferred to human T cells for therapy through the transfer of sequences encoding antigen-specific T-cell receptors (TCRs) or through the antigen-specific antibody-based chimeric antigen receptors (CARs). By either method, the advantage of the adoptive transfer of modified T cells over immunotherapies developed to date such as vaccines or cytokines alone is that T cells have the ability to directly traffic to the site of tumor or infection, to multiply at that site, and to persist in a memory state for months to years.

Methods of Gene Transfer to T Cells

The feasibility of TCR gene transfer was first reported in 1986, with TCR genes from one mouse T cell being transferred via protoplast fusion into another mouse T-cell clone with a different specificity, and this demonstrated that transfer of TCRs was necessary and sufficient to redirect T-cell activity [4]. Similarly, TCR gene transfer via cell fusion or electroporation also was shown to confer MHC restricted specificity to nonspecific T cells [5, 6]. However these initial processes were generally inefficient for preparation of meaningful numbers of T cells for clinical application. The advent of retroviral vectors not only made gene transfer more efficient at large scale but also enabled the stable integration of exogenous genes. In 1990, the clinical feasibility and safety of gamma retroviral gene transduced T-cell therapy was demonstrated by Rosenberg et al. using T cells modified to express the gene encoding neomycin resistance as a method of assessing T-cell biodistribution and survival after infusion [7]. Despite these demonstrations of scalability and stable expression, gamma retroviral vectors require full T-cell activation for efficient gene transduction that may, in certain culture systems, inhibit cell function [8, 9]. Gamma retroviral vectors also are susceptible to gene silencing [10, 11] and preferentially integrate near transcriptional start sites, increasing the potential risk for insertional mutagenesis [12, 13]. Nevertheless, to date, TCR gene therapy trials have relied entirely upon gamma retroviral gene transfer as a means of stable TCR gene integration. Considerable data shows that gamma retroviral vectors are safe when expressed in human T cells. However, gamma retroviral vectors appear less safe when used in human stem cells as observed in a clinical trial utilizing murine-derived gamma retroviral vectors to transfer the γ signaling chain to stem cells of infants with X-linked SCID [14, 15]. To date, four of nine children in that trial have developed integration site-induced cancers, and one of those children died after relapse. Of note, a parallel study to the French study where the same disease, payload, and target cell were used, no tumorigenesis was observed at first, suggesting that minor differences, such as vector envelope and stem cell growth factors, might be relevant in long-term safety of integrating vectors [16]. However, in December of 2007, one of the ten children treated in that study also developed insertion-mediated oncogenesis [17]. Today, the causes of insertional oncogenesis remain poorly understood, but in the X-linked SCID studies, a common integration site, LMO-2, is observed in at least three of the five patients. In general, the target cell, vector, and disease payload are considered major factors contributing to the risks of vector-induced tumors. However, there is insufficient data to date to support the relative contribution of each of these factors in assuring safety of gamma retroviral gene transfer.

Lentiviral vectors have the potential to be safer from the perspective of insertional mutagenesis, and they have substantially higher transduction efficiency for genetically engineering human T cells [18–21]. In 2006, the first clinical use of an engineered lentiviral vector in humans was reported. This study demonstrated the feasibility, safety, and tolerability of administering autologous T cells modified with a conditionally replicating lentiviral vector encoding an antisense gene against the HIV envelope in patients with HIV [22]. A longitudinal analysis of samples from these patients showed no evidence for abnormal expansions of cells due to vector-mediated insertional activation of proto-oncogenes as of 32 weeks postinfusion [23]. Recently, lentiviral vectors have emerged as a favorable vector system for stably expressing two-gene TCRs, which require dual gene expression. Bicistronic lentiviral vectors that provide for high-level tumor or viral antigen-specific TCR gene expression in T cells have been developed that show strong promise for clinical application [24, 25]. Although viral vectors provide efficient and stable transgene expression, the process is limited by the time and cost required for clinical grade vector production and restrictions on the size and number of genes that can be packaged by these vectors. Potential safety issues could also limit the use of viral vectors in certain clinical applications [26]. A study in tumor prone mice comparing the tumorigenicity of gamma retroviral vectors to lentiviral vectors demonstrated that lentiviral vector gene transfer into hematopoietic stem cells of up to an average of six copies per cell was not tumorigenic in contrast to gamma retroviral vectors at an average copy number of three per cell [12]. While instructive, this study still does not preclude the potential for insertional mutagenesis with lentiviral vectors, particularly when used to transduce stem cells in immunosuppressed populations. It is notable that T-cell leukemia is not a recognized side effect of HIV infection, although a high proportion of proviruses are defective and therefore would not mask a leukemic event.

In light of the potential serious adverse events associated with gamma retroviral gene transfer, the Food and Drug Administration has attempted to arrive at a consensus recommendation for long-term follow-up of patients for delayed adverse events associated with these studies. Follow-up as long as the patient lifetime has been proposed, but a compromise was reached that combined feasibility with safety concerns. In the November, 2006 FDA Guidance for Industry "Gene Therapy Clinical Trials-Observing Subjects for Delayed Adverse Events," long-term follow-up is restricted to vectors that integrate or that have the potential for latency and reactivation. It will take establishment of a larger safety database of patient treatment before these vectors will be accepted for clinical application in nonlife-threatening diseases, other than for local gene transfer. Additional research studies in serial transplantation of stem cells and T cells in immunodeficient mice will help to elucidate the effects of lentiviral integration on clonal dominance, and such studies will help to further establish the safety/risk profile of individual vectors.

Recent advances have shown electroporation-mediated mRNA transfection is an alternative and promising approach without many of the limitations associated with gene transfer vectors that insert, by and large, randomly in the genome. mRNA manufacture and delivery does not require the significant costs of manufacture and testing required for the use of exogenous viral vectors. Large quantities of mRNA can be rapidly prepared and evaluated. Importantly, mRNA expression is transient so that the transfection process does not result in permanent genetic modification of cells. Preclinical studies based on mRNA electroporation of human T lymphocytes have been reported recently. Using this approach for transfer of tumor antigen-specific TCR genes, Zhao et al. demonstrated a >90% transduction efficiency that conferred tumor reactivity to nonreactive, unstimulated human T cells [27]. Yoon et al. showed in a xenograft mouse model that the adoptive transfer of Her2/neu chimeric antigen receptor mRNA electroporated into peripheral blood lymphocytes (PBL) led to significant inhibition of tumor growth compared with transfer of mock-transfected PBL. Significantly, the inhibition observed with mRNA-transfected cells was higher than that observed with Herceptin antibody [28]. The transient expression of exogenous protein following RNA electroporation may at first seem a limitation to long term persistence of infused antigen-specific T cells, a factor associated with clinical response [29]. However, transient TCR or CAR expression may be desirable when T cells are redirected against self-tumor antigens that have the potential to elicit treatment related autoimmunity.

T-Cell Receptor Gene Therapy

Natural antigen recognition by T cells is conferred by the heterodimeric TCR, comprised of an α/β chain dimer, which recognizes intracellular protein derivatives in an MHC restricted and antigen processing dependent manner (Fig. 12.1). With this knowledge of T-cell biology, one strategy to harness the strengths of T cells for immunotherapy is to identify therapeutically effective T-cell clones, clone the heterodimeric TCR, and express it in polyclonal peripheral blood T cells isolated from the patient, creating bispecific T cells with specificity conferred by the endogenous TCR and by the cloned TCR. Important factors for consideration during the selection of TCRs for potential use in TCR gene therapy include, the degree to which the cognate antigen is expressed among a particular cancer or virus type, the affinity of the TCR for cognate peptide/MHC complex, its ability of the TCR to bind independently of CD8 or CD4 coreceptors, and its ability to properly assemble and pair on the T-cell surface. Significant effort has been made to isolate antigen-specific TCRs that exhibit high affinity binding to cognate peptide antigens that are highly expressed among virally infected cell populations or particular cancer types and then transfer these high functional avidity TCRs to patient T cells [30–32]. Affinity, as defined here, refers to the strength of the sole interaction between TCR and peptide/MHC complex, and excludes accessory molecule contributions. The combined strength of multiple interactions is defined as avidity. Naturally occurring T cells expressing high affinity TCRs specific for self-tumor antigens are difficult to obtain due to intrathymic deletion. However, in the setting of melanoma, rare high-affinity TCRs, derived from TILs that mediate strong *in vivo* tumor regression, have been identified, cloned, and transduced into the T cells of patients with cancer [30, 33]. TCR-transduced T cells demonstrate cytolytic activity and secrete cytokines *in vitro* after stimulation with melanoma cell lines and have been clinically evaluated [34].

In 1998, the first use of retroviral vector driven TCR gene transfer was reported by the Pease laboratory with the LTR directing expression of TCRα and an internal CMV promoter driving TCRα expression in mouse T cells [35, 36]. The following year, Clay used a gamma retroviral vector to transfer a human, HLA-A2 restricted TCR with specificity for MART-1, derived from tumor infiltrating lymphocytes from a melanoma patient, to T cells from three healthy donors [37]. To date, the identification of high affinity TCRs bearing T cells from the natural human T-cell repertoire has been performed by brute-force T-cell cloning [34] or via selection using alpha 3 domain-modified peptide/MHC tetramers that lack CD8 coreceptor binding capacity and therefore exclusively interrogate TCR binding to peptide/MHC complex [38]. Natural TCRs can be further optimized for tumor antigen reactivity through focused amino acid substitutions in TCR complementarity-determining regions [32, 39].

Alternative nonautologous approaches of TCR isolation can now bypass natural TCR affinity issues resulting from thymic selection. For example, high avidity, tumor antigen-specific, HLA-A*0201-positive T cells can be generated through *in vitro* stimulation of allogeneic HLA-A*0201-negative T cells with tumor antigen loaded HLA-A*0201-positive APCs, as a source for TCR isolation [40]. Another approach is to generate TCRs in the absence of *in vivo* selection pressure by phage or yeast display [31, 32, 41]. Here, high affinity TCRs are generated *in vitro* by random mutation of the TCR degenerate complementarity-determining regions, followed by subsequently screening for increased affinity via tetramer staining analysis, and lastly transduction of high affinity TCR cDNA candidates into T cells for functional avidity testing. The generation of high affinity TCRs for some self-tumor antigens can be performed *in vivo* in immunized HLA-*0201 transgenic mice where tolerance to human proteins is incomplete due to differences in protein sequences between mice and human [42, 43]. Because of the nonhomologous nature of this approach, high affinity TCR αβ chain pairs generated from HLA-*0201 mice have the added benefit of being coreceptor independent and preferentially pairing with mouse TCR αβ chain, but not endogenous human TCRs, after transfer to human T cells [42]. However, xenogeneic proteins expressed in human T cells for infusion also have the potential to induce transgene immunogenicity and result in rejection of the exogenous protein expressing T cells [44–47].

Transfer of both the α and β chains of the human TCR directs expression of the intact TCR, however, mis-pairing with endogenous TCR a and b chains can occur, thereby reducing the surface density of antigen-specific TCR and increasing the potential for inducing unpredictable, novel, self-autoreactive TCR α and β pairs. One strategy to overcome this barrier is to allow for the natural development of human T cells *in vivo* in immunodeficient mice where tolerance to human proteins does not occur [48]. Another approach is to perform TCR gene transduction on preselected cellular subsets. TCR transduction of hematopoietic stem cells (HSCs) followed by T-cell-lineage differentiation through *in vitro* Notch signaling [49, 50] can provide forced expression of transduced TCRs by repression of expression of the Rag genes, such that endogenous TCRβ are not expressed [51]. Alternatively, transduction TCR α and β genes into gdT cells, which lack natural expression of TCRαβ chains, can circumvent TCR mis-pairing [52]. A second strategy to overcome the problem of mispairing with endogenous TCR α and β chains is to preselect TCR αβ pairs that naturally pair preferentially or to redesign existent TCRs for preferred pairing. Recently, Heemskerk showed that select antigen-specific TCRs naturally possess a high TCR-CD3 intrinsic affinity that allows for competitive, preferential expression of the paired TCR on the cell surface, suggesting that candidate TCRs can be prescreened for their pairing properties [53]. To better enforce preferred pairing, others have manipulated the transmembrane association domains of TCR α and TCR β by chimeric single chain construction [54]. More recently, specific amino acid alterations in the TCRs themselves have been used to restrict TCR αβ pairing. Replacement of human TCR constant regions with mouse TCR constant regions improves antigen-specific TCR pairing and overall antigen-specific function of the transduced T cell [42, 55, 56], however, the use of mouse TCR regions provides a risk of inducing transgene immunogenicity [57]. Minimal manipulation of human TCR genes to include a novel secondary disulfide bond reinforces selective exogenous TCR pairing with reduced likelihood for antitransgene immune induction [55, 58]. Sebestyen et al. have reported on construction of a two-chain chimeric CD3zeta-modified TCR α and β chain resulted in preferred pairing between the exogenous CD3zeta-modified chains, exclusion of endogenous TCR chains, and offered antigen-specific signaling that was independent of additional TCR signaling components [59]. Such molecules may be less immunogenic since they do not require novel, nonphysiologic amino acid substitutions within the TCR chains.

Engineered T-Cell Receptor Clinical Trials

Perhaps the strongest rationale to date for using engineered antigen-specific T cells in cancer immunotherapy is the clinical studies performed with MART-1 TILs. MART-1 is a differentiation antigen overexpressed in melanoma and a common target of tumor infiltrating $CD8^+$ T cells derived from these tumors. Autologous transfer of MART-1 reactive HLA-A*0201-restricted TILs has resulted in cancer regression in multiple patients [2]. The MART-1-specific DMF4 TCR was isolated from the TIL of an HLA-A*0201$^+$ patient with melanoma who experienced cancer regression and following infusion of autologous TILs. Two distinct MART-1-specific T-cell populations, a $VB12^+$ subset that exhibited long-term persistence at a high frequency in the blood after infusion (clone DMF4), and a VB12-negative subset that rapidly declined after infusion (clone DMF5) were observed. The DMF4 clone was selected for TCR isolation and use in gene therapy based upon these qualities. Genetic transfer of DMF4 TCR with gamma retroviral delivery conveyed antigen specificity and biologic activity to nonreactive T lymphocytes [60]. The therapeutic potential of DMF4 TCR gene therapy was reported by Morgan, Rosenberg, and colleagues who showed that autologous transfer of DMF4 TCR engineered T cells to 15 patients with metastatic melanoma after lymphodepletion resulted in a 13% objective cancer response rate [61]. Following lymphodepletion, patients received T cells at doses ranging between 1×10^9 to 86×10^9 at a mean TCR transduction efficiency of 42%, with IL-2 coadministered to tolerance. Therapy was well tolerated, with no toxicities attributed to gene marked cells, and resulted in the stable engraftment of DMF4 TCR expressing T cells in the circulation even 1 year after cell transfer. Despite the low rate of objective responses and lack of complete responses in this trial, this work represents the first demonstration of TCR gene therapy in the clinic.

The low response rate of DMF4 TCR gene therapy, relative to TIL-based therapy, suggested that substantial improvement was possible. One element limiting TCR gene therapy was hypothesized to be the affinity of the TCR selected for clinical application. To identify the optimal TCR for therapy, 24 different MART-1 specific TCR clones selected from individual MART-1-reactive TIL clones were screened for their capacity to confer high functional avidity to nonreactive T cells [30]. Among all clones tested, the DMF4 TCR showed low to intermediate avidity to MART-1 peptide or MART-1 expressing, HLA-matched melanoma cells, and CD8 coreceptor dependence. Interestingly, TCR clone DMF5, derived from the same patient as DMF4, showed high affinity binding to MART-1 specific tetramers mutated to eliminate CD8 binding capacity, and exhibited the greatest CD8-independent recognition of MART-1 antigen expressing tumors when transferred to CD4 T cells. Johnson and colleagues recently reported on the treatment of 20 patients with metastatic melanoma with high-avidity DMF5 TCR transduced T cells. T-cell doses between 1.5 and 107×10^9 T cells with a 71% mean transduction efficiency were administered with exogenous IL-2, and yielded a 30% (6/20) objective response rate [34]. Unlike DMF4 TCR therapy, treatment with DMF5 TCR led to a high incidence of melanocyte directed autoimmunity (80%), including induction of vitiligo, ocular uveitis, and hearing loss. DMF5 TCR expressing T cells were detectable in autoimmune tissues and persisted at high levels, comprising up to 80% of T lymphocytes, in the peripheral blood. In the same report, 16 additional patients received autologous T cells engineered to express a highly reactive mouse TCR called gp100(154) against the human melanocyte gp100:154–162 epitope that was isolated from HLA-A*0201 transgenic mice immunized with this peptide, which differs from the mouse sequence by a single amino acid. Similar to DMF5, gp100 TCR therapy induced cancer regression in a subset of patients (19%) and autoimmunity in nearly all (94%). Of note, expression of HLA and target antigens by tumor was confirmed in all patients prior to therapy, but not after.

Elucidation of the mechanisms which account for the disparity between recognition of tumor cells and normal quiescent cells expressing the targeted cancer antigen has clear clinical implications for the delivery of safe adoptive immunotherapy. Disparate factors include differential lymphocyte homing to normal versus tumor tissue, differences in antigen:MHC density, heterogeneity of antigen or HLA expression by tumor after treatment (antigen loss), quantitative and qualitative difference in antigen presenting cells, or elevated immune suppression and evasion in the tumor microenvironment. Each of these may need to be considered to improve tumor rejection and reduce toxicity in patients. Although not randomized, findings from these trials emphasize the importance of TCR avidity in the elimination of tumors *in vivo*. This underscores the need for caution when targeting self-tumor antigens with redirected T-cell-based therapy. Accordingly, the selection of cancer testis antigens, which are abundantly expressed in cancer cells but limited in somatic tissues, as T-cell targets should be considered in future trials. The pace of clinical research testing engineered TCRs has accelerated tremendously in recent years. A compilation of T-cell adoptive transfer clinical trials testing engineered TCR is shown in Table 12.2.

Chimeric Antigen Receptors

An alternative strategy to produce genetically engineered T cells is the chimeric antigen receptor (CAR) approach, in which a chimeric receptor composed of signaling receptor domains is combined with antibody antigen binding domains. In this manner, the effector functions of T lymphocytes are enhanced with the ability of antibodies to recognize predefined surface antigens with high specificity in a non-MHC restricted manner [62, 63]. These receptors have the ability to recognize intact membrane proteins independent of antigen processing. In principle, universal targeting vectors can be constructed because the single-chain variable fragment (scFv) of the antibody binds to native cell surface epitopes and bypass the requirement for MHC restriction. The tumor binding function of CAR is usually accomplished by the inclusion of a scFv antibody, containing the V_H and V_L chains joined by a peptide linker of about 15 residues in length [64]. Early versions of CARs pioneered by Eshhar combined the antigen recognition domain of antibody with the intracellular domain of the T-cell receptor-zeta (TCR-zeta) chain or Fc gammaRI protein [65]. There is now a much greater understanding of T-cell signaling pathways and the importance of optimization of chimeric receptors with combinations of signaling domains that may differ in different settings [66, 67]. The addition of costimulatory domains, particularly the intracellular domain of CD28, can significantly augment the ability of these receptors to stimulate cytokine secretion and enhance antitumor efficacy in preclinical animal models using both solid tumors and leukemias that lack the expression of the CD28 receptor ligands CD80 and CD86 [68–70]. Inclusion of domains from receptors such as the tumor necrosis factor receptor family members,

Table 12.2 Clinical trials of T-cell receptor gene modified T cells

Target antigen	TCR type	Indication	Investigators, center	Reference	ClinicalTrials.gov Identifier
CEA	anti-CEA TCR	Metastatic cancer expressing CEA antigen	S.A. Rosenberg, NCI/NIH		NCT00809978
CEA	anti-CEA TCR	Metastatic cancer expressing CEA antigen	S.A. Rosenberg, NCI/NIH		NCT00923806
CEA	anti-CEA TCR	Adenocarcinoma	R.P. Junghans, Beth Israel Deaconess Medical Center		NCT00004178
gp100	TCR gene-engineered TILs	Melanoma	S.A. Rosenberg, NCI/NIH		NCT00085462
gp100	anti-gp100:154-162 TCR	Melanoma	S.A. Rosenberg, NCI/NIH	4452 (Johnson)	NCT00509496
gp100	anti-gp100:154-162 TCR, ALVAC virus	Melanoma	S.A. Rosenberg, NCI/NIH		NCT00610311
MART-1	anti-MART-1 F5 TCR, ALVAC virus	Melanoma	S.A. Rosenberg, NCI/NIH		NCT00612222
MART-1	TCR anti-MART-1 F5	High-risk melanoma	M. C. Jensen, Beckman Research Institute	4097 (Morgan)	NCT00706992
MART-1	CTL line expressing TCR ant-MART-1	Melanoma	H. von der Maase, Aarhus University Hospital	4491 (Duval)	

(continued)

Table 12.2 (continued)

Target antigen	TCR type	Indication	Investigators, center	Reference	ClinicalTrials.gov Identifier
MART-1 (DMF4)	anti- MART-1 TCR-engineered tumor-infiltrating lymphocytes or PBLs	Melanoma	S.A. Rosenberg, NCI/NIH	4097 (Morgan)	NCT00091104
MART-1 (DMF5)	anti-MART-1 F5 TCR	Melanoma	S.A. Rosenberg, NCI/NIH	4452 (Johnson)	NCT00509288
MART-1 and gp100	anti-MART-1 and anti-gp100 TCR	Melanoma	S.A. Rosenberg, NCI/NIH		NCT00814684
NY ESO-1	anti-NY ESO-1 TCR	Kidney cancer, melanoma, metastatic cancer expressing NY ESO-1	S.A. Rosenberg, NCI/NIH		NCT00670748
p53	anti-p53 TCR	Metastatic cancer with p53 overexpression	S.A. Rosenberg, NCI/NIH		NCT00393029
p53	anti-p53 TCR	Kidney cancer, melanoma, metastatic cancer expressing p53	S.A. Rosenberg, NCI/NIH		NCT00704938

CD134 (OX-40) and CD137 (4-1BB) into CARs has also been shown to augment CAR-mediated T-cell responses [71, 72]. Likewise, the efficiency of *ex vivo* T-cell stimulation and gene transfer can affect the engraftment and persistence of adoptively transferred engineered T cells [22, 73–76].

There are several potential limitations to the CAR T cells: (1) the tumor must express the target antigen on the cell surface; (2) large amounts of shed or soluble antigen can inhibit the CAR T cells; and (3) the chimeric receptor may be immunogenic, resulting in the elimination of the redirected T cells by the host immune system. While the cell surface expression of an antigen can be examined in tissue specimens and in primary cell lines, the degree of shedding *in vivo* and in the tumor mileu is much more difficult to determine. Immunogenicity may arise if foreign sequences such as antibiotic selection genes or mouse antibody sequences are expressed, or because of novel epitopes that are created at the fusion joint of human signaling domains that are not normally juxtaposed. The basis for this supposition is that human retrovirally modified CTLs expressing a fusion protein consisting of hygromycin:HSV thymidine kinase were eliminated by host CTLs in patients with advanced HIV infection [77]. Importantly, this immune mediated elimination was not accompanied by adverse effects and required 6–8 weeks to occur. As of this writing, there is one clinical trial reported where CAR containing a scFv with mouse sequences have been given to cancer patients. Following a single dose (0.6 to 4×10^9 T cells), the CAR T cells were detected in circulation from 23, 32, and 53 days after infusion in three patients with renal cell carcinoma [46, 78]. All three patients developed low levels of anti-scFv antibodies between 37 and 100 days after the CAR T-cell infusion.

CAR Clinical Trials

Over the past decade, CARs directed against a wide variety of tumor antigens have been developed [reviewed in 79]. One of these, CD19, is a 95 kDa glycoprotein present on B cells from early development until differentiation into plasma cells [80–82]. CD19 is also expressed by most B cell lymphomas, mantle cell lymphoma, ALLs, CLLs, hairy cell leukemias, and a subset of acute myelogenous leukemias [80, 83, 84]. CD19 thus represents a highly attractive target for immunotherapy. Furthermore, CD19 is not present on most normal tissues, other than normal B cells, including pluripotent blood stem cells [81], which makes CD19 a relatively safe target presenting a minimal risk of autoimmune disease or irreversible myelotoxicity. Anti-CD19 antibodies and scFvs either native or conjugated to radioisotopes or toxins are currently being developed and have demonstrated promise in both mouse models [85, 86] and human and nonhuman primates [87–90]. Gamma retrovirally transduced primary human T cells containing a TCR-zeta cytosolic signaling domain were shown to engraft and to eradicate Raji Burkitt lymphoma tumor cells in immunodeficient mice [89]. Human T cells that express the CD19 scFv:TCRz or CD19 scFv:TCRz:4-1BB can kill B-ALL and lymphoma cells *in vitro* [67, 72, 90–92].

Recently Brentjens, Sadelain, Riviere and colleagues initiated a clinical trial evaluating a chimeric immune receptor targeting CD19 and linked to the CD28 and CD3ζ signaling chains, for treatment of CLL. This is a two cohort study, with the first cohort receiving a low dose of cells to evaluate safety. The second cohort is designed to better evaluate potential efficacy, and the patients receive cytoreductive chemotherapy and a higher dose of cells. Cells are modified using a gamma retroviral vector, and undergo rapid expansion using anti-CD28 mAb and anti-CD3 mAb coated beads [93]. We recently opened a clinical trial at the University of Pennsylvania of adoptive immunotherapy using CART-19 expressing T cells with a 4-1BB:TCRζ, which have not been previously tested in humans. A recent report testing CART-20 cells with a TCRζ reported safety in patients with mantle cell lymphoma [94].

Lamers and colleagues recently reported the interim results of a trial testing CAR in three patients with metastatic renal cell carcinoma [46, 78]. The CAR were engineered with an scFv specific for carboxy-anhydrase-IX (the G250 antigen) that is overexpressed on clear cell RCC and in the biliary tract. The CAR was expressed in autologous T cells using gamma retroviral transduction. The subjects were treated with a dose-escalation scheme of intravenous doses of 2×10^7 T cells at day 1; 2×10^8 T cells at day 2; 2×10^9 T cells at days 3–5 (treatment cycle 1); and 2×10^9 cells at days 17–19, in combination with human recombinant IL-2, given subcutaneously at 5×10^5 U/m^2 twice daily administered at days 1–10 and days 17 to 26. Infusions of these G250 CAR T cells were initially well-tolerated. However, after four to five infusions, liver enzyme disturbances reaching NCI CTC grades 2–4 developed. These toxicities necessitated the cessation of treatment in subjects 1 and 3, corticosteroid treatment in patient 1, and reduction of the maximal T-cell dose in subjects 2 and 3. All three patients developed low levels of anti-scFv (G250) antibodies between 37 and 100 days after the start of T-cell therapy. The liver enzyme elevations resolved. The engineered cells circulated from 32 to 53 days after infusion, and the authors could demonstrate antitumor activity of the G250 CAR T cells in the peripheral blood of the patients after infusion. The authors concluded that the liver toxicity was most likely due to the reactivity of G250 T bodies against the target antigen expressed on normal tissue, that is, the epithelial cells lining the bile ducts. Thus, this is interpreted as a form of "on-target, off-organ" toxicity and argues for the need to identify and test antigens with even more highly restricted tissue distribution. It may also be advantageous in certain settings to consider the use of transiently expressed sequences or the incorporation of suicide switches to dampen an undesired autoimmune response.

In the case of ovarian cancer, T cells transduced with CARs that bind to antigens on ovarian tumors have been shown to have potent therapeutic activity in preclinical models [95–98]. Kershaw, Hwu and colleagues carried out a

14-patient trial for ovarian cancer with two cohorts evaluating a CAR targeting the folate receptor and linked to the Fcγ signaling chain [45]. This trial consisted of two cohorts, the first of which received three cycles of escalating doses of T bodies expanded with αCD3 Ab and IL-2 for doses of 3×10^9, 1×10^{10} and $3–5 \times 10^{10}$ respectively, combined with systemic IL-2. The second cohort received one–two cycles (depending upon the number of cells available) of cells expanded with allogeneic PBMCs and IL-2, followed by allogeneic stimulation *in vivo* with live unirradiated PBMCs. Doses ranged from 4 to 169×10^9 cells, and two of the six patients received two cycles. Cells were modified using a gamma retroviral vector containing the neomycin resistance gene, and were selected using G418 after transduction. Grade 3 and 4 adverse events were observed in cohort 1, but were attributed to IL-2 administration. No serious adverse events were attributed to CAR administration. Persistence of modified cells was poor and did not exceed 3 weeks. This is likely a result of generation of antibodies to the infused T cells, as serum inhibitory factor that could be removed by incubation with protein G was found in patient serum post but not preinfusion. No effect on tumor, as measured by imaging and CA-125 antigen was observed in patients.

Mesothelin is a tumor associated antigen that was originally identified by Pastan and colleagues using the antibody CAK-1 to stain mesothelial cells, mesotheliomas, and ovarian cancers [99, 100]. It is a secreted protein anchored at the cell membrane by glycosylphosphatidylinositol (GPI) linkage. Scholler and Bast have found that secreted, soluble or shed forms of mesothelin may provide useful biomarkers for diagnosis of ovarian carcinoma [101, 102]. Mesothelin has relatively limited expression in normal tissues, as no reactivity was detectable with numerous normal tissues such as liver, heart, brain, kidney, or bone marrow, including many other epithelial tissues such as cervix, prostate, stomach or esophagus, or skin [100]. In contrast, normal peritoneal, pleural and pericardial mesothelia do express mesothelin, albeit at lower levels than generally found on malignant tissue. The biologic functions of mesothelin are not known, although it appears to be involved in cell adhesion via its interaction with CA125/MUC16, and this has been proposed to play a role in cancer progression [103, 104]. Mesothelin is a target of a natural immune response in ovarian [105] and pancreatic [106] cancer, and has been proposed to be a target for cancer immunotherapy [105, 107–111]. The adoptive transfer of natural $CD8^+$ CTLs specific for MHC class I restricted mesothelin epitopes has been shown to have therapeutic activity for preclinical models of ovarian cancer as well [112]. CAR targeted to mesothelin have recently been shown in a xenograft model to significantly reduce tumor burden when mice were engrafted with large pre-established tumors. Incorporation of the CD137 signaling domain enhanced persistence of the CAR transduced T cells *in vivo* [66]. On the basis of this preclinical data, we are planning a clinical trial with CAR directed against mesothelin in ovarian cancer. A summary of T-cell adoptive transfer clinical trials testing CAR is shown in Table 12.3.

Table 12.3 Clinical trials of chimeric antigen receptor gene modified T cells

Target antigen	CAR type	Indication	Investigators, center	Reference	ClinicalTrials.gov Identifier
Carboxy-anhydrase IX	CAR IL-13 zetakine (with the Hy/TK selection/suicide fusion protein)	Renal cell carcinoma	J. Gratama, Erasmus University Medical Center,	(4087, 4088)	
CD19	CAR (with the Hy/TK selection/suicide fusion protein)	Relapsed or refractory follicular non-Hodgkin's lymphoma	M.C. Jensen, City of Hope		NCT00182650
CD19	CAR:CD28-CD3ζ	Chronic lymphocytic leukemia (CLL)	R. Brentjens, I. Riviere, Memorial Sloan-Kettering		NCT00466531
CD19	CAR: scFv-CD3ζ or scFv-CD28-CD3ζ	B-cell non-Hodgkin's lymphoma and chronic lymphocytic leukemia	M.K. Brenner, R. Kamble, Baylor		NCT00586391
CD19	CAR: scFv-CD3ζ or scFv-CD28-CD3ζ	Low-grade B-cell non-Hodgkin's lymphoma and chronic lymphocytic leukemia	M.K. Brenner, H.E. Heslop, Baylor		NCT00608270
CD19	CAR: scFv-CD3ζ or scFv-CD28-CD3ζ in ATC* or EBV-specific T-cell	B-cell non-Hodgkin's lymphoma and chronic lymphocytic leukemia	C. Ramos, Baylor		NCT00709033
CD20	CAR: scFvFc-CD3ζ	Mantle cell lymphoma or indolent B-cell non-Hodgkin's lymphoma	O.W. Press, Fred Hutchinson Cancer Research Center	4318 (Till)	NCT00012207
CD20	CAR: scFvFc-CD28-CD137-CD3ζ	Mantle cell lymphoma or indolent B-cell non-Hodgkin's lymphoma	B. Till, Fred Hutchinson Cancer Research Center		NCT00621452

CEA	CAR scFv-CD28-CD3ζ	Gastric cancer	R.P. Junghans, Roger Williams Hospital		NCT00429078
CEA	CAR scFv-CD28-CD3ζ	Colorectal cancer	R.P. Junghans, Roger Williams Hospital		NCT00673322
CEA	CAR scFv-CD28-CD3ζ	Metastatic breast cancer	R.P. Junghans, Roger Williams Hospital		NCT00673829
GD2	CAR scFv-CD3ζ in ATC* or EBV-specific T-cell + CD45 ab	Advanced neuroblastoma	C. Louis, Baylor	4400 (Pule)	NCT00609206
α-Folate Receptor	CAR Mov-scFv-CD3ζ	Ovarian epithelial cancer	S.A. Rosenberg, NCI/NIH	4082 (Kershaw)	NCT00019136
Her-2	CAR ErbB2 scFv-CD3ζ/CD28	Metastatic cancer that expresses Her-2	S.A. Rosenberg, NCI/NIH		NCT00924287
IL13R α-2	CAR IL-13 zetakine (with the Hy/TK selection/suicide fusion protein)	High-grade malignant glioma	M.C. Jensen, Beckman Research Institute		NCT00730613
L1-cells adhesion molecule (CD171)	CAR CE7R sc-Fv-CD3ζ (with the Hy/TK selection/suicide fusion protein)	Recurrent or refractory disseminated neuroblastoma	J.R. Park, Fred Hutchinson Cancer Research Center	4103 (Park)	NCT00006480
PSMA	CAR scFv-CD3ζ	Hormone refractory prostate cancer	R.P. Junghans, Roger Williams Hospital		NCT00664196

Considerations in T Cell *Ex vivo* Engineering: Gene Delivery, Choice of Antigens, and Balancing Efficacy and Safety

Recent scientific and technologic advances have enabled the creation of genetically engineered T-cell therapies with specificity for tumor antigens or pathogens in addition to enhanced or novel functions. Considerations of the best method to produce a large population of functional antigen-specific T cells *ex vivo* have centered on the source of T cells followed by repetitive stimulations in extended *ex vivo* culture that generates effector T cells, but at the risk of long term engraftment. This requirement for extended culture time may now be surmounted by redirection of T cell specificity through gene transfer of chimeric receptor or engineered TCR constructs. Central to the success of future clinical trials of engineered T lymphocytes is the determination of whether one approach or a combination of approaches should be employed and how to engineer and manufacture the respective T lymphocytes for human testing in a variety of disease settings and preparative regimens. In settings of immune deficiency or tumor tolerance, vaccination with peptides, proteins, DNA or peptide pulsed dendritic cells, prior to the collection, modification, expansion and reinfusion of the primed T cells may be the best method to achieve clinically significant immune responses [113, 114].

Gene transfer to T cells via currently available gamma retroviral vectors provides high-level expression of transgenes in T cells *in vitro*. A potential safety concern when infusing individuals with engineered T cells is viral insertional mutagenesis potentially leading to cellular transformation. In four patients in one study of genetically engineered hematopoietic stem cells (HSCs) for severe combined immunodeficiency [115], and recently one patient in a second similar study [116–118], this serious adverse event has been observed. The introduction of engineered lentiviral vectors has greatly increased the efficiency of gene transfer to human hematopietic cells, and a recent pilot study with lentivirally transduced T cells that expressed an antisense HIV vector showed promise in patients infected with HIV [22]. As mentioned above, insertional mutagenesis is a safety concern with any integrating viral vector. It is reassuring that the natural history of HIV does not include an increased incidence of T-cell leukemia; this provides empirical data that lentiviral vector transduction in a differentiated cell such as a T cell rather than HSC's might be a safer path towards retargeting immune cells compared to gamma retroviral (also known as oncoretroviral) vectors. Furthermore, side-by-side tests in preclinical models [12] and initial clinical evidence indicate that lentiviral vectors are less prone to insertional mutagenesis [23]. Electroporation-mediated DNA or mRNA transfection is an alternative and represents a promising approach without many of the limitations associated with vectors that insert by and large randomly in the genome [28]. Importantly, mRNA expression is transient so that the transfection process does not result in permanent genetic modification of cells.

Unlike most infectious pathogen antigens, tumor antigens as self antigens pose the potential risk for off-target effects for long term persisting cells. For this reason, knowledge of off-tumor expression of candidate target antigens is particularly important.

With redirected TCR and CAR targeted T cells, off-target autoimmunity may be observed in pre-clinical animal models. In some cases, due to the fidelity or unavailability of an appropriate animal model, definitive knowledge of off target effects may not be possible. Clinical trial design therefore must balance this risk of autoimmunity with long-term persistence of the genetically retargeted T cells that can proliferate following enagagement with target antigen. In addition to transient expression, engineering a "self-destruct" or suicide gene into constructs targeting self antigens would provide an additional level of protection. At a minimum, early trials of novel retargeted T-cell therapies should proceed cautiously in determining the dose of engineered T cells to be administered. The implications of these findings need to be incorporated into the translation of therapeutic approaches from animal models to the clinic. In addition, the ideal target antigen, *ex vivo* gene transfer and culture process in a research setting must be adapted to clinical scale and clinically compatible reagents to achieve regulatory compliance, meet requirements for safety and feasibility, and ultimately clinical benefit.

Acknowledgements The authors would like to acknowledge helpful discussions and assistance from Dr. Carl June, Dr. James Riley, Dr. Richard Carroll, Dr. Anne Chew, Dr. Gwendolyn Binder, and Dr. Yangbing Zhao.

References

1. Billingham, R. E., L. Brent, and P. B. Medawar. 1954. Quantitative studies on tissue transplantation immunity. II. The origin, strength and duration of actively and adoptively acquired immunity. Proc R Soc Lond B Biol Sci 143:58.
2. Dudley, M. E., J. R. Wunderlich, P. F. Robbins, J. C. Yang, P. Hwu, D. J. Schwartzentruber, S. L. Topalian, R. Sherry, N. P. Restifo, A. M. Hubicki, M. R. Robinson, M. Raffeld, P. Duray, C. A. Seipp, L. Rogers-Freezer, K. E. Morton, S. A. Mavroukakis, D. E. White, and S. A. Rosenberg. 2002. Cancer regression and autoimmunity in patients after clonal repopulation with antitumor lymphocytes. Science 298:850.
3. Dudley, M. E., J. R. Wunderlich, J. C. Yang, R. M. Sherry, S. L. Topalian, N. P. Restifo, R. E. Royal, U. Kammula, D. E. White, S. A. Mavroukakis, L. J. Rogers, G. J. Gracia, S. A. Jones, D. P. Mangiameli, M. M. Pelletier, J. Gea-Banacloche, M. R. Robinson, D. M. Berman, A. C. Filie, A. Abati, and S. A. Rosenberg. 2005. Adoptive cell transfer therapy following non-myeloablative but lymphodepleting chemotherapy for the treatment of patients with refractory metastatic melanoma. J. Clin. Oncol. 23:2346.
4. Dembic, Z., W. Haas, S. Weiss, J. McCubrey, H. Kiefer, H. von Boehmer, and M. Steinmetz. 1986. Transfer of specificity by murine alpha and beta T-cell receptor genes. Nature 320:232.
5. Gabert, J., C. Langlet, R. Zamoyska, J. R. Parnes, A. M. Schmitt-Verhulst, and B. Malissen. 1987. Reconstitution of MHC class I specificity by transfer of the T cell receptor and Lyt-2 genes. Cell 50:545.
6. Saito, T., A. Weiss, J. Miller, M. A. Norcross, and R. N. Germain. 1987. Specific antigen-Ia activation of transfected human T cells expressing murine Ti alpha beta-human T3 receptor complexes. Nature 325:125.
7. Rosenberg, S. A., P. Aebersold, K. Cornetta, A. Kasid, R. A. Morgan, R. Moen, E. M. Karson, M. T. Lotze, J. C. Yang, S. L. Topalian, and. 1990. Gene transfer into humans–immunotherapy of patients with advanced melanoma, using tumor-infiltrating lymphocytes modified by retroviral gene transduction. N. Engl. J. Med. 323:570.

8. Cooper, L. J., M. S. Topp, C. Pinzon, I. Plavec, M. C. Jensen, S. R. Riddell, and P. D. Greenberg. 2004. Enhanced transgene expression in quiescent and activated human CD8+ T cells. Hum. Gene Ther. 15:648.
9. Sauce, D., M. Bodinier, M. Garin, B. Petracca, N. Tonnelier, A. Duperrier, J. V. Melo, J. F. Apperley, C. Ferrand, P. Herve, F. Lang, P. Tiberghien, and E. Robinet. 2002. Retrovirus-mediated gene transfer in primary T lymphocytes impairs their anti-Epstein-Barr virus potential through both culture-dependent and selection process-dependent mechanisms. Blood 99:1165.
10. Ellis, J. 2005. Silencing and variegation of gammaretrovirus and lentivirus vectors. Hum. Gene Ther. 16:1241.
11. Pfeifer, A., M. Ikawa, Y. Dayn, and I. M. Verma. 2002. Transgenesis by lentiviral vectors: lack of gene silencing in mammalian embryonic stem cells and preimplantation embryos. Proc. Natl. Acad. Sci. USA 99:2140.
12. Montini, E., D. Cesana, M. Schmidt, F. Sanvito, M. Ponzoni, C. Bartholomae, S. L. Sergi, F. Benedicenti, A. Ambrosi, S. C. Di, C. Doglioni, K. C. von, and L. Naldini. 2006. Hematopoietic stem cell gene transfer in a tumor-prone mouse model uncovers low genotoxicity of lentiviral vector integration. Nat. Biotechnol. 24:687.
13. Mitchell, R. S., B. F. Beitzel, A. R. Schroder, P. Shinn, H. Chen, C. C. Berry, J. R. Ecker, and F. D. Bushman. 2004. Retroviral DNA integration: ASLV, HIV, and MLV show distinct target site preferences. PLoS. Biol. 2:E234.
14. Hacein-Bey-Abina, S., C. von Kalle, M. Schmidt, F. Le Deist, N. Wulffraat, E. McIntyre, I. Radford, J. L. Villeval, C. C. Fraser, M. Cavazzana-Calvo, and A. Fischer. 2003. A serious adverse event after successful gene therapy for X-linked severe combined immunodeficiency. N. Engl. J. Med. 348:255.
15. Hacein-Bey-Abina, S., C. von Kalle, M. Schmidt, M. P. McCcormack, N. Wulffraat, P. Leboulch, A. Lim, C. S. Osborne, R. Pawliuk, E. Morillon, R. Sorensen, A. Forster, P. Fraser, J. I. Cohen, G. de Saint Basile, I. Alexander, U. Wintergerst, T. Frebourg, A. Aurias, D. Stoppa-Lyonnet, S. Romana, I. Radford-Weiss, F. Gross, F. Valensi, E. Delabesse, E. Macintyre, F. Sigaux, J. Soulier, L. E. Leiva, M. Wissler, C. Prinz, T. H. Rabbitts, F. Le Deist, A. Fischer, and M. Cavazzana-Calvo. 2003. LMO2-associated clonal T cell proliferation in two patients after gene therapy for SCID-X1. Science 302:415.
16. Gaspar, H. B., K. L. Parsley, S. Howe, D. King, K. C. Gilmour, J. Sinclair, G. Brouns, M. Schmidt, K. C. von, T. Barington, M. A. Jakobsen, H. O. Christensen, G. A. Al, H. N. White, J. L. Smith, R. J. Levinsky, R. R. Ali, C. Kinnon, and A. J. Thrasher. 2004. Gene therapy of X-linked severe combined immunodeficiency by use of a pseudotyped gammaretroviral vector. Lancet. 364:2181.
17. Howe, S. J., M. R. Mansour, K. Schwarzwaelder, C. Bartholomae, M. Hubank, H. Kempski, M. H. Brugman, K. Pike-Overzet, S. J. Chatters, R. D. de, K. C. Gilmour, S. Adams, S. I. Thornhill, K. L. Parsley, F. J. Staal, R. E. Gale, D. C. Linch, J. Bayford, L. Brown, M. Quaye, C. Kinnon, P. Ancliff, D. K. Webb, M. Schmidt, K. C. von, H. B. Gaspar, and A. J. Thrasher. 2008. Insertional mutagenesis combined with acquired somatic mutations causes leukemogenesis following gene therapy of SCID-X1 patients. J. Clin. Invest. 118:3143.
18. Naldini, L., U. Blomer, P. Gallay, D. Ory, R. Mulligan, F. H. Gage, I. M. Verma, and D. Trono. 1996. *In vivo* gene delivery and stable transduction of nondividing cells by a lentiviral vector. Science. 272:263.
19. Sinn, P. L., S. L. Sauter, and P. B. McCray, Jr. 2005. Gene therapy progress and prospects: development of improved lentiviral and retroviral vectors – design, biosafety, and production. Gene Ther. 12:1089.
20. Verhoeyen, E., V. Dardalhon, O. Ducrey-Rundquist, D. Trono, N. Taylor, and F. L. Cosset. 2003. IL-7 surface-engineered lentiviral vectors promote survival and efficient gene transfer in resting primary T lymphocytes. Blood. 101:2167.
21. Humeau, L. M., G. K. Binder, X. Lu, V. Slepushkin, R. Merling, P. Echeagaray, M. Pereira, T. Slepushkina, S. Barnett, L. K. Dropulic, R. Carroll, B. L. Levine, C. H. June, and B. Dropulic. 2004. Efficient lentiviral vector-mediated control of HIV-1 replication in CD4 lymphocytes from diverse HIV+infected patients grouped according to CD4 count and viral load. Mol. Ther. 9:902.

22. Levine, B. L., L. M. Humeau, J. Boyer, R. R. MacGregor, T. Rebello, X. Lu, G. K. Binder, V. Slepushkin, F. Lemiale, J. R. Mascola, F. D. Bushman, B. Dropulic, and C. H. June. 2006. Gene transfer in humans using a conditionally replicating lentiviral vector. Proc. Natl. Acad. Sci. USA. 103:17372.
23. Wang, G. P., B. L. Levine, G. K. Binder, C. C. Berry, N. Malani, G. McGarrity, P. Tebas, C. H. June, and F. D. Bushman. 2009. Analysis of lentiviral vector integration in HIV+study subjects receiving autologous infusions of gene modified CD4+ T cells. Mol. Ther. 17:844.
24. Varela-Rohena, A., P. E. Molloy, S. M. Dunn, Y. Li, M. M. Suhoski, R. G. Carroll, A. Milicic, T. Mahon, D. H. Sutton, B. Laugel, R. Moysey, B. J. Cameron, A. Vuidepot, M. A. Purbhoo, D. K. Cole, R. E. Phillips, C. H. June, B. K. Jakobsen, A. K. Sewell, and J. L. Riley. 2008. Control of HIV-1 immune escape by CD8 T cells expressing enhanced T-cell receptor. Nat. Med. 14:1390.
25. Yang, S., C. J. Cohen, P. D. Peng, Y. Zhao, L. Cassard, Z. Yu, Z. Zheng, S. Jones, N. P. Restifo, S. A. Rosenberg, and R. A. Morgan. 2008. Development of optimal bicistronic lentiviral vectors facilitates high-level TCR gene expression and robust tumor cell recognition. Gene Ther. 15:1411.
26. Ferber, D. 2001. Gene therapy. Safer and virus-free? Science 294:1638.
27. Zhao, Y., Z. Zheng, C. J. Cohen, L. Gattinoni, D. C. Palmer, N. P. Restifo, S. A. Rosenberg, and R. A. Morgan. 2006. High-efficiency transfection of primary human and mouse T lymphocytes using RNA electroporation. Mol. Ther. 13:151.
28. Yoon, S. H., J. M. Lee, H. I. Cho, E. K. Kim, H. S. Kim, M. Y. Park, and T. G. Kim. 2009. Adoptive immunotherapy using human peripheral blood lymphocytes transferred with RNA encoding Her-2/neu-specific chimeric immune receptor in ovarian cancer xenograft model. Cancer Gene Ther. 16:489.
29. Robbins, P. F., M. E. Dudley, J. Wunderlich, M. El-Gamil, Y. F. Li, J. Zhou, J. Huang, D. J. Powell Jr., and S. A. Rosenberg. 2004. Cutting edge: persistence of transferred lymphocyte clonotypes correlates with cancer regression in patients receiving cell transfer therapy. J. Immunol. 173:7125.
30. Johnson, L. A., B. Heemskerk, D. J. Powell, Jr., C. J. Cohen, R. A. Morgan, M. E. Dudley, P. F. Robbins, and S. A. Rosenberg. 2006. Gene transfer of tumor-reactive TCR confers both high avidity and tumor reactivity to nonreactive peripheral blood mononuclear cells and tumor-infiltrating lymphocytes. J. Immunol. 177:6548.
31. Li, Y., R. Moysey, P. E. Molloy, A. L. Vuidepot, T. Mahon, E. Baston, S. Dunn, N. Liddy, J. Jacob, B. K. Jakobsen, and J. M. Boulter. 2005. Directed evolution of human T-cell receptors with picomolar affinities by phage display. Nat. Biotechnol. 23:349.
32. Dunn, S. M., P. J. Rizkallah, E. Baston, T. Mahon, B. Cameron, R. Moysey, F. Gao, M. Sami, J. Boulter, Y. Li, and B. K. Jakobsen. 2006. Directed evolution of human T cell receptor CDR2 residues by phage display dramatically enhances affinity for cognate peptide-MHC without increasing apparent cross-reactivity. Protein Sci. 15:710.
33. Zhao, Y., Z. Zheng, P. F. Robbins, H. T. Khong, S. A. Rosenberg, and R. A. Morgan. 2005. Primary human lymphocytes transduced with NY-ESO-1 antigen-specific TCR genes recognize and kill diverse human tumor cell lines. J. Immunol. 174:4415.
34. Johnson, L. A., R. A. Morgan, M. E. Dudley, L. Cassard, J. C. Yang, M. S. Hughes, U. S. Kammula, R. E. Royal, R. M. Sherry, J. R. Wunderlich, C. C. Lee, N. P. Restifo, S. L. Schwarz, A. P. Cogdill, R. J. Bishop, H. Kim, C. C. Brewer, S. F. Rudy, C. Vanwaes, J. L. Davis, A. Mathur, R. T. Ripley, D. A. Nathan, C. M. Laurencot, and S. A. Rosenberg. 2009. Gene therapy with human and mouse T cell receptors mediates cancer regression and targets normal tissues expressing cognate antigen. Blood. 114:535.
35. Pogulis, R. J., M. J. Hansen, and L. R. Pease. 1998. Retroviral-mediated expression of an MHC class I-restricted T cell receptor in the CD8 T cell compartment of bone marrow-reconstituted mice. Hum. Gene Ther. 9:2285.
36. Pogulis, R. J., and L. R. Pease. 1998. A retroviral vector that directs simultaneous expression of alpha and beta T cell receptor genes. Hum. Gene Ther. 9:2299.
37. Clay, T. M., M. C. Custer, J. Sachs, P. Hwu, S. A. Rosenberg, and M. I. Nishimura. 1999. Efficient transfer of a tumor antigen-reactive TCR to human peripheral blood lymphocytes confers anti-tumor reactivity. J. Immunol. 163:507.

38. Pittet, M. J., V. Rubio-Godoy, G. Bioley, P. Guillaume, P. Batard, D. Speiser, I. Luescher, J. C. Cerottini, P. Romero, and A. Zippelius. 2003. Alpha 3 domain mutants of peptide/MHC class I multimers allow the selective isolation of high avidity tumor-reactive CD8 T cells. J. Immunol. 171:1844.
39. Robbins, P. F., Y. F. Li, M. El-Gamil, Y. Zhao, J. A. Wargo, Z. Zheng, H. Xu, R. A. Morgan, S. A. Feldman, L. A. Johnson, A. D. Bennett, S. M. Dunn, T. M. Mahon, B. K. Jakobsen, and S. A. Rosenberg. 2008. Single and dual amino acid substitutions in TCR CDRs can enhance antigen-specific T cell functions. J. Immunol. 180:6116.
40. Stanislawski, T., R. H. Voss, C. Lotz, E. Sadovnikova, R. A. Willemsen, J. Kuball, T. Ruppert, R. L. Bolhuis, C. J. Melief, C. Huber, H. J. Stauss, and M. Theobald. 2001. Circumventing tolerance to a human MDM2-derived tumor antigen by TCR gene transfer. Nat. Immunol. 2:962.
41. Holler, P. D., P. O. Holman, E. V. Shusta, S. O'Herrin, K. D. Wittrup, and D. M. Kranz. 2000. *In vitro* evolution of a T cell receptor with high affinity for peptide/MHC. Proc. Natl. Acad. Sci. USA. 97:5387.
42. Cohen, C. J., Z. Zheng, R. Bray, Y. Zhao, L. A. Sherman, S. A. Rosenberg, and R. A. Morgan. 2005. Recognition of fresh human tumor by human peripheral blood lymphocytes transduced with a bicistronic retroviral vector encoding a murine anti-p53 TCR. J. Immunol. 175:5799.
43. Kuball, J., F. W. Schmitz, R. H. Voss, E. A. Ferreira, R. Engel, P. Guillaume, S. Strand, P. Romero, C. Huber, L. A. Sherman, and M. Theobald. 2005. Cooperation of human tumor-reactive CD4+ and CD8+ T cells after redirection of their specificity by a high-affinity p53A2.1-specific TCR. Immunity. 22:117.
44. Berger, C., C. A. Blau, M. L. Huang, J. D. Iuliucci, D. C. Dalgarno, J. Gaschet, S. Heimfeld, T. Clackson, and S. R. Riddell. 2004. Pharmacologically regulated Fas-mediated death of adoptively transferred T cells in a nonhuman primate model. Blood. 103:1261.
45. Kershaw, M. H., J. A. Westwood, L. L. Parker, G. Wang, Z. Eshhar, S. A. Mavroukakis, D. E. White, J. R. Wunderlich, S. Canevari, L. Rogers-Freezer, C. C. Chen, J. C. Yang, S. A. Rosenberg, and P. Hwu. 2006. A phase I study on adoptive immunotherapy using gene-modified T cells for ovarian cancer. Clin. Cancer Res. 12:6106.
46. Lamers, C. H., S. C. Langeveld, C. M. Groot-van Ruijven, R. Debets, S. Sleijfer, and J. W. Gratama. 2007. Gene-modified T cells for adoptive immunotherapy of renal cell cancer maintain transgene-specific immune functions *in vivo*. Cancer Immunol. Immunother. 56:1875.
47. Park, J. R., D. L. DiGiusto, M. Slovak, C. Wright, A. Naranjo, J. Wagner, H. B. Meechoovet, C. Bautista, W. C. Chang, J. R. Ostberg, and M. C. Jensen. 2007. Adoptive transfer of chimeric antigen receptor re-directed cytolytic T lymphocyte clones in patients with neuroblastoma. Mol. Ther. 15:825.
48. Shultz, L. D., F. Ishikawa, and D. L. Greiner. 2007. Humanized mice in translational biomedical research. Nat. Rev. Immunol. 7:118.
49. Clark, R. A., K. Yamanaka, M. Bai, R. Dowgiert, and T. S. Kupper. 2005. Human skin cells support thymus-independent T cell development. J. Clin. Invest. 115:3239.
50. Schmitt, T. M., R. F. de Pooter, M. A. Gronski, S. K. Cho, P. S. Ohashi, and J. C. Zuniga-Pflucker. 2004. Induction of T cell development and establishment of T cell competence from embryonic stem cells differentiated *in vitro*. Nat. Immunol. 5:410.
51. Schlissel, M. S. 2003. Regulating antigen-receptor gene assembly. Nat. Rev. Immunol. 3:890.
52. van der Veken, V., R. S. Hagedoorn, M. M. van Loenen, R. Willemze, J. H. Falkenburg, and M. H. Heemskerk. 2006. Alphabeta T-cell receptor engineered gammadelta T cells mediate effective antileukemic reactivity. Cancer Res. 66:3331.
53. Heemskerk, M. H., R. S. Hagedoorn, M. A. van der Hoorn, d. van, V, M. Hoogeboom, M. G. Kester, R. Willemze, and J. H. Falkenburg. 2007. Efficiency of T-cell receptor expression in dual-specific T cells is controlled by the intrinsic qualities of the TCR chains within the TCR-CD3 complex. Blood. 109:235.
54. Willemsen, R. A., M. E. Weijtens, C. Ronteltap, Z. Eshhar, J. W. Gratama, P. Chames, and R. L. Bolhuis. 2000. Grafting primary human T lymphocytes with cancer-specific chimeric single chain and two chain TCR. Gene Ther. 7:1369.

55. Kuball, J., M. L. Dossett, M. Wolfl, W. Y. Ho, R. H. Voss, C. Fowler, and P. D. Greenberg. 2007. Facilitating matched pairing and expression of TCR chains introduced into human T cells. Blood. 109:2331.
56. Voss, R. H., J. Kuball, R. Engel, P. Guillaume, P. Romero, C. Huber, and M. Theobald. 2006. Redirection of T cells by delivering a transgenic mouse-derived MDM2 tumor antigen-specific TCR and its humanized derivative is governed by the CD8 coreceptor and affects natural human TCR expression. Immunol. Res. 34:67.
57. Greenberg, P. D. 1991. Adoptive T cell therapy of tumors: mechanisms operative in the recognition and elimination of tumor cells. Adv. Immunol. 49:281–355.
58. Cohen, C. J., Y. F. Li, M. El-Gamil, P. F. Robbins, S. A. Rosenberg, and R. A. Morgan. 2007. Enhanced antitumor activity of T cells engineered to express T-cell receptors with a second disulfide bond. Cancer Res. 67:3898.
59. Sebestyen, Z., E. Schooten, T. Sals, I. Zaldivar, J. E. San, B. Alarcon, S. Bobisse, A. Rosato, J. Szollosi, J. W. Gratama, R. A. Willemsen, and R. Debets. 2008. Human TCR that incorporate CD3zeta induce highly preferred pairing between TCRalpha and beta chains following gene transfer. J. Immunol. 180:7736.
60. Hughes, M. S., Y. Y. Yu, M. E. Dudley, Z. Zheng, P. F. Robbins, Y. Li, J. Wunderlich, R. G. Hawley, M. Moayeri, S. A. Rosenberg, and R. A. Morgan. 2005. Transfer of a TCR gene derived from a patient with a marked antitumor response conveys highly active T-cell effector functions. Hum. Gene Ther. 16:457.
61. Morgan, R. A., M. E. Dudley, J. R. Wunderlich, M. S. Hughes, J. C. Yang, R. M. Sherry, R. E. Royal, S. L. Topalian, U. S. Kammula, N. P. Restifo, Z. Zheng, A. Nahvi, C. R. de Vries, L. J. Rogers-Freezer, S. A. Mavroukakis, and S. A. Rosenberg. 2006. Cancer regression in patients after transfer of genetically engineered lymphocytes. Science. 314:126.
62. Pinthus, J. H., T. Waks, K. Kaufman-Francis, D. G. Schindler, A. Harmelin, H. Kanety, J. Ramon, and Z. Eshhar. 2003. Immuno-gene therapy of established prostate tumors using chimeric receptor-redirected human lymphocytes. Cancer Res. 63:2470.
63. Gross, G., T. Waks, and Z. Eshhar. 1989. Expression of immunoglobulin-T-cell receptor chimeric molecules as functional receptors with antibody-type specificity. Proc. Natl. Acad. Sci. USA. 86:10024.
64. Mullaney, B. P., and M. G. Pallavicini. 2001. Protein-protein interactions in hematology and phage display. Exp. Hematol. 29:1136.
65. Eshhar, Z., T. Waks, G. Gross, and D. G. Schindler. 1993. Specific activation and targeting of cytotoxic lymphocytes through chimeric single chains consisting of antibody-binding domains and the gamma or zeta subunits of the immunoglobulin and T-cell receptors. Proc. Natl. Acad. Sci. USA. 90:720.
66. Carpenito, C., M. C. Milone, R. Hassan, J. C. Simonet, M. Lakhal, M. M. Suhoski, A. Varela-Rohena, K. M. Haines, D. F. Heitjan, S. M. Albelda, R. G. Carroll, J. L. Riley, I. Pastan, and C. H. June. 2009. Control of large, established tumor xenografts with genetically retargeted human T cells containing CD28 and CD137 domains. Proc. Natl. Acad. Sci. USA. 106:3360.
67. Milone, M. C., J. D. Fish, C. Carpenito, R. G. Carroll, G. K. Binder, D. Teachey, M. Samanta, M. Lakhal, B. Gloss, G. net-Desnoyers, D. Campana, J. L. Riley, S. A. Grupp, and C. H. June. 2009. Chimeric receptors containing CD137 signal transduction domains mediate enhanced survival of T Cells and increased antileukemic efficacy *in vivo*. Mol. Ther. 17:1453.
68. Finney, H. M., A. D. Lawson, C. R. Bebbington, and A. N. Weir. 1998. Chimeric receptors providing both primary and costimulatory signaling in T cells from a single gene product. J. Immunol. 161:2791.
69. Maher, J., R. J. Brentjens, G. Gunset, I. Riviere, and M. Sadelain. 2002. Human T-lymphocyte cytotoxicity and proliferation directed by a single chimeric TCRzeta/CD28 receptor. Nat. Biotechnol. 20:70.
70. Friedmann-Morvinski, D., A. Bendavid, T. Waks, D. Schindler, and Z. Eshhar. 2005. Redirected primary T cells harboring a chimeric receptor require costimulation for their antigen-specific activation. Blood. 105:3087.

71. Finney, H. M., A. N. Akbar, and A. D. Lawson. 2004. Activation of resting human primary T cells with chimeric receptors: costimulation from CD28, inducible costimulator, CD134, and CD137 in series with signals from the TCR zeta chain. J. Immunol. 172:104.
72. Imai, C., K. Mihara, M. Andreansky, I. C. Nicholson, C. H. Pui, T. L. Geiger, and D. Campana. 2004. Chimeric receptors with 4-1BB signaling capacity provoke potent cytotoxicity against acute lymphoblastic leukemia. Leukemia. 18:676.
73. Levine, B. L., W. B. Bernstein, M. Connors, N. Craighead, T. Lindsten, C. B. Thompson, and C. H. June. 1997. Effects of CD28 costimulation on long-term proliferation of CD4+ T cells in the absence of exogenous feeder cells. J. Immunol. 159:5921.
74. Paulos, C. M., M. M. Suhoski, G. Plesa, T. Jiang, S. Basu, T. N. Golovina, S. Jiang, N. A. Aqui, D. J. Powell, Jr., B. L. Levine, R. G. Carroll, J. L. Riley, and C. H. June. 2008. Adoptive immunotherapy: good habits instilled at youth have long-term benefits. Immunol. Res. 42:182.
75. Walker, R. E., C. M. Bechtel, V. Natarajan, M. Baseler, K. M. Hege, J. A. Metcalf, R. Stevens, A. Hazen, R. M. Blaese, C. C. Chen, S. F. Leitman, J. Palensky, J. Wittes, R. T. Davey, J. Falloon, M. A. Polis, J. A. Kovacs, D. F. Broad, B. L. Levine, M. R. Roberts, H. Masur, and H. C. Lane. 2000. Long-term *in vivo* survival of receptor-modified syngeneic T cells in patients with human immunodeficiency virus infection. Blood 96:467.
76. Mitsuyasu, R. T., P. A. Anton, S. G. Deeks, D. T. Scadden, E. Connick, M. T. Downs, A. Bakker, M. R. Roberts, C. H. June, S. Jalali, A. A. Lin, R. Pennathur-Das, and K. M. Hege. 2000. Prolonged survival and tissue trafficking following adoptive transfer of CD4zeta gene-modified autologous CD4(+) and CD8(+) T cells in human immunodeficiency virus-infected subjects. Blood 96:785.
77. Riddell, S. R., M. Elliott, D. A. Lewinsohn, M. J. Gilbert, L. Wilson, S. A. Manley, S. D. Lupton, R. W. Overell, T. C. Reynolds, L. Corey, and P. D. Greenberg. 1996. T-cell mediated rejection of gene-modified HIV-specific cytotoxic T lymphocytes in HIV-infected patients. Nat. Med. 2:216.
78. Lamers, C. H., S. Sleijfer, A. G. Vulto, W. H. Kruit, M. Kliffen, R. Debets, J. W. Gratama, G. Stoter, and E. Oosterwijk. 2006. Treatment of metastatic renal cell carcinoma with autologous T-lymphocytes genetically retargeted against carbonic anhydrase IX: first clinical experience. J. Clin. Oncol. 24:e20-e22.
79. Sadelain, M., I. Riviere, and R. Brentjens. 2003. Targeting tumours with genetically enhanced T lymphocytes. Nat. Rev. Cancer. 3:35.
80. Freedman, A. S., A. W. Boyd, F. R. Bieber, J. Daley, K. Rosen, J. C. Horowitz, D. N. Levy, and L. M. Nadler. 1987. Normal cellular counterparts of B cell chronic lymphocytic leukemia. Blood. 70:418.
81. Uckun, F. M., W. Jaszcz, J. L. Ambrus, A. S. Fauci, K. Gajl-Peczalska, C. W. Song, M. R. Wick, D. E. Myers, K. Waddick, and J. A. Ledbetter. 1988. Detailed studies on expression and function of CD19 surface determinant by using B43 monoclonal antibody and the clinical potential of anti-CD19 immunotoxins. Blood. 71:13.
82. Fearon, D. T., and M. C. Carroll. 2000. Regulation of B lymphocyte responses to foreign and self-antigens by the CD19/CD21 complex. Annu. Rev. Immunol. 18:393–422.
83. Scheuermann, R. H., and E. Racila. 1995. CD19 antigen in leukemia and lymphoma diagnosis and immunotherapy. Leuk. Lymphoma. 18:385.
84. Schwonzen, M., C. Pohl, T. Steinmetz, W. Seckler, B. Vetten, J. Thiele, D. Wickramanayake, and V. Diehl. 1993. Immunophenotyping of low-grade B-cell lymphoma in blood and bone marrow: poor correlation between immunophenotype and cytological/histological classification. Br. J. Haematol. 83:232.
85. Li, Q., W. Hudson, D. Wang, E. Berven, F. M. Uckun, and J. H. Kersey. 1998. Pharmacokinetics and biodistribution of radioimmunoconjugates of anti-CD19 antibody and single-chain Fv for treatment of human B-cell malignancy. Cancer Immunol. Immunother. 47:121.
86. Ek, O., P. Gaynon, T. Zeren, L. M. Chelstrom, D. E. Myers, and F. M. Uckun. 1998. Treatment of human B-cell precursor leukemia in SCID mice by using a combination of the anti-CD19 immunotoxin B43-PAP with the standard chemotherapeutic drugs vincristine, methylprednisolone, and L-asparaginase. Leuk. Lymphoma. 31:143.

87. Bejcek, B. E., D. Wang, E. Berven, C. A. Pennell, S. C. Peiper, S. Poppema, F. M. Uckun, and J. H. Kersey. 1995. Development and characterization of three recombinant single chain antibody fragments (scFvs) directed against the CD19 antigen. Cancer Res. 55:2346.
88. Nicholson, I. C., K. A. Lenton, D. J. Little, T. Decorso, F. T. Lee, A. M. Scott, H. Zola, and A. W. Hohmann. 1997. Construction and characterisation of a functional CD19 specific single chain Fv fragment for immunotherapy of B lineage leukaemia and lymphoma. Mol. Immunol. 34:1157.
89. Brentjens, R. J., J. B. Latouche, E. Santos, F. Marti, M. C. Gong, C. Lyddane, P. D. King, S. Larson, M. Weiss, I. Riviere, and M. Sadelain. 2003. Eradication of systemic B-cell tumors by genetically targeted human T lymphocytes co-stimulated by CD80 and interleukin-15. Nat. Med. 9:279.
90. Cooper, L. J., M. S. Topp, L. M. Serrano, S. Gonzalez, W. C. Chang, A. Naranjo, C. Wright, L. Popplewell, A. Raubitschek, S. J. Forman, and M. C. Jensen. 2003. T-cell clones can be rendered specific for CD19: toward the selective augmentation of the graft-versus-B-lineage leukemia effect. Blood. 101:1637.
91. Serrano, L. M., T. Pfeiffer, S. Olivares, T. Numbenjapon, J. Bennitt, D. Kim, D. Smith, G. McNamara, Z. Al-Kadhimi, J. Rosenthal, S. J. Forman, M. C. Jensen, and L. J. Cooper. 2006. Differentiation of naive cord-blood T cells into CD19-specific cytolytic effectors for post-transplantation adoptive immunotherapy. Blood. 107:2643.
92. Cooper, L. J., Z. Al-Kadhimi, L. M. Serrano, T. Pfeiffer, S. Olivares, A. Castro, W. C. Chang, S. Gonzalez, D. Smith, S. J. Forman, and M. C. Jensen. 2005. Enhanced antilymphoma efficacy of CD19-redirected influenza MP1-specific CTLs by cotransfer of T cells modified to present influenza MP1. Blood. 105:1622.
93. Hollyman, D., J. Stefanski, M. Przybylowski, S. Bartido, O. Borquez-Ojeda, C. Taylor, R. Yeh, V. Capacio, M. Olszewska, J. Hosey, M. Sadelain, R. J. Brentjens, and I. Riviere. 2009. Manufacturing validation of biologically functional T cells targeted to CD19 antigen for autologous adoptive cell therapy. J. Immunother. 32:169.
94. Till, B. G., M. C. Jensen, J. Wang, E. Y. Chen, B. L. Wood, H. A. Greisman, X. Qian, S. E. James, A. Raubitschek, S. J. Forman, A. K. Gopal, J. M. Pagel, C. G. Lindgren, P. D. Greenberg, S. R. Riddell, and O. W. Press. 2008. Adoptive immunotherapy for indolent non-Hodgkin lymphoma and mantle cell lymphoma using genetically modified autologous CD20-specific T cells. Blood. 112:2261.
95. Westwood, J. A., M. J. Smyth, M. W. Teng, M. Moeller, J. A. Trapani, A. M. Scott, F. E. Smyth, G. A. Cartwright, B. E. Power, D. Honemann, H. M. Prince, P. K. Darcy, and M. H. Kershaw. 2005. Adoptive transfer of T cells modified with a humanized chimeric receptor gene inhibits growth of Lewis-Y-expressing tumors in mice. Proc. Natl. Acad. Sci. USA. 102:19051.
96. Parker, L. L., M. T. Do, J. A. Westwood, J. R. Wunderlich, M. E. Dudley, S. A. Rosenberg, and P. Hwu. 2000. Expansion and characterization of T cells transduced with a chimeric receptor against ovarian cancer. Hum. Gene Ther. 11:2377.
97. Hwu, P., J. C. Yang, R. Cowherd, J. Treisman, G. E. Shafer, Z. Eshhar, and S. A. Rosenberg. 1995. *In vivo* antitumor activity of T cells redirected with chimeric antibody/T-cell receptor genes. Cancer Res. 55:3369.
98. Hwu, P., G. E. Shafer, J. Treisman, D. G. Schindler, G. Gross, R. Cowherd, S. A. Rosenberg, and Z. Eshhar. 1993. Lysis of ovarian cancer cells by human lymphocytes redirected with a chimeric gene composed of an antibody variable region and the Fc receptor gamma chain. J. Exp. Med. 178:361.
99. Chang, K., and I. Pastan. 1996. Molecular cloning of mesothelin, a differentiation antigen present on mesothelium, mesotheliomas, and ovarian cancers. Proc. Natl. Acad. Sci. USA. 93:136.
100. Chang, K., L. H. Pai, J. K. Batra, I. Pastan, and M. C. Willingham. 1992. Characterization of the antigen (CAK1) recognized by monoclonal antibody K1 present on ovarian cancers and normal mesothelium. Cancer Res. 52:181.
101. Scholler, N., N. Fu, Y. Yang, Z. Ye, G. E. Goodman, K. E. Hellstrom, and I. Hellstrom. 1999. Soluble member(s) of the mesothelin/megakaryocyte potentiating factor family are detectable in sera from patients with ovarian carcinoma. Proc. Natl. Acad. Sci. USA. 96:11531.

102. Badgwell, D., Z. Lu, L. Cole, H. Fritsche, E. N. Atkinson, E. Somers, J. Allard, R. G. Moore, K. H. Lu, and R. C. Bast Jr. 2007. Urinary mesothelin provides greater sensitivity for early stage ovarian cancer than serum mesothelin, urinary hCG free beta subunit and urinary hCG beta core fragment. Gynecol. Oncol. 106:490.
103. Rump, A., Y. Morikawa, M. Tanaka, S. Minami, N. Umesaki, M. Takeuchi, and A. Miyajima. 2004. Binding of ovarian cancer antigen CA125/MUC16 to mesothelin mediates cell adhesion. J. Biol. Chem. 279:9190.
104. Gubbels, J. A., J. Belisle, M. Onda, C. Rancourt, M. Migneault, M. Ho, T. K. Bera, J. Connor, B. K. Sathyanarayana, B. Lee, I. Pastan, and M. S. Patankar. 2006. Mesothelin-MUC16 binding is a high affinity, N-glycan dependent interaction that facilitates peritoneal metastasis of ovarian tumors. Mol. Cancer. 5:50.
105. Ho, M., R. Hassan, J. ZHANG, Q. C. Wang, M. Onda, T. Bera, and I. Pastan. 2005. Humoral immune response to mesothelin in mesothelioma and ovarian cancer patients. Clin. Cancer Res. 11:3814.
106. Thomas, A. M., L. M. Santarsiero, E. R. Lutz, T. D. Armstrong, Y. C. Chen, L. Q. Huang, D. A. Laheru, M. Goggins, R. H. Hruban, and E. M. Jaffee. 2004. Mesothelin-specific CD8(+) T cell responses provide evidence of *in vivo* cross-priming by antigen-presenting cells in vaccinated pancreatic cancer patients. J. Exp. Med. 200:297.
107. Yokokawa, J., C. Palena, P. Arlen, R. Hassan, M. Ho, I. Pastan, J. Schlom, and K. Y. Tsang. 2005. Identification of novel human CTL epitopes and their agonist epitopes of mesothelin. Clin. Cancer Res. 11:6342.
108. Breidenbach, M., D. T. Rein, M. Everts, J. N. Glasgow, M. Wang, M. J. Passineau, R. D. Alvarez, N. Korokhov, and D. T. Curiel. 2005. Mesothelin-mediated targeting of adenoviral vectors for ovarian cancer gene therapy. Gene Ther. 12:187.
109. Hassan, R., and M. Ho. 2008. Mesothelin targeted cancer immunotherapy. Eur. J. Cancer. 44:46.
110. Hassan, R., W. Ebel, E. L. Routhier, R. Patel, J. B. Kline, J. ZHANG, Q. Chao, S. Jacob, H. Turchin, L. Gibbs, M. D. Phillips, S. Mudali, C. Iacobuzio-Donahue, E. M. Jaffee, M. Moreno, I. Pastan, P. M. Sass, N. C. Nicolaides, and L. Grasso. 2007. Preclinical evaluation of MORAb-009, a chimeric antibody targeting tumor-associated mesothelin. Cancer Immun. 7:20.
111. Chang, C. L., T. C. Wu, and C. F. Hung. 2007. Control of human mesothelin-expressing tumors by DNA vaccines. Gene Ther. 14:1189.
112. Hung, C. F., Y. C. Tsai, L. He, and T. C. Wu. 2007. Control of mesothelin-expressing ovarian cancer using adoptive transfer of mesothelin peptide-specific CD8+ T cells. Gene Ther. 14:921.
113. Powell, D. J., Jr., M. E. Dudley, K. A. Hogan, J. R. Wunderlich, and S. A. Rosenberg. 2006. Adoptive transfer of vaccine-induced peripheral blood mononuclear cells to patients with metastatic melanoma following lymphodepletion. J. Immunol. 177:6527.
114. Rapoport, A. P., E. A. Stadtmauer, N. Aqui, A. Badros, J. Cotte, L. Chrisley, E. Veloso, Z. Zheng, S. Westphal, R. Mair, N. Chi, B. Ratterree, M. F. Pochran, S. Natt, J. Hinkle, C. Sickles, A. Sohal, K. Ruehle, C. Lynch, L. Zhang, D. L. Porter, S. Luger, C. Guo, H. B. Fang, W. Blackwelder, K. Hankey, D. Mann, R. Edelman, C. Frasch, B. L. Levine, A. Cross, and C. H. June. 2005. Restoration of immunity in lymphopenic individuals with cancer by vaccination and adoptive T-cell transfer. Nat. Med. 11:1230.
115. Cavazzana-Calvo, M., S. Hacein-Bey, B. de Saint, F. Gross, E. Yvon, P. Nusbaum, F. Selz, C. Hue, S. Certain, J. L. Casanova, P. Bousso, F. L. Deist, and A. Fischer. 2000. Gene therapy of human severe combined immunodeficiency (SCID)-X1 disease. Science 288:669.
116. Thrasher, A. J., H. B. Gaspar, C. Baum, U. Modlich, A. Schambach, F. Candotti, M. Otsu, B. Sorrentino, L. Scobie, E. Cameron, K. Blyth, J. Neil, S. H. Abina, M. Cavazzana-Calvo, and A. Fischer. 2006. Gene therapy: X-SCID transgene leukaemogenicity. Nature. 443:E5-E6.
117. Hacein-Bey-Abina, S., C. von Kalle, M. Schmidt, M. P. McCormack, N. Wulffraat, P. Leboulch, A. Lim, C. S. Osborne, R. Pawliuk, E. Morillon, R. Sorensen, A. Forster,

P. Fraser, J. I. Cohen, B. G. de Saint, I. Alexander, U. Wintergerst, T. Frebourg, A. Aurias, D. Stoppa-Lyonnet, S. Romana, I. Radford-Weiss, F. Gross, F. Valensi, E. Delabesse, E. Macintyre, F. Sigaux, J. Soulier, L. E. Leiva, M. Wissler, C. Prinz, T. H. Rabbitts, F. Le Deist, A. Fischer, and M. Cavazzana-Calvo. 2003. LMO2-associated clonal T cell proliferation in two patients after gene therapy for SCID-X1. Science 302:415.

118. Hacein-Bey-Abina, S., K. C. von, M. Schmidt, D. F. Le, N. Wulffraat, E. McIntyre, I. Radford, J. L. Villeval, C. C. Fraser, M. Cavazzana-Calvo, and A. Fischer. 2003. A serious adverse event after successful gene therapy for X-linked severe combined immunodeficiency. N. Engl. J. Med. 348:255.

Part IV
Non-Cellular Aspects of Cancer Immunotherapy

Chapter 13
Cytokine Immunotherapy

Megan Nelles, Vincenzo Salerno, Yixin Xu, and Christopher J. Paige

Abstract The immune system protects the body not only from invasion by foreign infectious agents but also from abnormal self-cells with the capacity to cause disease. Several lines of evidence suggest that the immune system can effectively rid the body of cells with malignant potential under normal physiologic conditions and tumor development results from a failure of the immune system. There are also rare cases of spontaneous regression of an established tumor, suggesting that the immune system can regain control if stimulated in the appropriate manner, despite the potentially immunosuppressive nature of the tumor microenvironment. Therapeutic manipulation of the cytokine balance may be such an appropriate stimulation. Cytokines are key immunomodulatory agents that shape responses by the immune system and, conversely, are also involved in the suppression of a response. By manipulating the cytokine milieu, endogenous protection may be reestablished or even enhanced. It is therefore no surprise that cytokine immunotherapy holds great theoretical promise for the treatment of cancer. This theoretical promise has been borne out in a wide variety of preclinical models but unfortunately, clinical trials have to date failed to recapitulate these results.

In this chapter, we discuss the activities of the more promising cytokines as monotherapies, multiple cytokine therapies or paired with other treatment modalities. Our growing understanding of each cytokine alone and within the complex microenvironment of the tumor will lead to the refinement of protocols and improve their therapeutic efficacy. We present the outcomes of some clinical trials and the preclinical models that informed their design; highlighting what the achievements as well as the failures can teach us going forward. We also discuss the many challenges faced by this field and the areas of inquiry in which focused efforts will bear the most fruit. Ultimately, understanding which differences between preclinical and clinical protocols account for the discrepancy in outcomes will help us in designing more effective treatments for those cancers that remain refractory to therapy.

Keywords Cancer • Clinical trials • Cytokine • Immunotherapy • Preclinical models

---

M. Nelles (✉)
Department of Medical Biophysics, University of Toronto, Toronto, ON, Canada
e-mail: mnelles@uhnres.utoronto.ca

J. Medin and D. Fowler (eds.), *Experimental and Applied Immunotherapy*,
DOI 10.1007/978-1-60761-980-2_13,

Introduction

Cytokines in the tumor microenvironment can profoundly influence cancer pathogenesis. Cytokines participate in complex interactions with cells and molecules to promote tumor growth, metastasis and invasion or, alternatively, engender a robust immune response that inhibits cancer progression. The balance of immunostimulatory and immuno-suppressive cytokines may determine the outcome of a developing cancer.

There is ample evidence that the immune system is capable of recognizing and eliminating cells undergoing transformation. In some cases, genetic and epigenetic events that lead to malignant transformation may also result in the expression of tumor-specific or tumor-associated antigens that mark malignant cells for destruction by the immune system. There is also evidence that the products of some oncogenes can summon immune cells to the tumor microenvironment by inciting strong inflammatory signals [1]. Such signals can be produced downstream of stress-induced proteins, such as Hsp70, that are released during the tissue remodeling that results from neoplastic growth and invasion [2]. Furthermore, the observation that there is a higher cancer incidence among immunosuppressed patients supports the theory that immunosurveillance can effectively rid the body of pre-malignant cells before any notable tumor growth occurs under normal physiological conditions and in the absence of intervention. Graft-versus-leukemia (GVL) responses in patients following allogeneic hematopoietic stem cell transplantation further demonstrate that cancer cells that develop in a suppressed patient can be sensitive to recognition and elimination by an intact immune system transplanted from a healthy donor. Additionally, in rare cases, cancer patients experience spontaneous regression, indicating that the immune system is also capable of regaining control over established disease and is capable of mediating tumor clearance if stimulated appropriately.

These stimuli, however, do not always lead to a protective immune response; indeed, development of a malignancy is evidence of immune insufficiency. The immune system either fails to recognize the developing malignancy, mounts a suboptimal response, or is actively suppressed by the tolerogenic mechanisms normally activated to prevent auto-immunity. These tolerogenic programs can be co-opted to prevent anti-cancer responses and allow for tumor outgrowth. In this way, early inflammatory responses shape tumor development. Immunosurveillance applies selective pressure on the tumor cell population and immuno-editing is the resulting dynamic process whereby a neoplasm is either eliminated, reaches equilibrium or escapes immune control [3]. In the case of escape, the tumors that emerge are those with reduced immunogenicity or those that have exploited the inflammatory milieu for their own benefit [3]. Indeed, many tumors over-produce immunosuppressive cytokines, such as IL-10, or growth promoting cytokines, such as IL-1β; furthermore, immune effector cells of cancer patients sometimes display evidence of signaling defects that make them unresponsive to appropriate activation stimuli. It has also been observed that chronic inflammation can lead to effector cell anergy or exhaustion that restricts subsequent anti-tumor responses. Therefore, the mixture of cytokines produced in the tumor microenvironment may prove decisive in molding the host response and could be exploited to alter the balance between immune suppression and stimulation [4].

In addition to immuno-editing, the tumor is also shaped by therapeutic interventions. Current standard of care (SOC) protocols can often reduce tumor burden and induce remission but disease relapse is common upon termination of treatment. Such tumor outgrowth is often highly resistant to further radiation and chemotherapy. Some argue that this phenomenon is due to a minority population of persistent cancer initiating cells (CICs) that possess stem cell-like characteristics (also referred to as cancer stem cells, CSCs) [5]. This population demonstrates inherent resistance to therapy due to lower cycling rates and increased expression of multi-drug resistance proteins, along with the propensity to repopulate the tumor [6, 7]. There is some evidence that CICs produce higher levels of cytokines that support their growth and survival as compared to the cells that make up the bulk of the tumor. For example, the over-expression of IL-4 by CICs acts to negatively regulate apoptosis and cancer-directed immunosurveillance, thereby conferring resistance to chemotherapy-induced cell death [8]. IL-4 blockade has resulted in reduced production of prosurvival molecules and, consequently, strong sensitization to chemotherapeutic drugs. However, little work has been done to show the effect of cytokine therapy on this distinct and difficult to treat population.

Cytokine therapy in general has been reserved for those patients refractory to radiation therapy and for whom the chemotherapeutic armamentarium has already been exhausted. In principle, the induction of immunity by cytokine therapy is appealing for several reasons: (1) the systemic nature of immunity makes it ideal for eradicating widespread disease; (2) the immune system should be effective against CICs because it does not require dividing cellular targets; (3) the immune response would be expected to lead not only to short-term disease clearance, but also to long-term immunological memory of the disease, thereby preventing relapse; and (4) targeted immunity should be relatively non-toxic as long as auto-immune reactions are avoided. Accelerated cytokine discovery and our growing understanding of their modes of action will help in harnessing their potential to overcome or reverse tolerance; cytokine therapy seeks to pharmacologically enhance the cytokine's normal physiological function [9] to stimulate immune cells into becoming active under conditions where they have remained inert. Many challenges remain as the pleiotropy and redundancy inherent to cytokines frustrate efforts to delineate their activity; as well as the further challenge of determining how cytokines relate with tumor cells and identifying how cytokine combinations interact [4]. However, past and ongoing clinical trials are providing important insights and clarifying the therapeutic potential of cytokines.

In this chapter, we will discuss some of the preclinical studies and clinical trials that have employed various cytokines as monotherapies, as well as the rational combination of various cytokines for increased benefit. We will also discuss obstacles that this field still faces (Table 13.1).

Table 13.1 Applications for the cytokines discussed in this chapter

Cytokine	Action	Preclinical Models	Clinical Applications & Trial Phase	Administration	References
IFN-α	• Enhances tumor antigen presentation, and expression of death receptors and adhesion molecules • Promotes activity of macrophages, DCs, B and T cells • Induces apoptosis • Anti-angiogenic	• *In vitro* anti-proliferative and apoptotic assays • Human xenograft in nude mice • Wide range of murine malignancies	• Urological malignancies (III) • Melanoma (III) • ML (III) • Lymphoma (III) • Multiple myeloma (III) • Primary brain tumor (II) • Ovarian carcinoma (III)	• IV • Gene therapy – liposome • SC • IT • IP	• [4, 10–16, 114–117, 120–122]
IL-2	• Expands T cell populations (Tregs also) • Increases permeability of vasculature	• *In vitro* T cell proliferation • Wide range of murine malignancies	• Urological malignancies (III) • Melanoma (III) • Leukemia (III) • Lymphoma (III) • Colorectal carcinoma (III) • Ovarian carcinoma (II)	• IV • SC • IT • IP • PEGylated recombinant protein • Autologous fibroblast/tumor cell vaccine • Cell culture for adoptive transfer	• [11, 17–29], [107–112, 114–119]
IL-12	• Promotes DC maturation • Activates APC, increases MHC expression and antigen presentation • Th1 polarizing • Enhances IFN-γ production • Anti-angiogenic	• Wide range of murine malignancies • *In vitro* cytotoxicity assays	• Melanoma (II) • Leukemia (I) • Lymphoma (II) • AIDS-related Kaposi sarcoma (II) • Gastrointestinal/urological carcinoma (I)	• IV • IT • Intravesical • Autologous fibroblast/DC/tumor cell vaccine • Plasmid electroporation	• [30, 32–38, 40–45, 107–113]

IFN-γ	• Antiproliferative effect on neoplastic cells • Stimulates NK cells and macrophages • Increases MHC expression and antigen presentation • Downregulates proto-oncogene expression	• Growth inhibition assays • Wide range of murine malignancies	• Ovarian cancer (III) • Urological carcinoma (II) • Melanoma (III) • Lung cancer (III)	• IP • Intravesical • IV • Allogeneic/ autologous tumor cell vaccine • *In vitro* culture of DCs/ macrophages	• [46–57]
GM-CSF	• Increases myeloid progenitor populations • Stimulates growth, differentiation, migration and maturation of monocytes/macrophages and DCs • Enhances antigen presentation	• *In vitro* cytotoxicity assays • Wide range of murine malignancies	• Immune reconstitution post transplantation • Lymphoma (III) • Lung cancer (III) • Prostate cancer (III) • Melanoma (III)	• SC • IT • Vaccinia virus gene therapy • Autologous tumor cell vaccine	• [59–70, 118–122, 124–126]
IL-10	• Th2 polarizing • Suppresses cytokine secretion by Th1 and NK cells • Supports humoral immunity • Enhances B cell survival, proliferation and antibody production	• Murine breast and colon cancer models • *In vivo* and *in vitro* metastatic models • Differential gene expression assays		• Retroviral tumor cell vaccine • SC • IV	• [71–75]
TNF-α	• Activates Caspase-8 to promote apoptotic death • Causes destruction of stroma • Leads to production of ROS and hyperpermeability of vasculature	• Various rodent malignancy models	• Soft tissue sarcoma • Melanoma (II) • Pancreatic (III) • Esophageal (II)	• Limb perfusion • IT • IV • Radio- and chemo-inducible adenovector	• [76–83]

(continued)

Table 13.1 (continued)

Cytokine	Action	Preclinical Models	Clinical Applications & Trial Phase	Administration	References
IL-7	• Maintains homeostasis of resting naïve and memory T cells • Augments expansion of cycling T cells	• *In vitro* maturation and survival assays • Wide range of murine malignancies	• Incurable non-haematoloic malignancy (I) • Serum assays in various disease states • *In vitro* cytotoxicity assays	• SC	• [84, 85]
IL-15	• Promotes proliferation of memory $CD8^+$ T cells • Protects cells from AICD	• Various murine malignancy models • *In vitro* expansion assays		• IV	• [86, 87, 107]
IL-18	• Stimulates proliferation of T and NK cells • Elevated serum levels associated with metastatic disease	• *In vitro* proliferation and signaling assays • Various murine malignancy models	• Renal cell carcinoma (I) • Melanoma (II) • Lymphoma (I)	• IV	• [92–100, 107, 113]
IL-21	• Maintains activated T cells in less mature state – avoids exhaustion	• Various murine malignancy models • Adoptive therapy	• Renal cell carcinoma (I) • Melanoma (I)	• IV • SC	• [101–103, 107]

Cytokines as Monotherapies

Cytokines have been used in a number of treatment platforms over the years: they have been utilized singularly, in combination with other cytokines, combined with SOC treatments such as chemotherapeutics or radiation, or with experimental treatments that combine cytokines with tumor-specific antigens in vaccination platforms. Cytokines have been administered locally or systemically, or used in the culture and expansion of specific anti-tumor lymphocytes for adoptive transfer therapy. Recombinant proteins have been administered by various routes; further, cytokines have been produced as the downstream product of gene therapy approaches that have included DNA-based intratumoral injection, or somatic cell gene therapy that uses dendritic cells (DCs) or autologous tumor cells for cytokine delivery. Each platform offers certain advantages and suffers from certain limitations. For example, gene therapy is more feasible and cost-effective than using recombinant proteins, which can be difficult and expensive to produce in quantities sufficient for *in vivo* applications; however, dosing can be difficult to control with gene therapy approaches. Systemic cytokine administration can be more effective against disseminated disease but is often more toxic. Autologous tumor cell vaccines ensure the availability of appropriate antigens in the context of immune stimulation but cytokine-secreting DCs may more accurately mimic the natural process.

Many studies have used cytokines as single agents. Among the more extensively studied are interferon-alpha (IFN-α), interleukin-2 (IL-2), IL-12, granulocyte macrophage colony-stimulating factor (GM-CSF), IFN-γ, and tumor necrosis factor alpha (TNF-α).

IFN-α

IFN-α actually comprises a family of proteins but, other than a limited number of phase I trials using IFN-α1 that showed equivalent biological response with reduced side-effects, only IFN-α2 has been broadly assessed [10]; as such, it is this latter variant to which we will henceforth refer. IFN-α has the longest history of therapeutic use of any cytokine in clinical oncology and is perhaps the most effective agent yet identified [4]. It promotes cytotoxicity by enhancing tumor-antigen presentation, the expression of death receptors and adhesion molecules, and by promoting B cell, T cell, macrophage and DC activity. Besides acting directly on cancer cells to induce apoptosis, IFN-α also exhibits anti-angiogenic properties. The anti-cancer activity of IFN-α has been well-established using *in vitro* assays and preclinical model systems (reviewed in [10]) that provided the rationale for numerous subsequent clinical trials. Although trials of renal cell carcinoma (RCC) consistently demonstrated response rates in the 5–10% range, IFN-α has none-theless become standard treatment as a second-line palliative effort for more advanced patients, especially those who cannot tolerate IL-2 [11]. In 1995, a phase III clinical trial was conducted that showed significant improvements in the disease free-survival of melanoma patients, compared to the observation arm, as well as a survival benefit at follow-up time points up to 9 years out. This led to IFN-α being the first cytokine to receive FDA approval for this indication. However, subsequent trials yielded contradictory results, raising controversy about the overall benefit of IFN-α for the treatment of solid tumors. IFN-α therapy has been applied to more than a dozen types of cancer [12] and better results have been consistently observed for patients with hairy cell leukemia [13] and chronic myelogenous leukemia (CML) [14, 15]. Although it has now been replaced by other chemotherapeutic agents for these cancer types [12], IFN-α continues to be used in the treatment of other cancers, specifically in patients at high risk of recurrence [16]. Finally, the differences observed between IFN-α1 and IFN-α2 suggest that further research focused on the structure of cytokines, and how alterations to specific domains might preserve the desired activity while mitigating toxicity, might be warranted.

IL-2

In 1998, IL-2 became the second cytokine to be approved by the FDA for the treatment of cancer [17]. IL-2 is of particular interest for the supportive role it plays in the expansion of conventional T cell populations. However, due to the greater expression of CD25, the IL-2Rα chain, by regulatory T cells (Treg), limiting concentrations of IL-2 may actually lead to preferential expansion of the Treg population.

In clinical trials conducted between 1985 and 1993, of the 270 patients treated with high-dose intravenous IL-2, 16% demonstrated an objective response [18–22]. More recently, high-dose intravenous IL-2 has also been used in the treatment of patients with metastatic RCC, achieving response rates of up to 24% [23]. However, giving IL-2 at high doses has been compared to inducing a controlled state of septic shock [11]. "Capillary leak syndrome" is the most common side effect of IL-2 therapy and leads to symptoms such as edema, weight gain, and pulmonary congestion resulting from increased permeability of capillaries. Although these symptoms are largely reversible and quickly resolve upon treatment termination [24], patients require intensive management during the course of therapy and this limits applicability. Administration of low-dose IL-2 has effectively avoided some of this toxicity, while still achieving objective response in 23% of patients [25, 26]; however, studies of different malignancies using different routes of administration, have mostly been unsuccessful even if such side effects were reduced.

A synthetic derivative of IL-2, with a two amino acid modification, is being tested in preclinical models and appears to have the same activity as unmodified IL-2 without any observable toxicity [11]. That said, a phase I trial utilizing this compound was not as promising [27] and clinical anti-tumor activity was shown to be negligible. There is also anecdotal evidence for the effectiveness of cell-mediated gene therapy using autologous fibroblasts transduced to express IL-2 and CD40L, injected along with leukemia blasts into patients with high-risk acute leukemia. Nine of the 10 patients treated with this protocol survived past 5 years without experiencing significant toxicity [28]. However, due to the small sample size in this trial, further studies are required to confirm the utility of this approach.

Possibly the most common application for IL-2 is in the culture of cells for adoptive therapy. In fact, the availability of recombinant IL-2 has been a key advancement enabling the large-scale culture of lymphocytes required for adoptive transfer therapy. One study that administered IL-2-expanded tumor-infiltrating lymphocytes (TILs) combined with an IL-2 dosing regimen to patients with metastatic melanoma, following lymphodepletion, noted tumor regression in more than 50% of the patients [29]. However, it is technically and logistically difficult to derive tumor-reactive TILs from all patients and much work remains to be done to optimize this technique.

IL-12

Cellular rather than humoral responses to cancer are generally considered preferable, which makes IL-12 particularly attractive for use in immunotherapy, as it is the archetypal Th1-polarizing cytokine. As a key mediator of both innate and adaptive immunity, IL-12 acts on DCs to increase maturation, MHC class I and II expression, and antigen presentation, thus allowing for the initiation of tumor-specific cytotoxic responses. Furthermore, IL-12 has anti-angiogenic activity. IL-12 can also promote the activation of innate immune cells such as macrophages and eosinophils through its induction of IFN-γ and other cytokines. Indeed, the anti-tumor activity of IL-12 is largely mediated by IFN-γ. In particular, Gollob et al. have demonstrated in year 2000 that the induction and maintenance of IL-12-induced IFN-γ expression is highly correlated with favorable outcomes in patients with metastatic RCC [30]. Yet high levels of IFN-γ have also been associated with the induction of antagonistic molecules such as IL-10 and the depletion of signaling molecules downstream of IL-12 such as STAT4 [31]. Both toxicity and the induction of such antagonistic mechanisms continue to pose a challenge in terms of further development of IL-12. Overcoming these formidable obstacles is the impetus for testing the efficacy of IL-12 therapy following different dose and scheduling protocols as well as various routes and methods of administration.

A large body of data points to the potent anti-cancer activity of IL-12 and initial phase I human trials demonstrated adequate treatment safety to warrant further investigation. Cutaneous T cell lymphoma [32, 33], AIDS-related Kaposi sarcoma [34], and non-Hodgkin's lymphoma [35] are among the cancer variants for which IL-12 therapy has had the greatest success. A phase I trial treating metastatic melanoma lesions with intratumoral electroporation of an IL-12 plasmid achieved either a partial response or disease stabilization in 42% of the patients treated. Another 10% of patients achieved complete tumor regression, even of untreated lesions, and systemic toxicity was minimal in all cases [36]. Possibly the most advantageous application of IL-12 therapy thus far has been as an adjuvant to vaccines. In one protocol, stage III and IV melanoma patients that had undergone debulking surgery were then given peptide vaccines derived from tyrosinase and gp100; 87% experienced a boost in vaccine-specific peripheral immune responses when the vaccine was administered with IL-12 [37].

In contrast to these isolated successes, most clinical investigations involving IL-12 have failed to recapitulate the promise of pre-clinical studies. The first phase II trial of systemic IL-12 given intravenously, according to the schedule established in phase I studies, resulted in the unexpected death of two patients due to severe toxicity [38]. The dose limiting toxicities experienced in the phase I study included: hyperbilirubinemia, neutropenia, oral stomatitis, liver function test abnormalities and predominantly elevated transaminases; importantly, the 2 patients that died in the phase II trial did so due to toxicity from massive organ failure [39]. However, other clinical studies have proven safe, if not very effective. For example, patients with transitional cell carcinoma of the bladder were administered intravesical IL-12 therapy that was well tolerated but showed no evidence of clinical response [40]. IL-12 can also be used to artificially propagate the function of DCs and potentiate cellular responses. Gastrointestinal carcinoma patients experienced only mild side effects after intratumoral injection of IL-12-secreting DCs. The treatment was well tolerated overall, but minimal clinical efficacy was demonstrated [41].

In place of genetically engineering DCs to express IL-12, Dohnal et al. have optimized a manufacturing protocol whereby peripheral blood mononuclear cells (PBMC) are collected from the leukocyte apheresis product, enriched for monocytes, and cultured with lipopolysaccharide (LPS) and IFN-γ. The resulting DCs are harvested at a semi-mature (smDC) stage during which they are the most potent producers of IL-12. This may prove to be a very valuable and effective therapeutic approach. In one intriguing *in vitro* study with human cells from melanoma patients, it was shown that Th1-polarized DCs (prepared as above), but not unpolarized DCs (generated as above but in the absence of LPS and IFN-γ), were able to rescue patient type-1 anti-melanoma $CD4^{+}$ T cell responses [42]. The type-1 DCs produced high levels of IL-12; the study concluded that properly prepared DCs might be able to "correct" type-1 insufficiency and establish anti-cancer responses. These observations are important because cell-based vaccines are used once cancers are already established and may have induced a state of non-responsiveness in the host that needs to be reset. A murine sarcoma model was used to test the *in vivo* anti-tumor capacity of DCs prepared as above. A similar deceleration in the growth of tumors was seen in only those mice administered smDCs; with no effect seen if the mice were treated using fully mature DCs [43]. Along these lines, a number of clinical trials have also been conducted. One phase I trial determined that the protocol was well tolerated and delayed type hypersensitivity (DTH) responses were seen in 65% of the patients. Unfortunately, this did not translate to any significant survival advantage [44].

Finding the reason for the apparent disconnect between preclinical and clinical experiences is of paramount importance, as this will inform the rational design of future clinical protocols. Our own recent results indicate that one reason for failures to date may be that inadequate amounts of IL-12 are delivered to the effector cells within the specific locale of the tumor microenvironment [45]. We have shown in a murine system that the amount of IL-12 provided at the local site of interaction is the difference between cancer progression and immune elimination [45]. A sufficient local dose of IL-12, or even other cytokine effectors, may overcome immunosuppressive or tolerogenic signals that can preclude an effective response without leading to systemic levels that can be toxic and counter-productive. This observation may prove relevant to both our understanding of cytokine function within the complex *in vivo* environment as well as the design of future studies. At a minimum, the findings indicate that there exists a need for different measurement and selection techniques in human vaccination preparations.

IFN-γ

Besides its role in mediating the anti-tumor effects of IL-12, IFN-γ has also been tested as a monotherapeutic for its antiproliferative effect on neoplastic cells and its ability to stimulate NK cells and macrophages. IFN-γ has also been studied for its capacity to both up-regulate the expression of tumor antigens and increase the expression of MHC molecules on APCs, thereby effectively enhancing tumor immunogenicity [46]. Furthermore, IFN-γ also leads to the down-regulation of HER-2/neu proto-oncogene protein and mRNA message, which was the rationale for usage of this cytokine in the treatment of ovarian cancer. In general, intraperitoneal administration is favored because ovarian cancer is confined to the peritoneal cavity, even at later stages, allowing for local delivery of high concentrations [47]. The results of early trials were not consistent but showed objective responses in sub-groups of the patient population and IFN-γ was proven effective as a second-line therapy for epithelial ovarian cancer [47–50]. A subsequent phase III trial showed a progression-free survival benefit as compared to the control group; however, overall survival did not reach statistical significance due to the small sample size [46]. Intravesical administration of IFN-γ has also been examined for therapy of superficial bladder cancer and has shown a protective role, preventing short-term disease recurrence [51].

A small number of trials have employed systemic IFN-γ but dosing proves difficult, giving bimodal results: that is, 1 mg had no effect while 0.5 mg, in the same patients, restored cytotoxic monocyte function. Moreover, systemic application of this cytokine leads to significant toxicity, as is the case with many other cytokines; by comparison, cell-mediated approaches have generally been well-tolerated but produced insignificant clinical responses [52, 53]. The inclusion of INF-γ in the culture media of autologous DCs and macrophages boosts their efficacy and *in vivo* activation [54–57], resulting in improved safety but, again, no real clinical improvement.

GM-CSF

GM-CSF was named for its ability to stimulate colony formation of myeloid progenitors; this capacity was originally exploited to reduce the incidence of infection and minimize hospital stay times by boosting neutrophil recovery in patients undergoing stem cell transplantation [58]. Leukine®, recombinant human GM-CSF, is still used clinically for this indication and is also under study for use as an adjuvant in cancer immunotherapy. By stimulating the growth, differentiation, migration, and maturation of monocytes/ macrophages and DCs, and stimulating enhanced antigen presentation by APCs, GM-CSF bridges innate and adaptive immunity. This, as well as its propensity to cause vasculopathy and destroy the vascular network on which a growing tumor depends for nutrients, makes GM-CSF an attractive candidate for cancer immunotherapy. *In vitro* studies demonstrated stimulation of cytotoxic effects on cancer cells [59, 60] due to induction of a subset of DCs that express high levels of costimulatory molecules and are superior at phagocytosing apoptotic tumor cells. GM-CSF also triggers DC expression of CD1d, a MHC class I-like molecule that presents lipid antigens to NKT cells [61]. Based on these observations, many clinical trials have followed.

The survival of stage III and IV melanoma patients was prolonged by subcutaneous administration of GM-CSF [62]. Other trials employed gene therapy approaches, such as intratumoral injections of a GM-CSF-encoding vaccinia virus. This approach negated the toxic effects of the treatment and generated regression of injected tumors in up to 71% of patients, with concomitant regression of non-injected lesions in some recipients [63]. Autologous tumor cells have also been explored as gene therapy effectors [64]. Preclinical mouse data shows that this approach can result in dense DC infiltrates at the site of injection, which is not seen with injection of tumor cells alone or tumor cells plus concurrent injection of recombinant GM-CSF at the same site [65]. In one human trial, patients with non-small cell lung carcinoma were treated with GM-CSF-secreting autologous tumor cells with negligible resulting toxicity and an impressive response rate that was highly related to the vaccine dose. Seven of the 25 assessable patients achieved clinical responses; two patients remained disease free after the surgical resection that was performed upon study induction and five patients showed stable disease with prolonged survival [66]. The biggest drawback to this approach is the inconsistency in vaccine preparation. Vaccine preparations were successfully prepared for 34 of the 35 enrolled patients but the number of transduced cells, and their level of expression, varied and the clinical response was linked to this variation [66]. To improve consistency, subsequent trials used a preparation of autologous primary tumor cells mixed with a bystander cell line that was engineered and selected for high levels of GM-CSF secretion. This approach caused increased toxicity; both in incidence and severity, but no clinical responses were noted [67]. GVAX® is another vaccine platform that consists of two particularly advanced and malignant lines of hormone refractory prostate cancer cells that secrete GM-CSF and are irradiated before injection. This formula has undergone extensive testing.

A human phase I/II trial [68] looked promising but both phase III trials were terminated. One trial was terminated due to a greater number of deaths in the vaccine treatment arm as compared to the control arm; importantly, no specific reason was identified to account for the imbalance in deaths between the study arms [69]. The other trial was terminated because it was deemed highly unlikely that the primary endpoint criteria of the study would be fulfilled [70].

IL-10

IL-10 is produced predominately by monocytes and Th2-type T cells. IL-10 leads to suppression of cytokine secretion by Th1 and NK cells, thus inhibiting cellular immunity. Rather, IL-10 supports humoral immunity by enhancing B cell survival, proliferation, and antibody production [71, 72]. IL-10 is generally considered to be an immunosuppressive cytokine that inhibits APCs from robustly stimulating $CD4^+$ T cells; as such, it came as a surprise when a small number of preclinical studies of IL-10 demonstrated anti-tumor and anti-metastatic activity. Several groups showed this to be NK cell- and IFN-γ-dependent or to result from vasculopathy or the prevention of neoangiogenesis. Furthermore, these responses appeared durable and even led to immune memory [71, 73–75]. However, it must be mentioned that these studies are in the minority and detection of IL-10 in the serum of cancer patients undergoing therapy tends to be a negative prognostic indicator. Hence, the primary research focus is on the application of IL-10 to mitigate inappropriate activation of the immune system for the prevention and treatment of auto-immunity.

TNF-α

The clinical use of TNF-α for the treatment of soft tissue sarcoma and melanoma metastases localized to the limb is associated with a high rate of limb salvage [76]; however, overwhelming toxicity obviates the use of the recombinant protein in systemic immunotherapy. Isolated limb perfusion is the current application method, thus restricting effectiveness against metastatic disease. Although TNF-α can activate Caspase-8, resulting in apoptotic death, the primary mechanism of anti-tumor activity appears to be through destruction of the stromal environment, with an absence of direct activity against cancer cells. Stimulation of TNFR1 on endothelial cells of the tumor vasculature results in production of reactive oxygen species (ROS) that cause vascular thrombosis, leading to hyperpermeability and massive hemorrhagic necrosis of the tumor – hence the name. This hyperpermeability allows for synergistic effects of TNF-α combined with SOC chemotherapeutics as it facilitates the accumulation and biodistribution of chemotherapeutic drugs into the tumor [77]. The specific target of ROS-mediated destruction appears to be proliferating endothelial cells, which suggests that this drug may be most effective against nascent lesions. TNF-α also plays a number of roles in the regulation of the immune system, such as granulocyte activation and extravasation; however, it remains to be seen if this biology is of clinical significance.

Several clinical trials using TNF-α gene therapy [78–80] have been conducted and a phase III [80] trial for pancreatic cancer is still underway which, if successful, may lead to approval of the first gene therapy treatment for cancer in the United States. A replication-incompetent adenovector, TNFerade, was designed to express the human TNF-α gene under control of a radio- and chemo-inducible promoter sequence that allows for spatial and temporal control of expression. Patients were given intratumoral injections of TNFerade along with SOC for their particular cancer, including esophageal, head and neck, pancreatic, and rectal cancers. Outcomes have demonstrated objective responses or stable disease in 70–85% of patients, with complete pathological response in 15–30%. One Phase I/II study for esophageal cancer employing this approach resulted in 48% of patients surviving to 5 years; by comparison, patient survival in other comparable studies was 12–18 months [81]. Interim analysis of the phase III pancreatic cancer trial shows a 12-month survival rate that is twice that of the control group (40% as compared with 20%) [81].

A preclinical model of spontaneous melanoma [82] suggests that TNFerade may be effective against metastatic disease as there was an observed reduction in the number of lymph node metastases. Furthermore, sparing of healthy tissue was noted in a preclinical model of malignant glioma where tumor-bearing mice experienced the vascular thrombotic effect typical of TNFerade treatment but radiation-induced activation of vector injected into normal muscle tissue caused only minor fibrosis and no abnormalities to the vasculature [83]. In summary, TNFerade is a perfect example of the sort of multi-combinatorial approach that is likely to make real advances in the field of immunotherapy. However, unless there is an immunostimulatory component to this therapy, as may be suggested by the clearance of distant metastases in some models, it is difficult to envisage how this approach will successfully clear minimal residual disease (MRD) and prevent relapse.

Other Cytokines

IL-7 is a cytokine with immuno-restorative and immuno-enhancing qualities. IL-7 is imperative for development of various components of the immune system and for homeostasis of resting naïve and memory T cells. Although IL-7 is a prototypical homeostatic cytokine, as a member of the common γ-chain family, IL-7 also augments expansion of the cycling T cell population. Due to lower IL-7R expression on Tregs as compared with $CD4^+$ T cells, it is proposed that therapy could expand the potential helper and effector populations without expanding the Treg population, where the exact opposite can be true of IL-2. A recent study showed a role for this cytokine that is unique among cytokines under investigation for therapeutic activity against cancer [84]. The substantial T cell expansion in response to IL-7 led to broadening in the diversity of the circulating TCR repertoire with a concomitant decrease in senescent $CD8^+$ effector and Treg cells. This profile marks IL-7 for use in patients who are T cell depleted for a number of reasons including iatrogenic causes such as chemotherapy or myeloablative radiation prior to bone marrow transplantation. Despite this promise, a phase I trial demonstrated anti-tumor activity in only one of the 16 patients treated with a well tolerated dose of IL-7 [85] administered subcutaneously.

IL-15, another member of the common γ-chain family, shares some of the biological activity of IL-2 but uniquely, it promotes the proliferation of memory $CD8^+$ T cells [86] and also protects T cells from activation induced cell death (AICD) [87]. Klebanoff et al. [88] used a model of $CD8^+$ antigen-specific T cell adoptive transfer therapy to compare the benefit of combined IL-15 immunotherapy with their previously published results using IL-2. IL-15, at potentially saturating doses, was equally beneficial with regards to tumor size and animal survival but the anti-tumor response initiated with IL-15, administered at physiologic levels, was sustained to a greater degree than that initiated by IL-2. When used as a single agent, IL-15 has demonstrated various beneficial outcomes including: regression of established melanoma tumors [89], metastasis inhibition in a metastatic lymphoma model [90], reversal of T cell anergy [91] and the ability to overcome the need for $CD4^+$ T cell help by $CD8^+$ CTLs [91]. However, to date, no clinical data are available. One reason for the slow clinical development of this promising cytokine is its poor pharmacokinetics *in vivo*, which necessitates the use of large doses. Further study into the signaling of IL-15 suggested that the dominant mechanism *in vivo* is through trans-presentation; that is, IL-15 bound to IL-15Rα is expressed on APCs and this pre-formed complex then binds to the other components of the receptor, IL-15Rβ/γ, which is expressed on NK and T cells. Accordingly, work is underway to study the potential clinical benefit of a fusion protein to extend the *in vivo* half-life as well as to more accurately mimic the endogenous pathway of IL-15 activity [90].

IL-18 plays disparate roles in cancer therapy. It is of interest for its ability to stimulate the proliferation of T and NK cells capable of tumor clearance [92–96], yet squamous cell carcinoma, melanoma, and various other tumor cell lines secrete high levels of IL-18. As well, elevated serum levels of IL-18 are associated with metastatic disease [97, 98]. Despite this duplicitous character, a phase I evaluation was carried out; intravenous administration of IL-18 in patients with RCC, melanoma, and Hodgkin's lymphoma was determined to be safe [99]. However, the follow-up phase II trial showed significant toxicity and only one patient in 64 achieved a partial response and four others had stable disease. This trial was terminated due to the apparent low level of clinical efficacy [100].

IL-21 is a more recently discovered common γ-chain family member that caused tumor regression in a mouse model of RCC after intravenous administration [101]. In a subsequent phase I trial of patients with metastatic RCC and melanoma [102], of the six patients treated, 2 achieved stable disease and 1 achieved a partial response by RECIST (Responsive Evaluation Criteria in Solid Tumors). Perhaps a more unique and intriguing study is one conducted in a preclinical model of adoptive transfer therapy [103]. The antigen-induced maturation of naïve $CD8^+$ T cells into cytolytic effector cells expressing granzyme-B and CD44 for adoptive transfer was inhibited by culture with IL-21; these results were in direct contrast to the results found with IL-2 or IL-15 culture. Yet, counter intuitively, subsequent transfer of the variously cultured T cells into tumor-bearing mice showed superior tumor regression mediated by the less mature IL-21 primed T cells as compared to cells primed with either IL-2 or IL-15.

Along the way, some unforeseen and unfortunate outcomes have also been experienced in cytokine therapy. One such example is IL-6, which demonstrated potential as an immunotherapeutic agent but in fact acted as a growth factor for myeloma cells [11]. A better understanding must still be reached about how cytokines act in a therapeutic setting before we can take full advantage of their curative potential. Optimization of protocols will, however, need to consider the way a given cytokine interacts with other treatment platforms as well as with other cytokines.

Cytokines in Combination Therapies

Immunomodulation by IL-12 is highly dependent upon the production of supplementary cytokines; and it has been observed that those patients that maintain elevated IFN-γ levels throughout the course of treatment are those that benefit from IL-12 therapy. *In vitro* studies have shown that lymphocytes can respond to IL-12 only in the presence of IL-2 and an optimal response, efficient IFN-γ production, requires both IL-15 and IL-18 [104–106]. This is true for other cytokines also, making a strong argument for combination therapy protocols. IL-12 shows particular promise for combination with other cytokines because it plays a critical role in regulating both the innate and adaptive immune systems and synergizes with several other cytokines to enhance their anti-tumor potential (reviewed in [107]).

Many of the common γ-chain family interleukins have been combined with IL-12 in therapeutic protocols. IL-2 and IL-12 ultimately lead to complementary biological effects but signal through distinct pathways. They also upregulate the expression of each other's receptors in a reciprocal fashion [107]. This is the premise for a number of murine models that have demonstrated the utility of dual IL-12/IL-2 therapy [108–111] where tumor regression was either additively or synergistically enhanced by the combination over either cytokine alone. Sufficient safety was demonstrated [112] in human subjects, leading to a follow-up trial by the same group. IFN-γ production was significantly augmented and there was a three-fold expansion in the NK cell population. Combining IL-15 with IL-12 has some theoretical advantages in that IL-15, in contrast to IL-2, does not support Treg populations. Preclinical models have shown significant tumor regression by several distinct mechanisms: dependent on T and NK cells and mediated by IFN-γ, or alternatively, independent of any of these factors but dependent on the synthesis of nitric oxide and demonstrating a role for macrophages in the effector phase. Furthermore, anti-tumor activity downstream of this combination differs with the type of tumor and the mode of delivery [107]. Considering the unique biological characteristics of each of these cytokines, it may be that the most beneficial application of this combination is to use IL-12 in the priming phase and IL-15 to boost the response and support the development of the memory T cell pool. A model of neuroblastoma was used to test this premise [107]: the authors reported enhanced $CD8^+$ CTL activity and improved overall outcome from sequential vaccination as compared to either vaccination alone or both administered concurrently. This benefit was also seen when they used the sequential regimen to treat established tumors. The characteristics of IL-12 and IL-21 also warrant investigation into the value of integrated therapy with these two cytokines. IL-21 results in significantly increased production of IL-12Rβ2 and IFN-γ mRNA message, as well as increased expression of several other cytokines. The attenuation of IFN-γ production during IL-12 monotherapy has been largely responsible for nullifying potential benefit; it is possible that concurrent expression of IL-12 and IL-21 may remedy this limitation [107].

IL-18 has been coupled with IL-12 to prolong IFN-γ production. When combined with IL-2, IL-18 induces the production of Th2 cytokines and is associated with immune suppression. However, in combination with IL-12, Th1 cytokines predominate with a concomitant attenuation of IL-10 production [107]. This cooperation has been borne out in murine models as well [113].

IL-2 has also been combined with a number of other cytokines: IL-2 and IFN-α have been combined, with and without chemotherapy, in several trials [114–116] but this resulted in severe toxicity and some treatment-related mortality. Lower doses abated the severity of side effects and even led to improved clinical outcomes for lung metastasis in patients with RCC [117]. However, IL-2 combined with GM-CSF has not demonstrated any advantage over monotherapy in any study, most of which were performed in cutaneous melanoma and advanced RCC patients [118, 119].

Research has shown that IFN-α, when combined with GM-CSF, is able to drive the differentiation of blood monocytes into DCs in culture [120]. Subsequent studies suggest that this is a critical step toward induction of anti-tumor immunity [121]. One study, using DCs that had been obtained in this manner and pulsed with tumor antigen, reported long-term survival for 100% of their metastatic medullary thyroid carcinoma patients at a mean follow-up time of 37 months. Unfortunately, this is a very small (five patients) and uncontrolled study [122] but warrants continued investigation.

Concluding Remarks

While SOC protocols can prolong survival and improve quality of life, MRD or the existence of resistant CICs often result in disease relapse and the outgrowth of even more aggressive malignancies. Hence, there is an urgent need for the development of novel curative therapies. Ample evidence demonstrates that cytokines can stimulate potent and effective immune responses and that the immune system is capable of affecting specific and complete cancer clearance without requiring the identification of target antigens. However, the complex interplay between tumor cell, immune cell, and stromal cell in the tumor microenvironment can frustrate and subvert therapeutic attempts. Achieving or restoring the appropriate cytokine balance is crucial. In the past few decades, our understanding of the role that cytokines play in both cancer pathogenesis and cancer therapy has increased profoundly, as has our understanding of their individual modes of action. However, cytokine immunotherapy has thus far failed to reflect preclinical expectation. Cytokines function in a symphony of factors and there remains much to learn about cytokine-cytokine, as well as cytokine-cell interactions, before they can be exploited for maximum therapeutic benefit.

We have proposed that one key reason for the lack of success in human trials is that inadequate amounts of the therapeutic cytokine are available within the tumor microenvironment [45]. In animal systems of cell-based therapy, more extensive selection of highly expressing variants is possible. For human application, however, cytokine production is measured on bulk populations, making it impossible to know if the requisite number of cells are making adequate cytokine to induce immune responses in their local environments. More specifically, we are suggesting that cell-mediated delivery of a sufficient stimulant changes the APC-T cell dynamics on the micro scale of the synapse such that non-responsiveness can be overcome or bypassed and a robust immune response ensue. This localized response can then efficiently clear disease, while avoiding elevated systemic cytokine levels and the associated toxicity. Once an appropriate local response is initiated, the systemic nature of immunity will combat disseminated and metastatic disease.

Stagg et al. posit that inducing an effective clinical response will require, at a minimum, an agent or combination of agents that work in concert to enhance antigen presentation and overcome regulatory checkpoints that restrict tumor immunity, while rescuing intrinsic tumor suppressor pathways and affecting cell death in a manner that increases the pool of potential tumor antigens [3]. Although we have seen that it is unlikely that cytokine monotherapy will be successful, combining immunotherapy with cell death-inducing radiation or chemotherapy theoretically fulfills these proposed requirements. TNFerade certainly serves as an example of this. However, it is conceivable that the most beneficial application of cytokine therapy will be to treat MRD following remission-induction chemotherapy. This will require an understanding of the effect that therapy has on CICs. To this end, future preclinical studies will need to focus more on the use of spontaneous tumor systems to better model this scenario.

Disease progression depends on both the nature of the host and the nature of the malignancy. The stage and characteristics of disease have myriad effects on the functioning of the host immune system that, in turn, influence cytokine activity. Therefore, consideration must also be given to the particular strengths of each cytokine for the treatment of: early stage or established cancers, solid tumors or hematological malignancies, and localized or metastatic disease. A need exists to optimize therapeutic protocols to harness the differences observed for different doses and different routes or methods of administration for maximum benefit. This will require consideration of the role of the tumor microenvironment, which is replete with non-transformed cells that support the outgrowth, invasion and metastasis of malignant cells. Furthermore, while some studies appear to offer promise that ground is being gained, the inconsistency of results from one study to another has made objective evaluation difficult. There is an over-arching need in the field for standards of both conduct and evaluation in order to actualize the potentially curative role of cytokines in cancer therapy. For example, a number of phase II and III trials have failed according to RECIST criteria but are able to demonstrate anecdotal evidence of significant objective responses that are not appreciated by existing guidelines. Therefore, a more appropriate set of benchmarks must be developed that allow for the unique characteristics of immunotherapy and capture a true measure of treatment success [123]. While the promise of curative immunotherapy remains to be realized, the steady progress in our understanding of immune regulation, including cytokine biology, is providing a rational basis for developing new approaches to induce an effective anti-cancer immune response.

References

1. Finn OJ (2008) Cancer immunology. N Engl J Med 358: 2704–2715
2. Vega VL, Maio AD (2007) Macrophages and the stress response. In: Asea AAA, Maio AD Heat shock proteins: potent mediators of inflammation and immunity, Springer, Dordrecht
3. Stagg J, Johnstone RW, Smyth MJ (2007) From cancer immunosurveillance to cancer immunotherapy. Immunol Rev 220: 82–101
4. Dranoff G (2004) Cytokines in cancer pathogenesis and cancer therapy. Nat Rev Cancer 4: 11–22
5. Dick JE (2008) Stem cell concepts renew cancer research. Blood 112: 4793–4807
6. Bao S, Wu Q, McLendon RE et al (2006) Glioma stem cells promote radioresistance by preferential activation of the DNA damage response. Nature 444: 756–760
7. Levina V, Marrangoni AM, DeMarco R et al (2008) Drug-selected human lung cancer stem cells: cytokine network, tumorigenic and metastatic properties. PLoS ONE 3: e3077
8. Francipane MG, Alea MP, Lombardo Y et al (2008) Crucial role of interleukin-4 in the survival of colon cancer stem cells. Cancer Res 68: 4022–4025
9. Mach N, Dranoff G (2000) Cytokine-secreting tumor cell vaccines. Curr Opin Immunol 12: 571–575
10. Borden EC, Lindner D, Dreicer R et al (2000) Second-generation interferons for cancer: clinical targets. Semin Cancer Biol 10: 125–144
11. Jeffrey Weber M, PhD. Cytokines and Cancer Therapy. http://www.meds.com/immunotherapy/intro.html
12. Ferrantini M, Capone I, Belardelli F (2008) Dendritic cells and cytokines in immune rejection of cancer. Cytokine Growth Factor Rev 19: 93–107
13. Quesada JR, Hersh EM, Manning J et al (1986) Treatment of hairy cell leukemia with recombinant alpha-interferon. Blood 68: 493–497
14. Allan NC, Richards SM, Shepherd PC (1995) UK Medical Research Council randomised, multicentre trial of interferon-alpha n1 for chronic myeloid leukaemia: improved survival irrespective of cytogenetic response. The UK Medical Research Council's Working Parties for Therapeutic Trials in Adult Leukaemia. Lancet 345: 1392–1397
15. Ohnishi K, Ohno R, Tomonaga M et al (1995) A randomized trial comparing interferon-alpha with busulfan for newly diagnosed chronic myelogenous leukemia in chronic phase. Blood 86: 906–916
16. Kirkwood JM, Strawderman MH, Ernstoff MS et al (1996) Interferon alfa-2b adjuvant therapy of high-risk resected cutaneous melanoma: the Eastern Cooperative Oncology Group Trial EST 1684. J Clin Oncol 14: 7–17
17. Kirkwood JM, Tarhini AA, Panelli MC et al (2008) Next generation of immunotherapy for melanoma. J Clin Oncol 26: 3445–3455
18. Atkins MB, Lotze MT, Dutcher JP et al (1999) High-dose recombinant interleukin 2 therapy for patients with metastatic melanoma: analysis of 270 patients treated between 1985 and 1993. J Clin Oncol 17: 2105–2116

19. Atkins MB, Kunkel L, Sznol M et al (2000) High-dose recombinant interleukin-2 therapy in patients with metastatic melanoma: long-term survival update. Cancer J Sci Am 6 Suppl 1: S11–14
20. Dutcher JP, Creekmore S, Weiss GR et al (1989) A phase II study of interleukin-2 and lymphokine-activated killer cells in patients with metastatic malignant melanoma. J Clin Oncol 7: 477–485
21. Parkinson DR, Abrams JS, Wiernik PH et al (1990) Interleukin-2 therapy in patients with metastatic malignant melanoma: a phase II study. J Clin Oncol 8: 1650–1656
22. Rosenberg SA, Yang JC, Topalian SL et al (1994) Treatment of 283 consecutive patients with metastatic melanoma or renal cell cancer using high-dose bolus interleukin 2. J Am Med Assoc 271: 907–913
23. McDermott DF, Regan MM, Clark JI et al (2005) Randomized phase III trial of high-dose interleukin-2 versus subcutaneous interleukin-2 and interferon in patients with metastatic renal cell carcinoma. J Clin Oncol 23: 133–141
24. Schwartzentruber DJ (2001) Guidelines for the safe administration of high-dose interleukin-2. J Immunother 24: 287–293
25. Lissoni P, Barni S, Ardizzoia A et al (1997) Immunotherapy for metastatic renal carcinoma with interleukin-2 in a subcutaneous administration schedule of short duration. Subcutaneous IL-2 in renal carcinoma. Arch Ital Urol Androl 69: 159–162
26. Tourani JM, Lucas V, Mayeur D et al (1996) Subcutaneous recombinant interleukin-2 (rIL-2) in out-patients with metastatic renal cell carcinoma. Results of a multicenter SCAPP1 trial. Ann Oncol 7: 525–528
27. Margolin K, Atkins MB, Dutcher JP et al (2007) Phase I trial of BAY 50–4798, an interleukin-2-specific agonist in advanced melanoma and renal cancer. Clin Cancer Res 13: 3312–3319
28. Rousseau RF, Biagi E, Dutour A et al (2006) Immunotherapy of high-risk acute leukemia with a recipient (autologous) vaccine expressing transgenic human CD40L and IL-2 after chemotherapy and allogeneic stem cell transplantation. Blood 107: 1332–1341
29. Dudley ME, Rosenberg SA (2007) Adoptive cell transfer therapy. Semin Oncol 34: 524–531
30. Gollob JA, Mier JW, Veenstra K et al (2000) Phase I trial of twice-weekly intravenous interleukin 12 in patients with metastatic renal cell cancer or malignant melanoma: ability to maintain IFN-gamma induction is associated with clinical response. Clin Cancer Res 6: 1678–1692
31. Portielje JE, Gratama JW, van Ojik HH et al (2003) IL-12: a promising adjuvant for cancer vaccination. Cancer Immunol Immunother 52: 133–144
32. Rook AH, Wood GS, Yoo EK et al (1999) Interleukin-12 therapy of cutaneous T-cell lymphoma induces lesion regression and cytotoxic T-cell responses. Blood 94: 902–908
33. Duvic M, Sherman ML, Wood GS et al (2006) A phase II open-label study of recombinant human interleukin-12 in patients with stage IA, IB, or IIA mycosis fungoides. J Am Acad Dermatol 55: 807–813
34. Little RF, Pluda JM, Wyvill KM et al (2006) Activity of subcutaneous interleukin-12 in AIDS-related Kaposi sarcoma. Blood 107: 4650–4657
35. Younes A, Pro B, Robertson MJ et al (2004) Phase II clinical trial of interleukin-12 in patients with relapsed and refractory non-Hodgkin's lymphoma and Hodgkin's disease. Clin Cancer Res 10: 5432–5438
36. Daud AI, DeConti RC, Andrews S et al (2008) Phase I trial of interleukin-12 plasmid electroporation in patients with metastatic melanoma. J Clin Oncol 26: 5896–5903
37. Lee P, Wang F, Kuniyoshi J et al (2001) Effects of interleukin-12 on the immune response to a multipeptide vaccine for resected metastatic melanoma. J Clin Oncol 19: 3836–3847
38. Atkins MB, Robertson MJ, Gordon M et al (1997) Phase I evaluation of intravenous recombinant human interleukin 12 in patients with advanced malignancies. Clin Cancer Res 3: 409–417
39. Cohen J (1995) IL-12 deaths: explanation and a puzzle. Science 270: 908
40. Weiss GR, O'Donnell MA, Loughlin K et al (2003) Phase 1 study of the intravesical administration of recombinant human interleukin-12 in patients with recurrent superficial transitional cell carcinoma of the bladder. J Immunother 26: 343–348

41. Mazzolini G, Alfaro C, Sangro B et al (2005) Intratumoral injection of dendritic cells engineered to secrete interleukin-12 by recombinant adenovirus in patients with metastatic gastrointestinal carcinomas. J Clin Oncol 23: 999–1010
42. Wesa A, Kalinski P, Kirkwood JM et al (2007) Polarized type-1 dendritic cells (DC1) producing high levels of IL-12 family members rescue patient TH1-type antimelanoma CD4+ T cell responses *in vitro*. J Immunother 30: 75–82
43. Huttner KG, Breuer SK, Paul P et al (2005) Generation of potent anti-tumor immunity in mice by interleukin-12-secreting dendritic cells. Cancer Immunol Immunother 54: 67–77
44. Dohnal AM, Witt V, Hugel H et al (2007) Phase I study of tumor Ag-loaded IL-12 secreting semi-mature DC for the treatment of pediatric cancer. Cytotherapy 9: 755–770
45. Labbe A, Nelles M, Walia J et al (2008) IL-12 Immunotherapy of murine leukemia: comparison of systemic versus gene modified cell therapy. J Cell Mol Med 13(8): 1962–1976
46. Windbichler GH, Hausmaninger H, Stummvoll W et al (2000) Interferon-gamma in the first-line therapy of ovarian cancer: a randomized phase III trial. Br J Cancer 82: 1138–1144
47. Allavena P, Peccatori F, Maggioni D et al (1990) Intraperitoneal recombinant gamma-interferon in patients with recurrent ascitic ovarian carcinoma: modulation of cytotoxicity and cytokine production in tumor-associated effectors and of major histocompatibility antigen expression on tumor cells. Cancer Res 50: 7318–7323
48. Colombo N, Peccatori F, Paganin C et al (1992) Anti-tumor and immunomodulatory activity of intraperitoneal IFN-gamma in ovarian carcinoma patients with minimal residual tumor after chemotherapy. Int J Cancer 51: 42–46
49. D'Acquisto R, Markman M, Hakes T et al (1988) A phase I trial of intraperitoneal recombinant gamma-interferon in advanced ovarian carcinoma. J Clin Oncol 6: 689–695
50. Pujade-Lauraine E, Guastalla JP, Colombo N et al (1996) Intraperitoneal recombinant interferon gamma in ovarian cancer patients with residual disease at second-look laparotomy. J Clin Oncol 14: 343–350
51. Stavropoulos NE, Hastazeris K, Filiadis I et al (2002) Intravesical instillations of interferon gamma in the prophylaxis of high risk superficial bladder cancer – results of a controlled prospective study. Scand J Urol Nephrol 36: 218–222
52. Brill TH, Kubler HR, von Randenborgh H et al (2007) Allogeneic retrovirally transduced, IL-2- and IFN-gamma-secreting cancer cell vaccine in patients with hormone refractory prostate cancer – a phase I clinical trial. J Gene Med 9: 547–560
53. Nemunaitis J, Bohart C, Fong T et al (1998) Phase I trial of retroviral vector-mediated interferon (IFN)-gamma gene transfer into autologous tumor cells in patients with metastatic melanoma. Cancer Gene Ther 5: 292–300
54. Monnet I, Breau JL, Moro D et al (2002) Intrapleural infusion of activated macrophages and gamma-interferon in malignant pleural mesothelioma: a phase II study. Chest 121: 1921–1927
55. Schwaab T, Heaney JA, Schned AR et al (2000) A randomized phase II trial comparing two different sequence combinations of autologous vaccine and human recombinant interferon gamma and human recombinant interferon alpha2B therapy in patients with metastatic renal cell carcinoma: clinical outcome and analysis of immunological parameters. J Urol 163: 1322–1327
56. Hildenbrand B, Sauer B, Kalis O et al (2007) Immunotherapy of patients with hormone-refractory prostate carcinoma pre-treated with interferon-gamma and vaccinated with autologous PSA-peptide loaded dendritic cells – a pilot study. Prostate 67: 500–508
57. Dohnal AM, Graffi S, Witt V et al (2008) Comparative evaluation of techniques for the manufacturing of dendritic cell-based cancer vaccines. J Cell Mol Med 13: 125–135
58. Greenberg P, Advani R, Keating A et al (1996) GM-CSF accelerates neutrophil recovery after autologous hematopoietic stem cell transplantation. Bone Marrow Transplant 18: 1057–1064
59. Grabstein KH, Urdal DL, Tushinski RJ et al (1986) Induction of macrophage tumoricidal activity by granulocyte-macrophage colony-stimulating factor. Science 232: 506–508
60. Wing EJ, Magee DM, Whiteside TL et al (1989) Recombinant human granulocyte/macrophage colony-stimulating factor enhances monocyte cytotoxicity and secretion of tumor necrosis factor alpha and interferon in cancer patients. Blood 73: 643–646
61. Nemunaitis J (2005) Vaccines in cancer: GVAX, a GM-CSF gene vaccine. Expert Rev Vaccines 4: 259–274

62. Spitler LE, Grossbard ML, Ernstoff MS et al (2000) Adjuvant therapy of stage III and IV malignant melanoma using granulocyte-macrophage colony-stimulating factor. J Clin Oncol 18: 1614–1621
63. Mastrangelo MJ, Maguire HC, Jr., Eisenlohr LC et al (1999) Intratumoral recombinant GM-CSF-encoding virus as gene therapy in patients with cutaneous melanoma. Cancer Gene Ther 6: 409–422
64. Soiffer R, Lynch T, Mihm M et al (1998) Vaccination with irradiated autologous melanoma cells engineered to secrete human granulocyte-macrophage colony-stimulating factor generates potent antitumor immunity in patients with metastatic melanoma. Proc Natl Acad Sci USA 95: 13141–13146
65. Simmons AD, Li B, Gonzalez-Edick M et al (2007) GM-CSF-secreting cancer immunotherapies: preclinical analysis of the mechanism of action. Cancer Immunol Immunother 56: 1653–1665
66. Salgia R, Lynch T, Skarin A et al (2003) Vaccination with irradiated autologous tumor cells engineered to secrete granulocyte-macrophage colony-stimulating factor augments antitumor immunity in some patients with metastatic non-small-cell lung carcinoma. J Clin Oncol 21: 624–630
67. Nemunaitis J, Jahan T, Ross H et al (2006) Phase 1/2 trial of autologous tumor mixed with an allogeneic GVAX vaccine in advanced-stage non-small-cell lung cancer. Cancer Gene Ther 13: 555–562
68. Higano CS, Corman JM, Smith DC et al (2008) Phase 1/2 dose-escalation study of a GM-CSF-secreting, allogeneic, cellular immunotherapy for metastatic hormone-refractory prostate cancer. Cancer 113: 975–984
69. GVAX Cancer Vaccine Fails in Phase III Study (2008) http://www.fdanews.com/newsletter/article?articleId=109976&issueId=11915. September 3, 2008
70. Cell Genesys Halts Another Late-Stage GVAX Trial (2008) http://www.fdanews.com/newsletter/article?articleId=111480&issueId=12068. October 22, 2008
71. Zheng LM, Ojcius DM, Garaud F et al (1996) Interleukin-10 inhibits tumor metastasis through an NK cell-dependent mechanism. J Exp Med 184: 579–584
72. Adris S, Klein S, Jasnis M et al (1999) IL-10 expression by CT26 colon carcinoma cells inhibits their malignant phenotype and induces a T cell-mediated tumor rejection in the context of a systemic Th2 response. Gene Ther 6: 1705–1712
73. Kundu N, Beaty TL, Jackson MJ et al (1996) Antimetastatic and antitumor activities of interleukin 10 in a murine model of breast cancer. J Natl Cancer Inst 88: 536–541
74. Sun H, Jackson MJ, Kundu N et al (1999) Interleukin-10 gene transfer activates interferon-gamma and the interferon-gamma-inducible genes Gbp-1/Mag-1 and Mig-1 in mammary tumors. Int J Cancer 80: 624–629
75. Gerard CM, Bruyns C, Delvaux A et al (1996) Loss of tumorigenicity and increased immunogenicity induced by interleukin-10 gene transfer in B16 melanoma cells. Hum Gene Ther 7: 23–31
76. van Horssen R, Ten Hagen TL, Eggermont AM. (2006) TNF-alpha in cancer treatment: molecular insights, antitumor effects, and clinical utility. Oncologist 11: 397–408
77. van der Veen AH, de Wilt JH, Eggermont AM et al (2000) TNF-alpha augments intratumoural concentrations of doxorubicin in TNF-alpha-based isolated limb perfusion in rat sarcoma models and enhances anti-tumour effects. Br J Cancer 82: 973–980
78. McLoughlin JM, McCarty TM, Cunningham C et al (2005) TNFerade, an adenovector carrying the transgene for human tumor necrosis factor alpha, for patients with advanced solid tumors: surgical experience and long-term follow-up. Ann Surg Oncol 12: 825–830
79. Mundt AJ, Vijayakumar S, Nemunaitis J et al (2004) A Phase I trial of TNFerade biologic in patients with soft tissue sarcoma in the extremities. Clin Cancer Res 10: 5747–5753
80. Senzer N, Mani S, Rosemurgy A et al (2004) TNFerade biologic, an adenovector with a radiation-inducible promoter, carrying the human tumor necrosis factor alpha gene: a phase I study in patients with solid tumors. J Clin Oncol 22: 592–601

81. Weichselbaum RR, Kufe D (2009) Translation of the radio- and chemo-inducible TNFerade vector to the treatment of human cancers. Cancer Gene Ther
82. MacGill RS, Davis TA, Macko J et al (2007) Local gene delivery of tumor necrosis factor alpha can impact primary tumor growth and metastases through a host-mediated response. Clin Exp Metastasis 24: 521–531
83. Staba MJ, Mauceri HJ, Kufe DW et al (1998) Adenoviral TNF-alpha gene therapy and radiation damage tumor vasculature in a human malignant glioma xenograft. Gene Ther 5: 293–300
84. Sportes C, Hakim FT, Memon SA et al (2008) Administration of rhIL-7 in humans increases *in vivo* TCR repertoire diversity by preferential expansion of naive T cell subsets. J Exp Med 205: 1701–1714
85. Sportes C, Gress RE (2007) Interleukin-7 immunotherapy. Adv Exp Med Biol 601: 321–333
86. Zhang X, Sun S, Hwang I et al (1998) Potent and selective stimulation of memory-phenotype CD8+ T cells *in vivo* by IL-15. Immunity 8: 591–599
87. Marks-Konczalik J, Dubois S, Losi JM et al (2000) IL-2-induced activation-induced cell death is inhibited in IL-15 transgenic mice. Proc Natl Acad Sci USA 97: 11445–11450
88. Klebanoff CA, Finkelstein SE, Surman DR et al (2004) IL-15 enhances the *in vivo* antitumor activity of tumor-reactive CD8+ T cells. Proc Natl Acad Sci USA 101: 1969–1974
89. Ugen KE, Kutzler MA, Marrero B et al (2006) Regression of subcutaneous B16 melanoma tumors after intratumoral delivery of an IL-15-expressing plasmid followed by *in vivo* electroporation. Cancer Gene Ther 13: 969–974
90. Bessard A, Sole V, Bouchaud G et al (2009) High antitumor activity of RLI, an interleukin-15 (IL-15)-IL-15 receptor alpha fusion protein, in metastatic melanoma and colorectal cancer. Mol Cancer Ther 8: 2736–2745
91. Cheever MA (2008) Twelve immunotherapy drugs that could cure cancers. Immunol Rev 222: 357–368
92. Tominaga K, Yoshimoto T, Torigoe K et al (2000) IL-12 synergizes with IL-18 or IL-1beta for IFN-gamma production from human T cells. Int Immunol 12: 151–160
93. Ahn HJ, Maruo S, Tomura M et al (1997) A mechanism underlying synergy between IL-12 and IFN-gamma-inducing factor in enhanced production of IFN-gamma. J Immunol 159: 2125–2131
94. Tomura M, Zhou XY, Maruo S et al (1998) A critical role for IL-18 in the proliferation and activation of NK1.1+ CD3- cells. J Immunol 160: 4738–4746
95. Micallef MJ, Tanimoto T, Kohno K et al (1997) Interleukin 18 induces the sequential activation of natural killer cells and cytotoxic T lymphocytes to protect syngeneic mice from transplantation with Meth A sarcoma. Cancer Res 57: 4557–4563
96. Song H, Hur DY, Kim KE et al (2006) IL-2/IL-18 prevent the down-modulation of NKG2D by TGF-beta in NK cells via the c-Jun N-terminal kinase (JNK) pathway. Cell Immunol 242: 39–45
97. Park H, Byun D, Kim TS et al (2001) Enhanced IL-18 expression in common skin tumors. Immunol Lett 79: 215–219
98. Park S, Cheon S, Cho D (2007) The dual effects of interleukin-18 in tumor progression. Cell Mol Immunol 4: 329–335
99. Robertson MJ, Mier JW, Logan T et al (2006) Clinical and biological effects of recombinant human interleukin-18 administered by intravenous infusion to patients with advanced cancer. Clin Cancer Res 12: 4265–4273
100. Tarhini AA, Millward M, Mainwaring P et al (2009) A phase 2, randomized study of SB-485232, rhIL-18, in patients with previously untreated metastatic melanoma. Cancer 115: 859–868
101. Hughes S, Chin L, Waggie K et al (2004) Interleukin 21 efficacy in a mouse model of metastatic renal cell carcinoma. J Clin Oncol 22: 2004ASCO. Annual Meeting
102. Curti BD (2006) Immunomodulatory and antitumor effects of interleukin-21 in patients with renal cell carcinoma. Expert Rev Anticancer Ther 6: 905–909
103. Hinrichs CS, Spolski R, Paulos CM et al (2008) IL-2 and IL-21 confer opposing differentiation programs to CD8+ T cells for adoptive immunotherapy. Blood 111: 5326–5333

104. Quill H, Bhandoola A, Trinchieri G et al (1994) Induction of interleukin 12 responsiveness is impaired in anergic T lymphocytes. J Exp Med 179: 1065–1070
105. Fehniger TA, Carson WE, Caligiuri MA (1999) Costimulation of human natural killer cells is required for interferon gamma production. Transplant Proc 31: 1476–1478
106. Jung T, Witzak K, Dieckhoff K et al (1999) IFN-gamma is only partially restored by co-stimulation with IL-12, IL-2, IL-15, IL-18 or engagement of CD28. Clin Exp Allergy 29: 207–216
107. Weiss JM, Subleski JJ, Wigginton JM et al (2007) Immunotherapy of cancer by IL-12-based cytokine combinations. Expert Opin Biol Ther 7: 1705–1721
108. Addison CL, Bramson JL, Hitt MM et al (1998) Intratumoral coinjection of adenoviral vectors expressing IL-2 and IL-12 results in enhanced frequency of regression of injected and untreated distal tumors. Gene Ther 5: 1400–1409
109. Wigginton JM, Komschlies KL, Back TC et al (1996) Administration of interleukin 12 with pulse interleukin 2 and the rapid and complete eradication of murine renal carcinoma. J Natl Cancer Inst 88: 38–43
110. Wigginton JM, Gruys E, Geiselhart L et al (2001) IFN-gamma and Fas/FasL are required for the antitumor and antiangiogenic effects of IL-12/pulse IL-2 therapy. J Clin Invest 108: 51–62
111. Wigginton JM, Park JW, Gruys ME et al (2001) Complete regression of established spontaneous mammary carcinoma and the therapeutic prevention of genetically programmed neoplastic transition by IL-12/pulse IL-2: induction of local T cell infiltration, Fas/Fas ligand gene expression, and mammary epithelial apoptosis. J Immunol 166: 1156–1168
112. Gollob JA, Veenstra KG, Parker RA et al (2003) Phase I trial of concurrent twice-weekly recombinant human interleukin-12 plus low-dose IL-2 in patients with melanoma or renal cell carcinoma. J Clin Oncol 21: 2564–2573
113. Subleski JJ, Hall VL, Back TC et al (2006) Enhanced antitumor response by divergent modulation of natural killer and natural killer T cells in the liver. Cancer Res 66: 11005–11012
114. Ravaud A, Trufflandier N, Ferriere JM et al (2003) Subcutaneous interleukin-2, interferon alpha-2b and 5-fluorouracil in metastatic renal cell carcinoma as second-line treatment after failure of previous immunotherapy: a phase II trial. Br J Cancer 89: 2213–2218
115. Marincola FM, White DE, Wise AP et al (1995) Combination therapy with interferon alfa-2a and interleukin-2 for the treatment of metastatic cancer. J Clin Oncol 13: 1110–1122
116. Sparano JA, Dutcher JP, Kaleya R et al (1991) Colonic ischemia complicating immunotherapy with interleukin-2 and interferon-alpha. Cancer 68: 1538–1544
117. Akaza H, Tsukamoto T, Onishi T et al (2006) A low-dose combination therapy of interleukin-2 and interferon-alpha is effective for lung metastasis of renal cell carcinoma: a multicenter open study. Int J Clin Oncol 11: 434–440
118. Elias EG, Zapas JL, McCarron EC et al (2008) Sequential administration of GM-CSF (Sargramostim) and IL-2 +/– autologous vaccine as adjuvant therapy in cutaneous melanoma: an interim report of a phase II clinical trial. Cancer Biother Radiopharm 23: 285–291
119. Amato RJ, Malya R, Rawat A (2008) Phase II study of combination thalidomide/interleukin-2 therapy plus granulocyte macrophage-colony stimulating factor in patients with metastatic renal cell carcinoma. Am J Clin Oncol 31: 237–243
120. Paquette RL, Hsu NC, Kiertscher SM et al (1998) Interferon-alpha and granulocyte-macrophage colony-stimulating factor differentiate peripheral blood monocytes into potent antigen-presenting cells. J Leukoc Biol 64: 358–367
121. Santini SM, Lapenta C, Santodonato L et al (2009) IFN-alpha in the generation of dendritic cells for cancer immunotherapy. Handb Exp Pharmacol (188): 295–317
122. Papewalis C, Jacobs B, Wuttke M et al (2008) IFN-alpha skews monocytes into CD56+ -expressing dendritic cells with potent functional activities *in vitro* and *in vivo*. J Immunol 180: 1462–1470
123. Peggs KS, Quezada SA, Korman AJ et al (2006) Principles and use of anti-CTLA4 antibody in human cancer immunotherapy. Curr Opin Immunol 18: 206–213

124. Rasmussen T, Hansson L, Osterborg A et al (2003) Idiotype vaccination in multiple myeloma induced a reduction of circulating clonal tumor B cells. Blood 101: 4607–4610
125. Abdalla AO, Hansson L, Eriksson I et al (2007) Idiotype protein vaccination in combination with adjuvant cytokines in patients with multiple myeloma – evaluation of T-cell responses by different read-out systems. Haematologica 92: 110–114
126. Hansson L, Abdalla AO, Moshfegh A et al (2007) Long-term idiotype vaccination combined with interleukin-12 (IL-12), or IL-12 and granulocyte macrophage colony-stimulating factor, in early-stage multiple myeloma patients. Clin Cancer Res 13: 1503–1510

Chapter 14
Transcriptional Modulation Using Histone Deacetylase Inhibitors for Cancer Immunotherapy

Takashi Murakami

Abstract Epigenetic processes, in addition to well-characterized genetic abnormalities, play a critical role in cancer initiation and progression. The acetylation status of histones affects levels of gene expression, and causes aberrant transcriptional repression in cancer cells. As a result, this is also implicated with unresponsiveness to immune-based therapies as well as conventional chemotherapies against cancer. To sensitize cancer cells to those therapies, histone deacetylase inhibitors (HDACi) have attracted much attention as epigenetic-modulating agents, because these compounds possess the pleiotropic effects on malignant cells such that they are more prone to differentiation, growth arrest, and apoptosis. This chapter will highlight the pleiotropic antitumor effects of HDACi when combined with antigen-specific tumor immunotherapy, and describe the potential clinical implications for the improved cancer immunotherapy.

Keywords Adoptive immune cell-transfer • Gene expression • Histone acetylation • Histone deacetylase inhibitors • Tumor-sensitization

Introduction

Great efforts and innovations in biomedical science have been made in the field of cancer immunology since the latter part of the twentieth century, and attempts to enhance cellular immune responses have used various cancer antigens and immunizing vectors (e.g., [1, 2]). However, the effectiveness of cancer immunotherapy in humans remains limited [2, 3]. For example, IL-2, a cytokine that nonspecifically stimulates T cells, is approved by the US Food and Drug Administration for the treatment of patients with metastatic melanoma or renal cancer, and can mediate

T. Murakami (✉)
Division of Bioimaging Sciences, Center for Molecular Medicine, Jichi Medical University, 3311-1Yakushiji, Shimotsuke, Tochigi, 329-0498, Japan
e-mail: takmu@jichi.ac.jp

J. Medin and D. Fowler (eds.), *Experimental and Applied Immunotherapy*,

DOI 10.1007/978-1-60761-980-2_14,

cancer regression in around 15–20% of treated patients [4]. Cancer vaccine strategies also allow for the generation of immune T cells that recognize antigenic peptides present on tumor cells, whereas the regression of growing tumors in patients treated with active immunization has been very sporadic and rare [2]. In contrast, cell transfer approaches, such as those using adoptive cell transfer or adoptive immunotherapy, may be relatively effective and have been demonstrated in one case to mediate cancer regression in 50–70% of patients with metastatic melanoma [2]. Although this cell transfer approach has provided important clues concerning requirements for the successful immunotherapy of cancers in general [5, 6], the fact that tumors have primarily remained so resistant to T-cell attack requires further investigation and explanation. This chapter will focus on genetic alterations in cancer cells and the potential modification of such cells for effective cancer immunotherapy.

In order to mediate antitumor effects, T cells of sufficient avidity for the recognition of tumor antigens must be present in sufficient quantities, pass to the tumor site by extravasating from the circulation, and then mediate effector functions to destroy cancer cells. Indeed, all of these criteria may be required for effective treatment. However, the cellular unresponsiveness of solid tumors through aberrant transcriptional regulation remains a critical barrier that limits the therapeutic potential of adoptively transferred T cells in patients with cancer. Herein, I will highlight aberrant transcriptional regulation in cancer cells and describe potential tumor sensitization strategies employing the use of histone deacetylase (HDAC) inhibitors to enhance adoptive immunotherapy.

Histone Acetylation in Cancer

Histone Acetylation and Gene Expression

Complexes of genomic DNA and histones form the nucleosome of chromatin in eukaryotic cells, where 146 base pairs of double-stranded DNA are wrapped around a central core of four basic histones (H3, H4, H2A, and H2B). Each nucleosomal structure has eight histone proteins arranged in a tripartite structure with one (H3H4) tetramer and two H2AH2B dimers. Nucleosomes are separated by linker DNA and compacted into higher-ordered structures by histone H1 [7]. While this provides a mechanism of inserting several meters of DNA into a single nucleus, structural compactions can also restrict the access of regulatory proteins to the DNA. The N-terminal tails of all four histones protrude outward from the core histones and are thus accessible to histone-modifying enzymes [7]. Many studies over the past decade have demonstrated that multiple covalent modifications (acetylation, phosphorylation, methylation, ubiquitination, sumoylation, and ADP-ribosylation) take place on histone tail residues and, as more recent data has demonstrated, within the body of the histone proteins. The histone code hypothesis suggests that a dynamic constellation of these posttranslational modifications determines the binding of chromatin remodeling factors to the nucleosome [8]. These factors regulate the accessibility of transcription factors, cofactors, and the general transcriptional machinery to DNA by altering the chromatin structure, which ultimately determines gene expression. All of the epigenetic alterations currently recognized on histones are reversible, and separate sets of enzymes for removing these modifications have been identified [9]. The histone acetyltransferase (HAT) and histone deacetylase (HDAC) enzymes determine the status of nucleosome histone acetylation, one of the most extensively studied of the chromatin modifications. It is the balance between the opposing activities of HATs and HDACs that determines chromatin structure at the gene level and, therefore, at the gene expression level. Additionally, the relative amount of total cellular HATs and HDACs determines the global status of acetylation in the genome. This balance is important in regulating the cellular response to endogenous and exogenous stimuli. Histone acetylation is generally correlated with gene activation, whereas histone deacetylation is correlated with gene repression. HDAC enzymes also modulate the acetylation of numerous other nonhistone proteins such as p53, cytoskeleton protein α-tubulin, and the molecular chaperone Hsp90 [10–12]. Direct acetylation of transcription factors and cofactors by HATs is another important mechanism by which acetylation regulates gene expression [13].

Classification of HDAC Enzymes and Activity in Normal and Cancer Tissues

There is increasing evidence to suggest that the 18 HDAC enzymes (HDACs) in humans are not redundant in function. These HDACs are currently divided into four classes based on phylogenetic and functional properties [14, 15]. Class I HDACs (1–3, 8) are structurally similar to yeast transcription factor RPD3. Class II HDACs (4–7, 9, 10) possess homology to yeast HDA1 deacetylase and are subdivided into two classes: class IIa HDACs (4, 5, 7, 9) and class IIb HDACs (6 and 10). Class III HDACs (sirtuins) are structurally similar to yeast SirT2, and the only known class IV HDAC is HDAC 11.

Class I and class II HDACs are zinc-dependent enzymes containing catalytic sequences that can be inhibited by zinc chelating compounds [15]. Class IV HDAC (HDAC11) has conserved residues in the catalytic center that are shared by both class I and class II HDACs [14, 16], although the target substrates remain to be identified [17]. Class III HDACs require NAD+ as a cofactor for enzymatic activity (they do not contain zinc in the catalytic site) [18]. Given their zinc-dependent enzymatic activity, class I, class II, and class IV enzymes are referred to as classical HDACs, and represent the main targets in current clinical drug development.

While there are structural similarities among HDACs, these enzymes possess nonredundant roles during embryogenesis. For example, mice lacking HDAC 1, 2, 3, or 7 show aberrant cell cycle regulation or abnormal blood vessel development, and lead to embryonic lethality (reviewed in [17]). In contrast, knockout mice of

HDAC 4, 5, 6, or 9 are not lethal, but show markedly abnormal cardiovascular, bone, and muscle development [17]. These findings suggest that individual HDAC enzymes possess unique roles in embryonic development and organ-specific functions, and highlight the potential adverse effects associated with the systemic targeted inhibition of HDACs.

Given the specific roles of HDACs in normal tissues, it is perhaps not surprising that the effects of targeted inhibition of HDACs in cancer cells seem to be tissue-dependent. For instance, knockdown of HDAC 1 inhibits proliferation of colon cancer cells in culture, and induces apoptosis in osteosarcoma and breast cancer cells [19, 20]. Inhibition of HDAC 2 can induce growth arrest in colon cancer cells, but not in osteosarcoma and breast cancer cells [19, 20]. However, knockdown of HDAC 2 down-regulates estrogen receptor (ER) expression, and potentiates tamoxifen-induced apoptosis in ER-positive breast cancer cells [21]. Knockdown of HDAC 2 enhances p53-dependent gene activation and repression, and can inhibit the proliferation of cultured breast cancer cells [22]. These examples suggest that HDAC enzymes in cancers may be associated with tissue-specific programming.

Histone Modifications in Cancer

The transformation of a cell from a normal to a cancerous state involves multiple processes, which include both genetic and epigenetic pathways. Chromatin modification and DNA methylations are epigenetic changes that can regulate gene expression, thereby influencing cell growth and differentiation without altering DNA sequences. Aberrant DNA methylation, which is a well-characterized gene silencing mechanism, is a hallmark of cancer [23]. Furthermore, alterations in both histone acetyltransferases and HDACs have been found in many human cancers [14, 16, 24–32]. For example, a mutation in the cAMP response element-binding protein (CREB) binding protein leads to the inactivation of its histone acetyltransferase activity (Rubinstein-Taybi syndrome) [33], and loss of heterozygosity in the CREB binding protein locus has been observed in a subset of lung cancers [34]. Thus, dysregulation of histone modifications also appears to induce inappropriate gene expression or repression, thereby contributing to the pathogenesis of many forms of cancers. To date, there are three major mechanisms known to be involved in epigenetic modifications found in cancer: (1) activation of silenced genes, particularly oncogenes; (2) silencing of normally active tumor suppressors; and (3) silencing of certain immune genes involved in the antitumor response. There is substantial accumulating evidence that oncogenes and tumor suppressor genes are regulated by HDACs, and that both HAT and HDAC mutations potentially occur during carcinogenesis (reviewed by [35]). Although it has been established that tumor suppressor and immune-associated genes are often silenced by HDACs in cancer cells, the events leading to epigenetic silencing remains poorly understood.

HDAC Inhibitors and Cancer Cell Death-Sensitivity

Inhibition of HDACs and Gene Expression in Cancer

As described earlier, aberrant transcriptional repression of genes that control cell growth and differentiation is associated with cancer development, and alteration of HDACs underlies the transcriptional repression associated with many cancer types [36, 37]. Thus, blockade of HDACs might restore global gene expression in cancer cells, making these sensitive to cell cycle arrest, differentiation, and apoptotic cell death [38, 39]. The mode of action of HDAC inhibitors, as transcriptional modulators, differs from that of other anticancer agents [16]. Indeed, HDAC inhibitors are expected to be effective for many cancer types that are unresponsive to conventional chemotherapy [16, 39]. Table 14.1 presents a summary of how HDAC inhibitors may function in cancer therapy. Gene expression profiles in cancer cells mediated by HDAC inhibitors of diverse structural classes are time and dose dependent [40, 41]. However, although many similarities have been observed concerning the effects of various HDAC inhibitors on gene expression, it is notable that some profiles are agent-specific [42].

HDAC inhibitors have shown cytotoxicity against a variety of human and rodent cancer cells *in vitro* and *in vivo* [43], and some of these inhibitors are being tested in clinical studies [39] (Table 14.2). For example, depsipeptide (also referred to as FK228 or FR901228) is regarded as a promising HDAC inhibitor for use in the treatment of human melanoma. Depsipeptide was originally isolated from *Chromobacterium violaceum* (No 968) as a compound that reversed the malignant phenotype of H-ras-transformed fibroblasts by blocking the p21 ras-mediated signal transduction pathway [44, 45]. Depsipeptide suppressed cell proliferation and induced apoptosis in human uveal melanoma cell lines at relatively high doses [36], and produced substantial therapeutic effects against various malignancies [39].

Table 14.1 Representative biological effects of HDAC inhibitors in cancer therapy

Inhibition of cell proliferation
Induction of p21, G1 arrest, and cellular differentiation
Augmentation of nuclear receptor response driving terminal cell differentiation
Reactivation of silenced tumor suppressor genes (in combination with DNA methyl transferase inhibitors)
Suppression of telomerase gene expression
Induction of apoptosis
Activation of mitochondria-dependent apoptosis
Activation and/or sensitization of death-receptor mediated killing
Induction of topoisomerase II (potential alteration of sensitivity to DNA-damaging agents)
Other mechanisms
Alteration of angiogenic signaling
Alteration of microtubule function
Induction of MHC antigens on the cell surface (to augment immune responses)
Suppression of IL-2-mediated gene expression

Table 14.2 Representative histone deacetylase inhibitors in clinical development

Group	Compound	HDAC target (effective concentration)	Clinical phase trial
Hydroxamates	Trichostatin A (TSA)	Class I, II, IV (nM)	–
	Vorinostat, Suberoyl anilide hydroxamic acid (SAHA)	Class I, II, IV (μM)	III
	M-carboxycinnamic acid bishydroxamate (CBHA)	N/A (μM)	–
	LAQ-824/LBH589	Class I, II, IV (nM)	I
	PXD-101 (Bellinostat)	Class I, II, IV (μM)	II
Cyclic peptide	Depsipeptide (FK228)	Class I (HDAC1, 2)(nM)	II
Aliphatic acids	Valproic acid	Class I, IIa (nM)	II
	Phenyl butyrate	Class I, IIa (nM)	II
	AN-9 (Pivanex)	N/A (μM)	II
Benzamides	MS-275	Class I (HDAC1, 2, 3)(μM)	II
	MGCD0103	Class I (μM)	II

N/A not available

In a clinical trial, Schrump et al. [46] examined global gene expression profiles in laser-captured tumor cells from pre- and posttreatment biopsies from lung cancer patients receiving depsipeptide infusions. Pretreatment RNA was used as the reference for each respective posttreatment array study. Considerable heterogeneity was detected in baseline as well as posttreatment gene expression profiles. Only 16 genes were induced by at least twofold in one or more patients following depsipeptide treatment. In contrast, more than 1,000 genes were repressed by at least twofold in one or more patients following depsipeptide infusion [46]. Interestingly, the genes that were induced or repressed by at least twofold following depsipeptide treatment seemed to be down-regulated or overexpressed, respectively, in resected primary lung cancers relative to adjacent, histologically normal bronchial epithelial cells. In other words, depsipeptide appears to normalize gene expression in lung cancer cells.

In another *in vitro* study to determine the molecular basis of the cytotoxic effect of depsipeptide against melanoma, depsipeptide treatment resulted in suppression of the Ras-MAPK signaling pathway through Rap1 up-regulation, thereby leading to apoptosis in human melanoma cells [37]. Recent investigations have revealed that abnormalities in the Ras–B-Raf–MAP kinase pathway are observed in most patients with melanoma [47, 48], and that this aberrant signaling plays a critical role in melanoma development [49]. Thus, gene up-regulation induced by depsipeptide renders melanoma cells more susceptible to apoptotic cell death [37]. Additionally, since HDAC inhibitors potentially enhance Fas death receptor and MHC class I expression on cancer cells (see Table 14.1), the use of these inhibitors may prove to be a useful adjunct in human immunotherapy strategies against cancer [50, 51].

HDAC Inhibitors and Cancer Cell Death-Sensitivity

Emerging evidence suggests that a variety of factors are present that limit tumor regression in the host-tumor interaction. For example, cancer progression often takes place despite the presence of circulating cancer-specific cytotoxic T cells (CTLs). Even with patients in whom large numbers of highly activated tumor-specific CTLs have been infused, clinical improvement has been difficult to achieve [2, 52]. Recent evidence concerning host factors suggests that regulatory elements of the immune responses, including $CD4^{+}CD25^{+}$ regulatory T cells (Tregs), inhibit the ability of CTLs to produce effective antitumor responses [53, 54]. In regard to tumor escape factors, many aggressive tumors do not express tumor antigens or MHC antigens [55, 56]. Moreover, many cancer types lack sufficient apoptotic cell death pathways due to aberrant transcription [57, 58]. It is essential that these tumor escape factors be countered to facilitate the establishment of efficient cancer immunotherapy strategies. One approach might be to manipulate the tumor as well as the host prior to adoptive immune cell transfer. In this regard, the use of HDAC inhibitors is promising (see Table 14.1).

While HDAC inhibition is capable of inducing varying degrees of growth arrest, differentiation, or apoptosis of cancer cells [39, 57], cultured normal fibroblasts and melanocytes are almost always considerably more resistant than tumor cells [37, 39]. As such, cellular modulation through the use of HDAC inhibitors may be relatively specific to cultured melanoma cells [37]. This tumor-selective effect following HDAC inhibitor treatment was observed in the up-regulation of MHC class I molecules *in vitro* and *in vivo* [50, 51, 59]. Since the MHC class I molecule is released from the endoplasmic reticulum only after binding peptide and then moving to the cell surface [60], this up-regulation could provide substantial benefits for immunological recognition. However, there are abnormalities in the expression and/or function of various components of the MHC class I antigen-processing pathway in human malignant cells [61, 62], and it remains to be determined whether HDAC inhibitors can also affect various components of the protein-processing machinery.

Combining HDAC Inhibitors and Immunotherapy

Although the use of HDAC inhibitors has induced partial and complete responses in patients with hematological malignancies when administered alone [63–65], only a partially objective response was observed in solid tumor cancer patients [66, 67]. These findings suggest that monotherapy with the HDAC inhibitor alone will not prove sufficiently effective for the majority of cancer patients. However, it is possible that HDAC inhibitor therapy may provide greater benefits when used primarily to modulate tumor factors. It is also possible that simultaneous targeting by employing a variety of HDACs could provide greater benefits, particularly for selective immunotherapy against cancer. Intriguingly, HDAC inhibitors could activate components of death receptors including Fas and tumor necrosis factor-related apoptosis-inducing ligand (TRAIL) receptor [68–71]. In fact, murine B16/F10 cells are less immunogenic and highly resistant to a variety of apoptotic stimuli when they are not manipulated [72–74]. However, it has been demonstrated that exposure of B16 tumor cells to a limited dose of depsipeptide (FK228) induced cell surface expression of Fas receptor and MHC class I, and the enforced Fas engagement synergistically increased caspase-3/7 activity of B16/F10 cells in the presence of depsipeptide *in vitro* [50]. These changes should provide CTLs with an enhanced ability to recognize and destroy target tumor cells (Fig. 14.1).

Furthermore, it has been demonstrated that HDACi synergizes with exogenously added TRAIL to induce apoptosis of various human solid tumor cell lines *in vitro* [70, 71, 75, 76]. While the TRAIL system seems to account for a significant portion of the HDACi effect on the human death receptor pathway, it has been suggested that the effect of HDAC inhibitors on the death receptor pathway may not occur universally in all tumor types [39]. For example, epigenetic silencing of caspase-8 in human cancer cells causes apoptosis resistance against TRAIL treatment and even the use of pan-HDAC inhibitors failed to reverse the silencing [77]. However, interferon (IFN)-γ can restore caspase-8 expression, and the use of HDAC inhibitors with IFN-γ can successfully overcome the resistance to TRAIL *in vitro* and *in vivo* [77]. The Fas–Fas ligand system is well known as the major pathway involved in

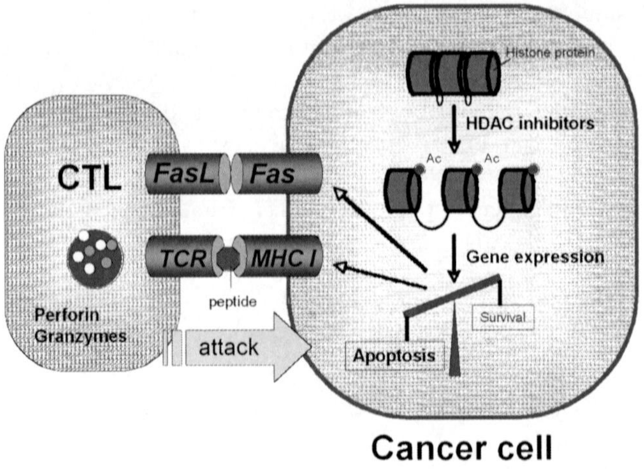

Fig. 14.1 *The rationale of cancer immunotherapy using HDAC inhibitors*. Acetylation of histone (and non-histone) proteins by HDAC inhibitors results in change in the expression and activity of apoptotic proteins in cancer cells that favors a proapoptotic response and a lowering of the cellular apoptotic threshold. Pro-apoptotic proteins are up-regulated and antiapoptotic proteins are down-regulated. Moreover, MHC antigens are induced on the cell surface and immune recognition and responses are augmented. Therefore, these cancer cells have been sensitized to apoptosis induced by death receptor ligands (e.g., Fas ligand and TRAIL). These cells are also recognized by antigen-specific CTLs that are adoptively transferred into the patients. Potentially toxic granules in CTLs (i.e., perforin and granzymes) are released into the target cancer cells. *Ac* acetylated, *CTLs* cytotoxic T cells, *TCR* T cell receptor, *FasL* Fas ligand

CTL-mediated tumor destruction in rodent models [78–80], and therefore some differences between species (mouse and human) may be considered for the major death receptor pathway. However, recent animal studies have demonstrated that the augmented tumoricidal effects of tumor-specific CTLs induced as a consequence of sensitization by HDAC inhibitors resulted in melanoma cells with enhanced CTL-mediated cytotoxicity *in vivo* [50, 51]. Limited administration of HDACi (depsipeptide/FK228) sufficiently modulated the acetylation level of histone H3 at the tumor site [50]. Substantial enhanced expression of MHC class I and gp100 melanoma antigen by HDAC inhibitor (LAQ-824) was observed at B16 tumor sites [51]. These findings support the rationale for protocols that pretreat cancer patients with HDACi to potentiate adoptive immune therapy (see Fig. 14.1).

Emerging experimental data indicate that lympho-depletion using cyclophosphamide prior to adoptive cell transfer of tumor-specific T-lymphocytes plays a key role in enhancing treatment efficacy by eliminating $CD4^{+}CD25^{+}$ Tregs and other competing elements of the immune system in patients with cancer [81–83]. Such the depletion of endogenous cells that compete for the activation of cytokines (known as the "cytokine sink") is now required to maximize the exposure of homeostatic cytokines on the transferred CTLs [1].

In addition to this type of chemotherapy pretreatment of the host, it may be necessary to eliminate a variety of tumor factors that limit tumor regression despite the administration of antitumor CTLs [2]. It appears that clinical tumors possess the following characteristics that impact resistance to endogenously generated CTLs [3]: (1) loss of MHC expression; (2) insufficient antigen expression; (3) lack of sufficient apoptotic or other cell destruction pathways; and (4) production of local immunosuppressive factors. However, it is known that the use of HDAC inhibitors can counter the effects of the first three aforementioned characteristics (see Table 14.1). Therefore, it is possible that HDAC inhibitors such as vorinostat [84] may safely be used to reduce tumor factors that currently limit the therapeutic potential of adoptively transferred CTLs.

HDAC Inhibitors and Immune Responses

While HDAC inhibitors are often highlighted as anticancer agents themselves, several recent experimental studies have demonstrated that HDAC inhibitors can modulate immune responses at concentrations much lower than those needed to effect antitumor activity [85–89]. For example, Reddy et al. [90] demonstrated that pretreatment of dendritic cells (DCs) with HDAC inhibitors significantly reduced Toll-like receptor-induced secretion of proinflammatory cytokines, suppressed the expression of CD40 and CD80, and reduced the *in vitro* and *in vivo* allo-stimulatory responses induced by the DCs. In fact, injection of DCs treated *ex vivo* with HDAC inhibitors reduced experimental graft-versus-host disease (GVHD) in a murine allogeneic bone-marrow transplantation (BMT) model [90]. The mechanism by which DC functions are inhibited following exposure to HDAC inhibitors involves increased expression of indoleamine 2,3-dioxygenase (IDO), a suppressor of DC function [90]. Direct injection of HDAC inhibitors early after allogeneic BMT into chimeric animals whose bone marrow-derived cells lacked IDO failed to protect against GVHD [90]. Therefore, it is suggested that HDAC inhibitors can regulate DC functions through the induction of IDO [90]. Moreover, Tao et al. [91] showed that treatment of mice with the HDACi trichostatin-A (TSA) increased the number of $FoxP3^{+}CD4^{+}CD25^{+}$ Tregs, and TSA suppressed T-cell–dependent alloresponses *in vivo* in fully MHC-mismatched cardiac and islet transplant models [91]. Villagra et al. [92] showed that class IV HDAC11 regulates IL-10 expression levels on antigen-presenting cells (APCs), thereby supporting the view that particular HADC enzymes are involved in the immune regulation of tolerance.

As HDAC inhibitors are being considered for the treatment of leukemia/lymphoma and solid tumors (reviewer by [93]), it is important to understand the complexity of their effect, especially on immunity-related cell responses, for the design of successful clinical regimens for cancer patients. Adverse effects in the immune system through the induction of tolerance may compromise the health condition of cancer patients undergoing HDAC inhibitor therapy. To date, the emerging clinical evidence of HDAC inhibitors in phase I and II trials show that HDAC inhibitors are well tolerated with minimal adverse effects in terms of immunology (mild myelosuppression: thrombocytopenia) [84], However, cardiotoxicity, and particularly arrhythmia, may need to be evaluated further in larger populations of treated patients [93]. Thus, the adverse effects of HDAC inhibitors on immunity-related cell responses have not yet been raised.

Some reports have suggested that HDAC inhibitors possess immunosuppressive effects [88, 94], including seemingly discordant results relating to the effects on Tregs [91, 95]. However, the accumulating data support the previously demonstrated findings relating to HDAC inhibitor-induced immune potentiation, such as MS-275 increasing the antitumor activity of high-dose IL-2 against the Renca murine kidney cancer model [95], the effect of the HDAC inhibitor TSA against modified lung cancer cell line TC-1 [96], and the effect of HDAC inhibitor depsipeptide (FK228) against B16 melanoma [50]. Moreover, adoptive transfer therapy strategies using tumor-specific CTLs possess a potential advantage by employing prior exposure to HDAC inhibitors. Murakami et al. [50] showed that the HDAC inhibitor depsipeptide increased the level of perforin in activated T cells to varying degrees, where the number of perforin-expressing CTLs increased in mice, and an accumulation of perforin was observed in humans. Perforin release by T cells in conjunction with granzymes induces an apoptotic cascade in target cells [79]. In fact, Palmer et al. [97] showed a B16 melanoma cell-dependent release of perforin following adoptive cell transfer of pmel-1 gp100-specific CTLs, and caspase-3 activation was also shown at the tumor site as a consequence of the downstream activation of perforin. Therefore, residual HDAC inhibitor in the plasma of pretreated hosts could be expected to result in the release large amounts of toxic granules at the sites of the target tumor. As such, the use of HDAC inhibitors should be considered for host and tumor modulation prior to adoptive transfer of tumor-specific CTLs (Fig. 14.1). Therefore, it is likely that the immunomodulating effects of HDAC inhibitors on Tregs and other immunosuppressive cell subsets are not dominant in the context of powerful tumor immunotherapy strategies.

Conclusions and Perspectives: Future for Cancer Immunotherapy

Based on the many preclinical studies investigating the biology and therapeutic value of HDAC inhibitors, the use of these compounds with other cancer therapies has been examined. Since HDAC inhibitors can preferentially alter the balance of apoptotic cell pathways, the use of HDAC inhibitors has been considered in combination with conventional chemotherapeutic agents in solid tumors [98–101]. Additionally, HDAC inhibitors enhance tumor cell radiosensitivity and are being tested in conjunction with ionizing radiation in solid tumors [102]. These strategies are providing promising data for the use of HDAC inhibitors in cancer immunotherapy.

Although there are currently several HDAC inhibitors available or in clinical development, it may also be necessary to consider the differences in the effects of individual HDAC inhibitors [93]. In this context, generalizations pertaining to HDAC inhibitors-induced responses require caution. Current clinical studies have yet to employ substantial biomarkers to predict responses to HDAC inhibitor treatment [93]. Therefore, in order to maximize the clinical responses to combination therapies that use HDAC inhibitors, it is necessary to determine whether a tumor-specific HDAC iso-enzyme profile can predict a response to individual HDAC inhibitors. The development of these tumor-specific biomarkers to predict HDAC inhibitor responses is a necessity. In the near future, it is expected that preclinical studies using targeted knockdown of individual HDAC iso-enzymes will shed light on any subtle biological differences that may be present [93].

The accumulated evidence in clinical oncology has clearly demonstrated that HDAC inhibitors can alter the balance of cancer cells such that they are more prone to differentiation, growth arrest, and apoptosis. The emerging data supports the view that HDAC inhibitors could sensitize and make tumor cells responsive to immunotherapy, and it is conceivable that successful modification by the use of HDAC inhibitors could result in tumor cells being susceptible to attack by CTLs. Although studies investigating the potential of HDAC inhibitors to improve vaccine-based and adoptive cell therapy-based immunotherapies are limited [50, 51, 95, 96], the use of promising HDAC inhibitors should improve cancer immunotherapy strategies in translational clinical studies.

Acknowledgement I would like to thank Dr. Eiji Kobayashi (Center for Molecular Medicine, Jichi Medical University, Tochigi, Japan) for helpful comments and suggestions.

References

1. Gattinoni L, Powell DJ Jr, Rosenberg SA, et al. (2006) Adoptive immunotherapy for cancer: building on success. Nat Rev Immunol 6: 383–393
2. Rosenberg SA, Yang JC, Restifo NP (2004) Cancer immunotherapy: moving beyond current vaccines. Nat Med 10: 909–915
3. Rosenberg SA, Dudley ME (2004) Cancer regression in patients with metastatic melanoma after the transfer of autologous antitumor lymphocytes. Proc Natl Acad Sci USA 101 Suppl 2: 14,639–14,645
4. Rosenberg SA, Yang JC, Topalian SL, et al. (1994) Treatment of 283 consecutive patients with metastatic melanoma or renal cell cancer using high-dose bolus interleukin-2. J Am Med Assoc 271: 907–913
5. Dudley ME, Wunderlich JR, Robbins PF, et al. (2002) Cancer regression and autoimmunity in patients following clonal repopulation with anti-tumor lymphocytes. Science 298: 850–854
6. Dudley ME, Wunderlich JR, Yang JC, et al. (2005) Adoptive cell transfer therapy following non-myeloablative but lymphodepleting chemotherapy for the treatment of patients with refractory metastatic melanoma. J Clin Oncol 23: 2346–2357
7. Luger K, Mader AW, Richmond RK, et al. (1997) Crystal structure of the nucleosome core particle at 2.8 A resolution. Nature 389: 251–260
8. Jenuwein T, Allis CD (2001) Translating the histone code. Science 293: 1074–1080
9. Tomasi TB, Magner WJ, Khan ANH (2006) Epigenetic regulation of immune escape genes in cancer. Cancer Immunol Immunother 55: 1159–1184
10. Gu W, Roeder RG (1997) Activation of p53 sequence-specific DNA binding by acetylation of the p53 C-terminal domain. Cell 90: 595–606
11. Haggarty SJ, Koeller KM, Wong JC, et al. (2003) Domain-selective small-molecule inhibitor of histone deacetylase 6 (HDAC6)-mediated tubulin deacetylation. Proc Natl Acad Sci USA 100: 4389–4394
12. Kovacs JJ, Murphy PJ, Gaillard S, et al. (2005) HDAC6 regulates Hsp90 acetylation and chaperone-dependent activation of glucocorticoid receptor. Mol Cell 18: 601–607
13. Roth SY, Denu JM, Allis CD (2001) Histone acetyltransferases. Annu Rev Biochem 70: 81–120
14. Bolden JE, Peart MJ, Johnstone RW (2006) Anticancer activities of histone deacetylase inhibitors. Nat Rev Drug Discov 5: 769–784
15. DeRuijten AJM, van Gennip AH, Caron HN, et al. (2003) Histone deacetylases (HDACs): characterization of the classical HDAC family. Biochem J 370: 737–49
16. Marks P, Rifkind RA, Richon VM, et al. (2001) Histone deacetylases and cancer: causes and therapies. Nat Rev Cancer 1: 194–202
17. Witt O, Deubzer HE, Milde T, et al. (2009) HDAC family: what are the cancer relevant targets? Cancer Lett 277: 8–21
18. Vaquero A, Sternglanz R, Reinberg D (2007) NAD+ -dependent deacetylation of H4 lysine 16 by class III HDACs. Oncogene 26: 5505–5520
19. Senese S, Zaragoza K, Minardi S, et al. (2007) Role for histone deacetylase 1 in human tumor cell proliferation. Mol Cell Biol 27: 4784–4795
20. Weichert W, Roske A, Niesporek S, et al. (2008) Class I histone deacetylase expression has independent prognostic impact in human colorectal cancer: specific role of class I histone deacetylases *in vitro* and *in vivo*. Clin Cancer Res 14: 1669–1677
21. Bicaku E, Marchion DC, Schmitt ML, et al. (2008) Selective inhibition of histone deacetylase 2 silences progesterone receptor-mediated signaling. Cancer Res 68: 1513–1519
22. Harms KL, Chen X (2007) Histone deacetylase 2 modulates p53 transcriptional activities through regulation of p53-DNA binding activity. Cancer Res 67: 3145–52
23. Laird PW (2005) Cancer epigenetics. Hum Mol Genet 14: R65–76
24. Choi Y, Elagib KE, Goldfarb AN (2005) AML-1-ETO-Mediated erythroid inhibition: new paradigms for differentiation blockade by a leukemic fusion protein. Crit Rev Eukaryot Gene Expr 15: 207–216

25. Cress WD, Seto E (2000) Histone deacetylases, transcriptional control, and cancer. J Cell Physiol 184: 1–16
26. Crowley JA, Wang Y, Rapoport AP, et al. (2005) Detection of MOZ-CBP fusion in acute myeloid leukemia with 8;16 translocation. Leukemia 19: 2344–2345
27. Linggi BE, Brandt SJ, Sun ZW, et al. (2005) Translating the histone code into leukemia. J Cell Biochem 96: 938–950
28. Murati A, Adelaide J, Mozziconacci MJ, et al. (2004) Variant MYST4-CBP gene fusion in a t(10;16) acute myeloid leukaemia. Br J Haematol 125: 601–604
29. Panagopoulos I, Fioretos T, Isaksson M, et al. (2001) Fusion of the MORF and CBP genes in acute myeloid leukemia with the t(10;16)(q22;p13) Hum Mol Genet 10: 395–404
30. Ropero S, Fraga MF, Ballestar E, et al. (2006) A truncating mutation of HDAC2 in human cancers confers resistance to histone deacetylase inhibition. Nat Genet 38: 566–569
31. Sugita K, Taki T, Hayashi Y, et al. (2000) MLL-CBP fusion transcript in a therapy-related acute myeloid leukemia with the t(11;16)(q23;p13) which developed in an acute lymphoblastic leukemia patient with Fanconi anemia. Genes Chromosom Cancer 27: 264–269
32. Wilson AJ, Byun DS, Popova N, et al. (2006) Histone deacetylase 3 (HDAC3) and other class I HDACs regulate colon cell maturation and p21 expression and are deregulated in human colon cancer. J Biol Chem 281: 13,548–13,558
33. Alarcón JM, Malleret G, Touzani K, Vronskaya S, Ishii S, Kandel ER, Barco A (2004) Chromatin acetylation, memory, and LTP are impaired in CBP+/- mice: a model for the cognitive deficit in Rubinstein-Taybi syndrome and its amelioration. Neuron 42: 947–959
34. Kishimoto M, Kohno T, Okudela K, et al. (2005) Mutations and deletions of the CBP gene in human lung cancer. Clin Cancer Res 11: 512–519
35. Hake SB, Xiao A, Allis CD (2004) Linking the epigenetic language of covalent histone modifications to cancer. Br J Cancer 90: 761–769
36. Klisovic DD, Katz SE, Effron D, et al. (2003) Depsipeptide (FR901228) inhibits proliferation and induces apoptosis in primary and metastatic human uveal melanoma cell lines. Investig Ophthalmol Vis Sci 44: 2390–2398
37. Kobayashi Y, Ohtsuki M, Murakami T, et al. (2006) Histone deacetylase inhibitor FK228 suppresses the Ras-MAP kinase signaling pathway by upregulating Rap1 and induces apoptosis in malignant melanoma. Oncogene 25: 512–524
38. Johnstone RW, Licht JD (2003) Histone deacetylase inhibitors in cancer therapy: is transcription the primary target? Cancer Cell 4: 13–18
39. Minucci S, Pelicci PG (2006) Histone deacetylase inhibitors and the promise of epigenetic (and more) treatments for cancer. Nat Rev Cancer 6: 38–51
40. Crabb SJ, Howell M, Rogers H, et al. (2008) Characterisation of the *in vitro* activity of the depsipeptide histone deacetylase inhibitor spiruchostatin A. Biochem Pharmacol 76: 463–475
41. Peart MJ, Smyth GK, van Laar RK, et al. (2005) Identification and functional significance of genes regulated by structurally different histone deacetylase inhibitors. Proc Natl Acad Sci USA 102: 3697–702
42. Schrump DS (2009) Cytotoxicity mediated by histone deacetylase inhibitors in cancer cells: mechanisms and potential clinical implications. Clin Cancer Res 15: 3947–3957
43. Hoshikawa Y, Kwon HJ, Yoshida M, et al. (1994) Trichostatin A induces morphological changes and gelsolin expression by inhibiting histone deacetylase in human carcinoma cell lines. Exp Cell Res 214: 189–197
44. Ueda H, Manda T, Matsumoto S, et al. (1994a) FR901228, a novel antitumor bicyclic depsipeptide produced by *Chromobacterium violaceum* No. 968. III. Antitumor activities on experimental tumors in mice. J Antibiot (Tokyo) 47: 315–323
45. Ueda H, Nakajima H, Hori Y, et al. (1994b) Action of FR901228, a novel antitumor bicyclic depsipeptide produced by *Chromobacterium violaceum* no. 968, on Ha-ras transformed NIH3T3 cells. Biosci Biotechnol Biochem 58: 1579–1583
46. Schrump DS, Fischette MR, Nguyen DM, et al. (2008) Clinical and molecular responses in lung cancer patients receiving Romidepsin. Clin Cancer Res 14: 188–198
47. Davies H, Bibnell GR, Cox C, et al. (2002) Mutations of the BRAF gene in human cancer. Nature 417: 949–954

48. Madhunapantula SV, Robertson GP (2008) Is B-Raf a good therapeutic target for melanoma and other malignancies? Cancer Res 68: 5–8
49. Chin L, Garraway LA, Fisher DE (2006) Malignant melanoma: genetics and therapeutics in the genomic era. Genes Dev 20: 2149–2182
50. Murakami T, Sato A, Chun NAL, et al. (2008) Transcriptional modulation using HDACi depsipeptide promotes immune cell-mediated tumor destruction of murine B16 melanoma. J Invest Dermatol 28: 1506–1516
51. Vo DD, Prins RM, Begley JL, et al. (2009) Enhanced antitumor activity induced by adoptive T-cell transfer and adjunctive use of the histone deacetylase inhibitor LAQ824. Cancer Res 69: 8693–8699
52. Dudley ME, Wunderlich J, Nishimura MI, et al. (2001) Adoptive transfer of cloned melanoma-reactive T lymphocytes for the treatment of patients with metastatic melanoma. J Immunother 24: 363–373
53. Antony PA, Restifo NP (2005) $CD4^+CD25^+$ T regulatory cells, immunotherapy of cancer, and interleukin-2. J Immunother 28: 120–128
54. Dannull J, Su Z, Rizzieri D, et al. (2005) Enhancement of vaccine-mediated antitumor immunity in cancer patients after depletion of regulatory T cells. J Clin Invest 115: 3623–3633
55. Cabrera T, Lopez-Nevot MA, Gaforio JJ, et al. (2003) Analysis of HLA expression in human tumor tissues. Cancer Immunol Immunother 52: 1–9
56. Ferrone S, Marincola FM (1995) Loss of HLA class I antigens by melanoma cells: molecular mechanisms, functional significance and clinical relevance. Immunol Today 16: 487–494
57. Johnstone RW, Ruefli AA, Lowe SW (2002) Apoptosis: a link between cancer genetics and chemotherapy. Cell 108: 153–164
58. Maecker HL, Yun Z, Maecker HT, et al. (2002) Epigenetic changes in tumor Fas levels determine immune escape and response to therapy. Cancer Cell 2: 139–148
59. Khan AN, Gregorie CJ, Tomasi TB (2008) Histone deacetylase inhibitors induce TAP, LMP, Tapasin genes and MHC class I antigen presentation by melanoma cells. Cancer Immunol Immunother 57: 647–654
60. Williams A, Peh CA, Elliott T (2002) The cell biology of MHC class I antigen presentation. Tissue Antigens 59: 3–17
61. Chang CC, Ferrone S (2007) Immune selective pressure and HLA class I antigen defects in malignant lesions. Cancer Immunol Immunother 56: 227–236
62. Seliger B, Maeurer MJ, Ferrone S (2000) Antigen-processing machinery breakdown and tumor growth. Immunol Today 21: 455–464
63. Byrd JC, Marcucci G, Parthun MR, et al. (2005) A phase 1 and pharmacodynamic study of depsipeptide (FK228) in chronic lymphocytic leukemia and acute myeloid leukemia. Blood 105: 959–967
64. Piekarz RL, Robey R, Sandor V, et al. (2001) Inhibitor of histone deacetylation, depsipeptide (FR901228), in the treatment of peripheral and cutaneous T-cell lymphoma: a case report. Blood 98: 2865–2868
65. Piekarz RL, Robey RW, Zhan Z, et al. (2004) T-cell lymphoma as a model for the use of histone deacetylase inhibitors in cancer therapy: impact of depsipeptide on molecular markers, therapeutic targets, and mechanisms of resistance. Blood 103: 4636–4643
66. Sandor V, Bakke S, Robey RW, et al. (2002) Phase I trial of the histone deacetylase inhibitor, depsipeptide (FR901228, NSC 630176), in patients with refractory neoplasms. Clin Cancer Res 8: 718–728
67. Stadler WM, Margolin K, Ferber S, et al. (2006) A phase II study of depsipeptide in refractory metastatic renal cell cancer. Clin Genitourin Cancer 5: 57–60
68. Earel JK Jr, VanOosten RL, Griffith TS (2006) Histone deacetylase inhibitors modulate the sensitivity of tumor necrosis factor-related apoptosis-inducing ligand-resistant bladder tumor cells. Cancer Res 66: 499–507
69. Frew AJ, Lindemann RK, Martin BP, et al. (2008) Combination therapy of established cancer using a histone deacetylase inhibitor and a TRAIL receptor agonist. Proc Natl Acad Sci USA 105: 11,317–11,322

70. Nakata S, Yoshida T, Horinaka M, et al. (2004) Histone deacetylase inhibitors upregulate death receptor 5/TRAIL-R2 and sensitize apoptosis induced by TRAIL/APO2-L in human malignant tumor cells. Oncogene 23: 6261–6271
71. Singh TR, Shankar S, Srivastava RK (2005) HDAC inhibitors enhance the apoptosis-inducing potential of TRAIL in breast carcinoma. Oncogene 24: 4609–4623
72. Avent J, Vervaert C, Seigler HF (1979) Non-specific and specific active immunotherapy in a B16 murine melanoma system. J Surg Oncol 12: 87–96
73. Kalechman Y, Strassmann G, Albeck M, et al. (1998) Up-regulation by ammonium trichloro(dioxoethylene-0,0') tellurate (AS101) of Fas/Apo-1 expression on B16 melanoma cells: implications for the antitumor effects of AS101. J. Immunol 161: 3536–3542
74. Tsai V, Southwood S, Sidney J, et al. (1997) Identification of subdominant CTL epitopes of the GP100 melanoma-associated tumor antigen by primary *in vitro* immunization with peptide-pulsed dendritic cells. J Immunol 158: 1796–1802
75. Inoue S, MacFarlane M, Harper N, et al. (2004) Histone deacetylase inhibitors potentiate TNF-related apoptosis-inducing ligand (TRAIL)-induced apoptosis in lymphoid malignancies. Cell Death Differ 11 Suppl 2: S193–206
76. Lundqvist A, Abrams SI, Schrump DS, et al. (2006) Bortezomib and depsipeptide sensitize tumors to tumor necrosis factor-related apoptosis-inducing ligand: a novel method to potentiate natural killer cell tumor cytotoxicity. Cancer Res 66: 7317–7325
77. Häcker S, Dittrich A, Mohr A, et al. (2009) Histone deacetylase inhibitors cooperate with IFN-gamma to restore caspase-8 expression and overcome TRAIL resistance in cancers with silencing of caspase-8. Oncogene 28: 3097–3110
78. Caldwell SA, Ryan MH, McDuffie E, et al. (2003) The Fas/Fas ligand pathway is important for optimal tumor regression in a mouse model of CTL adoptive immunotherapy of experimental CMS4 lung metastases. J Immunol 171: 2402–2412
79. Kagi D, Vignaux F, Ledermann B, et al. (1994) Fas and perforin pathways as major mechanisms of T cell-mediated cytotoxicity. Science 265: 528–530
80. Lee CH, Kakinuma T, Wang J, et al. (2006) Sensitization of B16 tumor cells with a CXCR4 antagonist increases the efficacy of immunotherapy for established lung metastases. Mol Cancer Ther 5: 2592–2599
81. Ghiringhelli F, Larmonier N, Schmitt E, et al. (2004) $CD4^{+}CD25^{+}$ regulatory T cells suppress tumor immunity but are sensitive to cyclophosphamide which allows immunotherapy of established tumors to be curative. Eur J Immunol 34: 336–344
82. Lutsiak ME, Semnani RT, De Pascalis R, et al. (2005) Inhibition of CD4(+)25(+) T regulatory cell function implicated in enhanced immune response by low-dose cyclophosphamide. Blood 105: 2862–2868
83. Rosenberg SA, Restifo NP, Yang JC, et al. (2008) Adoptive cell transfer: a clinical path to effective cancer immunotherapy. Nat Rev Cancer 8: 299–308
84. Mann BS, Johnson JR, Cohen MH, et al. (2007) FDA approval summary: vorinostat for treatment of advanced primary cutaneous T-cell lymphoma. Oncologist 12: 1247–1252.
85. Leoni F, Zaliani A, Bertolini G, et al. (2002) The antitumor histone deacetylase inhibitor suberoylanilide hydroxamic acid exhibits antiinflammatory properties via suppression of cytokines. Proc. Natl. Acad. Sci. USA 99: 2995–3000
86. Mishra N, Reilly CM, Brown DR, et al. (2003) Histone deacetylase inhibitors modulate renal disease in the MRL-lpr/lpr mouse. J Clin Invest 111: 539–552
87. Reddy P, Maeda Y, Hotary K, et al. (2004) Histone deacetylase inhibitor suberoylanilide hydroxamic acid reduces acute graft-versus-host disease and preserves graft-versus-leukemia effect. Proc. Natl. Acad. Sci. USA 101: 3921–3926
88. Reilly CM, Mishra N, Miller JM, et al. (2004) Modulation of renal disease in MRL/lpr mice by suberoylanilide hydroxamic acid. J Immunol 173 4171–4178
89. Skov S, Rieneck K, Bovin LF, et al. (2003) Histone deacetylase inhibitors: a new class of immunosuppressors targeting a novel signal pathway essential for CD154 expression. Blood 101: 1430–1438

90. Reddy P, Sun Y, Toubai T, et al. (2008) Histone deacetylase inhibition modulates indoleamine 2,3-dioxygenase-dependent DC functions and regulates experimental graft-versus-host disease in mice. J Clin Invest 118: 2562–2573
91. Tao R, de Zoeten EF, Ozkaynak E, et al. (2007) Deacetylase inhibition promotes the generation and function of regulatory T cells. Nat Med 13: 1299–1307
92. Villagra A, Cheng F, Wang HW, et al. (2009) The histone deacetylase HDAC11 regulates the expression of interleukin 10 and immune tolerance. Nat Immunol 10: 92–100
93. Lane AA, Cabner BA (2009) Histone deacetylase inhibitors in cancer therapy. J Clin Oncol 27: 5459–5468
94. Brogdon JL, Xu Y, Szabo SJ, et al. (2007) Histone deacetylase activities are required for innate immune cell control of Th1 but not Th2 effector cell function. Blood 109: 1123–1130
95. Kato Y, Yoshimura K, Shin T, et al. (2007) Synergistic *in vivo* antitumor effect of the histone deacetylase inhibitor MS-275 in combination with interleukin 2 in a murine model of renal cell carcinoma. Clin Cancer Res 13: 4538–4546
96. Setiadi AF, Omilusik K, David MD, et al. (2008) Epigenetic enhancement of antigen processing and presentation promotes immune recognition of tumors. Cancer Res 68: 9601–9607
97. Palmer DC, Balasubramaniam S, Hanada K, et al. (2004) Vaccine-stimulated, adoptively transferred $CD8^+$ T cells traffic indiscriminately and ubiquitously while mediating specific tumor destruction. J Immunol 173: 7209–7216
98. Bishton M, Kenealy M, Johnstone R, et al. (2007) Epigenetic targets in hematological malignancies: combination therapies with HDAC is and demethylating agents. Expert Rev Anticancer Ther 7: 1439–1449
99. Kuendgen A, Lubbert M (2008) Current status of epigenetic treatment in myelodysplastic syndromes. Ann Hematol 87: 601–611
100. Marchion D, Munster P (2007) Development of histone deacetylase inhibitors for cancer treatment. Expert Rev Anticancer Ther 7: 583–598
101. Rasheed W, Bishton M, Johnstone RW, et al. (2008) Histone deacetylase inhibitors in lymphoma and solid malignancies. Expert Rev Anticancer Ther 8: 413–432
102. Camphausen K, Tofilon PJ (2007) Inhibition of histone deacetylation: a strategy for tumor radiosensitization. J Clin Oncol 25: 4051–4056

Chapter 15
Combining Cancer Vaccines with Conventional Therapies

Natalie Grinshtein and Jonathan Bramson

Abstract Cancer vaccines offer the promise of a noninvasive method to attack tumors using the patient's immune system. Although such strategies have shown robust activity in preclinical models, the results from clinical trials have been modest. Although the exact reason for the lack of clinical success remains to be determined, it is clear that the immunosuppressive environment of large tumors represents a hindrance to vaccine outcome. Therefore, it is reasonable to expect that combining vaccination with conventional debulking therapies should lead to an improved outcome. In this chapter, we discuss these issues and results from recent experiments demonstrating the combining vaccination with surgical resection or chemotherapy can greatly enhance the activity of cancer vaccines in situations of large tumor burden.

Keywords Cancer vaccine • Chemotherapy • Immune suppression • Surgery • T lymphocyte

Introduction

The identification of tumor-associated antigens (TAAs) has provided a platform for the development of cancer vaccines with the promise of eradicating antigen-positive tumor cells. In murine models, tumor vaccines can routinely provide complete protection from tumor challenge when employed in a prophylactic setting, however they exhibit reduced efficacy when applied in a therapeutic setting [1–3]. A 2004 review from investigators at the National Cancer Institute, NIH, summarizing their experience with 440 patients who received a variety of vaccines concluded that vaccines on themselves do not work as there were only four complete and nine partial responses observed, resulting in an overall response rate of ~3% [4].

N. Grinshtein (✉)
Department of Pathology and Molecular Medicine, McMaster University, Hamilton, ON, Canada
e-mail: ngrinshtein@hotmail.com

J. Medin and D. Fowler (eds.), *Experimental and Applied Immunotherapy*,
DOI 10.1007/978-1-60761-980-2_15, © Springer Science+Business Media, LLC 2011

Other reviews have been more hopeful and suggest the actual response rates may be as high as 10% [5]. Nevertheless, there is a general agreement that cancer vaccines have not performed as expected. These outcomes are not entirely surprising, however, as most clinical studies have been carried out in patients with advanced disease who are likely experiencing varying degrees of immune dysfunction.

T-cell-mediated tumor rejection is a complex process comprised of multiple steps. Tumor-specific T cells must first be activated to expand and differentiate into effector cells which then traffic to the tumor site and execute their effector function. Perturbation of any of these steps will result in a defective T cell response against the tumor. Interestingly, TAA-specific T cells can be detected in both healthy and tumor-bearing subjects prior to vaccination, albeit at low frequencies [6–9]. TAA-specific T cells can be expanded and activated by multiple rounds of vaccination to reach frequencies of 1–10% of the total $CD8^+$ T-cell pool [7, 8, 10–12]. The modest objective clinical response rates observed in vaccine trials [4, 5, 13] indicates that the mere presence of circulating T cells reactive against specific TAAs is insufficient to induce tumor regression [7, 10], suggesting a seemingly paradoxical coexistence of the antitumor T cells with an actively growing tumor. This apparent paradox can be explained simply enough: despite significant expansion of TAA-specific T cells following vaccination, the T cells still do not achieve a frequency or effector function sufficient to clear an established tumor. In support of this hypothesis, it has been demonstrated that the intensity of the immune response directly correlates with tumor clearance in a murine model [3, 14]. Moreover, adoptive transfer studies have demonstrated that infusing massive numbers of *ex vivo*-expanded autologous lymphocytes ($\sim 10^{11}$) into lymphopenic patients with late-stage melanoma results in 50–70% objective clinical responses [15]. However, these studies reflect a highly selected group of individuals because not all tumor-infiltrating lymphocyte (TIL) cultures achieve sufficient numbers to be used in adoptive transfer. Therefore, the T cells present in cultures that can be expanded to sufficiently high levels may have some inherent properties that make them better antitumor effectors. Furthermore, despite the high numbers of circulating TAA-specific T cells achieved using intensive preconditioning followed by adoptive transfer of expanded TILs, a large fraction of patients (30–50%) fails to respond to this line of therapy. Thus, factors other than the mere quantity of T-cells must influence tumor rejection. In that regard, two of the most likely scenarios for the limited efficacy of vaccination in a therapeutic setting are: (1) impaired effector T-cell function due to tumor-associated immunosuppression and (2) alterations within the tumor phenotype due to immune selection.

Immune Evasion

Tumor-Induced Immunosuppression

The phenomenon of local tumor-induced suppression and its impact on the phenotype and function of TILs has been extensively investigated. Importantly, with the exception of one report suggesting that TILs may not be functionally impaired [16], numerous groups have demonstrated that TILs are defective in their activation status [10, 17], cytolytic function [3, 10, 18–20], effector cytokine production [3, 9, 10, 21], and tetramer binding [18]. These defects in the TILs may also extend to circulating T cells. Murine studies have demonstrated that systemic tumor-specific $CD8^+$ T cells become activated at the early stages of tumor initiation; however, these cells progressively lose their cytolytic function at the late stage of tumor growth, suggesting specific impairment in antitumor T cells in the periphery [22, 23]. Furthermore, several reports have documented that TAA-specific circulating T cells isolated from tumor-bearing hosts become anergic early on during the course of tumor progression and exhibit a quiescent phenotype that cannot be rescued by vaccination [8, 24, 25].

Several clinical studies have demonstrated that vaccine-elicited circulating $CD8^+$ T cells are intact in terms of their phenotype and function, as they exhibit a characteristic effector/effector memory phenotype [8, 10, 12] and produce IFN-γ on restimulation *in vitro* [26, 27]. Thus, it is indeed possible to elicit robust antitumor immunity as evidenced by the circulating populations. However, the existing data suggest that the cells in the circulation may not adequately reflect the populations in the tumor. An examination of prostate cancer samples from nonvaccinated patients [17] revealed that T cells present in human prostatic carcinomas failed to up-regulate activation markers following polyclonal stimulation, whereas T cells within normal prostates were suitably responsive. Likewise, analysis of T cells within the peripheral blood and metastatic lesions from stage III/IV melanoma patients [9] demonstrated that circulating $CD8^+$ T cells responded appropriately to MART-1 peptide *in vitro* whereas MART-1 specific TILs exhibited a defective response to peptide stimulation. Analogous observations were made by analyzing samples from two melanoma patients who were immunized with a MART-1 peptide [10]: MART-1 specific TILs expressed lower levels of perforin and granzyme and failed to secret IFN-γ on restimulation *in vitro*, as compared to MART-1 $CD8^+$ T cells isolated from the peripheral blood. This dysfunctional phenotype was not a general effect of the tumor environment as restimulation of $CD8^+$ T cells with cytomegalovirus antigen resulted in comparable IFN-γ secretion from all the tissues, suggesting that the defect in cytokine production was antigen-specific. Similarly, we have observed in a murine melanoma model that vaccine-induced TILs specific for a tumor antigen exhibited impaired cytokine production whereas TILs specific for nontumor-associated antigens showed no defect in cytokine production [3].

Based on these data, it seems that various impairments in T-cell phenotype/function are observed within the TIL population that are not manifested within the circulating population of tumor-specific T cells. These observations complicate the interpretation of cancer vaccine studies that are typically dependent on peripheral blood sampling for measurement of tumor-specific immunity.

The mechanisms by which tumors exert their immune suppressive effects are multi-fold. Both tumor intrinsic effects have been described, such as production of regulatory cytokines (TGF-β and IL-10), elaboration of catabolic enzymes that deplete essential amino acids (arginase, indoleamine 2,3-dioxygenase) and inhibition of DC differentiation and maturation; in addition, tumor extrinsic effects have been identified, such as recruitment of suppressor cells (regulatory T cells and/or myeloid-derived suppressor cells) [28–32]. Although blockade of each of these pathways individually has shown beneficial impact on vaccination outcomes in specific tumor models, it remains unclear whether it will be possible to reverse the immunosuppressive environment of large, established tumors in the clinic.

Selection of Tumor-Escape Variants Following Vaccination

Spontaneous cancer cell variants arise in any large tumor due to unique mutations within their genomes. These variants may be resistant to immune-mediated rejection and, thusly, represent a significant theoretical concern for successful immune intervention [33]. Analysis of tumor biopsies from patients with metastatic melanoma has revealed that expression of both melanoma-associated antigens and MHC class I molecules can be down-regulated following immunotherapy or because of tumor progression [34–37]. Interestingly, these changes appear to result from reversible epigenetic events rather than from mutations. As such, expression of antigens, MHC molecules, and the antigen processing machinery can be recovered by treatment with epigenetic modifiers such as demethylating agents and histone deacetylase inhibitors [37, 38].

Studies have demonstrated that the potency of the CTL response following vaccination is inversely correlated with antigen expression, which suggests immune selection of antigen-loss variants *in vivo* [35, 37]. In that regard, Ohnmacht et al. analyzed antigen expression in metastases isolated from 52 melanoma patients before and after immunization with a modified gp100 peptide vaccine [39]. A decrease in gp100 expression was observed in metastases that regressed following immunization, but no change was detected in lesions that did not regress. These observations indicate that antigen loss was associated with effective immune attack; and, tumors are clearly able to evade immune attack without down-regulation of antigen expression. When targeting highly expressed antigens, tumor escape variants appear to be less of a concern because stromal cells can take up antigen and cross present to infiltrating CTL, thereby resulting in tumor regression due to stromal destruction [40]. Van Hall and colleagues have also identified a specific subset of CTLs capable of reacting against “transporter associated with antigen processing” (TAP)-deficient tumor cells, thus preventing them from escaping immune recognition [41]. Based on these results, we conclude that tumor escape variants do represent a concern but cannot explain the lack of clinical efficacy observed in vaccine trials.

Combining Surgical Resection and Vaccination

A central feature of all studies investigating tumor-associated immune suppression relates to the relationship between tumor burden and immune suppression. As the tumor grows larger, so do the defects in immune function. Indeed, we have observed that vaccine-induced TILs show defective cytokine production and cytolytic properties, which are directly related to tumor size; that is, TILs in smaller tumors displayed the least dysfunctional phenotype [3]. Therefore, a simple strategy to improve the outcomes of cancer vaccination would be to combine vaccination with tumor debulking therapy. Indeed, the simplest scenario would involve the use of adjuvant vaccination following surgical resection of primary tumors. Under these conditions, patients would not have been exposed to multiple rounds of chemotherapy that can deplete host immune effectors nor would they be experiencing significant generalized immune suppression that is associated with late-stage disease. The vaccine would be applied primarily to clear residual tumor within the resection site and any metastatic cells. In support of such as strategy, surgical debulking has been shown to reverse tumor-induced suppression and restore both humoral and cellular responses [42–44].

Several murine studies have assessed the value of immunization following surgical resection. Immunization with plasmid DNA encoding a TAA following resection of a primary B16F10 melanoma yielded a reduction in the number of lung metastases that formed [45]. Using a similar melanoma model, Gabri et al. demonstrated that immunization with GM3 ganglioside could provide complete protection against tumor relapse following surgery [44]. Likewise, Ohashi and colleagues observed that adjuvant immunization with autologous whole cell vaccines improved survival in a murine neuroblastoma model [46]. In contrast, we observed that immunization with a recombinant adenovirus vaccine failed to improve disease-free survival following surgery in the B16F10 melanoma model [47] even though this vaccine can provide complete protection against high doses of tumor cells when used in a prophylactic setting. Therefore, although adjuvant immunization has shown some promise, there also appear to be some limitations.

The results of clinical studies investigating adjuvant immunization have been mixed. Several small-scale studies in melanoma, renal cancer, colorectal cancer and glioma have reported positive clinical responses following vaccination [48–51]. However, the few large-scale, randomized studies that have been conducted have yielded more results that are disappointing. For example, a trial comparing vaccination with GM2 ganglioside (an over-expressed shared TAA) to treatment with high-dose interferon-α2b in 774 patients with resected stage IIb-III melanoma was halted early because the preliminary results were strongly in favor of interferon-α2b treatment [52]. Furthermore, two large randomized trials were conducted investigating the utility of adjuvant treatment with an allogeneic melanoma vaccine, Melacine, with and without interferon-α2b treatment [53, 54]. More than 600 patients were included in each treatment arm and no statistical benefit was observed for the vaccinated arms. Recently, two randomized trials compared the efficacy of whole cell melanoma vaccine CanVaxin administered in conjunction with Bacillus Calmette-Guerin (BCG) to that of BCG alone in patients with resected stage III and stage IV melanoma [55]. Both of those trials were also halted early because of lack of vaccine efficacy. Therefore, these clinical data suggests that vaccination has a limited efficacy in an adjuvant setting; however, it is equally possible that the vaccines used in those studies were simply inadequate and unable to elicit a response suitable for tumor rejection.

Despite these apparent negative outcomes, subset analysis has suggested that some patients did benefit from the adjuvant vaccination. As an example, a subset analysis revealed that, in fact, adjuvant immunization with the Melacine vaccine augmented the overall survival of patients expressing either HLA-A2 or HLA-C3 [56]. Thus, it appears that traditional methods, which have been employed for measuring outcomes of clinical trials, need to be reconsidered in the context of immunotherapies. The complex nature of the immune system and the multi-factorial process that underlies immune-mediated tumor rejection necessitates intensive immune monitoring to correctly interpret the outcome of clinical trials. To properly correlate the secondary endpoints with clinical outcomes, it will be necessary to consider the immunogenetics of the host, which will include HLA typing and, possibly, examination of functional polymorphisms within immune ligands and receptors [57].

Neoadjuvant Immunization May Prove More Effective than Adjuvant Immunization

As stated above, we observed that adjuvant immunization with a recombinant adenovirus vaccine for melanoma was ineffective in a murine model even when administered as little as 1 day following tumor resection [47]. These results could not be explained by rapid tumor regrowth because the relapsed tumors did not appear until 3–4 weeks following resection, which is ample time for the development of an immune response. In consideration of the data discussed in the previous section showing that circulating tumor-specific cells appear to function normally, we reasoned that immunization prior to surgery (termed *neo-adjuvant immunization*) would provide a source of functional, tumor-specific T cells at the time of resection that could be available to accomplish immediate rejection of residual tumor cells. Strikingly, this immunization strategy provided almost complete protection against tumor relapse. Immunological investigations revealed that neo-adjuvant immunization was associated with the presence of tumor-reactive $CD8^+$ T cells circulating throughout the animal (in the tumor, skin, and the draining lymph nodes). Thus, we speculate that the increased efficacy of vaccination in the neo-adjuvant setting relates to the availability of high frequencies of tumor-reactive $CD8^+$ T cells on the day of resection. In the murine models of adjuvant vaccination, the immune response does not reach a peak until 7–10 days following resection, which likely gives the tumor cells sufficient time to grow such that they can evade immunity.

It is interesting to note that neo-adjuvant vaccination appeared to prevent tumors from metastasizing to the draining lymph nodes in the mice receiving neo-adjuvant immunization (Fig. 15.1). Several reports have described marked immune suppression within sentinel lymph nodes associated with tumor metastases [58, 59]. Thus, the improved effectiveness of the neo-adjuvant immunization may relate to the ability

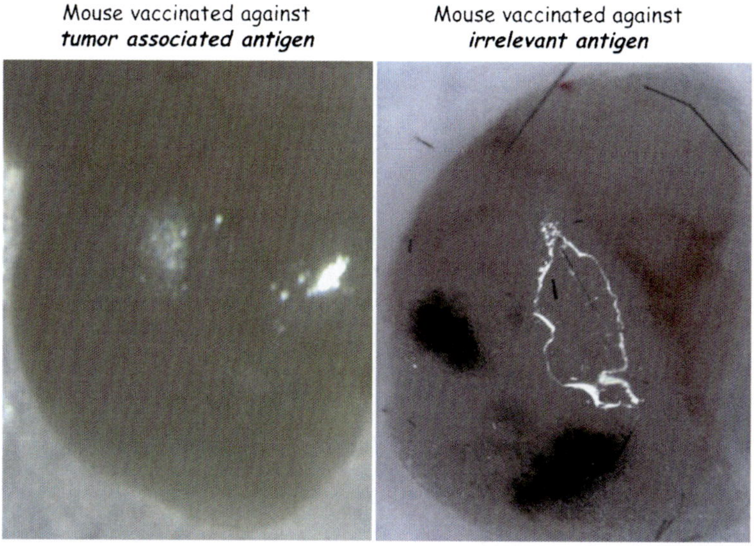

Fig. 15.1 C57Bl/6 mice bearing B16F10 tumors were immunized with either a recombinant adenovirus expressing DCT, a tumor-associated antigen (*left panel*) or a recombinant adenovirus expressing LCMV GP, a viral glycoprotein (*right panel*), as described in Grinshtein et al. [47]. Nine days later, the draining lymph nodes were removed. Deposits of melanoma cells are easily seen in the tumor draining lymph nodes from mice immunized against LCMV GP (*right panel* – dark patches) whereas no melanoma deposits were observed in the tumor draining lymph nodes from mice immunized against DCT (*left panel*)

of the vaccine-induced immune response to clear metastatic deposits before they get too large. Whether the same principle will prove to be true concerning human tumors remains to be determined. Neo-adjuvant immunization has been evaluated to date in small-scale clinical trials [60, 61]. Although these reports demonstrated evidence of vaccine immunogenicity, they were not sufficiently powered to make conclusions regarding efficacy.

Cytotoxic Therapies and Antitumor Vaccination

Although vaccination in combination with surgical resection has shown promise in preclinical models and some clinical studies, it is almost certain that physicians will continue to prescribe adjuvant chemotherapy as a first-line defense due to the proven efficacy of this strategy for treating numerous malignancies. Therefore, vaccines will likely be provided in combination with chemotherapy, so it is necessary to consider the impact of combination chemotherapy on vaccine outcomes. Adjuvant chemotherapy offers the added benefit of further reducing tumor burden; however, chemotherapy agents represent a possible drawback due to elimination of T and B-lymphocytes as well as APCs [62, 63]. Therefore, it is necessary to consider carefully which chemotherapies are being used in the adjuvant setting to make best use of the combination treatments.

Lymphopenia-Induced Homeostatic T-Cell Proliferation Enhances Autoimmunity

As stated above, the lymphopenia associated with some cytotoxic cancer agents is a theoretical concern for vaccination. However, preclinical studies have revealed that recovery from lymphopenia may actually enhance antitumor immunity. When the frequency of lymphocytes falls below a given threshold, naïve T cells begin to proliferate through a process termed "homeostatic expansion." This process is mediated by low-affinity interactions with self-peptides presented by self-MHC molecules and cytokines like IL-7 [64] but is independent of costimulatory signals provided through CD28, 4-1BB, or CD40 [65, 66]. Strikingly, naïve T cells undergoing homeostatic proliferation up-regulate cell surface molecules such as CD44, CD122 and Ly6C, which are typically associated with antigen-experienced cells. These "memory-like" T cells were found to be able to rapidly produce IFN-γ and exhibit cytolytic functions following antigen encounters [67–69]. Consistent with the notion that naïve T cells undergoing homeostatic expansion acquire such memory functions, microarray analyses revealed that the transcriptional signature of naive $CD8^+$ T cells in a lymphopenic environment closely resembled the signature of antigen-experienced memory cells, but not effector cells [70].

Naïve T cells, which convert to effector cells under lymphopenic conditions, seem to be beneficial to the host as they are highly functional and can protect against infections [71, 72]. Because homeostatic expansion is driven by interactions with self-peptides, T cells can potentially become auto reactive. This theory is supported by the notion that a variety of autoimmune diseases in humans, including rheumatoid arthritis, type I diabetes, and systemic lupus erythematosus are associated with lymphopenia [73]. In a similar vein, recipients of autologous BMT are prone to development of autoimmunity [74]. Because effective antitumor immunity may require activation of autoimmune T cells, it is possible that the lymphopenic environment will, in fact, provide a more favorable milieu for T cells to become activated against weakly immunogenic shared TAAs [75]. Indeed, several groups have demonstrated that adoptive transfer of minimal numbers of T cells into lymphopenic hosts can promote antitumor immunity [76, 77]. Therefore, it seems like an intriguing opportunity to exploit the period of reconstitution following chemotherapy for augmenting cancer immunotherapy.

Chemotherapeutic Agents Can Enhance Vaccination

The DNA alkylating agent cyclophosphamide (CTX), is a strong candidate for combination with vaccination as it is widely used for the treatment of both hematological and solid malignancies and has long been associated with immunomodulatory activity [78]. A number of mechanisms have been proposed for the immunostimulatory effects of CTX, including: selective elimination of regulatory T cells [79], disruption of tumor-stromal cell interactions [80], induction of bystander effects on T cells [81], stimulation of type I interferon [82], and initiation of a "cytokine storm" [83].

A number of groups have investigated the feasibility of combining CTX with various tumor vaccines [84–89]. In general, administration of CTX prior to vaccination was associated with an improved outcome whereas administration of CTX after vaccination impaired the immune response, consistent with the notion that CTX is toxic to effector cells. Similarly, we have observed that administration of CTX followed by immunization with a recombinant adenovirus increased long-term survival in a murine melanoma model [3]. These beneficial effects of CTX in our experience were associated with an increase in the frequency of tumor-specific T cells rather than a decrease in circulating Tregs.

Although CTX may function well as a treatment prior to vaccination in the adjuvant setting, administering the drug after immunization could prove problematic as CTX exhibits selective toxicity toward effector T cells [90]. This issue becomes a concern in the neo-adjuvant setting where immunization would likely precede adjuvant chemotherapy. An interesting exception is a report from Song et al. who administered CTX 3 days after vaccination with a recombinant vaccinia virus. Using this approach, they observed significantly elevated frequencies of antigen-specific $CD8^+$ T cells and complete tumor regression [85]. It should be noted that Song et al. employed a single low-dose of CTX (50 mg/kg) whereas other groups have used multiple higher-doses. Thus, it is likely that optimal timing of CTX treatment (before or after vaccination) will be dependent on whether the desired effect relates to an antitumor effect, Treg cell depletion, or the development of lymphopenia.

The taxane chemotherapy class (paclitaxel and docetaxel) represent another group of commonly used agents that could be used in combination with vaccinations. Similar to CTX, these agents invoke immunostimulatory consequences and it has been suggested that their antitumor activity may be a result of combined cytotoxic and immunologic effects [91]. Although the immune stimulating properties of the taxanes have not been investigated to the same extent as CTX, reports have shown that treatment with taxanes can augment various aspects of cellular immunity, including LAK activity, NK function, and lymphocyte proliferation [92, 93]. In one study, depressed cellular functions in breast cancer patients recovered to levels comparable to age-matched controls following six cycles of taxane therapy [92]. Taxane treatment is only associated with mild lymphopenia and, as such, would be expected to be compatible with vaccination strategies [94, 95]. Indeed, recent reports from murine and human studies demonstrated that concomitant treatment with docetaxel did not impair cancer vaccination [96, 97] and could actually improve therapeutic outcomes [84, 98].

Chemotherapeutic Agents Can Augment Tumor Immunogenicity

The previously described data provide good examples of the compatibility of chemotherapeutic agents and vaccines. Although we focused on cyclophosphamide and taxanes, other agents, such as the anthracyclines and platinum drugs, have also been proven compatible with vaccines when applied under defined conditions [99]. It should be noted that the beneficial impact of chemotherapeutic agents is not limited to their effects on T lymphocytes. That is, it has also been reported that chemotherapy can directly improve the immunogenicity of tumors. A number of molecules released by dying tumor cells can lead to antigen delivery and presentation by APCs including: proteins bound to calreticulin [100], proteins bound to heat-shock proteins [101], and HMGB1 [102]. Strikingly, in at least one murine model, the antitumor activity of mitoxantrone *in vivo* was dependent on calreticulin whereas *in vitro* sensitivity was calreticulin-independent [100, 103]. Studies in T-cell-deficient mice confirmed that the *in vivo* antitumor activity of mitoxantrone in this model was T-cell-dependent; thereby demonstrating that, in some cases, the activity of chemotherapy *in vivo* is actually dependent on the induction of antitumor immunity. Although this bodes well, in general, for efforts to combine vaccination with chemotherapy, it should be noted that not all chemotherapeutic agents were found to produce an "immunogenic" cell death [100]. Therefore, further investigation of this intriguing property of chemotherapeutic agents is required. Nonetheless, these combined data support the concept that implementation of cancer vaccine strategies may actually enhance the outcome of chemotherapy.

Conclusions

The data described in this chapter strongly support carefully designed studies that combine conventional treatments, such as surgery and chemotherapy, with vaccination. The beneficial effects of the conventional treatments stem from their ability to reduce tumor burden, augment the ability to elicit self-reactive T cells, and increase the immunogenicity of the tumor. Such combinatorial approaches need to be carefully considered. Should the combined treatment prove to be markedly robust, the resultant autoimmune sequellae could prove to be as unpleasant as the original disease. Indeed, this has proven to be the case with *allo*-transplantation where the graft-versus-leukemia effect can be curative but graft-versus-host disease (GVHD) can be debilitating. Nevertheless, the severity of GVHD can be managed to minimize patient discomfort. Indeed, a manageable autoimmune syndrome may be a reasonable outcome of curative cancer immunotherapy. Interestingly, we have observed that, in murine models, autoimmune sequellae are not a requisite outcome of an effective antitumor response. Rather, it appears that autoimmune pathology may be dependent on inflammatory stimulus within the target organ [104]. Furthermore, we have discovered that biasing the immune response enables the generation of antitumor immunity that does not cause autoimmune pathology even in the presence of overt inflammation within the target tissue [105]. Thus, the development of vaccination strategies that circumvent the risks of autoimmunity while maintaining antitumor immunity would mitigate the impact of immune enhancement by combination with chemotherapy and represents a realistic goal.

Designing the optimal vaccine strategy will likely prove to be patient specific, as treatment and disease history is typically a unique feature of each individual. A key concern that has been suggested, although not thoroughly discussed in this chapter, is the relevance of the vaccine to the tumor. In most cases, protein expression within the tumor tissue is deemed sufficient justification for selecting vaccine antigens. However, we have observed that while the B16F10 melanoma expresses both gp100 and TRP-2/DCT, only recombinant adenoviruses expressing TRP-2/DCT produced protective immunity against the B16F10 tumor whereas immunization with viruses expressing gp100 failed to elicit protective immunity [104, 106]. The lack of protection was not due to insufficient immunogenicity of the vaccine expressing gp100 [106]. Rather, the lack of protection was due to inadequate expression of the target epitope on the tumor itself. Therefore, an additional aspect of the tumor that needs to be included in selection criteria is epitope presentation. Indeed, this issue is critical to the proper interpretation of the randomized phase III studies of cancer vaccines where clinical efficacy did not match expectations.

As described above, characterizing antitumor immunity produced by experimental vaccines is a key challenge for clinical evaluation of vaccination strategies. The majority of trials employ peripheral blood as a source of T cells for immune monitoring, yet, as described above, the T cells within the tumor can manifest significant functional impairments that are not observed within the peripheral blood. Therefore, it is quite likely that we are not adequately informed regarding individuals who have developed successful antitumor immunity and, as a result, we are likely making incorrect conclusions regarding the magnitude of the T-cell response and clinical outcome. Herein lies an opportunity for strategies employing neo-adjuvant vaccines. In addition to the potential beneficial effects of vaccinating prior to surgery, this strategy also provides access to the tumor mass, and possibly the draining nodes, thereby enabling a careful analysis of the T cells within the tumor. The information provided by such analyses will prove critical for accurate evaluation of cancer vaccines.

In summary, despite the promising results obtained in preclinical studies, cancer vaccines have only shown modest clinical activity. As these novel strategies move toward standard practice of care, they will need to be incorporated into conventional treatment strategies involving surgery and chemotherapy. This integration will be facilitated by careful consideration of the opportunities for immunostimulation provided by these conventional therapies.

References

1. Bowne WB, Srinivasan R, Wolchok JD et al (1999) Coupling and uncoupling of tumor immunity and autoimmunity. J Exp Med 190:1717–1722
2. McWilliams JA, McGurran SM, Dow SW et al (2006) A modified tyrosinase-related protein 2 epitope generates high-affinity tumor-specific T cells but does not mediate therapeutic efficacy in an intradermal tumor model. J Immunol 177:155–161

3. Grinshtein N, Ventresca M, Margl R et al (2009) High-dose chemotherapy augments the efficacy of recombinant adenovirus vaccines and improves the therapeutic outcome. Cancer Gene Ther 16:338–350
4. Rosenberg SA, Yang JC, Restifo NP (2004) Cancer immunotherapy: moving beyond current vaccines. Nat Med 10:909–915
5. Mocellin S, Mandruzzato S, Bronte V et al (2004) Cancer vaccines: pessimism in check. Nat Med 10:1278–1279; author reply 1279–1280
6. Lee PP, Yee C, Savage PA et al (1999) Characterization of circulating T cells specific for tumor-associated antigens in melanoma patients. Nat Med 5:677–685
7. Rosenberg SA, Sherry RM, Morton KE et al (2005) Tumor progression can occur despite the induction of very high levels of self/tumor antigen-specific $CD8^+$ T cells in patients with melanoma. J Immunol 175:6169–6176
8. Speiser DE, Lienard D, Pittet MJ et al (2002) *In vivo* activation of melanoma-specific CD8(+) T cells by endogenous tumor antigen and peptide vaccines. A comparison to virus-specific T cells. Eur J Immunol 32:731–741
9. Zippelius A, Batard P, Rubio-Godoy V et al (2004) Effector function of human tumor-specific CD8 T cells in melanoma lesions: a state of local functional tolerance. Cancer Res 64:2865–2873
10. Appay V, Jandus C, Voelter V et al (2006) New generation vaccine induces effective melanoma-specific $CD8^+$ T cells in the circulation but not in the tumor site. J Immunol 177:1670–1678
11. Lee KH, Wang E, Nielsen MB et al (1999) Increased vaccine-specific T cell frequency after peptide-based vaccination correlates with increased susceptibility to *in vitro* stimulation but does not lead to tumor regression. J Immunol 163:6292–6300
12. Monsurro V, Nagorsen D, Wang E et al (2002) Functional heterogeneity of vaccine-induced CD8(+) T cells. J Immunol 168:5933–5942
13. Finke LH, Wentworth K, Blumenstein B et al (2007) Lessons from randomized phase III studies with active cancer immunotherapies – outcomes from the 2006 meeting of the Cancer Vaccine Consortium (CVC). Vaccine 25 (Suppl 2):B97–B109
14. Perez-Diez A, Spiess PJ, Restifo NP et al (2002) Intensity of the vaccine-elicited immune response determines tumor clearance. J Immunol 168:338–347
15. Dudley ME, Yang JC, Sherry R et al (2008) Adoptive cell therapy for patients with metastatic melanoma: evaluation of intensive myeloablative chemo radiation preparative regimens. J Clin Oncol 26:5233–5239
16. Prevost-Blondel A, Zimmermann C, Stemmer C et al (1998) Tumor-infiltrating lymphocytes exhibiting high *ex vivo* cytolytic activity fail to prevent murine melanoma tumor growth *in vivo*. J Immunol 161:2187–2194
17. Bronte V, Kasic T, Gri G et al (2005) Boosting antitumor responses of T lymphocytes infiltrating human prostate cancers. J Exp Med 201:1257–1268
18. Blohm U, Roth E, Brommer K et al (2002) Lack of effector cell function and altered tetramer binding of tumor-infiltrating lymphocytes. J Immunol 169:5522–5530
19. Kilinc MO, Aulakh KS, Nair RE et al (2006) Reversing tumor immune suppression with intratumoral IL-12: activation of tumor-associated T effector/memory cells, induction of T suppressor apoptosis, and infiltration of $CD8^+$ T effectors. J Immunol 177:6962–6973
20. Radoja S, Saio M, Schaer D et al (2001) $CD8^{(+)}$ tumor-infiltrating T cells are deficient in perforin-mediated cytolytic activity due to defective microtubule-organizing center mobilization and lytic granule exocytosis. J Immunol 167:5042–5051
21. Anderson MJ, Shafer-Weaver K, Greenberg NM et al (2007) Tolerization of tumor-specific T cells despite efficient initial priming in a primary murine model of prostate cancer. J Immunol 178:1268–1276
22. Hargadon KM, Brinkman CC, Sheasley-O'neill SL et al (2006) Incomplete differentiation of antigen-specific CD8 T cells in tumor-draining lymph nodes. J Immunol 177:6081–6090
23. Huang Y, Obholzer N, Fayad R et al (2005) Turning on/off tumor-specific CTL response during progressive tumor growth. J Immunol 175:3110–3116

24. Sotomayor EM, Borrello I, Tubb E et al (1999) Conversion of tumor-specific $CD4^+$ T-cell tolerance to T-cell priming through *in vivo* ligation of CD40. Nat Med 5:780–787
25. Staveley-O'Carroll K, Sotomayor E, Montgomery J et al (1998) Induction of antigen-specific T cell anergy: an early event in the course of tumor progression. Proc Natl Acad Sci U S A 95:1178–1183
26. Nielsen MB, Monsurro V, Migueles SA et al (2000) Status of activation of circulating vaccine-elicited $CD8^+$ T cells. J Immunol 165:2287–2296
27. Pittet MJ, Valmori D, Dunbar PR et al (1999) High frequencies of naive Melan-A/MART-1-specific $CD8^{(+)}$ T cells in a large proportion of human histocompatibility leukocyte antigen (HLA)-A2 individuals. J Exp Med 190:705–715
28. Whiteside TL (2006) Immune suppression in cancer: effects on immune cells, mechanisms and future therapeutic intervention. Semin Cancer Biol 16:3–15
29. Marincola FM, Wang E, Herlyn M et al (2003) Tumors as elusive targets of T-cell-based active immunotherapy. Trends Immunol 24:335–342
30. Zou W (2005) Immunosuppressive networks in the tumor environment and their therapeutic relevance. Nat Rev Cancer 5:263–274
31. Frey AB, Monu N (2006) Effector-phase tolerance: another mechanism of how cancer escapes antitumor immune response. J Leukoc Biol 79:652–662
32. Lizee G, Cantu MA, Hwu P (2007) Less yin, more yang: confronting the barriers to cancer immunotherapy. Clin Cancer Res 13:5250–5255
33. Khong HT, Restifo NP (2002) Natural selection of tumor variants in the generation of "tumor escape" phenotypes. Nat Immunol 3:999–1005
34. Jager E, Ringhoffer M, Altmannsberger M et al (1997) Immunoselection *in vivo*: independent loss of MHC class I and melanocyte differentiation antigen expression in metastatic melanoma. Int J Cancer 71:142–147
35. Jager E, Ringhoffer M, Karbach J et al (1996) Inverse relationship of melanocyte differentiation antigen expression in melanoma tissues and $CD8^+$ cytotoxic-T-cell responses: evidence for immunoselection of antigen-loss variants *in vivo*. Int J Cancer 66:470–476
36. Riker A, Cormier J, Panelli M et al (1999) Immune selection after antigen-specific immunotherapy of melanoma. Surgery 126:112–120
37. Sanchez-Perez L, Kottke T, Diaz RM et al (2005) Potent selection of antigen loss variants of B16 melanoma following inflammatory killing of melanocytes *in vivo*. Cancer Res 65:2009–2017
38. Setiadi AF, Omilusik K, David MD et al (2008) Epigenetic enhancement of antigen processing and presentation promotes immune recognition of tumors. Cancer Res 68:9601–9607
39. Ohnmacht GA, Wang E, Mocellin S et al (2001) Short-term kinetics of tumor antigen expression in response to vaccination. J Immunol 167:1809–1820
40. Spiotto MT, Rowley DA, Schreiber H (2004) Bystander elimination of antigen loss variants in established tumors. Nat Med 10:294–298
41. van Hall T, Wolpert EZ, van Veelen P et al (2006) Selective cytotoxic T-lymphocyte targeting of tumor immune escape variants. Nat Med 12:417–424
42. Danna EA, Sinha P, Gilbert M et al (2004) Surgical removal of primary tumor reverses tumor-induced immunosuppression despite the presence of metastatic disease. Cancer Res 64:2205–2211
43. Salvadori S, Martinelli G, Zier K (2000) Resection of solid tumors reverses T cell defects and restores protective immunity. J Immunol 164:2214–2220
44. Gabri MR, Mazorra Z, Ripoll GV et al (2006) Complete antitumor protection by perioperative immunization with GM3/VSSP vaccine in a preclinical mouse melanoma model. Clin Cancer Res 12:7092–7098
45. Hawkins WG, Gold JS, Blachere NE et al (2002) Xenogeneic DNA immunization in melanoma models for minimal residual disease. J Surg Res 102:137–143
46. Ohashi K, Kobayashi G, Fang S et al (2006) Surgical excision combined with autologous whole tumor cell vaccination is an effective therapy for murine neuroblastoma. J Pediatr Surg 41:1361–1368

47. Grinshtein N, Bridle B, Wan Y et al (2009) Neoadjuvant vaccination provides superior protection against tumor relapse following surgery compared with adjuvant vaccination. Cancer Res 69:3979–3985
48. Hsueh EC, Essner R, Foshag LJ et al (2002) Prolonged survival after complete resection of disseminated melanoma and active immunotherapy with a therapeutic cancer vaccine. J Clin Oncol 20:4549–4554
49. Jocham D, Richter A, Hoffmann L et al (2004) Adjuvant autologous renal tumor cell vaccine and risk of tumor progression in patients with renal-cell carcinoma after radical nephrectomy: phase III, randomized controlled trial. Lancet 363:594–599
50. Steiner HH, Bonsanto MM, Beckhove P et al (2004) Antitumor vaccination of patients with glioblastoma multiforme: a pilot study to assess feasibility, safety, and clinical benefit. J Clin Oncol 22:4272–4281
51. Mazzaferro V, Coppa J, Carrabba MG et al (2003) Vaccination with autologous tumor-derived heat-shock protein gp96 after liver resection for metastatic colorectal cancer. Clin Cancer Res 9:3235–3245
52. Kirkwood JM, Manola J, Ibrahim J et al (2004) A pooled analysis of eastern cooperative oncology group and intergroup trials of adjuvant high-dose interferon for melanoma. Clin Cancer Res 10:1670–1677
53. Sondak VK, Liu PY, Tuthill RJ et al (2002) Adjuvant immunotherapy of resected, intermediate-thickness, node-negative melanoma with an allogeneic tumor vaccine: overall results of a randomized trial of the Southwest Oncology Group. J Clin Oncol 20:2058–2066
54. Mitchell MS, Abrams J, Thompson JA et al (2007) Randomized trial of an allogeneic melanoma lysate vaccine with low-dose interferon Alfa-2b compared with high-dose interferon Alfa-2b for Resected stage III cutaneous melanoma. J Clin Oncol 25:2078–2085
55. Sondak VK, Sabel MS, Mule JJ (2006) Allogeneic and autologous melanoma vaccines: where have we been and where are we going? Clin Cancer Res 12:2337s–2341s
56. Sosman JA, Unger JM, Liu PY et al (2002) Adjuvant immunotherapy of resected, intermediate-thickness, node-negative melanoma with an allogeneic tumor vaccine: impact of HLA class I antigen expression on outcome. J Clin Oncol 20:2067–2075
57. Poland GA, Ovsyannikova IG, Jacobson RM et al (2007) Heterogeneity in vaccine immune response: the role of immunogenetics and the emerging field of vaccinomics. Clin Pharmacol Ther 82:653–664
58. Negin B, Panka D, Wang W et al (2008) Effect of melanoma on immune function in the regional lymph node basin. Clin Cancer Res 14:654–659
59. Lee JH, Torisu-Itakara H, Cochran AJ et al (2005) Quantitative analysis of melanoma-induced cytokine-mediated immunosuppression in melanoma sentinel nodes. Clin Cancer Res 11:107–112
60. Noguchi M, Yao A, Harada M et al (2007) Immunological evaluation of neoadjuvant peptide vaccination before radical prostatectomy for patients with localized prostate cancer. Prostate 67:933–942
61. Ullenhag GJ, Spendlove I, Watson NF et al (2006) A neoadjuvant/adjuvant randomized trial of colorectal cancer patients vaccinated with an anti-idiotypic antibody, 105 AD7, mimicking CD55. Clin Cancer Res 12:7389–7396
62. Guillaume T, Rubinstein DB, Symann M (1998) Immune reconstitution and immunotherapy after autologous hematopoietic stem cell transplantation. Blood 92:1471–1490
63. Mackall CL (2000) T-cell immunodeficiency following cytotoxic antineoplastic therapy: a review. Stem Cells 18:10–18
64. Schluns KS, Kieper WC, Jameson SC et al (2000) Interleukin-7 mediates the homeostasis of naive and memory CD8 T cells *in vivo*. Nat Immunol 1:426–432
65. Ernst B, Lee DS, Chang JM et al (1999) The peptide ligands mediating positive selection in the thymus control T cell survival and homeostatic proliferation in the periphery. Immunity 11:173–181
66. Prlic M, Blazar BR, Khoruts A et al (2001) Homeostatic expansion occurs independently of costimulatory signals. J Immunol 167:5664–5668

67. Goldrath AW, Bogatzki LY, Bevan MJ (2000) Naive T cells transiently acquire a memory-like phenotype during homeostasis-driven proliferation. J Exp Med 192:557–564
68. Kieper WC, Jameson SC (1999) Homeostatic expansion and phenotypic conversion of naive T cells in response to self peptide/MHC ligands. Proc Natl Acad Sci U S A 96:13306–13311
69. Murali-Krishna K, Ahmed R (2000) Cutting edge: naive T cells masquerading as memory cells. J Immunol 165:1733–1737
70. Goldrath AW, Luckey CJ, Park R et al (2004) The molecular program induced in T cells undergoing homeostatic proliferation. Proc Natl Acad Sci U S A 101:16885–16890
71. Oehen S, Brduscha-Riem K (1999) Naive cytotoxic T lymphocytes spontaneously acquire effector function in lymphocytopenic recipients: a pitfall for T cell memory studies? Eur J Immunol 29:608–614
72. Hamilton SE, Wolkers MC, Schoenberger SP et al (2006) The generation of protective memory-like $CD8^+$ T cells during homeostatic proliferation requires $CD4^+$ T cells. Nat Immunol 7:475–481
73. Gleeson PA, Toh BH, van Driel IR (1996) Organ-specific autoimmunity induced by lymphopenia. Immunol Rev 149:97–125
74. Sherer Y, Shoenfeld Y (1998) Autoimmune diseases and autoimmunity post-bone marrow transplantation. Bone Marrow Transplant 22:873–881
75. Maine GN, Mule JJ (2002) Making room for T cells. J Clin Invest 110:157–159
76. Baccala R, Gonzalez-Quintial R, Dummer W et al (2005) Tumor immunity via homeostatic T cell proliferation: mechanistic aspects and clinical perspectives. Springer Semin Immunopathol 27:75–85
77. Theofilopoulos AN, Dummer W, Kono DH (2001) T cell homeostasis and systemic autoimmunity. J Clin Invest 108:335–340
78. Colvin OM (1999) An overview of cyclophosphamide development and clinical applications. Curr Pharm Des 5:555–560
79. Lutsiak ME, Semnani RT, De Pascalis R et al (2005) Inhibition of CD4(+) 25+ T regulatory cell function implicated in enhanced immune response by low-dose cyclophosphamide. Blood 105:2862–2868
80. Ibe S, Qin Z, Schuler T et al (2001) Tumor rejection by disturbing tumor stroma cell interactions. J Exp Med 194:1549–1559
81. Proietti E, Greco G, Garrone B et al (1998) Importance of cyclophosphamide-induced bystander effect on T cells for a successful tumor eradication in response to adoptive immunotherapy in mice. J Clin Invest 101:429–441
82. Salem ML, Kadima AN, El-Naggar SA et al (2007) Defining the ability of cyclophosphamide preconditioning to enhance the antigen-specific $CD8^+$ T-cell response to peptide vaccination: creation of a beneficial host microenvironment involving type I IFNs and myeloid cells. J Immunother 30:40–53
83. Bracci L, Moschella F, Sestili P et al (2007) Cyclophosphamide enhances the antitumor efficacy of adoptively transferred immune cells through the induction of cytokine expression, B-cell and T-cell homeostatic proliferation, and specific tumor infiltration. Clin Cancer Res 13:644–653
84. Machiels JP, Reilly RT, Emens LA et al (2001) Cyclophosphamide, doxorubicin, and paclitaxel enhance the antitumor immune response of granulocyte/macrophage-colony stimulating factor-secreting whole-cell vaccines in HER-2/neu tolerized mice. Cancer Res 61:3689–3697
85. Song CK, Han HD, Noh KH et al (2007) Chemotherapy enhances CD8(+) T cell-mediated antitumor immunity induced by vaccination with vaccinia virus. Mol Ther 15:1558–1563
86. Song W, Levy R (2005) Therapeutic vaccination against murine lymphoma by intratumoral injection of naive dendritic cells. Cancer Res 65:5958–5964
87. Taieb J, Chaput N, Schartz N et al (2006) Chemoimmunotherapy of tumors: cyclophosphamide synergizes with exosome based vaccines. J Immunol 176:2722–2729
88. Tong Y, Song W, Crystal RG (2001) Combined intratumoral injection of bone marrow-derived dendritic cells and systemic chemotherapy to treat pre-existing murine tumors. Cancer Res 61:7530–7535

89. Liu JY, Wu Y, Zhang XS et al (2007) Single administration of low dose cyclophosphamide augments the antitumor effect of dendritic cell vaccine. Cancer Immunol Immunother 56:1597–1604
90. Hermans IF, Chong TW, Palmowski MJ et al (2003) Synergistic effect of metronomic dosing of cyclophosphamide combined with specific antitumor immunotherapy in a murine melanoma model. Cancer Res 63:8408–8413
91. Chan OT, Yang LX (2000) The immunological effects of taxanes. Cancer Immunol Immunother 49:181–185
92. Tsavaris N, Kosmas C, Vadiaka M et al (2002) Immune changes in patients with advanced breast cancer undergoing chemotherapy with taxanes. Br J Cancer 87:21–27
93. Carson WE, 3rd, Shapiro CL, Crespin TR et al (2004) Cellular immunity in breast cancer patients completing taxane treatment. Clin Cancer Res 10:3401–3409
94. Kotsakis A, Sarra E, Peraki M et al (2000) Docetaxel-induced lymphopenia in patients with solid tumors: a prospective phenotypic analysis. Cancer 89:1380–1386
95. Souglakos J, Kotsakis A, Kouroussis C et al (2002) Nonneutropenic febrile episodes associated with docetaxel-based chemotherapy in patients with solid tumors. Cancer 95:1326–1333
96. Prell RA, Gearin L, Simmons A et al (2006) The anti-tumor efficacy of a GM-CSF-secreting tumor cell vaccine is not inhibited by docetaxel administration. Cancer Immunol Immunother 55:1285–1293
97. Arlen PM, Gulley JL, Parker C et al (2006) A randomized phase II study of concurrent docetaxel plus vaccine versus vaccine alone in metastatic androgen-independent prostate cancer. Clin Cancer Res 12:1260–1269
98. Garnett CT, Schlom J, Hodge JW (2008) Combination of docetaxel and recombinant vaccine enhances T-cell responses and antitumor activity: effects of docetaxel on immune enhancement. Clin Cancer Res 14:3536–3544
99. Emens LA, Jaffee EM (2005) Leveraging the activity of tumor vaccines with cytotoxic chemotherapy. Cancer Res 65:8059–8064
100. Obeid M, Tesniere A, Ghiringhelli F et al (2007) Calreticulin exposure dictates the immunogenicity of cancer cell death. Nat Med 13:54–61
101. Spisek R, Charalambous A, Mazumder A et al (2007) Bortezomib enhances dendritic cell (DC)-mediated induction of immunity to human myeloma via exposure of cell surface heat shock protein 90 on dying tumor cells: therapeutic implications. Blood 109:4839–4845
102. Kazama H, Ricci JE, Herndon JM et al (2008) Induction of immunological tolerance by apoptotic cells requires caspase-dependent oxidation of high-mobility group box-1 protein. Immunity 29:21–32
103. Panaretakis T, Joza N, Modjtahedi N et al (2008) The co-translocation of ERp57 and calreticulin determines the immunogenicity of cell death. Cell Death Differ 15:1499–1509
104. Lane C, Leitch J, Tan X et al (2004) Vaccination-induced autoimmune vitiligo is a consequence of secondary trauma to the skin. Cancer Res 64:1509–1514
105. Zhang S, Bernard D, Khan WI et al (2009) $CD4^{+}$ T-cell-mediated anti-tumor immunity can be uncoupled from autoimmunity via the STAT4/STAT6 signaling axis. Eur J Immunol 39:1252–1259
106. Leitch J, Fraser K, Lane C et al (2004) CTL-dependent and -independent antitumor immunity is determined by the tumor not the vaccine. J Immunol 172:5200–5205

Chapter 16
Combining Oncolytic Viruses with Cancer Immunotherapy

Kyle B. Stephenson, John Bell, and Brian Lichty

Abstract Although there are a number of treatment modalities currently available for the treatment of cancer, mortality rates have only shown marginal improvement of late. Novel therapeutics, with decreased side effects, are desperately needed in order to improve the current prognosis for this deadly disease. To this end viruses with either natural or engineered tropism to tumors as well as immunotherapeutic approaches are being investigated as possible mono-therapies for the treatment of cancer. Each of these therapies have shown some successes on their own, however the combination of these two modalities may improve on the efficacy of the individual treatments. This chapter will focus on the use of oncolytic viruses in cancer therapy, the importance of the immune response, and its ability to further improve on the successes of oncolytic viral therapy.

Keywords Cancer immunotherapy • Combination therapy • Oncolytic virus • Tumor immunity • Viral vaccine vectors

Introduction

Although there have certainly been many advances in the use of chemotherapy, radiotherapy, and surgery for the treatment of cancer, mortality rates have remained mostly unchanged and substantial sides effects persist [1]. As such, there exists a need to develop new targeted treatment modalities for cancer therapy that might be safer and more effective. Towards this aim, oncolytic viruses (OVs) have been studied for decades; more recently, recombinant technologies have been developed to allow for engineering of these viruses. OVs can be engineered to be extremely specific, thereby

B. Lichty (✉)
Center for Gene Therapeutics, Department of Pathology and Molecular Medicine, McMaster University, Hamilton, ON, Canada
e-mail: lichtyb@mcmaster.ca

J. Medin and D. Fowler (eds.), *Experimental and Applied Immunotherapy*,
DOI 10.1007/978-1-60761-980-2_16, © Springer Science+Business Media, LLC 2011

limiting viral infection solely to within the tumor. Infection of tumors with these viruses leads to tumor destruction through lytic infections. OVs have also been found to naturally induce antitumor immunity following oncolysis; importantly, this antitumor immunity can be further augmented through the addition of immunostimulatory transgenes into the vector design. This chapter will introduce the different strategies that have been tried in both preclinical and clinical testing of OVs and immunotherapy.

History of Oncolytic Viruses and Their Use in Immunotherapy of Cancer

Induction of Antitumor Immune Responses Following Oncolytic Virus Therapy

As the study of OVs has evolved, increased attention has been paid to the host immune response and its influence on oncolysis. Immune responses against viral vectors likely impair viral oncolysis, and thereby represent a barrier to clinical success. On the other hand, immune responses against the tumor should aid in tumor destruction of the tumor, thus preventing disease relapse. As tumor cells are killed, tumor debris becomes available for recognition by the immune system for generation of antitumor immune responses. The death of the tumor cells through the viral oncolysis process should be more immunogenic than other forms of apoptotic death induced by current therapies due to the presence of viral pathogen-associated molecular patterns (PAMPs), such as viral nucleic acid, release of endogenous "danger" molecules, and the production of cytokines and chemokines [2–4].

The antitumor effect of OVs is mediated by lytic infection and also by the resultant immune response. A number of studies have found that both innate and adaptive immune responses are generated following viral oncolysis mediated by HSV [5–10], VSV [11], adenovirus [3], and parvovirus [12]. The adaptive immune responses generated by these OVs are specific to the tumor treated and can effectively cure untreated tumors. Such immune responses are long-lasting, as evidenced by an inability to re-engraft cured mice with tumor cells [6, 10]. The production of cytokines and chemokines following viral infection facilitates immune cell homing to the tumor, and promotes immune activation rather than immune tolerance [13]. APCs, mainly dendritic cells, are then able to cross present TAA to the adaptive immune system, thereby leading to the induction of tumor specific T cells [4, 14]. The presence of viral PAMPs in the phagocytosed tumor cells, especially dsRNA, promotes cross-presentation by dendritic cells (DCs). Toll-like receptor (TLR)-3 is an innate sensor that is able to recognize dsRNA produced as a by-product of viral replication. When the infected tumor cells are phagocytosed, the recognition of this dsRNA by TLR 3 will help activate the DC and enhance cross-priming of viral and tumor antigens. [15]. The immune responses generated following viral oncolysis are an important aspect of this therapy as they will be able to recognize and destroy any remaining tumor cells that are not destroyed by the virus.

Viral Oncolysates

Since the 1960s, OVs have been used to generate therapeutic cell-based vaccines known as viral oncolysates [16]. A number of different viruses have been used to treat a variety of tumors [17–19]. Viral oncolysates are made using either autologous or allogeneic tumor cells infected with an OV; a high multiplicity of infection is used to ensure infection of every tumor cell. Preparation of the lysate varies between labs but usually consists of disrupting the infected cells and removing the cellular debris while keeping viral, cellular, and virally modified proteins. The injection site for OV lysates varies depending on the tumor being treated. For example, intradermal injection is typically used for the treatment of melanoma to induce immunity within the lymph system, thereby leading to systemic immunity [20]. For ovarian cancer, the oncolysate can be delivered intraperitoneally to induce localized immune responses at the site of tumor metastasis [17]. Since the goal of this therapy is to induce an immune response, *in vivo* replication of the OV is not necessarily required; as such, some groups have inactivated the virus prior to vaccination [18, 19].

It is important to consider the potential advantages of the OV approach relative to simply producing an oncolysate from uninfected tumor cells. First, infection of tumor cells prior to oncolysate production yields cells that are more immunogenic. Infected tumor cells activate innate antiviral pathways through numerous sensors, including Toll-like receptors (TLRs). The subsequent signaling effects up-regulate numerous molecules, including type-I Interferon (IFN) and other proinflammatory cytokines (see Fig. 16.1). Type 1 IFN acts as a survival and maturation signal for DCs and augments DC ability to induce immune responses through antigen presentation [21]. Viral infection has also been shown to induce immune responses against proteins unique to the viral oncolysate [19, 22]. Studies using viral oncolysates have shown that antibodies against proteins in the oncolysate can be induced following treatment. These antibodies are not able to recognize proteins from uninfected tumor cells or viral proteins; rather, they possess specificity for those proteins that have been modified by the viral infection [19, 22].

A number of clinical trials have been carried out using viral oncolysates. In all cases, the safety of this therapy was excellent with only mild adverse reactions observed. Furthermore, there has been no case of tumor growth at the OV injection site; this latter finding is critical, as any oncolysate vaccine used in the clinic must ensure that no intact tumor cells remain that would allow for tumor formation. The most common viruses used in human trials have been vaccinia virus, attenuated influenza A virus, and Newcastle disease virus (NDV) [16, 23–27]. These viruses are very safe for use in humans for a number of reasons. First, each virus used has

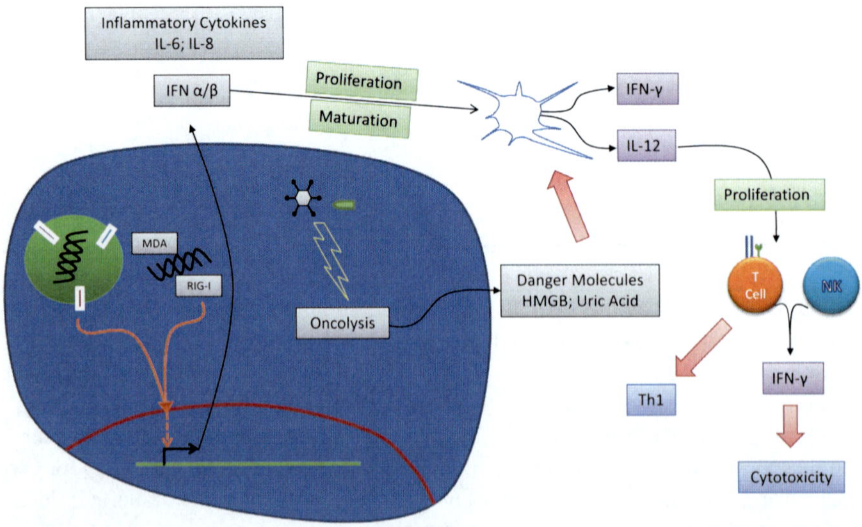

Fig. 16.1 How an oncolytic virus might impact antitumoral immunity. Upon viral infection toll-like receptor (*TLR*) will sense viral nucleic acid in endosomes, while RIG-I and MDA will sense cytoplasmic nucleic acids. Ligation of these molecules will lead to the production of proinflammatory cytokines and type I interferon, the latter of which is able to induce maturation and proliferation of DCs leading to IFN-γ and IL-12 production. The viral infection will lead to cell death releasing other danger molecules that will also impact on DC maturation. DC derived IL-12 is able to activate both T cells and NK cells leading to cytotoxic killing of tumor cells and Th1 polarization of T cell responses

been attenuated and is significantly less virulent relative to their wild-type counterparts. Vaccinia was the strain used for smallpox vaccination until smallpox was declared eradicated in 1980 [28]; as such, anyone old enough to have been vaccinated has immunity to this virus. An added benefit to using attenuated forms of influenza virus as an oncolytic virus is that it has been reported that people have been found to have antibodies that recognize the most common oncolytic influenza strain [17]. Finally, NDV is not a human pathogen and only causes minor infections leading to a self-limiting conjunctivitis [29].

In both syngeneic murine tumor models and human trials, viral oncolysates have been shown to induce tumor-specific immune responses and favorably influence tumor growth and survival. A number of murine studies have identified the induction of humoral responses that recognize antigens specific to the viral oncolysate that are not present in either virus or tumor cell lysates alone [19, 22]. In addition to the induction of humoral responses, T-cell-mediated responses have also been observed [23]. In these murine models, survival was increased in mice treated with viral oncolysates; in some cases, survival was further improved with the addition of exogenous cytokines. Similar to murine models, cancer patients that receive viral oncolysates on clinical trials are able to mount both humoral- and cell-mediated immune responses [17, 18, 25]. Treatment with these viral oncolysates has resulted

Table 16.1 Examples of preclinical testing of oncolytic viruses

		Findings	References
Antitumor immune induction following Oncolysis	HSV	Induction of both humoral and T cell (CD8) mediated immune responses required for efficacy	[6–9]
	VSV	VSV oncolysis induced antitumor immune responses in melanoma expressing a model TAA. Immune responses were required for full antitumor efficacy	[11]
	Parvovirus	Parvovirus H1 infection of tumor cells led to maturation of DCs, cross presentation and activation of CTL *in vitro*	[12]
Oncolytic virus with immunostimulatory transgene	IL-2	In both HSV and NDV, IL2 expression led to increased CD4 and CD8 cells in tumor, tumor draining lymph nodes and spleen	[39, 41, 42]
	GMCSF	JX-594 (vaccinia-GMCSF) delivered systemically had increased efficacy due to both viral replication and cytokine production	[47]
	IL-12	When expressed from VSV and HSV led to increased survival. HSV IL-12 induced immune responses also able to control non-infected contralateral tumor	[36, 38]
	B7.1 (CD80)	Soluble B7.1 expressed from HSV led to increased CD4/CD8 TILs and better efficacy, which was dependent on intact immune system	[48]
Oncolytic virus as a vaccine vector	NDV	Using a model CD8 TAA epitope, NDV vaccination against TAA increased survival, which was further augmented by NDV-IL2 co-administration	[78]
	Vaccinia	Vaccinia expressing a model tumor antigen was able to delay tumor growth when used as a prophylactic vaccine. This effect was improved on with the addition of IL-2	[79]

in objective responses in patients (see Table 16.2). Some promising results for the use of this therapy were observed through the prospective monitoring of patients with surgically resected metastatic melanoma treated postsurgically with NDV oncolysate. In these studies, it was found that viral oncolysate treatment doubled the 10- and 15-year survival rates compared to historical controls. It was also found that this increased survival was associated with increased numbers and clonality of $CD8^+$ T cells [20, 25, 30]. However, one limitation with these studies is the lack of control arms. Although the historical data is useful and well documented, these trials would have more power if they were completed as randomized controlled trials.

Immunostimulatory Oncolytic Viruses

Another strategy that has been investigated to increase the immunostimulatory properties of OVs is their combined use with chemokines, cytokines, and costimulatory molecules. Recombinant cytokines have been used in the clinic to treat cancer; however, toxicities associated with higher doses of systemically administered cytokines are substantial [31]. By incorporating cytokines and chemokines as transgenes into OVs, it is possible to safely increase immune stimulation through the local expression of cytokines.

Various immune cells can be targeted to increase antitumor immune responses following OV treatment. Both innate and adaptive cells can be manipulated, with the ultimate end goal to induce cytotoxic effector cells that can be either nonspecific NK cells or tumor antigen-specific T cells. The activation of APCs, including DCs, is important in both of these processes. DCs have an exquisite ability to take up released tumor antigens and cross-present them to $CD8^+$ T cells as well as activate $CD4^+$ T cells through MHC-II molecules. As the DCs take up this antigen they will mature as they travel to lymph nodes, and subsequently secrete cytokines that will impact both innate effector cells (NK) as well as adaptive immune cells ($CD4^+$ and $CD8^+$ T cells) [32, 33]. In an effort to activate and recruit APCs, a number of groups have added cytokines, chemokines, and costimulatory molecules to their OVs in the preclinical setting [34–42] and the clinical setting [43, 44]. RANTES is one such chemokine that is able to recruit numerous immune cell populations. Expression of RANTES leads to higher numbers of activated DCs within the tumor, thereby increasing tumor-specific T cells [45]. Alternatively, other groups have added GM-CSF to HSV and adenovirus to enhance maturation and activation of APCs [34, 46]. Treatment with these modified viruses led to increased efficacy with evidence of tumor-specific immune responses. Vaccinia virus expressing GM-CSF has also been show to increase vaccinia efficacy in preclinical models [47] and has shown efficacy in phase I clinical trials. Of the ten treated patients, one showed progressive disease while three patients had a partial response and six had stable disease according to RECIST criteria [44].

The majority of tumor cells have decreased expression of both MHC-I and costimulatory molecules and therefore do not efficiently present antigen directly to T cells [48]. The expression of costimulatory molecules, including B7-1, can help induce activation of T cells that otherwise may be anergic [49]. Expression of soluble B7-1 (CD80) from OVs slightly increased vaccine efficacy with concomitant induction of a systemic antitumor immune response [50]. The use of B7-1 expressing OVs can also be enhanced by coexpression of other cytokines such as IL-12, which is up-regulated in activated DCs. IL-12 promotes cellular proliferation and induction of IFN-γ from both T cells and NK cells, thereby inducing Th1 responses [51, 52]; IL-12 has been shown to have a significant impact on tumor growth when expressed from the liver following hydrodynamic injection of DNA [53]. However, IL-12 has been associated with substantial toxicity when given systemically; therefore, the use of viruses to restrict IL-12 expression may reduce toxicity associated with this cytokine [54, 55]. Importantly, the combination of IL-12 and B7-1 was shown to further enhance the efficacy of oncolytic viral therapy [31, 50].

Combining Oncolytic Virotherapy and Immunotherapy

Vaccine Approaches

Harnessing the immune system through vaccination has been a successful strategy in combating viral and bacterial infections. These vaccinations are used prophylactically to protect immunized individuals from specific pathogens. The development of cancer vaccines is more challenging due to the fact that successful vaccines will have to target tumor-associated antigens that the host may be tolerized against. And, since it is possible that every tumor is slightly different, prophylactic vaccination is unlikely to become a prominent therapy. One exception to these general principles is the recent advent of prophylactic vaccines for cervical cancer that target the causative virus HPV [56].

Many relevant tumor antigens have been identified and it is likely that more will continue to be found. These antigens are cellular proteins that are associated with tumors but are different from the proteins found in normal cells due to either aberrant expression of differentiation or embryonic antigens, or mutation or overexpression of cellular proteins [57]. By inducing immune responses against these proteins with vaccination, it may be possible to eradicate tumors in vaccinated individuals. There have been a large number of trials with very limited success and no approved therapies [57]; therefore, new strategies need to be evaluated, including the use of viral vectors.

Phase I and II clinical trials have been performed to assess the ability of viral vaccine vectors expressing TAA to induce immune responses in cancer patients. Some examples of tumor antigens that have been evaluated include: 5T4, carcinoembryonic antigen (CEA), MAGE, and NY-ESO-1 [58–64] (see Table 16.2). In these trials, some patients were found to develop an antibody and/or cell-mediated immune response against the immunized TAA. Since these studies involved small numbers of patients and did not incorporate control arms, it is difficult to assess the relevance of these immune responses; however, there was some indication that the immune responses correlated with clinical outcomes, mainly stabilization of disease. Complete responses were only observed in a very small number of instances; one of eight melanoma patients immunized with NY-ESO-1 expressing vaccinia and fowlpox vectors [64], two of thirty-eight patients treated with chemotherapy followed by ALVAC expressing CEA and B7-1 [62], and one of twenty-five patients treated with vaccinia expressing 5T4 combined with IL-2 [60]. In these trials, the same vector was repeatedly administered even though the patients developed antiviral immune responses early on in the course of treatment; as such, host immunity to the vaccine may have limited the therapeutic efficacy of subsequent vaccinations. In addition, it is possible that the overwhelming immune response to the viral antigens may have limited the expansion of specific immune responses to

Table 16.2 Examples of oncolytic virus and immunotherapy clinical trials

		Findings	References
Oncolytic virus		JX-594 (vaccinia virus-GMCSF) was found to be safe in cancer therapy with no dose limiting toxicity in a phase I trial. Some patients had objective responses at both the primary tumor and distant metastases	[44]
Viral oncolysates	NDV melanoma oncolysate	A 15-year follow up on patients treated with NDV oncolysate postsurgically indicated a 63% survival rate (compared to historical rate of ~40%)	[20, 25, 30]
	Influenza ovarian oncolysate	Treatment of ovarian cancer patients with allogeneic influenza oncolysate was safe and resulted in ~25% of patients showing clinical activity	[17]
Cancer vaccines	5T4	In phase I/II clinical trials modified Vaccinia Ankara expressing 5T4 induced humoral and cell mediated anti-5T4 responses with the former correlating with stable disease.	[59, 60]
	MAGE	Melanoma patients were vaccinated with canarypox (ALVAC) expressing an HLA-A1 restricted MAGE peptide. Some patients mounted a monoclonal T-cell response that correlated with tumor regression	[58]
	NY-ESO-1	Vaccinia and foulpox expressing NY-ESO-1 induced antibody, CD4 and CD8 responses	[64]
	CEA	Expression of CEA from ALVAC, along with costimulatory molecule B-7.1 induced anti-CEA T-cell responses that correlated with disease progression. This response could be boosted and is not impaired by combination with chemotherapy	[62, 63]
Adoptive cell therapy		TILs generated from excised melanoma tumors were expanded *in vitro* infused back into the patient following lymphodepletion by chemotherapy. Almost half of the patients treated had objective responses to the therapy	[92]

the TAA of interest [65]. One method to circumvent this biology is through the use of heterologous prime-boost strategies. In this scenario, the priming vaccination consists of a different strain of virus or a different virus relative to the agent used during the boost phase of vaccination. The large majority of studies assessing heterologous vaccination have used a variety of poxviruses, as well as adenovirus and Semliki Forest virus [66–71]. Using a heterologous boost led in most cases to an expansion of TAA-specific T cells in both murine models and clinical trials. However, different combinations of viral vectors were not always equal; that is,

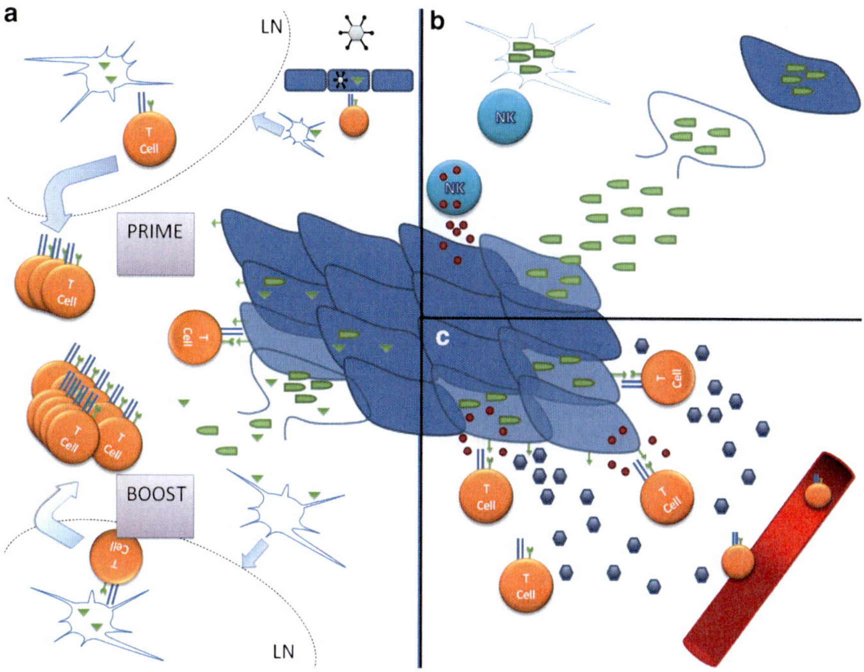

Fig. 16.2 Combining oncolytic viral therapy and immunotherapy. Combining OV therapy and immunotherapy lends numerous different ways to attack the same tumor. (**a**) Vaccination of cancer patients with TAA expressing viruses can prime antitumor immune responses, however these responses are rarely sufficient to lead to significant objective responses. OVs expressing the same TAA can be combined with therapy in a heterologous prime-boost regimen, which will lead to expansion of the initial pool of TAA specific T cells as well as de-bulking the tumor. (**b**) In the face of pre-existing immunity OVs delivered systemically have a large barrier to overcome before infecting the tumor. By using immune cells and tumor cells as "Trojan horses" OVs can be hidden from the immune system and a large dose of virus can be delivered to the tumor. The use of infected immune cells can also induce innate immunity and NK mediate tumor killing. (**c**) OV therapy should be able to augment adoptive cell therapy in two ways: Infection of the tumor will lead to induction of pro-inflammatory cytokines leading to enhanced homing of the adoptively transferred cells to the tumor; and the oncolysis resulting from viral infection will help de-bulk the tumor leaving less tumor for the adoptively transferred cells to clean up

some approaches led to enhanced immune responses, while others were no better than a homologous prime-boost strategy. Because of this, there exists a need to determine which viruses can be optimally used for priming or boosting vectors, and what combinations will lead to enhanced immune responses (Fig. 16.2).

For induction of immunity, the antigens being expressed need to be picked up and processed by APCs, mainly DCs, and then presented to T cells. Therefore, another strategy for the use of vaccination in cancer therapy is to use DCs loaded with TAAs. Upon immunization, loaded mature DCs can quickly activate tumor-specific T cells. Different methods exist to load the DCs, including ones that involve pulsing DCs with tumor lysates [72, 73], electroporation of TAA mRNA [74], or transductions using viral vectors [75, 76]. In each of these cases, whole TAAs have been successfully incorporated into the DCs. By including the entire antigen rather than MHC-restricted peptides, the DC vaccine can avoid limitations associated with HLA alleles because the DCs are able to process the antigen in the proper MHC context. Also, when peptides are used, they tend to be MHC class I restricted; in contrast, the use of full-length TAAs allows for the induction of helper epitopes through MHC class II and $CD4^+$ T cells, which play a major role in the induction of tumor-specific CTLs [75, 76]. Although all of these studies were able to show induction of T-cell responses, there was also an indication that NK cells were activated. Even though these NK cells on their own did not confer protection, without them there was a loss in efficacy [76]. DCs loaded with TAA were able to induce robust immune responses that correlated with tumor protection in a murine tumor model; however, with the addition of immunostimulatory cytokines, such immune responses could be further augmented [76].

DCs have the ability to pick up TAA and activate T and NK cells endogenously and therefore can also be used without loading of tumor antigen *in vitro*. Farrell et al. [77] have shown in 2008 that injecting immature DCs following OV infection leads to increased immune responses and efficacy. Although classical viral vaccine vectors are nonreplicating, the use of replicating OVs as vaccine vectors has begun to be interrogated. As with adenovirus and modified vaccinia Ankara, other OVs can be engineered to express TAAs. Along with expressing their TAA transgene to induce a specific immune response, these viruses will also infect and de-bulk the tumor. This process will not only leave less residual tumor for the immune system to eradicate but will also release a number of other tumor antigens that may allow for epitope spreading, thereby leading to antitumor immune responses against a number of other TAAs that have not been vaccinated against. Vaccinia virus, VSV, and NDV engineered to express model TAAs have been used in an attempt to lytically destroy the tumor while inducing a specific immune response [11, 78, 79]. OVs carrying transgenes can also be used to transduce DCs as a vaccine platform [80]. The expression of TAAs from VSV in transduced DCs is able to induce tumor-specific CTL responses and very efficiently activate NK cells [80]. It was shown that although the CTL have a role in protection, the NK cells were the main effectors responsible for tumor protection. Interestingly, even though no virus is produced from these infected DCs, infection of the DCs with VSV expressing no TAA was able to significantly impact tumor burden in a murine melanoma lung metastases model (although not to the same level as when a TAA was included) [80]. As well, we have recently shown that OVs expressing TAAs are excellent boosting vectors. Combining the benefits of viral oncolysis (tumor debulking and reversal of local immunosuppression) with that of heterologous prime:boost strategies (if the priming vaccine was a different vector) has led to significantly enhanced therapeutic benefit in animal models [81].

Immune Cells as OV Carriers

A number of OVs currently in preclinical and clinical testing are human viruses, including HSV1 and 2, reovirus, and measles virus. Because of this, additional barriers exist with respect to using these viruses to treat cancer systemically. Patients who have immunity against these viruses through previous exposure or immunization will have neutralizing antibodies that will have a negative impact on an ability to use these viruses systemically. The seroprevalance for HSV1 is around 65% [82]; most adults have pre-existing immunity to reovirus [83], and anyone who has had a normal childhood immunization schedule will have immunity against measles. To circumvent this limitation, these viruses have been administered by intratumoral delivery rather than systemically; consequently, there may be a limited ability of the virus to reach metastases. An alternative strategy is to use cells as a cloaking device, thereby hiding the OV from the immune system.

Immune cells have been used to circumvent host immunity to OVs, and are an ideal cell due to their ability to traffic to the tumor [84–86]. The cells and virus used depends on the particular study; however, the most common carriers employed to date are DCs and T cells. Although the ideal situation would be to match up a virus and immune cell that leads to a productive infection to concentrate the OV in the tumor, this has not always been the case; even a nonproductive infection is able to improve on monotherapy with cells or OVs [85, 86]. Measles virus naturally infects monocytic cells, and thus makes this cell population an ideal target to combine with MV therapy. MV is naturally fusogenic, infected carrier cells can fuse with tumor cells leading to infection of the tumor and protection of the virus from the immune system [87]. Although T cells and monocytes successfully deliver virus to the tumor, monocytes are better able to protect the virus from neutralizing antibodies, as such, monocytes are considered a better choice of cell carrier for MVs [87, 88].

Immune cells have also been used to deliver reovirus and VSV to tumors in a murine metastatic melanoma model. Both T cells and DCs have been used to deliver reovirus [89], while only T cells have been used as a carrier for VSV [85, 86]. The best combination for reducing tumor burden in the LN was using either T cells or mature DCs at low MOIs. However, the treatment of melanoma with reovirus-carrying cells did not lead to significant impact on the primary tumors [89], while VSV-carrying T cells were able to impact primary tumors and metastatic lymph nodes effectively [85, 86]. Finally, the use of cytokine-induced killer (CIK) cells has also been tested in murine xenograft models in combination with oncolytic vaccinia virus. CIK cells are an attractive carrier cell since they are able to traffic virus to the tumor while not presenting any viral antigens, further hiding the virus from the immune system [90]. Although CIK cells were able to kill some xenograft tumors, the combination of CIK and vaccinia virus was better than either intervention alone; this strategy was able to cure all mice engrafted with a CIK sensitive xenograft, while curing upwards of 70% of a resistant xenograft and a syngeneic murine tumor. An additional benefit to the use of CIK cells is their ability to penetrate the tumor mass and uniformly distribute the OV; by comparison, the virus alone would only be able to infect perivascular areas of the tumor [84, 90]. The combination of carrier cells and OV also leads to the induction of antitumor immune responses that are required for the treatment success [85, 89]. Finally, allogeneic tumor cells can also be used to deliver bulk payloads of OV to the tumor. Since OVs have the innate ability to replicate within tumor cells, due to their increased proliferation and impaired antiviral responses, these cells can be used as a "Trojan horse" to deliver large amounts of virus to the tumor. This property is likely to increase the therapeutic dose of virus that is able to reach and infect the tumor [91].

With the technological advances in producing clinical grade immune cells, the use of such cells as vehicles to deliver OVs to tumors while hiding them from the immune system will become important for the use of attenuated human viruses against which patients are seropositive. A great deal of preclinical research remains to determine optimal combinations of virus and immune cell. The cell vehicle used should ideally allow for a productive infection to deliver maximal doses of virus to the tumor causing greater viral replication and tumor cell destruction. However, if sufficient antitumor immune responses can be generated following treatment then less virus may be required as the immune response generated following treatment will destroy any tumor that the virus is unable to kill. It may also be possible to engineer the viruses to help the cells better home to tumors by expressing chemokine receptors to aid in trafficking or to better activate the immune cells with cytokines so that the carrier cell is able to better interact with the immune system.

Future Directions

As we learn more about viral-host interactions and tumor immunology, we should become better able to combine oncolytic viral therapy and other immunotherapies. These two therapeutic strategies are certainly related and in many cases already overlap, although they may do so in an unplanned, or even unexpected fashion. It makes sense to marry these approaches as both have features that compliment the other. Immunotherapies face the immunosuppressive properties of the tumor and may have difficulties in dealing with bulky tumors; by comparison, OVs are generally immunostimulatory and so may reverse the immunosuppressive microenvironment in the tumor while having the potential to debulk tumor burden. On the other hand, the immunogenic nature of these replicating viruses means that antiviral immunity may limit the duration of effective therapeutic impact. However, the generation of antitumor immunity during viral oncolysis may serve to extend the therapeutic effect; that is, the task of cancer destruction is at some point transferred from the OV to the ensuing, and potentially long-term antitumoral immune responses.

Although viral oncolysis appears capable of directly promoting antitumoral immunity in some instances, we do not yet fully understand the mechanism of action of this therapy. Importantly, the magnitude of the ensuing response is rarely impressive. In the near future, the most attractive immunotherapeutic strategies for combination with OV therapy are cell-based or vaccine-based therapies. Pairing OV therapy with adoptive T-cell therapy is an attractive proposition because infection of the tumor by the OV may serve to "call in the troops" by inducing an inflammatory milieu that will reduce or eliminate barriers to lymphocyte infiltration of tumors; in addition, OVs may generate chemotactic signals that enhance killing of tumors by these T cells. Likewise, it may be possible to use OVs to reduce immunosuppression while providing a rich source of destroyed tumor cells; in this manner, tumor antigens may be more optimally delivered *in vivo* to adoptively transferred DCs. Of course, using immune cells to carry OVs past an existing or mounting immune response limits the negative aspects of the patient's immune response while potentially enhancing OV therapy through targeted delivery of the viral vector into tumors, metastases, or involved lymph nodes.

We feel that a particularly attractive means to partner these strategies is through the combination of OVs with antitumor vaccinations. OVs expressing defined tumor antigens can be used to generate specific immune response against source tumors; however, these responses will likely be dominated by antiviral responses, as these vectors are so highly immunogenic. Nonetheless, once such an immune response is generated, subsequent administration of a heterologous OV expressing the same tumor antigen transgene will lead to a prime:boost effect, thereby resulting in an antitumoral immune response that can dominate the antiviral immune response. Indeed, we have recently demonstrated that infection of a tumor in such a primed animal leads to enhanced tumor infiltration by antigen-specific T cells as the immune system goes into the tumor to get at the virus [81]. These observations indicate that separate treatment modalities will have to be combined and that a single vector will not be nearly as efficacious as sequential vaccination/OV therapy using different treatment platforms.

The interaction between OVs and the patient's immune system will continue to be a bi-directional relationship where antiviral immunity continues to be a barrier to successful implementation of the OV therapeutic strategy. Yet, by embracing the immune system and using OVs to drive or enhance antitumor immunity, we may be able to turn the transient impact of the OV into a combined effect with a long-lasting, significant benefit for the cancer patient.

References

1. Jemal A, Siegel R, Ward E et al (2009) Cancer statistics, 2009. CA Cancer J Clin 59(4): 225–249
2. Benencia F, Courreges MC, Conejo-Garcia JR et al (2005) HSV oncolytic therapy upregulates interferon-inducible chemokines and recruits immune effector cells in ovarian cancer. Mol Ther 12: 789–802
3. Endo Y, Sakai R, Ouchi M et al (2008) Virus-mediated oncolysis induces danger signal and stimulates cytotoxic T-lymphocyte activity via proteasome activator upregulation. Oncogene 27: 2375–2381
4. Errington F, White CL, Twigger KR et al (2008) Inflammatory tumour cell killing by oncolytic reovirus for the treatment of melanoma. Gene Ther 15: 1257–1270
5. Endo T, Toda M, Watanabe M et al (2002) In situ cancer vaccination with a replication-conditional HSV for the treatment of liver metastasis of colon cancer. Cancer Gene Ther 9: 142–148
6. Hummel JL, Safroneeva E, Mossman KL (2005) The role of ICP0-Null HSV-1 and interferon signaling defects in the effective treatment of breast adenocarcinoma. Mol Ther 12: 1101–1110
7. Li H, Dutuor A, Fu X et al (2007) Induction of strong antitumor immunity by an HSV-2-based oncolytic virus in a murine mammary tumor model. J Gene Med 9: 161–169
8. Li H, Dutuor A, Tao L et al (2007) Virotherapy with a type 2 herpes simplex virus-derived oncolytic virus induces potent antitumor immunity against neuroblastoma. Clin Cancer Res 13: 316–322

9. Nakamori M, Fu X, Rousseau R et al (2004) Destruction of nonimmunogenic mammary tumor cells by a fusogenic oncolytic herpes simplex virus induces potent antitumor immunity. Mol Ther 9: 658–665
10. Toda M, Iizuka Y, Kawase T et al (2002) Immuno-viral therapy of brain tumors by combination of viral therapy with cancer vaccination using a replication-conditional HSV. Cancer Gene Ther 9: 356–364
11. Diaz RM, Galivo F, Kottke T et al (2007) Oncolytic immunovirotherapy for melanoma using vesicular stomatitis virus. Cancer Res 67: 2840–2848
12. Moehler MH, Zeidler M, Wilsberg V et al (2005) Parvovirus H-1-induced tumor cell death enhances human immune response *in vitro* via increased phagocytosis, maturation, and cross-presentation by dendritic cells. Hum Gene Ther 16: 996–1005
13. Khawli LA, Hu P, Epstein AL (2008) Cytokine, chemokine, and co-stimulatory fusion proteins for the immunotherapy of solid tumors. Handb Exp Pharmacol 181(5): 291–328
14. Greiner S, Humrich JY, Thuman P et al (2006) The highly attenuated vaccinia virus strain modified virus Ankara induces apoptosis in melanoma cells and allows bystander dendritic cells to generate a potent anti-tumoral immunity. Clin Exp Immunol 146: 344–353
15. Schulz O, Diebold SS, Chen M et al (2005) Toll-like receptor 3 promotes cross-priming to virus-infected cells. Nature 433: 887–892
16. Lindenmann J, Klein PA (1967) Viral oncolysis: increased immunogenicity of host cell antigen associated with influenza virus. J Exp Med 126: 93–108
17. Freedman RS, Edwards CL, Bowen JM et al (1988) Viral oncolysates in patients with advanced ovarian cancer. Gynecol Oncol 29: 337–347
18. Lotzova E, Savary CA, Freedman RS et al (1984) Natural killer cell cytotoxic potential of patients with ovarian carcinoma and its modulation with virus-modified tumor cell extract. Cancer Immunol Immunother 17: 124–129
19. Savage HE, Rossen RD, Hersh EM et al (1986) Antibody development to viral and allogeneic tumor cell-associated antigens in patients with malignant melanoma and ovarian carcinoma treated with lysates of virus-infected tumor cells. Cancer Res 46: 2127–2133
20. Cassel WA, Murray DR, Phillips HS (1983) A phase II study on the postsurgical management of Stage II malignant melanoma with a Newcastle disease virus oncolysate. Cancer 52: 856–860
21. Fitzgerald-Bocarsly P, Feng D (2007) The role of type I interferon production by dendritic cells in host defense. Biochimie 89: 843–855
22. Berthier-Vergnes O, Portoukalian J, Leftheriotis E et al (1994) Induction of IgG antibodies directed to a M(r) 31,000 melanoma antigen in patients immunized with vaccinia virus melanoma oncolysates. Cancer Res 54: 2433–2439
23. Arroyo PJ, Bash JA, Wallack MK (1990) Active specific immunotherapy with vaccinia colon oncolysate enhances the immunomodulatory and antitumor effects of interleukin-2 and interferon alpha in a murine hepatic metastasis model. Cancer Immunol Immunother 31: 305–311
24. Barnavon Y, Iwaki H, Bash JA et al (1988) Treatment of murine hepatic metastases with vaccinia colon oncolysates and IL-2. J Surg Res 45: 523–530
25. Batliwalla FM, Bateman BA, Serrano D et al (1998) A 15-year follow-up of AJCC stage III malignant melanoma patients treated postsurgically with Newcastle disease virus (NDV) oncolysate and determination of alterations in the CD8 T cell repertoire. Mol Med 4: 783–794
26. Tanaka N, Sivanandham M, Wallack MK (1994) Immunotherapy of a vaccinia colon oncolysate prepared with interleukin-2 gene-encoded vaccinia virus and interferon-alpha increases the survival of mice bearing syngeneic colon adenocarcinoma. J Immunother Emphasis Tumor Immunol 16: 283–293
27. Wallack MK (1981) Specific immunotherapy with vaccinia oncolysates. Cancer Immunol Immunother 12: 1–4
28. Belongia EA, Naleway AL (2003) Smallpox vaccine: the good, the bad, and the ugly. Clin Med Res 1: 87–92
29. Nelson CB, Pomeroy BS, Schrall K et al (1952) An outbreak of conjunctivitis due to Newcastle disease virus (NDV) occurring in poultry workers. Am J Public Health Nations Health 42: 672–678

30. Cassel WA, Murray DR (1992) A ten-year follow-up on stage II malignant melanoma patients treated postsurgically with Newcastle disease virus oncolysate. Med Oncol Tumor Pharmacother 9: 169–171
31. Lee YS, Kim JH, Choi KJ et al (2006) Enhanced antitumor effect of oncolytic adenovirus expressing interleukin-12 and B7-1 in an immunocompetent murine model. Clin Cancer Res 12: 5859–5868
32. Reschner A, Hubert P, Delvenne P et al (2008) Innate lymphocyte and dendritic cell cross-talk: a key factor in the regulation of the immune response. Clin Exp Immunol 152: 219–226
33. Tan JK, O'Neill HC (2005) Maturation requirements for dendritic cells in T cell stimulation leading to tolerance versus immunity. J Leukoc Biol 78: 319–324
34. Choi KJ, Kim JH, Lee YS et al (2006) Concurrent delivery of GM-CSF and B7-1 using an oncolytic adenovirus elicits potent antitumor effect. Gene Ther 13: 1010–1020
35. Fukuhara H, Ino Y, Kuroda T et al (2005) Triple gene-deleted oncolytic herpes simplex virus vector double-armed with interleukin 18 and soluble B7-1 constructed by bacterial artificial chromosome-mediated system. Cancer Res 65: 10,663–10,668
36. Shin EJ, Wanna GB, Choi B et al (2007) Interleukin-12 expression enhances vesicular stomatitis virus oncolytic therapy in murine squamous cell carcinoma. Laryngoscope 117: 210–214
37. Su C, Peng L, Sham J et al (2006) Immune gene-viral therapy with triplex efficacy mediated by oncolytic adenovirus carrying an interferon-gamma gene yields efficient antitumor activity in immunodeficient and immunocompetent mice. Mol Ther 13: 918–927
38. Varghese S, Rabkin SD, Liu R et al (2006) Enhanced therapeutic efficacy of IL-12, but not GM-CSF, expressing oncolytic herpes simplex virus for transgenic mouse derived prostate cancers. Cancer Gene Ther 13: 253–265
39. Vigil A, Park MS, Martinez O et al (2007) Use of reverse genetics to enhance the oncolytic properties of Newcastle disease virus. Cancer Res 67: 8285–8292
40. Zhao H, Janke M, Fournier P et al (2008) Recombinant Newcastle disease virus expressing human interleukin-2 serves as a potential candidate for tumor therapy. Virus Res 136: 75–80
41. Carew JF, Kooby DA, Halterman MW et al (2001) A novel approach to cancer therapy using an oncolytic herpes virus to package amplicons containing cytokine genes. Mol Ther 4: 250–256
42. Dasgupta S, Bhattacharya-Chatterjee M, O'Malley BW, Jr. et al (2006) Recombinant vaccinia virus expressing interleukin-2 invokes anti-tumor cellular immunity in an orthotopic murine model of head and neck squamous cell carcinoma. Mol Ther 13: 183–193
43. Liu TC, Hwang T, Park BH et al (2008) The targeted oncolytic poxvirus JX-594 demonstrates antitumoral, antivascular, and anti-HBV activities in patients with hepatocellular carcinoma. Mol Ther 16: 1637–1642
44. Park BH, Hwang T, Liu TC et al (2008) Use of a targeted oncolytic poxvirus, JX-594, in patients with refractory primary or metastatic liver cancer: a phase I trial. Lancet Oncol 9: 533–542
45. Lapteva N, Aldrich M, Weksberg D et al (2009) Targeting the intratumoral dendritic cells by the oncolytic adenoviral vaccine expressing RANTES elicits potent antitumor immunity. J Immunother 32: 145–156
46. Liu BL, Robinson M, Han ZQ et al (2003) ICP34.5 deleted herpes simplex virus with enhanced oncolytic, immune stimulating, and anti-tumour properties. Gene Ther 10: 292–303
47. Kim JH, Oh JY, Park BH et al (2006) Systemic armed oncolytic and immunologic therapy for cancer with JX-594, a targeted poxvirus expressing GM-CSF. Mol Ther 14: 361–370
48. Todo T, Martuza RL, Dallman MJ et al (2001) In situ expression of soluble B7-1 in the context of oncolytic herpes simplex virus induces potent antitumor immunity. Cancer Res 61: 153–161
49. Mueller DL, Jenkins MK, Schwartz RH (1989) Clonal expansion versus functional clonal inactivation: a costimulatory signalling pathway determines the outcome of T cell antigen receptor occupancy. Annu Rev Immunol 7: 445–480
50. Ino Y, Saeki Y, Fukuhara H et al (2006) Triple combination of oncolytic herpes simplex virus-1 vectors armed with interleukin-12, interleukin-18, or soluble B7-1 results in enhanced antitumor efficacy. Clin Cancer Res 12: 643–652
51. Trinchieri G, Pflanz S, Kastelein RA (2003) The IL-12 family of heterodimeric cytokines: new players in the regulation of T cell responses. Immunity 19: 641–644

52. de Jong EC, Smits HH, Kapsenberg ML (2005) Dendritic cell-mediated T cell polarization. Springer Semin Immunopathol 26: 289–307
53. Lui VW, He Y, Falo L et al (2002) Systemic administration of naked DNA encoding interleukin 12 for the treatment of human papillomavirus DNA-positive tumor. Hum Gene Ther 13: 177–185
54. Leonard JP, Sherman ML, Fisher GL et al (1997) Effects of single-dose interleukin-12 exposure on interleukin-12-associated toxicity and interferon-gamma production. Blood 90: 2541–2548
55. Wong RJ, Chan MK, Yu Z et al (2004) Angiogenesis inhibition by an oncolytic herpes virus expressing interleukin 12. Clin Cancer Res 10: 4509–4516
56. Rogers LJ, Eva LJ, Luesley DM (2008) Vaccines against cervical cancer. Curr Opin Oncol 20: 570–574
57. Lollini PL, Cavallo F, Nanni P et al (2006) Vaccines for tumour prevention. Nat Rev Cancer 6: 204–216
58. Lonchay C, van der Bruggen P, Connerotte T et al (2004) Correlation between tumor regression and T cell responses in melanoma patients vaccinated with a MAGE antigen. Proc Natl Acad Sci USA 101 Suppl 2: 14,631–14,638
59. Harrop R, Connolly N, Redchenko I et al (2006) Vaccination of colorectal cancer patients with modified vaccinia Ankara delivering the tumor antigen 5T4 (TroVax) induces immune responses which correlate with disease control: a phase I/II trial. Clin Cancer Res 12: 3416–3424
60. Amato RJ, Shingler W, Naylor S et al (2008) Vaccination of renal cell cancer patients with modified vaccinia ankara delivering tumor antigen 5T4 (TroVax) administered with interleukin 2: a phase II trial. Clin Cancer Res 14: 7504–7510
61. Kaufman HL, Taback B, Sherman W et al (2009) Phase II trial of Modified Vaccinia Ankara (MVA) virus expressing 5T4 and high dose Interleukin-2 (IL-2) in patients with metastatic renal cell carcinoma. J Transl Med 7: 2
62. Kaufman HL, Lenz HJ, Marshall J et al (2008) Combination chemotherapy and ALVAC-CEA/B7.1 vaccine in patients with metastatic colorectal cancer. Clin Cancer Res 14: 4843–4849
63. Horig H, Lee DS, Conkright W et al (2000) Phase I clinical trial of a recombinant canarypoxvirus (ALVAC) vaccine expressing human carcinoembryonic antigen and the B7.1 co-stimulatory molecule. Cancer Immunol Immunother 49: 504–514
64. Jager E, Karbach J, Gnjatic S et al (2006) Recombinant vaccinia/fowlpox NY-ESO-1 vaccines induce both humoral and cellular NY-ESO-1-specific immune responses in cancer patients. Proc Natl Acad Sci USA 103: 14,453–14,458
65. Harrop R, John J, Carroll MW (2006) Recombinant viral vectors: cancer vaccines. Adv Drug Deliv Rev 58: 931–947
66. Elzey BD, Siemens DR, Ratliff TL et al (2001) Immunization with type 5 adenovirus recombinant for a tumor antigen in combination with recombinant canarypox virus (ALVAC) cytokine gene delivery induces destruction of established prostate tumors. Int J Cancer 94: 842–849
67. Irvine KR, Chamberlain RS, Shulman EP et al (1997) Enhancing efficacy of recombinant anticancer vaccines with prime/boost regimens that use two different vectors. J Natl Cancer Inst 89: 1595–1601
68. Naslund TI, Uyttenhove C, Nordstrom EK et al (2007) Comparative prime-boost vaccinations using Semliki Forest virus, adenovirus, and ALVAC vectors demonstrate differences in the generation of a protective central memory CTL response against the P815 tumor. J Immunol 178: 6761–6769
69. Hodge JW, Poole DJ, Aarts WM et al (2003) Modified vaccinia virus ankara recombinants are as potent as vaccinia recombinants in diversified prime and boost vaccine regimens to elicit therapeutic antitumor responses. Cancer Res 63: 7942–7949
70. Hodge JW, McLaughlin JP, Kantor JA et al (1997) Diversified prime and boost protocols using recombinant vaccinia virus and recombinant non-replicating avian pox virus to enhance T-cell immunity and antitumor responses. Vaccine 15: 759–768
71. Marshall JL, Hoyer RJ, Toomey MA et al (2000) Phase I study in advanced cancer patients of a diversified prime-and-boost vaccination protocol using recombinant vaccinia virus and recombinant nonreplicating avipox virus to elicit anti-carcinoembryonic antigen immune responses. J Clin Oncol 18: 3964–3873

72. Palmer DH, Midgley RS, Mirza N et al (2009) A phase II study of adoptive immunotherapy using dendritic cells pulsed with tumor lysate in patients with hepatocellular carcinoma. Hepatology 49: 124–132
73. Shen X, Berger CL, Tigelaar R et al (2008) Development of immunogenic tumor-loaded dendritic cells through physical perturbation and apoptotic cell loading. Immunol Investig 37: 798–821
74. Bonehill A, Van Nuffel AM, Corthals J et al (2009) Single-step antigen loading and activation of dendritic cells by mRNA electroporation for the purpose of therapeutic vaccination in melanoma patients. Clin Cancer Res 15: 3366–3375
75. Kaplan JM, Yu Q, Piraino ST et al (1999) Induction of antitumor immunity with dendritic cells transduced with adenovirus vector-encoding endogenous tumor-associated antigens. J Immunol 163: 699–707
76. Ojima T, Iwahashi M, Nakamura M et al (2006) The boosting effect of co-transduction with cytokine genes on cancer vaccine therapy using genetically modified dendritic cells expressing tumor-associated antigen. Int J Oncol 28: 947–953
77. Farrell CJ, Zaupa C, Barnard Z et al (2008) Combination immunotherapy for tumors via sequential intratumoral injections of oncolytic herpes simplex virus 1 and immature dendritic cells. Clin Cancer Res 14: 7711–7716
78. Vigil A, Martinez O, Chua MA et al (2008) Recombinant Newcastle disease virus as a vaccine vector for cancer therapy. Mol Ther 16: 1883–1890
79. Bronte V, Tsung K, Rao JB et al (1995) IL-2 enhances the function of recombinant poxvirus-based vaccines in the treatment of established pulmonary metastases. J Immunol 154: 5282–5292
80. Boudreau JE, Bridle BW, Stephenson KB et al (2009) Recombinant vesicular stomatitis virus transduction of dendritic cells enhances their ability to prime innate and adaptive antitumor immunity. Mol Ther 17: 1465–1472
81. Bridle BW, Hanson S, Lichty BD (2010) Combining oncolytic virotherapy and tumour vaccination. Cytokine Growth Factor Rev. S1359-6101(10)00020-1 [pii] 10.1016/j.cytogfr.2010.02.009
82. Xu F, Sternberg MR, Kottiri BJ et al (2006) Trends in herpes simplex virus type 1 and type 2 seroprevalence in the United States. J Am Med Assoc 296: 964–973
83. Tai JH, Williams JV, Edwards KM et al (2005) Prevalence of reovirus-specific antibodies in young children in Nashville, Tennessee. J Infect Dis 191: 1221–4
84. Thorne SH, Negrin RS, Contag CH (2006) Synergistic antitumor effects of immune cell-viral biotherapy. Science 311: 1780–1784
85. Qiao J, Kottke T, Willmon C et al (2008) Purging metastases in lymphoid organs using a combination of antigen-nonspecific adoptive T cell therapy, oncolytic virotherapy and immunotherapy. Nat Med 14: 37–44
86. Qiao J, Wang H, Kottke T et al (2008) Loading of oncolytic vesicular stomatitis virus onto antigen-specific T cells enhances the efficacy of adoptive T-cell therapy of tumors. Gene Ther 15: 604–616
87. Iankov ID, Blechacz B, Liu C et al (2007) Infected cell carriers: a new strategy for systemic delivery of oncolytic measles viruses in cancer virotherapy. Mol Ther 15: 114–122
88. Ong HT, Hasegawa K, Dietz AB et al (2007) Evaluation of T cells as carriers for systemic measles virotherapy in the presence of antiviral antibodies. Gene Ther 14: 324–333
89. Ilett EJ, Prestwich RJ, Kottke T et al (2009) Dendritic cells and T cells deliver oncolytic reovirus for tumour killing despite pre-existing anti-viral immunity. Gene Ther 16: 689–699
90. Thorne SH, Contag CH (2008) Integrating the biological characteristics of oncolytic viruses and immune cells can optimize therapeutic benefits of cell-based delivery. Gene Ther 15: 753–758
91. Power AT, Wang J, Falls TJ et al (2007) Carrier cell-based delivery of an oncolytic virus circumvents antiviral immunity. Mol Ther 15: 123–130
92. Dudley ME, Wunderlich JR, Robbins PF et al (2002) Cancer regression and autoimmunity in patients after clonal repopulation with antitumor lymphocytes. Science 298: 850–854

Chapter 17
Radiation Therapy and Cancer Treatment: From the Basics to Combination Therapies that Ignite Immunity

Amanda Moretti, David A. Jaffray, and Jeffrey A. Medin

Abstract Since its inception in the late nineteenth century, radiation therapy has been a mainstay in the clinic to treat and palliate patients afflicted with cancer. With the rapid advancement in technology for radiation therapy equipment, delivery has improved and late sequalae have been minimized. Alongside the technological advancements, several groundbreaking biological discoveries have enabled scientists to delineate the cellular and molecular implications of radiation exposure. With a greater understanding of the cellular and coordinated tissue responses to irradiation, it became apparent that radiation can influence the activation state of the immune system. In recent years, strategies to enhance the biological response to tumors after radiation therapy have led to combination therapies that stimulate the immune system. Harnessing the specificity and efficiency of the immune system after radiation therapy offers potential for mobilizing the body`s innate defense system. This chapter will focus on the rationale for coupling radiation therapy and immunotherapy, and will provide evidence of its efficacy in preclinical and clinical studies.

Keywords Antibodies • Cell therapy • Cytokines • Gene therapy • Radiation therapy • Immunotherapy

Introduction

Current treatment modalities have proven insufficient to cure many forms of cancer. Patients may develop recurrent lesions or the disease may metastasize, thereby leading to low cure rates and high mortality, as well as significant treatment related side-effects. Consequently, cancer accounts for more than 25% of all deaths in the

---

D.A. Jaffray (✉)
Departments of Radiation Oncology and Medical Biophysics,
Princess Margaret Hospital/Ontario Cancer Institute,
University of Toronto, Toronto ON, Canada
e-mail: david.jaffray@rmp.uhn.on.ca

J. Medin and D. Fowler (eds.), *Experimental and Applied Immunotherapy*,
DOI 10.1007/978-1-60761-980-2_17,

United States and Canada, second only to heart disease [1]. Today, in the era of translational medicine, there is an urgent need to evolve our current paradigms for cancer treatment.

Immunotherapy has emerged as a treatment modality with great potential. Standard cancer treatments, such as chemotherapy, radiation therapy, and surgery aim to debulk tumors by cell killing or resection. Immunotherapy, conversely, aims to use the inherent capability of the immune system to specifically recognize and eliminate cancer cells. The unique nature of the immune system is further suited for such treatment purposes given its potential to survey and respond to recurrent and metastatic lesions without additional treatment. By strategically combining immunotherapy with other treatment modalities, there exists a potential to manipulate existing pathways of regulation and tolerance to yield therapeutic benefit [2–5]. This chapter will focus on the rationale for combining radiation therapy and immunotherapy, providing both preclinical and clinical evidence of its efficacy.

Background

The disciplines of radiation physics and biology have made tremendous progress in the last century. In 1895 Roentgen described the x-ray and the potential application of these agents was quickly realized. Only two months after this discovery, Grubbé, a medical student and novice experimenter, used x-rays therapeutically on a patient with locally advanced breast carcinoma. In doing so, Grubbé became one of the first people to treat cancer with radiation therapy. Five months later, Despeignes was the first to publish results using Roentgen's x-rays. Treating gastric carcinoma, he reported an improvement in tumor burden and pain relief in the patient [6].

In the early 1900s, radioactive isotopes, such as radium, were available and being used therapeutically to treat both benign and malignant disease. With its increasing popularity, concerns were raised regarding the adverse effects of radiation exposure. Insights into the biological effects of radiation emerged, and cellular dose-responses were quantified. Initial experiments investigating the effects of radiation exposure provided the foundation for our current understanding of radiation effects on tissues, and explained the observed disparities in results from early treatment protocols.

Standardization of treatment protocols became a necessity with the increasing usage of radiation therapy to treat cancer. Initial x-ray machines were unable to produce high-energy, penetrating rays to treat deep-seated tumors without severe skin reactions. With enhanced cooperation among medical and engineering professionals post-World War II, new sources of radiation, such as cesium and cobalt, along with new high-powered treatment machines were being used in the clinic. By the late 1950s, megavoltage equipment, such as linear accelerators, were being developed. In the 1980s, computers were used to calculate dose distribution, and treatment protocols were modified to increase reproducibility. With the advent of computer imaging, two-dimensional imaging was replaced by three-dimensional imaging for treatment planning. Three-dimensional (3D) imaging enabled radiologists to delineate the volume of the tumor and healthy tissues in close proximity with greater precision. This allowed conformation of the radiation beam to encompass the entire tumor while minimizing the inclusion of healthy tissues in the planned target volume for treatment. The development of intensity-modulated radiation therapy (IMRT) was an improvement to 3D conformal radiation therapy. IMRT offered optimal sculpting of the radiation beams to the shape of the tumor and allowed sharper dose gradients to be achieved to avoid healthy tissues. Despite these advances, normal organ movements and daily errors in patient positioning [7] led to inaccuracies in the precise doses delivered to the patient. This challenge led to the development of image-guided radiation therapy (IGRT). IGRT employs a linear accelerator integrated with an imaging system for guiding the placement of the radiation field relative to the body. This technology permits localization and targeting of the tumor at the time of treatment for accurate and precise delivery. IGRT further reduces the volume of normal tissue receiving radiation, thus reducing normal tissue toxicity, and allowing larger doses of radiation to be safely delivered. [8, 9] Most recently, combined use of IMRT and IGRT technologies has led to the emergence of stereotactic body radiation therapy (SBRT), wherein highly focused, high-dose treatments are delivered in fewer fractions [9]. This treatment schema is a departure from the established treatment paradigm of fractionated radiation therapy.

The practice of fractionated radiation therapy was initially used to allow the repair of sublethal damage and repopulation of cells in healthy tissues between fractions. Given that the planned target volume for irradiation included a significant volume of normal tissues prior to the development of detailed imaging modalities, damage to healthy tissues after large doses of radiation necessitated the delivery of treatment in multiple small fractions. Fractionated radiation therapy delivered at 1.8–2 Gy per fraction requires prolonged treatment schedules. These lengthy treatments may allow repair and accelerated repopulation of tumor cells between fractions, thus contributing to local tumor recurrence. As emergent technologies enable precise targeting of the radiation beams to the tumor while excluding healthy tissues, the optimal fraction dose and schedule requires re-evaluation.

SBRT aims to achieve the optimal therapeutic ratio by increasing the probability of tumor control while minimizing normal tissue toxicity. Reconstruction of the tumor volume using high quality images enables 3D analyses and precise treatment planning. SBRT radiation fields are only slightly larger than the gross tumor volume and steep dose gradients tightly conform to the tumor. Consequently, higher doses of radiation can be delivered to the tumor in a single treatment, and fewer fractions are required to achieve a biologically effective dose. SBRT typically uses ablative ranges of radiation doses (≥10 Gy/fraction) with a biologically effective dose ≥75–100 Gy [10]. Fowler *et al.* [11] compared the theoretical relative biological effectiveness of conventional fractionated dose regimes and SBRT regimes. An SBRT schedule in the range of 45–69 Gy in three to five fractions was expected to have at least twice the relative biological effectiveness in nonsmall cell lung cancer as a conventional fractionated schedule of 60–70 Gy in 30–35 fractions. Despite the expected increase in tumor control probability using SBRT, the radiobiology and the long-term implications of high-doses delivered to a small volume by hypofractionation are not well understood. Milano *et al.* [12] reviewed the literature on hypofractionation by SBRT, focusing on the safety and late tissue toxicities. Although early SBRT clinical data demonstrate efficacy and safety, this review highlights the importance of long-term follow-up, appropriate dosing schedules, and a better understanding of the limits of tolerance for SBRT in different types of tissues. It has been suggested by Fuks *et al.* [13] that conventional fractionated therapy and SBRT operate through different mechanisms to achieve tumor control. Fractionated low-dose radiation therapy (1.8–3 Gy/fraction) generates reactive oxygen species in tumor cells that subsequently results in the release of growth factors and cytokines that modify the radiation response. For instance, vascular endothelial growth factor (VEGF) and angiogenic factors produced post-irradiation attenuate endothelial cell apoptosis and support endothelial survival, thereby leading to tumor cell survival. Single-fraction high-dose radiation therapy (>8–10 Gy), on the other hand, can lead to lethal damage in tumor and endothelial cells. In both human and murine tumors, early death of endothelial cells preceded tumor cell apoptosis, thus suggesting an important link between endothelial and tumor cell survival. Endothelial cells exposed to single high-doses of radiation primarily undergo apoptosis due to ceramide-mediated signaling. High doses of radiation cause the accumulation of ceramide by acid spingomyelinase hydrolysis of sphingomyelin [14], or by the overcoming of ataxia telengiectasia mutated (ATM)-mediated inhibition of synthesis by ceramide synthetase [15]. Given that high-dose radiation treatment may have a greater tumor control probability due to endothelial and tumor cell apoptosis, it is appealing to postulate that SBRT may become a part of standard radiation therapy practice.

With such innovation in radiation treatment delivery and insight into the biological implications of high-dose radiation therapy, the stage has been set for combining other treatment modalities with radiation therapy.

Radiation Interactions with Matter

Radiation therapy delivers packets of energy released in the form of photons (gamma and x-rays) or particulate matter (electrons, neutrons, protons, or heavy charged ions). Although particulate radiation is used in radiation therapy, photons are used most frequently in treatment. When photons interact in matter, energy is transferred to secondary electrons. These electrons cause ionization of biological molecules either directly or indirectly via hydrolysis products and reactive radical intermediates. Indirect ionization damage occurs through chemical intermediates, and accounts for approximately 2/3 of the biological damage that is produced by x-rays [16]. Studies have estimated that a 1-Gy dose of x-ray radiation can result in 10^5 ionization events per cell, producing 1,000–2,000 single-stranded breaks (SSB) and 40 double-stranded breaks (DSB) [16, 17]. Although both DSB and SSB occur in DNA, DSB are thought to represent the principal lethal event [18]. DNA repair of SSB and DSB occurs quickly after radiation exposure; however, not all breaks are repaired with high fidelity. As such, the reproductive viability of cells can be altered and cell fate may be: apoptosis mitotic catastrophe leading to, necrosis, senescence, or terminal differentiation.

Factors that Shape the Cellular Radiation Response

Cellular responses to radiation strongly depend on the stage of the cell cycle and the presence or absence of free radical scavengers and biomolecules. Cells are most sensitive to radiation immediately before, during, and after mitosis. Given that the cellular machinery is mobilized to replicate DNA during S phase, radiation-induced DNA damage may be sensed and repaired more rapidly in this phase of the cell cycle. This biology may account for the increased radio-resistance observed during S phase. Rapidly dividing cells such as tumour cells are more radio-responsive than slowly dividing cells. As most irradiated cells undergo mitotic death, rapidly dividing populations are more rapidly depleted of cells than slowly dividing populations. However, irradiation does not always successfully eliminate all tumour cells. This is due to the unsynchronized cell division in the tumour and the nutrient and oxygenation status within the tumour bed. As the tumor grow, the disorganized vascular supply becomes insufficient to perfuse the expanding mass. As a result, many areas are outside of the diffusion limits of the vessels. In these areas, the hypoxic and nutrient-deprived cells adapt to the environment. Genetic stability and DNA repair capabilities are compromised in these cells and may allow the propagation of mutations in proliferating cells. These changes can influence the cellular responses to radiation damage [19]. The presence of oxygen at the time of irradiation fixes ionization damage, thus rendering more difficult to repair as such, cells exposed to partial pressures of oxygen lower than 10 mm Hg become two to three times more radio-resistant than normoxic cells [20]. Clinical studies have shown that patients irradiated in hyperbaric oxygen chambers benefit from improved local tumor control by 10% [21]. Despite the improved local tumor control obtained with hyperbaric oxygen chambers, treatment delivery using this method is difficult to administer. Given this, several radiation sensitizers have been designed to mimic the effect of oxygen to sensitize hypoxic cells in tumors. For example, agents such as nitroimidazoles have been used in the clinic, with encouraging results [22]. Additionally, fractionation of radiation therapy delivery has been used to sensitize radio-resistant cells. Fractionation re-sensitizes cells between dose fractions, thereby allowing for re-oxygenation of tumor cells and redistribution of cells in the cell cycle. Other agents used to sensitize tumor cells to radiation operate by different mechanisms. Nucleoside analogs, including 5`-fluorouracil, gemcitabine, and bromodeoxyuridine, act to sensitize cells to radiation by drug incorporation into DNA and/or RNA, with subsequent inhibition of nucleotide synthesis machinery [23]. Classical chemotherapeutic agents, such as cisplatin and paclitaxel, act to sensitize cells by a variety of different mechanisms [23]. For example, paclitaxel, which is a taxane type of chemotherapy agent, synchronizes cells in the G_2/M radio-sensitive phase of the cell cycle [24] and has been used in clinical trials in combination with radiotherapy [25].

Cellular Sensing and Responses to Radiation

The damage incurred by cells after radiation exposure initiates a variety of transcriptional responses in the cell that may alter cell cycle progression, induce DNA repair, and/or trigger cell death. SSBs and DSBs induced by radiation are sensed by proteins that activate surveillance proteins, including p53, ATM, and DNA-protein kinase. These surveillance proteins initiate cell-cycle arrest to allow repair of DNA either by homologous recombination or nonhomologous end-joining pathways [26, 27]. DNA damage that cannot be adequately repaired initiates signaling through ATM and p53 to initiate apoptosis via the mitochondrial pathway [26, 28]. In addition to irreparable DNA-damage, the radiation-induced activation of ceramide synthetase [15] and sphingomyelinase [27, 29] increases intracellular ceramide levels and promotes apoptosis via a cascade of second messengers [14, 15, 27, 29].

Membrane and cytoplasmic sensors of damage activate second messengers for communicating signals to the nucleus to coordinate the cellular response to radiation. MAPK, PI3K, and JAK/STAT are all downstream signaling pathways implicated in responses to radiation. The MAPK pathway consists of two distinct and differential cascades: (1) a mitogen and growth factor induced pathway (MAPK/ERK); and (2) an inflammatory and cellular stress (e.g., radiation) induced pathway (SAPK/JNK). Studies have shown the involvement of both of these pathways after radiation damage, with disparate end results of either survival and proliferation or apoptosis, respectively [26, 30]. Depending on the extent of DNA damage, signaling through the MAPK/ERK pathways may have dual effects, being activated with low levels of DNA damage and inactivated with high levels of nonrepairable damage [26].

Reactive oxygen and nitrogen species produced after irradiation can damage intracellular second messengers. This damage can alter the conformation of proteins and influence the signals sent to the nucleus [27]. In the nucleus, a variety of transcription factors downstream of the radiation-induced signaling cascades control gene expression. Genes that are affected by these cascades are involved in the cellular response to radiation exposure, and include genes involved in cell-cycle checkpoints, DNA repair, and inflammation.

Within hours after radiation exposure, cells produce factors that coordinate the tissues' response to radiation. Such factors include cytokines, chemokines, surface receptors, adhesion molecules, and enzymes. Cytokines, such as tumor necrosis factor- α (TNF-α), and death receptors, such as Fas, lead to downstream signaling after receptor ligation. These act in concert with radiation to induce apoptosis in exposed cells [31]. Transforming growth factor-β (TGF-β), however, can mediate cyto-protective effects in cells after radiation exposure. The mechanism of this protection varies with the cell type, being either MAPK- or PI3K-dependent [30]. Although several genes can be expressed in response to radiation damage, the ultimate effect is to restore tissue homeostasis. The outcome of the coordinated response to radiation-induced damage is determined by: the extent of damage; the signaling cascades initiated; and the stress-response genes expressed. The summation of these cellular signals will instruct the cell to either survive or undergo apoptosis. Cell death after radiation can occur premitotically, mitotically, or postmitotically via apoptosis or necrosis. The decision to undergo apoptosis versus necrosis is influenced by the integration of internal and external signals by the cell.

Cell Death

Under physiological conditions, cells undergoing apoptosis are rapidly cleared by phagocytes without activation of immune responses [32, 33]. Alternatively, necrosis is not immunologically silent [32, 33]. That is, necrosis is associated with the release of cellular debris into the extracellular space; then, proinflammatory mediators orchestrate the recruitment and activation of multiple innate and adaptive immune elements [33, 34]. Radiation can induce both necrotic and apoptotic cell death. The extent of each process depends on the state of the tumor, the natural propensity of the cells to undergo each process, and the damage incurred by the cells. Because most tumors are not exquisitely sensitive to radiation given their hypoxic, hypoglycemic, and mutated states, tumors do not often respond immediately to radiation therapy. This delayed death occurring after several mitoses is frequently considered necrotic due to the increase in cellular swelling and membrane permeability that occurs [35, 36]. Cell death after many mitotic cycles is generally attributed to unrepaired DNA breaks and chromosomal aberrations that cause genomic instability. Additionally, there is a smaller portion of radio-sensitive cells that die immediately after irradiation or in G2 arrest [35, 36].

The mode of cell death and the radio-sensitivity of tumors are influenced by the intratumoral levels of both glucose and oxygen. Although hypoxia may limit DNA damage and suppress radiation-induced SAPK activation [28], hypoglycemia in the tumor may sensitize cells to radiation. Glucose insufficiency limits the ATP-dependent phosphorylation cascades necessary to mediate the mitochondrial pathway of apoptosis [28] and limits the repair of DNA breaks by poly (ADP-ribose) polymerase (PARP) [37]. These observations suggest that DNA damage may not be adequately repaired in these cells after radiation-induced injury, and necrosis will occur as a result of failed apoptosis. Additionally, cells proceeding to secondary necrosis after delayed clearance by phagocytic cells may release danger signals into the microenvironment that will ignite the inflammatory reaction. Therefore, the type of death, timing of clearance, and factors released during cell death engender specialized responses from the host that determine the general activation state of the immune system after irradiation.

The Danger Hypothesis

The apoptotic and necrotic processes occurring in tumors act in opposing pathways of immunity. Treatments that alter the balance of apoptosis and necrosis occurring in the tumor may skew responses towards immune activation. As a tumor-specific immune response is desirable after radiation therapy, it is advantageous if radiation induces a form of cell death that is not immunologically inert. Matzinger [38] proposed a danger hypothesis, which states that dying cells release signals that prompt the immune system to recognize forms of nonphysiological cell death. Such endogenous danger signals released from necrotic or late apoptotic cells mediate inflammation and activate cells of the innate and adaptive immune system. Several danger signals have been identified, and include intracellular molecules such as heat-shock proteins (HSP), calreticulin, and high-mobility group box 1 proteins (HMGB1) [32, 39].

HSPs and calreticulin are intracellular proteins that are translocated to the plasma membrane on stressed or dying cells. HSPs bind toll-like receptor (TLR)-4 (TLR4) and CD14 on dendritic cells (DCs), thereby inducing maturation and the release of proinflammatory cytokines [33]. As HSPs are often noncovalently linked to cellular peptides, HSPs may facilitate cross-presentation of tumor antigens from dying tumor cells [40]. Calreticulin, on the other hand, stimulates the rapid uptake of tumor cell remnants by DCs [33]. HMGB1, which binds TLR2 and TLR4, is actively secreted by inflammatory cells and passively secreted from necrotic cells [32, 33, 39]. The binding of HMGB1 to TLR4 on DCs results in optimal antigen processing [33, 39]. It is thought that the DC-T cell cross-talk relies on calreticulin expression as an "eat me" signal to phagocytic cells, and HMGB1 secretion as a "danger" signal to license DCs for antigen uptake and processing [41]. In summary, we are currently pursuing the following hypothesis: the release of danger signals from dying tumor cells after irradiation plays a role in revving up the immune response to tip the balance in favor of the host defense against the tumor (Fig. 17.1).

The Tumor Microenvironment and Radiation

Cytokines

There exists a critical balance between proinflammatory and anti-inflammatory cytokines and chemokines induced by radiation. TNF-α is a proinflammatory cytokine that mediates many of its effects in concert with interleukin (IL)-6 (IL-6) and IL-1. These cytokines activate and mobilize cells to initiate an immune response and signal radiation injury to neighboring cells. Increases in TNF-α in several tissues post-irradiation have been reported [42–47]. This cytokine activates the vascular endothelium, inducing the expression of adhesion molecules and increasing vascular permeability. These effects will aid in the extravasation of leukocytes into the site of inflammation. IL-1 and TNF-α share many biological activities. IL-1α [44, 47] and IL-1β [44–46] are both increased early after radiation; however, IL-1β was found to be up-regulated more strongly than IL-1α early after exposure [44]. IL-6 expression is also increased after radiation therapy in several tissues [42, 45, 46, 48]. Increased levels of TGF-β, typically considered an

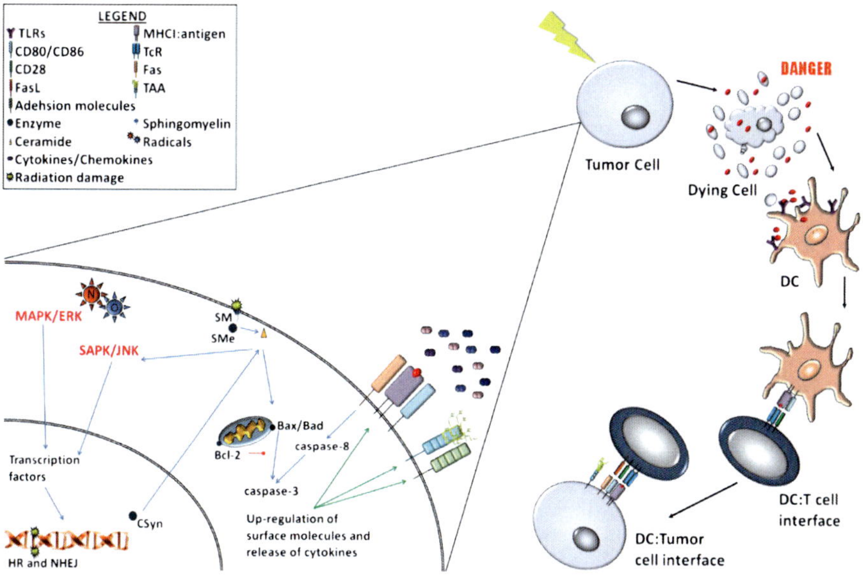

Fig. 17.1 *Radiation-induced modulation of tumor cells and the immune system.* After exposure to therapeutic doses of radiation, tumor cells undergo important changes which make them more amenable to immune recognition and kill. At a molecular level, the changes that occur within the tumor cell include: alterations in the genetic integrity of the cell; the activation of key signaling pathways which regulate survival, including the MAPK/ERK and/or the SAPK/JNK; the generation of free radicals which can propagate damage and activate signaling pathways; the release of cytokines and chemokines which can recruit and activate immune cells; and changes in cell surface markers, including major histocompatibility complex (MHC) and tumor associated antigens (TAAs). Cells which cannot faithfully repair the damage from ionizing radiation undergo cell death, either by apoptosis or necrosis. During the process of cell death, danger signals and tumor fragments are released into the microenvironment. The danger signals induce functional maturation of Dendritic cells (DCs) and tumor fragments are engulfed by DCs. TAAs from the tumor fragments are then processed and cross-presented onto MHCI with the appropriate co-stimulatory signals up-regulated during maturation. Upon migration to the tumor draining lymph node, these mature DCs can activate T cells into an effector phenotype so that after recognition, they can clear tumor cells. This process can be amplified at many different steps by supplementing danger signals, DCs or T cell populations, or by the addition of agents to support maturation and sustenance of key cell populations. HR homologous recombination, NHEJ non-homologous end joining, SM sphingomyelin, SMe sphingomyelinase, CSyn ceramide synthase, FasL Fas ligand, TLR toll-like receptor

anti-inflammatory mediator, have also been detected early after radiation exposure [49]. Other cytokines, chemokines, and their receptors, including IL-7 [50], IL-8 [45], IL-12 [51], IL-10 [45, 51, 52], VEGF [53], and the IL-6 receptor [45] have shown increased expression in various tissues post-irradiation. It is not surprising that negative regulators of inflammation are also released, such as TGF-β and IL-10, as they are likely involved in limiting the response to radiation. Analyses of gene expression in the lung after thoracic irradiation have even demonstrated a biphasic release of TNF-α, IL-6, and IL-1α [47]. This biphasic release is thought to mediate the late-effects of radiation in tissues, and may be beneficial for generating an immune response after hematopoietic recovery.

Taken together, it is well established that cytokines and chemokines are key players in the development of immune responses to tissue injury. Understanding the intricate signaling network induced after irradiation and the effects of these factors on the radiation response will be essential for determining optimal intervention methods for immunotherapeutic purposes.

Tumor Phenotype

In a classical study by Dranoff *et al.* [54], investigators showed that *in vivo* administration of lethally irradiated tumor cells dramatically increased their immunogenicity. Furthermore, sublethal irradiation has also been shown to have similar effects on tumor cells [55]. From these observations, it is evident that radiation causes important changes in the phenotype of tumor cells that have implications for its interaction with the immune system.

To subvert immune attack by lymphocytes, tumors have been shown to down-regulate the expression of major histocompatibility complex (MHC) molecules and receptors such as Fas [56]. After irradiation, however, both MHC class I [57–62] and Fas [57, 60, 63] have been shown to be up-regulated in a variety of tumors, both *in vitro* and *in vivo*. MHC class I expression is responsible for antigen presentation to cytotoxic T-lymphocytes (CTLs) and Fas expression on tumor cells can interact with Fas-ligand on CTLs. Thus, both of these molecules may enhance tumor cell recognition and killing following irradiation. In addition to MHC: peptide presentation on tumor cells, effector cells require costimulation in order to become fully activated. Radiation can enhance costimulation by increasing levels of both CD80 and CD86 on tumor cells [64, 65] and on DCs [51] to prevent T-cell anergy. Adhesion molecules, such as intercellular adhesion molecule-1 (ICAM-1) [58, 60, 62, 63, 66, 67] and vascular cell adhesion molecule-1 (VCAM-1) [68, 69] have been shown to be up-regulated on tumor cells or in the tumor vasculature after irradiation. The increased surface expression of adhesion molecules in the vessels can mediate extravasation of lymphocytes into the tumor site or support intercellular interactions between lymphocytes and tumor cells [70]. Therefore, radiation may render tumor cells more susceptible to CTL killing [57, 59, 60], prevent anergy, and promote tumor infiltration by lymphocytes [68].

Radiation can also increase the levels of tumor associated antigens (TAAs) expressed on tumors [60, 61]. Garnett *et al.* [60] studied the response of a panel of tumor cells to irradiation and noted that 74% of the cell types (17 of 23) up-regulated a TAA (CEA or MUC-1). As TAAs have been identified as a source of antigen for CTL recognition in many cancer patients, increasing expression of these antigens on tumor cells may increase their recognition by effector cells. Taken together, these results indicate that radiation therapy can alter the phenotype of tumor cells, making them more amenable to immune-mediated recognition and elimination.

Dendritic Cells

Dendritic cells (DCs) are sentinel cells in the body that constantly sample and present antigen to control immunity. However, in the immunosuppressive tumor microenvironment, functional maturation of DCs is inhibited. Tumor production of IL-10 and VEGF block DC maturation and results in defective differentiation and activation [71]. The presence of functionally immature DCs results in defective priming of tumor-reactive cells, thus leading to anergy or tolerance rather than activation. As outlined previously, radiation induces the expression of a variety of danger signals and proinflammatory cytokines. Their presence after irradiation may overcome the immunosuppression that occurs within the tumor bed and support the functional maturation of DCs.

DCs themselves are a generally radio-resistant cell population. Maturation and surface molecule expression on DCs are not significantly altered by radiation [72]. An increase in the intracellular peptide pool and an altered MHC:peptide repertoire has been reported after irradiation of DCs [73]. This change is consistent with a switch to the immunoproteosome, where the incorporation of interferon-γ-inducible elements influences its peptidase activities [74]. Therefore, the composition and activity of the proteosome affects the spectrum of peptides presented by DCs to the immune system after radiation exposure, and may alter the ensuing response [74].

Antigen processing may also be differentially modulated by exposure to ionizing radiation depending on whether peptides are generated from endogenous proteins or cross-presented from exogenous sources. Liao *et al.* [72] demonstrated that MART-1 transduced DCs failed to induce the specific immune response generated by MART-1_{26-35} peptide-pulsed DCs after irradiation. This observed defect in generating a CTL response to endogenous MART-1 was attributed to the presence of the immunoproteosome and a decreased loading of endogenous antigens onto MHC class I. The possibility that exogenous antigens are cross-presented more efficiently after radiation exposure is appealing given that DCs can uptake antigen from dying tumor cells. The increase in exogenous peptide loading onto MHC class I and the up-regulation of costimulatory molecules on DCs after irradiation may create the ideal DC: T cell interface to generate CTLs for tumor-specific immunity.

Depending on the mechanism of apoptosis induction, differential antigen processing, DC maturation modulation, and altered CTL responses have been observed. Phagocytosis of both necrotic and apoptotic cellular debris by DCs has been reported by investigators studying both human [75] and mouse [76] immature DCs. It has been suggested that necrotic and not apoptotic cells [75, 76] induce DC maturation and effective immunity induction; however there is contradictory evidence suggesting otherwise [77]. Therefore, it has been hypothesized that DCs may be able to distinguish between physiological and nonphysiological forms of cell death and respond in an appropriate manner [78]. These observations have implications for DC function and maturation after radiation-induced cell death.

Vasculature State and Leukocyte Localization

Localization of leukocytes to the sites of inflammation in the tumor vasculature requires the inducible expression of endothelial selectins (E- and P-selectin) and adhesion molecules (ICAM-1, and VCAM-1) [70]. In general, cytokines are key proteins regulating this process. Alterations in cytokines or adhesion molecules profoundly affect tumor infiltration and immune-mediated control of tumor growth. Local secretion of VEGF and anti-inflammatory cytokines, such as IL-10 and TGF-β, from the tumor can suppress the expression of endothelial adhesion molecules [70]. The reduction in expression of adhesion molecules on the endothelium may account for the observed reduction in leukocyte adherence to the endothelium in tumor vessels. Radiation, however, has been shown to overcome this suppression. Radiation alters the tumor vasculature by up-regulating adhesion molecules [79], including: ICAM-1 [67, 80, 81], VCAM-1 [69, 80, 81], and E-selectin [67, 81, 82] on tumor and/or endothelial cells. Expression of ICAM-1 on endothelial cells plays an essential role for leukocyte arrest on the endothelium [79]. The up-regulation of cytokines and chemokines after radiation also acts to increase adhesion molecules on leukocytes and the endothelium, and may render tumors more permissive for lymphocytic infiltration [69, 83]. For example, IFN-γ production after irradiation was found to be indispensible for the up-regulation of VCAM-1 on the vasculature, which mediates leukocyte rolling and arrest [69]. Therefore, there exists a complex interaction between radiation, endothelial cells, and tumor cells that determines the vessel activation state and its function for leukocyte interaction and extravasation.

Cytotoxic T-Lymphocyte(CTL) Responses

Tumor-specific CTLs exist in the tumor-bearing host and yet despite their presence, tumor progression still occurs. This occurrence is the result of effector-phase tolerance induction in the CTLs [84, 85]. The CTLs that accumulate in the tumor microenvironment are subjected to the immunosuppressive stimuli released from tumor cells [56], myeloid derived suppressor cells [71], and $CD4^+CD25^+$ T-regulatory cells [86]. Additionally, insufficient danger signals, along with inadequate priming and costimulation cause CTL dysfunction [87, 88]. The dysfunctional CTLs lack lytic and proximal signaling functions that are required for executing their effector functions [89, 90]. In the absence of tumor cells, however, signaling and lytic activity can be restored in the T-cells [89]. These findings suggest that the tumor microenvironment plays a pivotal role in shaping T-cell reactivity. By creating a proinflammatory environment where danger signals are abundant, effector lymphocytes may regain function and be rescued from their anergic state. Radiation can alter that balance in the tumor microenvironment. By inducing tumor cell death, promoting CTL infiltration, and changing transcriptional patterns in cells, it is conceivable that radiation can transform the tumor microenvironment into one that promotes an effective T-cell-mediated immune response.

Along these lines, a high level of integration exists between the cell- and tissue-oriented responses to radiation. Radiation-induced signaling cascades can establish a concentration gradient to modulate and direct the migration of cells into sites of inflammation where they can gain effector function. Understanding and exploiting the inflammatory environment and the changes that occur following radiation exposure (Table 17.1) will be essential for developing effective treatment strategies aimed at harnessing the immune system to generate a potent and long-lasting antitumor immune response.

Table 17.1 Radiation-induced modulation of relevant immune factors

Factor	Radiation-induced changes	Implication	References
Tumor cells	↑ TAA	Increases source of antigens for presentation on MHC and recognition by lymphocytes	[60, 61]
	↑ MHCI	Increases peptide presentation and recognition of tumor cells by lymphocytes	[57–62]
	↑ CD80 and CD86	Co-stimulates lymphocytes for effective activation	[64, 65]
	↑ Fas	Enhances Fas: FasL interaction and cell death	[57, 60, 63]
Vasculature	↑ VCAM-1	Leukocyte rolling and arrest on endothelium	[69, 80, 81]
	↑ ICAM-1	Leukocyte arrest on endothelium	[67, 80, 81]
	↑ E-selectin	Leukocyte recruitment and rolling on endothelium	[67, 81]
Cytokines and chemokines	↑ TNF-α	Activates vascular endothelium and mobilize cells	[42–47]
	↑ IL-1α	Activates vascular endothelium and mobilize cells	[44, 47]
	↑ IL-1β	Activates vascular endothelium and mobilize cells	[44–46]
	↑ IL-6	Activates lymphocytes	[42, 45, 46, 48]
	↑ IL-7	Sustains T cells	[50]
	↑ IL-8	Mobilizes and activates cells, including neutrophils and naive T cells	[45]
	↑ IL-12	Promotes Th1 and inflammatory reactions	[51]
	↑ IL-10	Suppresses DC activation and functions	[45, 51, 52]
	↑ TGF-β	Blocks differentiation of lymphocytes and monocytes; anti-inflammatory functions; induces apoptosis	[49]
	↑ VEGF	Angiogenesis; may play a role in re-oxygenating tissues after fractionated radiation therapy	[53]
	↑ CXCL16	Attracts effector T cells to the tumor	[145]

(continued)

Table 17.1 (continued)

Factor	Radiation-induced changes	Implication	References
APCs	Maturation of DCs	More effective priming and activation of T cells	Reviewed in [40]
	↑ Peptide repertoire and antigen cross-presentation onto MHCI	↑ Activation of tumor-specific $CD8^+$ T cells	[40, 73]
CTLs	↑ Suppressor cells	↑ Cell death of suppressor T cells and increased immune activation	[93]
	↑ Priming and activation	↑ CTL generation and tumor control	[68, 95]

APC antigen presenting cells, *CTLs* cytotoxic T lymphocytes, *MHCI* major histocompatibility complex I, *TAA* tumor associated antigen, *ICAM-1* intercellular adhesion molecule-1, *VCAM-1* vascular cell adhesion molecule-1, *DCs* dendritic cells, *IL* interleukin, *TGF-β* transforming growth factor-β, *TNF-α* tumor necrosis factor-α, *VEGF* vascular endothelial growth factor

Radiation Induced Immunosuppression: Radiation Versus Chemotherapy Coupling to Immunotherapy

Whole-body irradiation is known to cause immunosuppression, and radiation exposure is known to cause leucopenia [91–93]. Immunosuppression after whole-body irradiation is mainly ascribed to the depletion of lymphocytes. It is well established that lymphocytes are radio-sensitive; however, the radio-sensitivities of subsets of lymphocytes such as naive, effector, or regulatory populations remain unclear. It has been suggested that suppressor T cells are more sensitive to radiation [93]; in contrast, activated T cells appear to be more resistant to radiation-induced apoptosis [94]. Among lymphocytes, B cells are known to be the most radio-sensitive, followed by $CD4^+$ and $CD8^+$ T cells [91]. In contrast to whole-body irradiation, very low doses of radiation can mediate an immune-stimulatory effect [93]. Therefore, the dose and extent of exposure to radiation can have differential effects on the body.

Standard treatments using fractionated radiotherapy and chemotherapy may limit the radiation-mediated immune response that develops over time, and may contribute to malignant disease relapse. It has been suggested recently that ablative radiation therapy can act synergistically with immunotherapy to overcome tumor barriers and generate systemic immunity [95]. In a report by Lee *et al.* [95] it was demonstrated that tumor radio-sensitivity is T-cell mediated, and that APCs in the irradiated tumor and tumor draining lymph nodes present tumor antigen and efficiently activate T cells. The authors also found that ablative radiation therapy in combination with immunotherapy acted synergistically to eradicate tumors; in comparison, chemotherapy and fractionated radiotherapy were less effective. The authors hypothesized that chemotherapy and fractionated radiation therapy abolished the priming and expansion of specific $CD8^+$ T cells, thus depleting the effector T cells over time. Given the importance of the immune system for mediating antitumor effects, these results suggest that the schedule of standard radiation treatment should be reconsidered for preservation of the tumor-specific effector cells necessary for developing systemic immunity.

Technological advances in targeting the tumor have enabled the safe delivery of higher doses of radiation for better tumor control. Localized delivery which spares systemic, lymphoid, and bone marrow toxicity makes radiation therapy an attractive treatment to combine with immunotherapy. New strategies aimed at harnessing the specificity and efficiency of the immune system after radiation therapy offers potential for mobilizing the body's innate defense system for protection and potential cure from local and disseminated disease.

Pre-clinical Studies

Strategies to enhance the biological response to tumor irradiation have targeted the immune system. Initial investigations conducted in small animal models provide evidence that combining radiation therapy and immunotherapy may be a superior treatment strategy relative to either therapy alone.

Dendritic Cell (DC) Therapy

Several investigators have explored the capacity of ionizing radiation to augment the therapeutic efficacy of DCs. Reports have demonstrated that unpulsed DCs in combination with localized radiation therapy can induce tumor-specific CTL induction and antitumor activity [96–99]. This effect was not dependent on the degree of cell death induced by radiation, as it occurred in both radio-resistant [99], and radiosensitive cells [96, 97]. For example, Teitz-Tennenbaum *et al.* [99] demonstrated in 2003 that DC injection into irradiated tumors is superior to radiation, DCs, or tumor-pulsed DCs alone. This method of immunization was also able to protect mice from tumor rechallenge [97, 99]. The treatment response was attributed to the inhibition of tumor cell division [99] and the enhancement DC cross-presentation of antigen, trafficking to the lymph node, and activation of T cells [97, 99]. Although no significant cell death occurred after tumor irradiation in the study by Teitz-Tennenbaum *et al.* [99], DCs may have acquired tumor-specific antigens for cross-presentation from live cells in a cell contact-dependent mechanism recently described in the literature as "nibbling" [100]. As sentinel cells, DCs may play an instrumental role in mediating antitumor immune responses after radiation therapy; as such, strategies to enhance DC-based therapies have evolved. Gene-modification has emerged as an effective way to modulate the activity of DCs. Tatsuta *et al.* [101] demonstrated in 2009 that interferon-β gene transfer into naive DCs enhanced antitumor immunity after injection into irradiated tumors. In a different strategy to enhance the production of DCs, fms-like tyrosine kinase receptor 3 ligand (Flt3L) has also been used in combination with radiation. Flt3L in combination with radiation resulted in the induction of tumor-specific T-cell immunity and control of tumor growth outside of the irradiated field [102, 103].

Cytotoxic T-Cell Therapy

Radiation has been shown to increase the number of activated T cells due to increases in the antigen-presenting capabilities of APCs [68, 95]. Radiation-induced changes in the vascular network render tumors more accessible to infiltration by T cells [80], and leads to greater tumor control [68, 80]. Both single and fractionated doses of radiation increase the generation of tumor antigen-specific T cells and their localization to the site of the tumor; however, single-dose regimes have been shown in one study published in 2005 to lead to better tumor control [68]. Adoptive transfer of T cells has also been used to augment the number of CTLs in the host. To this end, the transfer of Th1 cells plus exogenous antigens after radiation treatment generated tumor specific $CD8^+$ T cells that eradicated tumors in the majority of mice and protected them against tumor rechallenge [104]. Adoptive transfer of specific T cells however, has shown divergent results after radiation therapy in different models. In one model, infusion of CEA-specific $CD8^+$ T cells effectively reduced CEA^+ tumor burden after radiation treatment [63]. Conversely, in a human tumor xenograft model, infusion of tumor-reactive CTLs did not enhance the efficacy of radiation treatment [105]. Although the discrepancy in these results cannot be accounted for by efficiency of homing to the tumor site, T cell survival after injection was not assessed, and optimal radiation schemes may not have been used.

T cell specificity, clonal expansion, and localization in the tumor bed are features of CTL therapy that make it appealing, especially in combination therapies for targeting tumors. Identifying methods to exploit CTLs to ensure tumor specificity and to quickly and reliably generate sufficient cell numbers for treatment will make this therapeutic strategy even more effective.

Antibody Therapy

Antibodies targeting molecules on both immune cells and tumors can be used to enhance the effectiveness of radiation treatment. Cytotoxic T lymphocyte associated antigen-4 (CTLA-4) has recently emerged as a promising therapeutic target for antibody-mediated therapy. As CTLA-4 is involved in regulating T-cell activities and plays a role in maintaining tolerance, monoclonal antibodies have been developed to block its functions. In a model of metastatic breast cancer, local radiation and CTLA-4 blockade induced a specific $CD8^+$ T-cell response that engendered systemic immunity that was able to inhibit metastases even outside of the field of radiation [106]. Similarly, in a bilateral tumor model with unilateral irradiation, CTLA-4 blockade was effective in mediating tumor regression outside of the irradiated field utilizing fractionated dose regimes [107]. Utilizing anti-CTLA-4 antibodies and radiation therapy may be particularly useful for generating immunity to poorly-immunogenic tumors that have been subjected to immune-editing and -selection.

Adjuvant Therapy

Stimulation of the innate immune system is necessary to achieve activation of the adaptive immune arm. By stimulating components of innate immunity, it may be possible to elicit better anti-tumor immune responses. β-glucans are naturally occurring polysaccharide components of the cell walls of certain pathogenic fungi and bacteria that can interact with surface receptors of immune cells, including complement and scavenger receptors. β-glucans have also been established as biological response modifiers given their ability to modulate hematopoiesis, stimulate phagocytosis, and induce the release of inflammatory cytokines. With such activities, it is not surprising that when treated with β-glucans, tumor growth can be inhibited and the leucopenia caused by radiation can be reversed [108]. After radiation, leucopenia and possible transient immunosuppression can pose a problem for generating a strong anti-tumor immune response. Therefore, combining radiation with β-glucans can be beneficial for preservation of the leukocyte population and enhancement of tumor control.

A potent strategy to boost immunity is through stimulation of toll-like receptors (TLRs) via exogenous ligand administration. The combination of radiation and unmethylated CpG motifs has demonstrated synergistic treatment responses in tumor models [109, 110]. In the presence of these oligodeoxynucleotides, type 1 responses are promoted and Th1 cytokines are released that support anti-tumor responses. In addition, CpG exposure protects T-cells and macrophages from ionizing radiation-induced cell death [111]. This may thus be an important adjuvant for use in fractionated radiotherapy to protect lymphocytes from deleterious effects and to preserve the immune response. Another strategy to exploit TLR ligands is through the use of synthetic double-stranded RNA: polyinosine-cytosine (poly(I:C)). Poly(I:C) has immunostimulatory properties, including the activation of interferon responses and enhancement of cross-presentation. In a recent study published in 2009, the combination of radiation and poly(I:C) enhanced tumor responses to treatment, and in some cases resulted in synergistic tumor regression [112].

Exploiting endogenous danger signals, HSPs have also been used to stimulate immunity. Akutsu *et al.* [113] demonstrated that DCs pulsed with HSP gp96 *in vitro* stimulated strong cytotoxicity; furthermore, *in vivo* administration resulted in tumor growth inhibition. Given these results, and the fact that radiation induces over-expression of HSP gp96, these investigators concluded that HSPs have the potential to boost systemic anti-tumor immunity, especially if used in combination with a cytotoxic treatment such as radiation therapy [113].

In a series of detailed experiments, administration of adenovirus (Ad) expressing Flt3L (Ad-Flt3L) and Ad expressing the herpes simplex virus-thymidine kinase (HSV-TK) gene (Ad-HSV-TK) was shown to be effective in treating gliomas. Treatment resulted in recruitment of bone-marrow derived myeloid DCs, activation of tumor- and antigen-specific T cells, induction of immunological memory, and increased long-term survival [113]. It was noted by the investigators that the release of HMGB1 from dying cells elicited TLR signaling in the myeloid DCs, which was essential for DC activation and induction of anti-tumor T-cell immunity [113]. Although HSV-TK was used in this model to induce tumor cell death, other forms of cytotoxic therapies, including radiation therapy, demonstrated significant release of HMGB1 from dying tumor cells [113]. Therefore, this study indicates that endogenous TLR ligands are important for eliciting strong adaptive immune responses; furthermore, these results indicate that combinatorial treatments that recruit DCs and stimulate TLRs together result in effective immunity.

Cytokine Therapy

Supplementation of radiation therapy with cytokines and growth factors has been investigated to optimize the generation of an anti-tumor response. To this end, exogenous cytokines [114–116], chemokines [117], or growth factors [102, 103], as well as cytokine-gene transduced tumor cells [118–120] have been used with some success. The use of systemically administered cytokines, however, should be approached with caution. Clinical experiences in humans with systemic IL-2 and TNF-α administration has revealed severe toxicities which were not observed in the pre-clinical murine studies. Consequently, methods to spatially and temporally control the expression of cytokines and other modulating factors after radiation have been developed to help control toxicities. Inducible promoters from genes that initiate transcription immediately after radiation have been used in this context. An Ad construct with the radiation-inducible promoter Egr-1 upstream of the cDNA encoding TNF-α (Ad-EGR1-TNFα) has demonstrated induction following radiation and localized production of TNF-α. The inducible production of TNF-α has demonstrated local destruction of the tumor vasculature causing overt tumor necrosis [121–123] and a reduction in the number of metastases [124]. The success of Ad-EGR1-TNFα in animal models warranted the therapeutic investigation of this construct in humans; towards this aim, several clinical studies have been conducted that will be discussed below.

Gene Therapy

Gene therapy strategies have also been coupled with radiation therapy. Gene delivery vehicles, such as recombinant viruses, have been used to deliver the cDNA sequences of TAAs, co-stimulatory molecules, and adhesion molecules for efficient expression in cells and recognition by the immune system. For example, the TRICOM vaccination uses a combination of CD80, ICAM-1, and lymphocyte function-associated antigen-3 (LFA-3) molecules expressed from a viral vector delivered in a prime-and-boost regimen. TRICOM vaccines coupled with TAAs have shown promise in pre-clinical studies. In both fractionated and single-dose radiation treatment schedules, CEA/TRICOM vaccination significantly reduced tumor burden [57]. With this combination of treatments, a significant increase in $CD4^+$ and $CD8^+$ tumor infiltrating lymphocytes was observed. The Fas pathway was also implicated in mediating tumor regression [57]. Using a similar model, multimodal therapy that included CEA/TRICOM vaccination, irradiation, and anti-$CD25^+$ monoclonal antibody-mediated suppressor cell depletion resulted in dramatic tumor regression and optimal induction of T-cell responsiveness against the self-antigen CEA [125]. Thus, employment of various immune-modulating therapies in the clinic may be required to obtain strong and durable responses, especially against self-antigens expressed on tumors. Clinical trials are warranted to investigate these types of combination treatments given their ability to mount specific immunity.

HSV-TK gene transfer followed by pro-drug administration (such as ganciclovir or acyclovir) results in the phosphorylation of pro-drugs into nucleotide analogs, which terminate DNA replication and lead to cell death. Delivery of the HSV-TK gene into cells is accomplished by genetic vehicles such as viruses. When HSV-TK and the pro-drug are present before radiation, they may sensitize cells to radiation-induced DNA damage and have greater cytotoxicity due to bystander effects. Murine studies examining the combination of an Ad-HSV-TK and radiation therapy have demonstrated that combined therapy results in an additive effect on tumor regression, prolonged survival, and increased protection from secondary tumor challenges [126]. Tumors which were treated with the combination therapy also had increased $CD4^+$ T-cell infiltrates, which suggests an ongoing immune reaction to the cell death induced by the two cytotoxic therapies [126].

Clinical Trials

Harnessing the immune system for optimal therapeutic intervention requires a comprehensive understanding of the tumor microenvironment and its interactions with the immune system (Fig. 17.1). Strategies to increase numbers of functional APCs, CTLs, and other stimulatory factors in the tumor have been met with some success in the clinic. Some of the promising results obtained using combination treatments in pre-clinical studies have translated into the clinic. Most of the clinical trials completed to date assess the safety and toxicity of these regimens, and evaluate preliminary immunological parameters pertinent for assessing the overall biological response to treatment (Table 17.2).

Table 17.2 Clinical trials of combined radio- and immunotherapy

Treatment	Radiation dose	Study type	Tumor type	Immune responses	Clinical responses	References
Autologous immature DCs injected intra-tumorally	Single fraction conformal 8 Gy	Phase I	Hepatoma	↑ AFP specific immune responses ↑ IFN-γ release and secretion from PBMC	21% of patients had stable disease, 14% had a partial response	[128]
β-Glucans	External (50 Gy) or intra-cavity (20 Gy)	Phase II/III	Cervical carcinoma	↑ In activated T cells subsets immediately after radiation More rapid recovery in $CD8^+$ T cells	5-year survival advantage, decreased recurrence in immune responders	[134, 135]
rV-PSA, rV-CD80 and booster rFP-PSA	Fractionated 1.8–2 Gy/dose = ≥70 Gy	Phase I (designed as Phase II)	Prostate cancer	↑ In PSA-specific T cells Antigenic cascade reactivity ↑ In NK cell percentage	Not powered to show a benefit in survival Low-dose metronomic IL-2 is associated with fewer toxicities	[138, 139]
IL-1β	Fractionated 1.8 Gy/dose = 54–59 Gy	Phase I/II	Non-small cell lung carcinoma	N.D.	81% response rate, 44% of responders had a complete response	[136, 137]
	Fractionated 2 Gy/dose = 60 Gy	Phase III			No survival advantage IL-1β toxicities caused non-compliance	

Ad-EGR1-TNFα	Fractionated 1.8–2 Gy/ dose = ~20–66.6 Gy	Phase I	Various origin	N.D.	43% objective tumor response 17% complete response (80% of which confirmed by pathology), 30% partial response No dose-limiting toxicities	[140]
	Fractionated 1.8–2 Gy/ dose = ~36–50.4 Gy	Phase I	Soft tissue sarcoma	N.D.	85% objective tumor response; 18% complete response, 82% partial response No dose-limiting toxicities	[141]
	Fractionated 1.8–2 Gy/ dose = 45–50.4 Gy Note: Concomitant 5-FU	Phase I, Phase II	Various origin	N.D.	Objective tumor responses	Reviewed [121]
Autologous IL-4 gene transfected fibroblasts and tumor lysate loaded DCs	Fractionated 2 Gy/ dose = 60 Gy	Phase I	Glioblastoma multiform or anaplastic astrocytoma	Type I DCs generated that produced high levels of IL-12 No increase in IFN-γ production	Laborious production of vaccine components Sub-optimal DC dose Lack of clinical and systemic immunological responses	[142]
Ad-HSV-TK	IMRT Fractionated 2 Gy/ dose = 76 Gy	Phase I/II	Prostate cancer	Long-term increases in activated $CD4^+$ and $CD8^+$ T cells	No dose limiting toxicities Combination improved patient prognostic factors	[143, 144]
	3D conformal radiation therapy	Phase I		N.D.		

AFP alpha-fetoprotein, *PBMC* peripheral blood mononuclear cells, *IMRT* intensity modulated radiation therapy, *IFN-γ* interferon-γ, *rV-PSA* recombinant vaccinia virus-prostate specific antigen, *rV-CD80* recombinant vaccinia virus-CD80, *rFP-PSA* recombinant fowlpox virus-prostate specific antigen, *Ad-EGR1-TNFα* adenovirus-EGR1 promoter-tumor necrosis factor-α, *5-FU* 5-fluorouracil, *Ad-HSV-TK* adenovirus-herpes simplex virus thymidylate kinase, *N.D.* no data

Dendritic Cell Therapy

As the most potent APCs, DCs have been used clinically to treat cancer with only modest success [127]. Despite the small numbers of patients who show a complete response in clinical trials, efforts are being made to improve response rates in humans. Towards achieving this aim, the combination of conformal radiation therapy and DC administrations has been tested in the clinic in 2005 [128]. In this phase I trial, autologous immature DCs were administered intra-tumorally after radiation therapy in patients with refractory hepatoma. Investigators reported that 70–80% of patients showed an increased α-fetoprotein-specific immune response after treatment, with 40% of participants showing significant increases in IFN-γ release and secretion from peripheral blood mononuclear cells. This study demonstrated the safety and feasibility of this treatment, and was well tolerated in patients. Optimizing DC culturing techniques, administration methods, and strategies to augment DC therapy after irradiation will be important to improve such clinical treatment protocols in future iterations.

Antibody Therapy

The use of anti-CTLA-4 antibodies in the clinic as a monotherapy has resulted in an objective response rate of up to 19%, with immune-related adverse events of grade 3 and 4 commonly occurring in the skin and gastrointestinal tract [129]. Although the immune-related adverse events were associated with a tumor response, the dose and treatment schedule requires optimization to minimize adverse events. Trials combining anti-CLTA-4 antibody and radiation therapy have not been published to date, however there is ongoing investigation of this coupling [129]. This combination may be particularly successful given the ability of CTLA-4 blockade to heighten T-cell responses. Taken together, these treatments, when combined, may result in prolonged tumor-antigen specific immune reactivity, and ultimately greater tumor control after radiation treatment.

Adjuvant Therapy

Approaches using innate immune stimulation, for example, through TLRs, have been used clinically in tumor immunization schemas. Autologous HSP gp96 purified from tumors and given as a vaccination to patients resulted in no major toxicities and significant increases in tumor-specific T cell responses [130, 131]. Importantly, patients that developed or enhanced an immune response after vaccination had a better prognosis than non-responders [131]. In addition to HSPs, other immunological adjuvants have been used clinically in cancer patients. Administration of

CpG intra-tumorally has demonstrated marked focal infiltration of DCs, T cells and B cells into the injection site [132], pronounced activation of DCs and increased cellularity in the draining lymph nodes [133]. Although not tested formally in clinical trials, the results of these reports provide rationale for combining these powerful adjuvants with radiation therapy to yield improved tumor control and survival.

As mentioned previously, β-glucans can have a role in both hematopoiesis and tumor control. Purified β-glucans have been used clinically in combination with radiation therapy. Several studies have been conducted using this agent, and all have demonstrated favorable outcomes in both tumor control and hematopoietic recovery. Miyazaki *et al.* [134] and Okamura *et al.* [135] both documented the immunostimulatory and recovery functions of β-glucans after radiation therapy, and emphasized the conferred survival advantage when used in combination.

Cytokine Therapy

Based on the pre-clinical evidence that exogenous IL-1β potentiated radiation-induced injury, McDonald *et al.* [136] conducted a phase I clinical trial demonstrating the safety in combining IL-1β and radiation in patients. As a follow up, Bradley *et al.* [137] conducted a phase II/III study using the same treatment schema. Results of this subsequent study indicated that no survival advantage was conferred to the patients treated with the combination therapy, and investigators suggested the use of other means to modify the radiation response. As there were no immunological analyses conducted in this study, no conclusions can be drawn on the systemic effects IL-1β on immunity.

In a similar manner, IL-2 and TNF-α, which showed particular promise in pre-clinical studies, demonstrated systemic toxicities in clinical trials. With local low-dose administrations of IL-2, however, lower toxicities have been noted with potential clinical application [138, 139]. To control the release of cytokines in patients in a clinical setting, radiation-inducible gene delivery constructs, such as Ad-EGR1-TNFα, have been used. Clinical trials assessing the combination of intra-tumoral Ad-EGR1-TNFα injection and radiation have corroborated several of the pre-clinical results [121]. In the first clinical trial performed, Ad-EGR1-TNFα was administered to patients with solid tumors of various origins [140], and in a second trial Ad-EGR1-TNFα was administered to patients with soft-tissue sarcoma [141]. In both studies, there were no dose limiting toxicities, with the most common adverse events being fever, chills and flu-like symptoms [140, 141]. Objective tumor responses were observed in 43% and 85% in the first and second trial, respectively [140, 141]. Results from the combination of radiation and Ad-EGR1-TNFα indicate that the treatment is well tolerated and effective in mediating tumor regression. Therefore, radiation-inducible constructs in combination with radiation therapy may be an effective way to exploit the downstream cellular effects of radiation, while obtaining potent anti-tumor responses.

Gene Therapy

Tumor cells genetically engineered to express cytokines or co-stimulatory molecules have been investigated as a novel cancer treatment platform in the clinic. Taking this idea a step further, Okada *et al.* [142] investigated autologous gene-transfected fibroblasts and tumor-lysate loaded DCs in a clinical trial in combination with radiation therapy. Despite radiographic evidence of improvement and preliminary results that suggested immune reactivity in a subset of patients, all patients eventually progressed. Investigators emphasized the need for more efficient protocols to generate the vaccine and a higher dose of DCs to demonstrate clinical benefit.

Combination of cancer gene therapy and radiation therapy has been investigated in other trials in the clinic with greater success. Studies combining direct injections of adenovirus encoding the suicide gene HSV-TK concurrent with radiation therapy have shown significant increases in activated $CD4^+$ and $CD8^+$ T-cells, which suggests potential immune activation after treatment [143]. The mechanism of this immune activation may be enhanced killing of the tumor cells after radiation therapy, which serves as an additional source of tumor antigens and danger signals. No dose-limiting toxicity was observed with the HSV-TK gene therapy vector [143, 144]. Interestingly, patients with intermediate risk appeared to benefit most from the combined treatment, being negative on biopsy and without PSA relapse. Investigators postulate, however, that the difference in effect between high and intermediate grade prostate cancers may relate to the permissivity of the cancer cells to adenovirus infection [144].

It has been shown in pre-clinical models that the combination of co-stimulatory molecules and TAA delivery via genetic vaccination yields increased survival and strong tumor-specific T-cell responses. The translation of this strategy to the clinic has yielded promising results as well, and it is now being combined with standard therapies, such as radiation therapy to boost clinical responses. In a study by Gulley *et al.* [138] published in 2005, the use of recombinant viral vaccines with standard radiotherapy in patients with localized prostate cancer was conducted. The trial was designed as a phase II study, with participants being randomized to receive radiation therapy, with or without the vaccine. The vaccinations consisted of recombinant viruses engineering expression of prostate-specific antigen (PSA) and the co-stimulatory molecule CD80, in a prime-boost regime with local GM-CSF and low-dose systemic IL-2. In the combination treatment arm, the majority of patients had at least a 3-fold increase in PSA-specific T-cells and reactivity against antigens not present in the vaccine. No such responses were detectable from patients receiving radiation alone. Given that toxicity was observed with systemic low-dose IL-2, a follow-up study using metronomic low-dose IL-2 administration was conducted. Results from this study confirmed previous results, and demonstrated a marked reduction in the toxicities associated with IL-2 [139]. Subsequent studies are needed to confirm the overall benefit in survival and progression to disease using this treatment.

Conclusions

The promising results obtained in early stage clinical trials offers hope for the combination of radiation therapy and immunotherapy in the treatment of cancer (Table 17.2). Although animal models provide the basis for many of the treatment regimens tried in the clinic, successful treatment regimens for cancer in animal models do not always translate into clinical success. Cancer can be cured in small animal models in part because artificial experimental models over-simplify the complexity of cancer and its interaction with the immune system. Recent advances in cellular and molecular biology, however, have provided tremendous insight into the biology of cancer cells and the mechanisms that govern immune cell activation and function. Future advancements in these areas will help to revolutionize our ability to treat cancer. This knowledge will be imperative for designing strategies to overcome immune regulation and heighten immune reactivity to tumors. It is well established that a variety of different cytokines, chemokines and growth factors are secreted post-irradiation; however, the exact roles of these factors have not been determined. A more thorough investigation of the cytokine storm post-irradiation may offer insight into the instructions the immune cells are receiving and may offer important insight into which cytokines mediate a productive anti-tumor response and which promote tissue injury. Clinical trials described at www.clinicaltrails.gov are investigating the magnitude and types of changes in cytokines, chemokines and growth factors after radiation treatment. Immunotherapeutic targets or rational combinations of targets may be necessary to optimally stimulate the immune system post-irradiation; investigation of potential combinations is warranted.

Immunological parameters aside, it will also be imperative to understand the radiobiology of SBRT and the long-term consequences of radiation therapy delivered in this manner. Several trials described at www.clinicaltrails.gov are actively investigating not only SBRT but also radiation and immunotherapy combination treatments. In a few of these trials, different cohorts have been set-up where timing of administration, dosing and types of treatments are being tested and directly compared on an immunological and tumor response basis. This design of clinical trial offers the opportunity to obtain vital information for understanding the impact of the treatment, and for planning future trials. With the knowledge and experience we gain from pre-clinical and clinical studies, we will be able to efficiently target the multiple components required for coordinating an effective attack against tumors, and elicit systemic immunity for protection against local and distant disease.

By combining immune-based and radiation therapies, a greater number of patients receiving radiation therapy may benefit from the treatment. Radiation, in addition to de-bulking the tumor, also provides the stimulus needed to ignite immune reactivity. Robust anti-tumor effects may be generated when such a combination is realized. The potential synergy between treatment modalities would offer initial tumor eradication, protection from recurrent disease, and ultimately prolonged survival in people afflicted with cancer.

References

1. Center for Disease Control and Prevention: National Center for Health Statistics (2009) Deaths, percent of total deaths, and death rates for the 15 leading causes of death in selected age groups, by race and sex: United States, 1999–2006. http://www.cdc.gov.myaccess.library.utoronto.ca/nchs/nvss/mortality_tables.htm#mortality. Accessed 26 March 2009
2. Blattman JN, Greenberg PD (2004) Cancer immunotherapy: a treatment for the masses. Science 305(5681):200–205
3. Formenti SC, Demaria S (2009) Systemic effects of local radiotherapy. Lancet Oncol 10(7):718–726
4. Pardoll DM (2002) Spinning molecular immunology into successful immunotherapy. Nat Rev Immunol 2(4):227–238
5. Waldmann TA (2003) Immunotherapy: past, present and future. Nat Med 9(3):269–277
6. Buschke F (1958) History of Radiation Therapy. In: Buschke F (ed) Progress in Radiation Therapy, 1st edn. Grune & Stratton, New York and London pp. 13–15
7. Langen KM, Jones DT (2001) Organ motion and its management. Int J Radiat Oncol Biol Phys 50(1):265–278
8. Jaffray DA (2005) Emergent technologies for 3-dimensional image-guided radiation delivery. Semin Radiat Oncol 15(3):208–216
9. Meyer JL, Verhey L et al (2007) New technologies in the radiotherapy clinic. Front Radiat Ther Oncol 40:1–17
10. Timmerman R, Bastasch M et al (2007) Optimizing dose and fractionation for stereotactic body radiation therapy. Normal tissue and tumor control effects with large dose per fraction. Front Radiat Ther Oncol 40:352–365
11. Fowler JF, Tome WA et al (2004) A challenge to traditional radiation oncology. Int J Radiat Oncol Biol Phys 60(4):1241–1256
12. Milano MT, Constine LS et al (2008) Normal tissue toxicity after small field hypofractionated stereotactic body radiation. Radiat Oncol 3:36
13. Fuks Z, Kolesnick R (2005) Engaging the vascular component of the tumor response. Cancer Cell 8(2):89–91
14. Kolesnick R, Fuks Z (2003) Radiation and ceramide-induced apoptosis. Oncogene 22(37):5897–5906
15. Liao WC, Haimovitz-Friedman A et al (1999) Ataxia telangiectasia-mutated gene product inhibits DNA damage-induced apoptosis via ceramide synthase. J Biol Chem 274(25):17908–17917
16. Hall EJ (2000) Radiobiology for the Radiologist, Lippincott Williama & Wilkins, Philadelphia
17. Lewanski CR, Gullick WJ (2001) Radiotherapy and cellular signalling. Lancet Oncol 2(6):366–370
18. Radford IR (1985) The level of induced DNA double-strand breakage correlates with cell killing after X-irradiation. Int J Radiat Biol Relat Stud Phys Chem Med 48(1):45–54
19. Bristow RG, Hill RP (2008) Hypoxia and metabolism. Hypoxia, DNA repair and genetic instability. Nat Rev Cancer 8(3):180–192
20. Tannock IF, Hill RP et al (ed) (2005) The Basic Science of Oncology. 4th edn. McGraw-Hill Ltd, New York
21. Overgaard J, Horsman MR (1996) Modification of hypoxia-induced radioresistance in tumors by the use of oxygen and sensitizers. Semin Radiat Oncol 6(1):10–21
22. Dayes IS, Abuzallouf S (2005) Local tumour control in women with carcinoma of the cervix treated with the addition of nitroimidazole agents to radiotherapy: a meta-analysis. Br J Radiol 78(933):777–782
23. DeVita VT, Hellman S et al (ed) (2005) Cancer: Principles and Practice of Oncology. 7th edn. Lippincott Williams & Wilkins, Philadelphia
24. Schiff PB, Horwitz SB (1980) Taxol stabilizes microtubules in mouse fibroblast cells. Proc Natl Acad Sci USA 77(3):1561–1565

25. Choy H (2000) Combining taxanes with radiation for solid tumors. Int J Cancer 90(3):113–127
26. Valerie K, Yacoub A et al (2007) Radiation-induced cell signaling: inside-out and outside-in. Mol Cancer Ther 6(3):789–801
27. Watters D (1999) Molecular mechanisms of ionizing radiation-induced apoptosis. Immunol Cell Biol 77(3):263–271
28. Hunter A, Hendrikse A et al (2006) Does the tumor microenvironment influence radiation-induced apoptosis? Apoptosis 11(10):1727–1735
29. Verheij M, Ruiter GA et al (1998) The role of the stress-activated protein kinase (SAPK/JNK) signaling pathway in radiation-induced apoptosis. Radiother Oncol 47(3):225–232
30. Dent P, Yacoub A et al (2003) Stress and radiation-induced activation of multiple intracellular signaling pathways. Radiat Res 159(3):283–300
31. Kimura K, Gelmann EP (2000) Tumor necrosis factor-alpha and Fas activate complementary Fas-associated death domain-dependent pathways that enhance apoptosis induced by gamma-irradiation. J Biol Chem 275(12):8610–8617
32. Rock KL, Kono H (2008) The inflammatory response to cell death. Annu Rev Pathol 3:99–126
33. Tesniere A, Panaretakis T et al (2008) Molecular characteristics of immunogenic cancer cell death. Cell Death Differ 15(1):3–12
34. Fonseca C, Dranoff G (2008) Capitalizing on the immunogenicity of dying tumor cells. Clin Cancer Res 14(6):1603–1608
35. Jonathan EC, Bernhard EJ et al (1999) How does radiation kill cells? Curr Opin Chem Biol 3(1):77–83
36. Illidge TM (1998) Radiation-induced apoptosis. Clin Oncol (R Coll Radiol) 10(1):3–13
37. Proskuryakov SY, Konoplyannikov AG et al (2003) Necrosis: a specific form of programmed cell death? Exp Cell Res 283(1):1–16
38. Matzinger P (1994) Tolerance, danger, and the extended family. Annu Rev Immunol 12:991–1045
39. Kono H, Rock KL (2008) How dying cells alert the immune system to danger. Nat Rev Immunol 8(4):279–289
40. Hatfield P, Merrick A et al (2005) Radiation-induced cell death and dendritic cells: potential for cancer immunotherapy? Clin Oncol (R Coll Radiol) 17(1):1–11
41. Ullrich E, Bonmort M et al (2008) Tumor stress, cell death and the ensuing immune response. Cell Death Differ 15(1):21–28
42. Fedorocko P, Egyed A et al (2002) Irradiation induces increased production of haemopoietic and proinflammatory cytokines in the mouse lung. Int J Radiat Biol 78(4):305–313
43. Hallahan DE, Spriggs DR et al (1989) Increased tumor necrosis factor alpha mRNA after cellular exposure to ionizing radiation. Proc Natl Acad Sci USA 86(24):10104–10107
44. Hong JH, Chiang CS et al (1999) Rapid induction of cytokine gene expression in the lung after single and fractionated doses of radiation. Int J Radiat Biol 75(11):1421–1427
45. Linard C, Ropenga A et al (2003) Abdominal irradiation increases inflammatory cytokine expression and activates NF-kappaB in rat ileal muscularis layer. Am J Physiol Gastrointest Liver Physiol 285(3):G556–G565
46. Marquette C, Linard C et al (2003) IL-1beta, TNFalpha and IL-6 induction in the rat brain after partial-body irradiation: role of vagal afferents. Int J Radiat Biol 79(10):777–785
47. Rube CE, Wilfert F et al (2004) Irradiation induces a biphasic expression of pro-inflammatory cytokines in the lung. Strahlenther Onkol 180(7):442–448
48. Van der Meeren A, Monti P et al (2005) Abdominal radiation exposure elicits inflammatory responses and abscopal effects in the lungs of mice. Radiat Res 163(2):144–152
49. Roedel F, Kley N et al (2002) Anti-inflammatory effect of low-dose X-irradiation and the involvement of a TGF-beta1-induced down-regulation of leukocyte/endothelial cell adhesion. Int J Radiat Biol 78(8):711–719
50. Toki J, Adachi Y et al (2003) Enhancement of IL-7 following irradiation of fetal thymus. Immunobiology 207(4):247–258

51. Liu SZ, Jin SZ et al (2001) Role of CD28/B7 costimulation and IL-12/IL-10 interaction in the radiation-induced immune changes. BMC Immunol 2:8
52. Broski AP, Halloran PF (1994) Tissue distribution of IL-10 mRNA in normal mice. Evidence that a component of IL-10 expression is T and B cell-independent and increased by irradiation. Transplantation 57(4):582–592
53. Ando S, Nojima K et al (1998) Evidence for mRNA expression of vascular endothelial growth factor by X-ray irradiation in a lung squamous carcinoma cell line. Cancer Lett 132(1-2):75–80
54. Dranoff G, Jaffee E et al (1993) Vaccination with irradiated tumor cells engineered to secrete murine granulocyte-macrophage colony-stimulating factor stimulates potent, specific, and long-lasting anti-tumor immunity. Proc Natl Acad Sci USA 90(8):3539–3543
55. Kliuchareva TE, Kashkina LM et al (1978) Comparative immunogenicity of irradiated and nonirradiated syngeneic and xenogeneic tumor cells containing a common specific transplantation antigen. Biull Eksp Biol Med 86(8):225–228
56. Rabinovich GA, Gabrilovich D et al (2007) Immunosuppressive strategies that are mediated by tumor cells. Annu Rev Immunol 25:267–296
57. Chakraborty M, Abrams SI et al (2004) External beam radiation of tumors alters phenotype of tumor cells to render them susceptible to vaccine-mediated T-cell killing. Cancer Res 64(12):4328–4337
58. Chiriva-Internati M, Grizzi F et al (2006) Gamma-radiation upregulates MHC class I/II and ICAM-I molecules in multiple myeloma cell lines and primary tumors. *In vitro* Cell Dev Biol Anim 42(3–4):89–95
59. Ciernik IF, Romero P et al (1999) Ionizing radiation enhances immunogenicity of cells expressing a tumor-specific T-cell epitope. Int J Radiat Oncol Biol Phys 45(3):735–741
60. Garnett CT, Palena C et al (2004) Sublethal irradiation of human tumor cells modulates phenotype resulting in enhanced killing by cytotoxic T lymphocytes. Cancer Res 64(21):7985–7994
61. Hareyama M, Imai K et al (1991) Effect of radiation on the expression of carcinoembryonic antigen of human gastric adenocarcinoma cells. Cancer 67(9):2269–2274
62. Santin AD, Hermonat PL et al (1997) Effects of irradiation on the expression of major histocompatibility complex class I antigen and adhesion costimulation molecules ICAM-1 in human cervical cancer. Int J Radiat Oncol Biol Phys 39(3):737–742
63. Chakraborty M, Abrams SI et al (2003) Irradiation of tumor cells up-regulates Fas and enhances CTL lytic activity and CTL adoptive immunotherapy. J Immunol 170(12):6338–6347
64. Morel A, Fernandez N et al (1998) Gamma-ray irradiation induces B7.1 costimulatory molecule neoexpression in various murine tumor cells. Cancer Immunol Immunother 46(5):277–282
65. Vereecque R, Buffenoir G et al (2000) Gamma-ray irradiation induces B7.1 expression in myeloid leukaemic cells. Br J Haematol 108(4):825–831
66. Gaugler MH, Squiban C et al (1997) Late and persistent up-regulation of intercellular adhesion molecule-1 (ICAM-1) expression by ionizing radiation in human endothelial cells *in vitro*. Int J Radiat Biol 72(2):201–209
67. Hallahan D, Kuchibhotla J et al (1996) Cell adhesion molecules mediate radiation-induced leukocyte adhesion to the vascular endothelium. Cancer Res 56(22):5150–5155
68. Lugade AA, Moran JP et al (2005) Local radiation therapy of B16 melanoma tumors increases the generation of tumor antigen-specific effector cells that traffic to the tumor. J Immunol 174(12):7516–7523
69. Lugade AA, Sorensen EW et al (2008) Radiation-induced IFN-gamma production within the tumor microenvironment influences antitumor immunity. J Immunol 180(5):3132–3139
70. Carlos TM (2001) Leukocyte recruitment at sites of tumor: dissonant orchestration. J Leukoc Biol 70(2):171–184
71. Kusmartsev S, Gabrilovich DI (2002) Immature myeloid cells and cancer-associated immune suppression. Cancer Immunol Immunother 51(6):293–298
72. Liao YP, Wang CC et al (2004) Ionizing radiation affects human MART-1 melanoma antigen processing and presentation by dendritic cells. J Immunol 173(4):2462–2469

73. Reits EA, Hodge JW et al (2006) Radiation modulates the peptide repertoire, enhances MHC class I expression, and induces successful antitumor immunotherapy. J Exp Med 203(5):1259–1271
74. McBride WH, Iwamoto KS et al (2003) The role of the ubiquitin/proteasome system in cellular responses to radiation. Oncogene 22(37):5755–5773
75. Sauter B, Albert ML et al (2000) Consequences of cell death: exposure to necrotic tumor cells, but not primary tissue cells or apoptotic cells, induces the maturation of immunostimulatory dendritic cells. J Exp Med 191(3):423–434
76. Gallucci S, Lolkema M et al (1999) Natural adjuvants: endogenous activators of dendritic cells. Nat Med 5(11):1249–1255
77. Gamrekelashvili J, Kruger C et al (2007) Necrotic tumor cell death *in vivo* impairs tumor-specific immune responses. J Immunol 178(3):1573–1580
78. Demaria S, Santori FR et al (2005) Select forms of tumor cell apoptosis induce dendritic cell maturation. J Leukoc Biol 77(3):361–368
79. Quarmby S, Kumar P et al (1999) Radiation-induced normal tissue injury: role of adhesion molecules in leukocyte-endothelial cell interactions. Int J Cancer 82(3):385–395
80. Ganss R, Ryschich E et al (2002) Combination of T-cell therapy and trigger of inflammation induces remodeling of the vasculature and tumor eradication. Cancer Res 62(5):1462–1470
81. Heckmann M, Douwes K et al (1998) Vascular activation of adhesion molecule mRNA and cell surface expression by ionizing radiation. Exp Cell Res 238(1):148–154
82. Hallahan D, Clark ET et al (1995) E-selectin gene induction by ionizing radiation is independent of cytokine induction. Biochem Biophys Res Commun 217(3):784–795
83. Ganss R, Hanahan D (1998) Tumor microenvironment can restrict the effectiveness of activated antitumor lymphocytes. Cancer Res 58(20):4673–4681
84. Frey AB, Monu N (2006) Effector-phase tolerance: another mechanism of how cancer escapes antitumor immune response. J Leukoc Biol 79(4):652–662
85. Frey AB, Monu N (2008) Signaling defects in anti-tumor T cells. Immunol Rev 222:192–205
86. Zou W (2006) Regulatory T cells, tumour immunity and immunotherapy. Nat Rev Immunol 6(4):295–307
87. Gallucci S, Matzinger P (2001) Danger signals: SOS to the immune system. Curr Opin Immunol 13(1):114–119
88. Matzinger P (2002) The danger model: a renewed sense of self. Science 296(5566):301–305
89. Monu N, Frey AB (2007) Suppression of proximal T cell receptor signaling and lytic function in $CD8^+$ tumor-infiltrating T cells. Cancer Res 67(23):11447–11454
90. Radoja S, Frey AB (2000) Cancer-induced defective cytotoxic T lymphocyte effector function: another mechanism how antigenic tumors escape immune-mediated killing. Mol Med 6(6):465–479
91. Belka C, Ottinger H et al (1999) Impact of localized radiotherapy on blood immune cells counts and function in humans. Radiother Oncol 50(2):199–204
92. Mori M, Desaintes C (2004) Gene expression in response to ionizing radiation: an overview of molecular features in hematopoietic cells. J Biol Regul Homeost Agents 18(3–4):363–371
93. North RJ (1984) Gamma-irradiation facilitates the expression of adoptive immunity against established tumors by eliminating suppressor T cells. Cancer Immunol Immunother 16(3):175–181
94. Lowenthal JW, Harris AW (1985) Activation of mouse lymphocytes inhibits induction of rapid cell death by x-irradiation. J Immunol 135(2):1119–1125
95. Lee Y, Auh SL et al (2009) Therapeutic effects of ablative radiation on local tumor require $CD8^+$ T cells: changing strategies for cancer treatment. Blood 114(3):589–595
96. Huang J, Wang Y et al (2007) Radiation-induced apoptosis along with local and systemic cytokine elaboration is associated with DC plus radiotherapy-mediated renal cell tumor regression. Clin Immunol 123(3):298–310
97. Kim KW, Kim SH et al (2004) Direct injection of immature dendritic cells into irradiated tumor induces efficient antitumor immunity. Int J Cancer 109(5):685–690

98. Nikitina EY, Clark JI et al (2001) Dendritic cells transduced with full-length wild-type p53 generate antitumor cytotoxic T lymphocytes from peripheral blood of cancer patients. Clin Cancer Res 7(1):127–135
99. Teitz-Tennenbaum S, Li Q et al (2003) Radiotherapy potentiates the therapeutic efficacy of intratumoral dendritic cell administration. Cancer Res 63(23):8466–8475
100. Harshyne LA, Zimmer MI et al (2003) A role for class A scavenger receptor in dendritic cell nibbling from live cells. J Immunol 170(5):2302–2309
101. Tatsuta K, Tanaka S et al (2009) Complete elimination of established neuroblastoma by synergistic action of gamma-irradiation and DCs treated with rSeV expressing interferon-beta gene. Gene Ther 16(2):240–251
102. Chakravarty PK, Alfieri A et al (1999) Flt3-ligand administration after radiation therapy prolongs survival in a murine model of metastatic lung cancer. Cancer Res 59(24):6028–6032
103. Demaria S, Ng B et al (2004) Ionizing radiation inhibition of distant untreated tumors (abscopal effect) is immune mediated. Int J Radiat Oncol Biol Phys 58(3):862–870
104. Yokouchi H, Chamoto K et al (2007) Combination tumor immunotherapy with radiotherapy and Th1 cell therapy against murine lung carcinoma. Clin Exp Metastasis 24(7):533–540
105. Krause M, Schmitz M et al (2007) Adoptive transfer of cytotoxic T-cells for treatment of residual disease after irradiation. Int J Radiat Biol 83(11–12):827–836
106. Demaria S, Kawashima N et al (2005) Immune-mediated inhibition of metastases after treatment with local radiation and CTLA-4 blockade in a mouse model of breast cancer. Clin Cancer Res 11(2):728–734
107. Dewan MZ, Galloway AE et al (2009) Fractionated but not single-dose radiotherapy induces an immune-mediated abscopal effect when combined with anti-CTLA-4 antibody. Clin Cancer Res 15(17):5379–5388
108. Vetvicka V, Dvorak B et al (2007) Orally administered marine (1-->3)-beta-D-glucan Phycarine stimulates both humoral and cellular immunity. Int J Biol Macromol 40(4):291–298
109. Chamoto K, Takeshima T et al (2009) Combination immunotherapy with radiation and CpG-based tumor vaccination for the eradication of radio- and immuno-resistant lung carcinoma cells. Cancer Sci 100(5):934–939
110. Milas L, Mason KA et al (2004) CpG oligodeoxynucleotide enhances tumor response to radiation. Cancer Res 64(15):5074–5077
111. Sohn WJ, Lee KW et al (2006) CpG-oligodeoxynucleotide protects immune cells from gamma-irradiation-induced cell death. Mol Immunol 43(8):1163–1171
112. Le UM, Kaurin DG et al (2009) Localized irradiation of tumors prior to synthetic dsRNA therapy enhanced the resultant anti-tumor activity. Radiother Oncol 90(2):273–279
113. Akutsu Y, Matsubara H et al (2007) Combination of direct intratumoral administration of dendritic cells and irradiation induces strong systemic antitumor effect mediated by GRP94/gp96 against squamous cell carcinoma in mice. Int J Oncol 31(3):509–515
114. Cameron RB, Spiess PJ et al (1990) Synergistic antitumor activity of tumor-infiltrating lymphocytes, interleukin 2, and local tumor irradiation. Studies on the mechanism of action. J Exp Med 171(1):249–263
115. Jurgenliemk-Schulz IM, Renes IB et al (1997) Anti-tumor effects of local irradiation in combination with peritumoral administration of low doses of recombinant interleukin-2 (rIL-2). Radiat Oncol Investig 5(2):54–61
116. Younes E, Haas GP et al (1995) Radiation-induced effects on murine kidney tumor cells: role in the interaction of local irradiation and immunotherapy. J Urol 153(6):2029–2033
117. Shiraishi K, Ishiwata Y et al (2008) Enhancement of antitumor radiation efficacy and consistent induction of the abscopal effect in mice by ECI301, an active variant of macrophage inflammatory protein-1alpha. Clin Cancer Res 14(4):1159–1166
118. Chiang CS, Hong JH et al (2000) Combining radiation therapy with interleukin-3 gene immunotherapy. Cancer Gene Ther 7(8):1172–1178
119. Newcomb EW, Demaria S et al (2006) The combination of ionizing radiation and peripheral vaccination produces long-term survival of mice bearing established invasive GL261 gliomas. Clin Cancer Res 12(15):4730–4737

120. Seetharam S, Staba MJ et al (1999) Enhanced eradication of local and distant tumors by genetically produced interleukin-12 and radiation. Int J Oncol 15(4):769–773
121. Mezhir JJ, Smith KD et al (2006) Ionizing radiation: a genetic switch for cancer therapy. Cancer Gene Ther 13(1):1–6
122. Mauceri HJ, Hanna NN et al (1996) Tumor necrosis factor alpha (TNF-alpha) gene therapy targeted by ionizing radiation selectively damages tumor vasculature. Cancer Res 56(19):4311–4314
123. Staba MJ, Mauceri HJ et al (1998) Adenoviral TNF-alpha gene therapy and radiation damage tumor vasculature in a human malignant glioma xenograft. Gene Ther 5(3):293–300
124. MacGill RS, Davis TA et al (2007) Local gene delivery of tumor necrosis factor alpha can impact primary tumor growth and metastases through a host-mediated response. Clin Exp Metastasis 24(7):521–531
125. Kudo-Saito C, Schlom J et al (2005) The requirement of multimodal therapy (vaccine, local tumor radiation, and reduction of suppressor cells) to eliminate established tumors. Clin Cancer Res 11(12):4533–4544
126. Chhikara M, Huang H et al (2001) Enhanced therapeutic effect of HSV-tk+GCV gene therapy and ionizing radiation for prostate cancer. Mol Ther 3(4):536–542
127. Rosenberg SA, Yang JC et al (2004) Cancer immunotherapy: moving beyond current vaccines. Nat Med 10(9):909–915
128. Chi KH, Liu SJ et al (2005) Combination of conformal radiotherapy and intratumoral injection of adoptive dendritic cell immunotherapy in refractory hepatoma. J Immunother 28(2):129–135
129. Wolchok JD, Saenger Y (2008) The mechanism of anti-CTLA-4 activity and the negative regulation of T-cell activation. Oncologist 13 Suppl 4:2–9
130. Belli F, Testori A et al (2002) Vaccination of metastatic melanoma patients with autologous tumor-derived heat shock protein gp96-peptide complexes: clinical and immunologic findings. J Clin Oncol 20(20):4169–4180
131. Mazzaferro V, Coppa J et al (2003) Vaccination with autologous tumor-derived heat-shock protein gp96 after liver resection for metastatic colorectal cancer. Clin Cancer Res 9(9):3235–3245
132. Haining WN, Davies J et al (2008) CpG oligodeoxynucleotides alter lymphocyte and dendritic cell trafficking in humans. Clin Cancer Res 14(17):5626–5634
133. Molenkamp BG, van Leeuwen PA et al (2007) Intradermal CpG-B activates both plasmacytoid and myeloid dendritic cells in the sentinel lymph node of melanoma patients. Clin Cancer Res 13(10):2961–2969
134. Miyazaki K, Mizutani H et al (1995) Activated (HLA-DR+) T-lymphocyte subsets in cervical carcinoma and effects of radiotherapy and immunotherapy with sizofiran on cell-mediated immunity and survival. Gynecol Oncol 56(3):412–420
135. Okamura K, Suzuki M et al (1989) Clinical evaluation of sizofiran combined with irradiation in patients with cervical cancer. A randomized controlled study; a five-year survival rate. Biotherapy 1(2):103–107
136. McDonald S, Chang AY et al (1993) Combined Betaseron R (recombinant human interferon beta) and radiation for inoperable non-small cell lung cancer. Int J Radiat Oncol Biol Phys 27(3):613–619
137. Bradley JD, Scott CB et al (2002) A phase III comparison of radiation therapy with or without recombinant beta-interferon for poor-risk patients with locally advanced non-small-cell lung cancer (RTOG 93-04). Int J Radiat Oncol Biol Phys 52(5):1173–1179
138. Gulley JL, Arlen PM et al (2005) Combining a recombinant cancer vaccine with standard definitive radiotherapy in patients with localized prostate cancer. Clin Cancer Res 11(9):3353–3362
139. Lechleider RJ, Arlen PM et al (2008) Safety and immunologic response of a viral vaccine to prostate-specific antigen in combination with radiation therapy when metronomic-dose interleukin 2 is used as an adjuvant. Clin Cancer Res 14(16):5284–5291

140. Senzer N, Mani S et al (2004) TNFerade biologic, an adenovector with a radiation-inducible promoter, carrying the human tumor necrosis factor alpha gene: a phase I study in patients with solid tumors. J Clin Oncol 22(4):592–601
141. Mundt AJ, Vijayakumar S et al (2004) A Phase I trial of TNFerade biologic in patients with soft tissue sarcoma in the extremities. Clin Cancer Res 10(17):5747–5753
142. Okada H, Lieberman FS et al (2007) Autologous glioma cell vaccine admixed with interleukin-4 gene transfected fibroblasts in the treatment of patients with malignant gliomas. J Transl Med 5:67
143. Fujita T, Teh BS et al (2006) Sustained long-term immune responses after in situ gene therapy combined with radiotherapy and hormonal therapy in prostate cancer patients. Int J Radiat Oncol Biol Phys 65(1):84–90
144. Freytag SO, Movsas B et al (2007) Phase I trial of replication-competent adenovirus-mediated suicide gene therapy combined with IMRT for prostate cancer. Mol Ther 15(5):1016–1023
145. Matsumura S, Wang B et al (2008) Radiation-induced CXCL16 release by breast cancer cells attracts effector T cells. J Immunol 181(5):3099–3107

Chapter 18
Assessing Immunotherapy Through Cellular and Molecular Imaging

John W. Barrett, Bryan Au, Ryan Buensuceso, Sonali de Chickera, Vasiliki Economopoulos, Paula Foster, and Gregory A. Dekaban

Abstract Molecular medicine is focusing its attention on developing immunotherapeutic strategies that engage the immune system to combat a number of human diseases, including cancer. As a result, great emphasis has been placed on enhancing existing imaging modalities and developing new imaging techniques in order to assess the *in vivo* consequences of a given immunotherapy. Recently, improvements in the resolution and sensitivity of existing *in vivo* imaging modalities, including computed tomography (CT), ultrasound (US), positron emission tomography (PET), single positron emission tomography (SPECT), optical imaging (OI), and magnetic resonance imaging (MRI), have evolved enormously. In this chapter, each modality, used either individually or together as multi-modal hybrid imaging techniques, will be evaluated in the context of how they contribute to assessing immunotherapies *in vivo* in preclinical and clinical settings.

Keywords CT • MRI • Multi-modal imaging • Optical imaging • PET • Ultrasound

Introduction

The application of imaging technologies to assess and validate the effect of cancer immunotherapies has evolved from its initial form, which was simply a qualitative analysis largely restricted in its application, into a form that can currently provide detailed two- or three-dimensional images yielding qualitative and often a quantitative assessment in both the preclinical and clinical setting. From its earliest inception, the immunotherapeutic treatment of cancer was largely based on pharmacological approaches centered on the use of various lymphokines such as IL-2 and type I interferons (IFN) to boost antitumor responses [1]. Historically,

J.W. Barrett (✉)
Robarts Research Institute, The University of Western Ontario, London, ON, Canada
e-mail: jbarrett@robarts.ca

J. Medin and D. Fowler (eds.), *Experimental and Applied Immunotherapy*,
DOI 10.1007/978-1-60761-980-2_18, © Springer Science+Business Media, LLC 2011

assessment of the effect of immunotherapeutics revolved around x-ray technology and included the early versions of the computed tomography (CT) scanners and microscopy-based analysis of tumor samples. As a result, the role that imaging played in these earlier studies was simply to follow and evaluate the tumor response. However, the role of imaging began to develop as cancer immunotherapy began to evolve into cell-based therapies. Early cell-based immunotherapy protocols generally required the isolation of autologous immune effector cells, such as tumor infiltrating lymphocytes and subsequent expansion and/or activation of these cells *ex vivo* before returning them to the patient [2–4]. While it remained important to monitor the tumor response in such studies, these types of cell-based immunotherapies warranted the requirement to track the *in vivo* fate of the therapeutic cells and their survival following their administration to the patient in order to confirm they were migrating to sites appropriate for their function and to ensure the patient's body was not rejecting them. The application of imaging modalities to track these cell-based therapies *in vivo* in humans relied heavily on the use of radioactive tracers like indium-111 or technetium-99 to label the cells prior to administration, in a process known as scintigraphy (Lee et al. [4, 5]). The limitation of this technique is the lack of anatomical detail and the lack of three-dimensional projections [6], features required for the accurate detection and localization of these cells *in vivo*. The radioactive tracers employed in scintigraphy also have a very short half-life, in the order of only a few hours, and thus did not allow for the longitudinal assessment of the *in vivo* fate of these labeled cells [7].

The advent of gene therapy and the use of viral or plasmid-based vectors to label *ex vivo*-prepared therapeutic cells circumvented the problem of the short half-life of scintigraphic tracers such as indium-111. The coevolution of vectors capable of delivering a therapeutic gene and a reporter gene became increasingly important as the use of viral or plasmid vectors to express immunomodulatory genes for the purpose of enhancing anticancer immune responses increased the need to track the bio-distribution and longevity of the vectors in target cells. Viral and plasmid vectors expressing a reporter gene whose expression was detectable by fluorescence or bioluminescence markers could theoretically label therapeutic cells, providing a method to detect and track the fate of these cells *in vivo* in longitudinal studies. However, in the early days of gene therapy and viral and plasmid vectors, bioluminescence and optical imaging methods either did not exist or were not sensitive enough to take advantage of this cell labeling technique. As a result, highly invasive biopsy or drawing of blood were required to detect and confirm the presence of these viral vector-transduced or plasmid vector transfected therapeutic cells *in vivo*. However, viral vectors, in conjunction with reporter genes such as β-galactosidase and luciferase did permit tracking in a more quantitative manner using biopsy or blood-derived material. This type of approach, while useful in certain preclinical studies, is not likely to be applicable for human use as both reporter genes are immunogenic [8–10]. Luciferin, the key reagent needed for luciferase to create luminescence, is not FDA approved for use in humans and its toxicity in humans still requires extensive investigation. Not until years after the advent of viral vectors have optical imaging techniques finally become sensitive enough for the detection of viral vector-transduced cells *in vivo*. Likewise, techniques which allow the application of positron emission tomography (PET) imaging (see below) have also become available in recent years, permitting the use of viral vector transduction in order to monitor the fate of cell-based and viral vector-based immunotherapies longitudinally.

The existence of imaging techniques remains an important part of the evaluation process of current and future genetically engineered cell-based immunotherapies. Recently, the application and versatility of *in vivo* imaging modalities has grown enormously as a result of improvements in these technologies, and now includes several different types of imaging techniques including CT, ultrasound (US), PET, single positron emission tomography (SPECT), optical imaging (OI), and magnetic resonance imaging (MRI). In the following sections, each of the above imaging modalities will be evaluated in the context of how they contribute to evaluating immunotherapies *in vivo* in preclinical and clinical settings. The application of multimodal hybrid imaging techniques which incorporate features from two different imaging techniques is soon becoming the way of the future, and will also be discussed.

Computed Tomography

Computed axial tomography (CAT or CT) scanning relies on the tissue-penetrating ability of x-rays. The fundamental principles of tomography, the imaging of single sections or slices of a body, began to evolve in the early 1900s after methods to do so were proposed by the Italian radiologist, Alessandro Vallebona [11]. CT scanning relies on an x-ray source and a corresponding detector. Visualization of various body tissues is based on the differential absorption of the x-rays. Dense tissues have higher absorption, and therefore appear darker on radiographs. In CT scans, the x-ray source and detectors rotate around the patient while obtaining serial x-ray images [12]. Movement of the x-ray source and detector modifies the focal plane, allowing only a given section or slice of the target to be visualized. This reduces the obscurities caused by out-of-focus objects. In newer machines, the patient slides through the machine while the x-ray source and detectors are rotated, permitting faster scanning [11, 13]. The series of images can be "stacked" upon one another to recreate a three-dimensional model of the target organ by a process called tomographic reconstruction. Due to the short wavelength and high energy of x-rays, CT scanning allows for high spatial resolution and deep tissue penetration. While these characteristics make CT well-suited for imaging dense tissue and tissues that differ in density from surrounding tissues, the ability to image soft tissues is limited [12].

CT scans are able to detect tumors smaller than 4 mm in diameter [14]. However, contrast agents are sometimes required for CT; the most common agents are barium and iodine. Both of these molecules possess large nuclei, and as such, are capable of obscuring the path of the x-rays to further improve tissue resolution. Currently, contrast agents are most often employed to examine disorders in the colon or vasculature. However, these agents are rapidly cleared by the kidney, thereby resulting in short imaging windows. CT is of use in screening for many different types of cancers owing to its high resolving capacity and the high penetration of x-rays, [14–16].

The use of CT for cell tracking and molecular imaging is in the early stage of development; the sensitivity of the technique has not been well established [12, 13]. Recently, strides have been made in the use of nanotechnology-based CT contrast agents. The use of gold nanoparticles was described in 2006 by Hainfeld et al. [17], who used these particles in the imaging of kidneys, tumors, and vasculature in mice [17]. Although these nanoparticles were still untargeted, the study gave rise to newer hybrid nanoparticles including polymer-coated gold nanoparticles [18], gadolinium-coated gold nanoparticles [19], and bismuth sulfide nanoparticles [20]. Popovtzer et al. described targeted gold nanoparticles for use in the CT imaging of cancer [21]. In their study, gold nanorods were conjugated to a head-and-neck cancer specific antibody, UM-A9. The application of nanoparticle-based contrast agents with CT has not yet been applied to tracking of therapeutic immune cells *in vivo*.

Despite the present shortcomings in the use of CT for cell tracking and molecular imaging, it remains an invaluable clinical tool. In combination with the high level of convenience associated with CT, further development of the targeted contrast agents may provide an effective tool for molecular imaging of immunotherapy. In the near term, combining CT with other imaging modalities (PET/SPECT and MRI; see below) that are already being used for cell tracking and assessing functional metabolic activity will likely reach clinical application first. In combination with other imaging techniques, CT can also provide important preclinical information in disease-specific animal models.

Ultrasound

Ultrasound (US) transduces high-frequency sound waves; as the waves bounce off of tissues, they create echoes that reflect back to the transducer. The transducer translates the vibrations into electrical pulses that are processed, and then transformed into an image. Two- and three-dimensional US have good sensitivity, are noninvasive, and do not use ionizing radiation. In addition, the equipment needed to obtain US images is modest compared to MRI, CT, and PET/SPECT scanners; as such, US scanners can be highly portable. As with other imaging modalities, US can be used to locate and characterize tumors. It can also provide additional information such as tumor volume, tumor vascularization, and the degree of necrosis that may be present [22, 23]. Anatomical US has also been used for the guided delivery of therapeutic cells. That is, the direct injection of *ex vivo*-prepared tumor-specific T cells, along with genetically engineered monocyte/macrophages and dendritic cells have all been performed using US-guided delivery [24, 25].

The main restriction of US is limited depth penetration (mm to cm). The sensitivity of US for cellular and molecular imaging is not yet well characterized. US can provide information, however, on molecular and cellular processes with the use of appropriate contrast agents [22]. For example, contrast-enhanced US (CEUS) typically uses small gas-filled bubbles surrounded by a defined protein-lipid formulation. These commercially available microbubbles range in size from several 100 nm to several microns in diameter; they contain a high molecular weight gaseous vapor, such as perfluorocarbons, that minimize gas loss through the shell. These shells are critical to provide sufficient bubble stability *in vivo* in order to produce contrast. Due to the size of the microbubbles, the CEUS contrast agent must be administered intravenously. Therefore, the microbubble contrast agent is retained almost exclusively within the intravascular space. Furthermore, microbubbles in the circulation have a very short half-life (<5 min) and can be cleared from the blood pool within 10 min. Thus, there is a relatively short period of time for the contrast agent to reach the target of interest. The CEUS agents can escape the blood vessels when there is a significant loss in blood vessel integrity, thereby allowing microbubbles to penetrate into extravascular spaces. Thus, at sites of tumor angiogenesis it is possible for microbubbles to penetrate the tumor, thereby contributing to tumor imaging and assessment of angiogenesis.

CEUS agents are largely designed to interact with receptors on activated leukocytes and endothelial cells at sites of disease and inflammation [26, 27]. Microbubble contrast agents can be targeted in three different ways. First, the chemical composition of the bubble shell can be designed to chemically interact with a target cell that expresses a specific type of surface chemistry through charge-mediated interactions [28]. Alternatively, the bubble shell can be coated with proteins that will be recognized by specific cellular receptors; for example, coating with albumin or other proteins have the capacity to interact with complement receptors. The most specific form of targeted CEUS is provided by microbubbles coated with a monoclonal antibody (mAb) or a ligand, such as an integrin adhesion molecule [29]. Association of a microbubble to a cell surface can impart greater long-term stability to the microbubble. In many cases when a mAb is used to target a microbubble to leukocytes, internalization of the microbubble occurs [30]. This can impart greater long-term stability to the microbubble, thereby improving imaging function.

While i.v. injection of the contrast agent is one way to label cells *in vivo* for US, it is also possible to label *ex vivo*-prepared therapeutic cells such as lymphocytes and dendritic cells. These microbubble-labeled cells could then potentially be injected and tracked by CEUS methods. The practical utility of this approach remains to be determined, especially with respect to the sensitivity and duration of detection *in vivo* in humans. A unique advantage of this type of cell label is that directed high energy ultrasound can induce the microbubbles to burst, thereby releasing the contrast agent from the therapeutic cell if necessary. Or, if the microbubbles are filled with a therapeutic drug, the application of high energy US can burst the bubble and release the drug such as a pro-inflammatory chemokine or cytokine to enhance recruitment and/or activation of immune effector cells to or at a tumor site. Similarly a microbubble containing a gene of interest at the target site and lead to an antitumor therapy or, at least theoretically, lead to immune activation or suppression within a particular targeted environment [26, 27].

As such, while traditional US imaging paradigms have already become an integral part of tumor diagnostics and therapy in general, US can also have a role in specific immunotherapy applications [31]. CEUS has the potential to become an important imaging tool, especially as new and more advanced targeting reagents are developed for microbubbles. The immunotherapeutic potential of CEUS will be greatly enhanced if targeted microbubbles can also be developed to successfully deliver drugs that can augment tumor-specific immune responses in a tumor-induced immunosuppressive environment.

Positron Emission Tomography (PET) and Single Photon Emission Computed Tomography (SPECT)

PET was developed in the mid-1970s as a method to generate functional information about the brain, and later, the heart. Over time, PET has evolved into a universal technique useful in preclinical and clinical environments that enables quantitative and tomographic assays of biological processes using positron-emitting radioisotope-labeled probes. PET moved to the forefront of molecular imaging techniques when it was adapted for whole body scans, successfully detecting both primary tumors and metastases [32], and later for monitoring chemotherapies [33]. A PET scan generates a digital image of chemical changes that occur in the tissue of interest, including malignancies, which are characterized by an increased rate of glycolysis.

PET radionuclides have short half-lives and are designed to mimic the turnover of biological markers [34]. The most common PET/SPECT tracer is fluorine-18 (half life ~110 min); however, a number of other probes exist (Table 18.1) and more are being developed and tested regularly. A patient undergoing a PET scan is given an injection that consists of a combination of a sugar (e.g., glucose) and a small amount of radioactively labeled sugar, for example ^{18}F-fluorodeoxyglucose (^{18}F-FDG). The uptake of ^{18}F-FDG at sites of increased glycolysis results in the accumulation of signal produced by the decay of the radiotracer and emission of positrons (positively charged antiparticle of the electron). The positron travels a few millimeters before colliding with an electron, thereby producing a pair of gamma photons that can be detected by a light-capturing instrument.

PET imaging of cancer has focused on the characterization of a range of oncogenic parameters, including: tumor metabolism, nucleoside metabolism, tumor hypoxia, and cancer cell-specific receptor expression. Specific radiolabels have been developed to target these individual phenomena (Table 18.1). The high sensitivity of PET allows the use of labeled tracers at concentrations that do not

Table 18.1 Representative PET radio-labeled tracers used for cancer evaluation (Information taken from [51])

Mechanism	Probe
Tumor hypoxia	^{18}F-fluoromisonidazol (^{18}F-FMISO)
	^{18}F-fluoroazomycin arabinoside (^{18}F-FAZA)
	^{62}Cu-diacetyl-bis(N^{4}-methylthiosemicarbazone) (^{62}Cu-ATSM)
Metabolic tracers	
Glucose	^{18}F-fluorodeoxyglucose (^{18}F-FDG)
Amino acid	^{11}C-methionine (^{11}C-MET)
Nucleoside	^{11}C-thymidine (^{11}C-Thd), ^{11}C-fluorothymidine (^{11}C-FLT)
Lipid	^{11}C-acetate, ^{11}C-choline
Receptor imaging	
Estrogen	^{18}F-fluoroestradiol (^{18}F-FES)
Somatostatin	^{111}In-pentetreotide (^{111}In-DTPAOC)

Table 18.2 Half-life and yield for selected PET/SPECT radio tracers (Data from [34])

Radionuclide	Half-life	Positron yield
^{15}O	2 min	
^{13}N	10 min	
^{11}C	20 min	
^{68}Ga	68 min	89%
^{18}F	110 min	97%
^{64}Cu	12.7 h	18%
^{86}Y	14.7 h	17.5%
^{76}Br	16.0 h	55%
^{89}Zr	78.4 h	22.7%
^{124}I	100.2 h	23%
^{131}I	8 days	

perturb tumor metabolism or overwhelm the molecular signatures that are up-regulated on tumor cells as part of the oncogenic transformation [35]. PET is also useful for detecting metabolic changes shortly after therapy and can provide functional information about activation of immune cells involved in immunotherapies [34, 36]. The detection of soft-tissue tumors accounts for approximately 90% of PET clinical applications [37].

SPECT is closely related to PET but uses lower energy gamma rays and isotopes with longer half-lives (for example, ^{131}I with a half-life of 8 days; Table 18.2). The technique was developed 10–20 years earlier than PET; however, the resolution of SPECT is lower than PET, yielding less detailed images. In contrast to PET's colliding positrons, SPECT relies on a single photon bundle of gamma radiation to generate images. Traditionally, the type of application of SPECT has been similar to those of PET. For imaging of cellular and molecular events, PET/SPECT is currently known to have the highest sensitivity of all *in vivo* imaging modalities. Another advantage of PET/SPECT is unlimited depth penetration.

Although both PET/SPECT are commonly used for preclinical as well as clinical investigations, there exist a number of nontrivial limitations. Compared to other *in vivo* imaging modalities, PET and SPECT currently have the lowest spatial resolution and therefore provide little anatomical information [38]. Other limitations include the radiation dose administered and the short half-lives of the radioisotopes. The most common tracers have half-lives that range from minutes to less than 2 h (Table 18.2); as such, access to a local cyclotron for production purposes is required. This requirement, and the high cost of tracer production, coupled with the aforementioned short half-life of most useful radioisotopes means that PET/SPECT is a costly technique.

The basic concept underlying nuclear imaging of cells and molecules is that an expressed protein can be detected with specific radiopharmaceuticals. The use of PET to image the location and expression of transgenes, to monitor activation of adoptive cell therapies, and to assess endogenous molecular events following treatment illustrates the goal of combining imaging techniques with reporter genes systems [39]. PET imaging strategies have also taken advantage of features of the herpes simplex virus (HSV) thymidine kinase (tk) suicide gene therapy alone or in combination with ganciclovir in a range of cancer models [40–43]. Future developments may include the incorporation of PET imaging in other suicide gene/ tion with high contrast imaging, promises new treatment of malignant gliomas and may be suitable for clinical applications [45]. It is critical that reporter gene proteins do not induce cell death or alter normal cellular functions at levels necessary for detection. Finally, another potential obstacle is that foreign gene expression can lead to a host immune response to transferred cells [36].

Dedicated PET scanners for small animal models have led to the rise of preclinical PET technologies and brought PET technology to a new set of investigators. New reporter systems have been developed that combine PET detection with immunotherapy strategies. These methods include indirect *in vivo* labeling using PET reporter genes, directly labeled adoptively transferred lymphocytes, and cell surface reporter strategies [46–50]. However, there are still limitations that must be addressed including issues of sensitivity, specificity, toxicity, and immunogenicity [36]. For example, lymphocyte immunotherapy requires PET detection of relatively small numbers of cells but with a resolution sufficient for anatomical elucidation. The spatial resolution of PET/SPECT, as applied to cancer imaging, is not sensitive enough to resolve some cancerous growths because target expression is often spatially limited and can be at levels below detectable resolution. Improving resolution to submillimeter levels is necessary [51]. Another major obstacle is the differential detection of reporter genes in immune cells; such genes must be distinguishable from endogenous levels of cellular expression to ensure that the signal-to-noise ratio is reasonable. Accelerated cellular metabolism is not limited to cancer tissues and from the point of view of PET, tumor hypoxia is no different from oxygen starvation in tissues by any other cause. Angiogenesis is common to tumors but increased blood vessel formation is also common during organ formation or wound repair. Therefore, development and identification of optimal probes, designed to meet the specific needs of patients will be essential for future PET/SPECT success [51].

Finally, PET depends on the differential uptake of the tracer between normal and cancerous cells. Unfortunately, the uptake mechanism of cancer cells and activated immune cells of ^{18}F can be difficult to differentiate, e.g., both melanoma cells and activated immune cells undergo a significant increase in glucose use and result in a high intensity PET signal [36]. The solution has been to investigate new radioisotopes and develop new PET strategies that can reveal and quantify minute differences, thereby allowing better separation of cell types. Future research for PET/SPECT improvement is directed at the development of new radio-probes for improved cancer target cell identification. Cancer-specific tracers that can differentiate between cancer type (colon, lung, breast etc.) between tumor cells and immune cells and cell type and functional state are key areas of future improvement [51–53].

Optical Imaging

The two main types of optical imaging for *in vivo* cell tracking are bioluminescence imaging (BLI) and fluorescence imaging (FI). Both of these techniques work by capturing light emissions. BLI uses visible light, whereas FI uses either visible or near-infrared light.

BLI often involves the labeling of cells through the delivery of the *luc* cDNA encoding the protein luciferase. The substrate, luciferin, is delivered i.v. and photons are generated through the process of luciferase-mediated oxidation of luciferin; therefore, no excitation of light is required. BLI has been used for cell tracking in small animal models [54, 55]. The main advantage of BLI is that it can be used to detect very low levels of signal because the light is emitted in a context that is virtually background free. BLI is quick and easy to perform and allows rapid testing in live experimental models. It is also uniquely suited for high-throughput imaging because of its ease of operation, short acquisition times (typically 10–60 s), and the possibility of simultaneous measurement of multiple mice. However, BLI has some drawbacks. First, the inefficiency of light transmission through an opaque animal leads to poor depth penetration. And second, the lack of an equivalent imaging modality applicable for human studies prevents translation into clinical use.

The main advantages of FI are its high sensitivity and the fact that no substrate is required for its visualization. Cells tagged with fluorescently labeled antibodies or those in which expression of the green fluorescent protein (GFP) cDNA is introduced can be followed by this technique. This simple, reflectance type of FI has been used extensively for cell imaging. However, FI is not quantitative, and the image information is surface-weighted; as such, anything close to the surface will appear brighter compared with deeper structures. A newer approach to FI of deeper structures is fluorescence-mediated tomography (FMT). The FMT subject is exposed to continuous wave or pulsed light from different sources, and detectors are arranged in a spatially defined order to capture the emitted light. Mathematical processing of this information results in a reconstructed tomographic image. Resulting images have a resolution of 1–2 mm, and the fluorochrome detection threshold is in the nanomolar range.

The most prominent class of fluorescent dyes are the near-infrared (NIR) cyanine dyes [56]. The development of NIR fluorescent probes has greatly improved the preclinical as well as the clinical feasibility of optical imaging by offering an increased penetration depth [57–59]. The longer wavelength of NIR also overcomes some of the limitations of using visible light for *in vivo* tracking, such as autofluorescence, limited tissue penetration, and scattering; as a result, the sensitivity of detection is greater than with other fluorescent dyes [60, 61]. NIR imaging in animal models allows for sensitive imaging of tumor growth and immune cell localization [38]. There is a wide range of nontargeting NIR fluorescent dyes available commercially for effective cell labeling to permit *in vivo* cell tracking using OI [62]. Many cell types, including human mesenchymal stem cells [63] and Ag-specific $CD4^+$ T cells [64] have been nonspecifically labeled using these dyes.

Simple targeting probes, cross-linking probes, and enzyme-activated probes can be used for targeted labeling of cells. Among these, enzyme-activated probes represent the most sensitive targeting strategy for optical imaging. These probes initially emit a low fluorescent signal in their intact state, and greatly increase in fluorescence after catalysis by enzymes specific to the target cell. One enzyme can activate multiple probes, thus resulting in amplification of the signal [61]. Increased protease activity is implicated in many cancers, thereby providing a target for diagnostic detection of cancer cells using corresponding enzyme-activated probes. Several studies have used an NIR probe for cathepsin B or cathepsin D to detect several tumor types; as the probe accumulates in tumor cells, cleavage by tumor proteases results in up to 12-fold increases in fluorescence [60, 65–69].

Quantum dots (QD) are inorganic fluorophores comprised of a semiconductor core (e.g., CdSe, PbSe, or InAs) that is enclosed in a semiconductor shell and surrounded by a chemical coating for stabilization and solubility [70]. QDs can be used for labeling tumor cells and immune cells by attaching biological probes to the QD coating. QDs offer controllable emission wavelengths, narrow emission profiles, continuous broad absorption spectra, and robust signal strength. As such, QD imaging reduces autofluorescence and permits multiple QDs to be visualized with a single light source ("multiplexing") [70–73]. The sensitivity of QD labeling even allows for single molecule detection. Furthermore, QDs demonstrate increased stability, along with reduced photo-bleaching and blinking which allow them to remain fluorescent in tissues *in vivo* for months [63, 74, 75]. Indeed, researchers were able to observe QDs in lymph nodes of mice for more than 4 months [74]. QD labeling has also been used for the *in vivo* tracking of a DC-based immunotherapy in mice [76]. However, the short and long-term toxicity of QDs will have to be thoroughly examined before the application can be tested in humans [73].

Bifunctional labeling is now being explored by developing contrast agents that can be detected by more than one imaging modality; this approach integrates the advantages and compensates for the limitations of each individual modality.

Imaging protocols that employ contrast agents that are detectable by OI and MRI benefit from the high sensitivity of optical imaging and the superior depth penetration and spatial resolution of MRI [77–79]. Such contrast agents have been used for imaging of Ag-specific CTL tracking to tumors [80] in preclinical models. Combinations of optical and nuclear imaging have also been attempted [81, 82]. These probes reap the depth penetration, three-dimensional resolution, and quantification benefits offered by techniques like PET and SPECT [83]. Furthermore, tri-functional approaches are being designed that integrate all three imaging modalities: optical, MRI, and radiotracers [63].

Magnetic Resonance Imaging

MRI is based on the principles of nuclear magnetic resonance (NMR), a spectroscopic technique used by scientists to obtain the chemical and physical properties of molecules. In 1971, Raymond Damadian discovered the basis for using MRI as a tool for medical diagnosis. He determined that there were differences between the signals emitted from normal, healthy tissues and cancerous tissues [84]. These visible differences motivated scientists to develop MRI as a method to detect disease.

Anatomical MRI relies on the properties of hydrogen atoms (protons) to generate an image. In MRI, protons are exposed to a strong external magnetic field, which causes them to align with the main magnetic field. A radiofrequency transmitter is then briefly turned on producing an electromagnetic field. The aligned protons absorb this energy. After the RF is turned off the protons decay to the original equilibrium state during relaxation, emitting energy that produces the signal which can be detected quantified, and translated into an image by advanced computer processing. Protons in different tissues of the body relax at different rates. Image contrast can be generated by exploiting these differences.

Anatomical MRI is routinely used for tumor detection using clinical MRI systems that currently have main magnetic field strengths of 1.5 or 3.0 T. Dedicated small animal MRI systems with stronger magnetic fields (7–14 T) and stronger gradients allow higher resolution and are widely used to evaluate preclinical murine models of disease.

MRI has a number of characteristics that make it well-suited for cell tracking. MRI can produce images with high spatial resolution and exquisite soft-tissue contrast. Current micro-MRI techniques can achieve *in vivo* resolution on the order of tens of micrometers. MRI is also free of ionizing radiation and is considered safe and noninvasive. The limitations of MRI are its expense and relatively low sensitivity to contrast agents used for cell detection.

Cellular MRI is an emerging field, which uses high resolution MRI along with cell labels to detect and track cells of interest. The most commonly used cell labels are superparamagnetic iron oxide nanoparticles. A variety of iron oxide-based labels are now available for cell labeling, including: standard superparamagnetic iron oxide particles (SPIO, 50–100 nm); ultra-small iron oxide agents (USPIO, 10–30 nm); and micron-sized iron oxide particles (MPIO, 0.75 μm and larger). The presence of the magnetic label causes a distortion in the magnetic field and leads to abnormal signal hypointensities in iron sensitive images. Areas containing iron labeled cells therefore appear as regions of low signal intensity on MRI images, thus creating negative contrast. A variety of cell types have been tracked *in vivo* with MRI using this approach including stem cells [85, 86], monocytes/macrophages [77], dendritic cells [87, 88] and lymphocytes [89].

A number of labs have now shown that MRI is a useful, noninvasive modality for tracking the *in vivo* migration of cell-based immunotherapies [90]. At this time, there are several papers published that have demonstrated the ability to track the migration of dendritic cells *in vivo* in mice using cellular MRI [91]. In addition, an important clinical study has shown that SPIO-labeled DCs can be tracked in humans by cellular MRI [92]. Kircher et al. have also tracked $CD8^{+}$ T cell migration to sites of tumors *in vivo* using cellular MRI [80]. This latter study showed that T lymphocytes were detectable to a near single-cell resolution *in vivo* with no significant effects on cellular function or phenotype. This work demonstrated, for the first time, the spatial and temporal recruitment of T cells to tumors *in vivo*.

There are certain limitations to the use of iron oxide-based reagents that must be taken into account when analyzing MRI image data. Contrast agents are diluted or lost when cells divide or die; as such, the timeframe for imaging cells labeled using these agents must be within days to several weeks, depending on the rates of cell proliferation and death. Another limitation is that other phenomenon can cause signal loss in MR images sensitive to iron. This includes anatomical features (such as air in sinuses, bone, blood vessels, and hemorrhage) and artifacts caused by local magnetic field inhomogeneities, all of which have the potential to cause false positives. This could lead to the incorrect identification of regions of signal loss as iron-labeled cells [78]. Similarly the nonspecific uptake of iron particles by macrophages that engulf dead labeled cells can be problematic [93].

To overcome these limitations, other contrast agents are being developed. ^{19}F MRI has been used to image perfluoropolyether (PFPE)–labeled DC *in vivo* with minimal adverse effects on DC function [94]. Due to the negligible endogenous ^{19}F background signal, this approach allows for more specific imaging of the labeled cells. The image obtained with the ^{19}F signal can be correlated with the anatomical information derived from a conventional ^{1}H image to obtain a more detailed image. Most recently, dendritic cells have been successfully labeled with a gadolinium contrast agent for detection by MRI [25]. The main limitation of gadolinium or PFPE labels is the much lower sensitivity compared with iron nanoparticles [95].

Although current MRI techniques allow for dynamic imaging of cell migration, accurate quantification of DC migration *in vivo* using MRI remains an ultimate goal. Noninvasive quantification would provide useful information, such as the ability to correlate the number of cells required to generate a suitable immune response by a cell-based immunotherapy.

Multimodal Imaging Techniques in Molecular and Cellular Imaging

Each of the above imaging modalities has advantages and limitations and each may provide certain specific information. However, multilayered information can be obtained by combining the best features of any two modalities; for example, high-resolution anatomical images (such as CT or MRI images) can be combined with functional or metabolic images (such as PET or SPECT images).

Currently the most promising hybrid multimodality imaging systems include integrated PET/CT, SPECT/CT, and PET/MRI. Integrated scanners can be used to track the distribution of different radioactive tracers *in vivo*. These tracers can provide metabolic and functional data in the clinic and be combined with a reporter gene system to enhance experimental modeling. These tracer-reporter gene systems allow implanted cells to be tracked over time. However, these systems also have some limitations. The tracer-reporter gene systems, such as the combination of the HSV/TK gene with the ^{131}I-Fialuridine (^{131}I-FIAU) tracer, have currently only been used in experimental models. The current clinical tracers are also not necessarily specific to a certain disease process or illness. For example, the tracer ^{18}F-Fluorodeoxyglucose (^{18}F-FDG), which is used to detect cancerous cell growth, will enhance any area of increased metabolic activity. This method thus cannot detect tumors that have spread to organs that naturally have an increased uptake of FDG and cannot detect tumors that do not display increased metabolic activity. This can lead to incorrect findings in some patients, causing them to receive inappropriate treatment. A PET/CT examination can also expose patients to higher doses of radiation (as high as 25 mSv [or 2,500 mrem]) [96], which is higher than the limit proposed by the Nuclear Regulatory Commission (100 mrem/year) [97]. In cell tracking studies, the higher radiation dose may have an effect on cell viability or functionality [98].

PET/MRI systems face major hurdles before they can become more readily available. PET examinations require an attenuation correction to display data accurately. In PET/CT, this attenuation correction is easily calculated from the CT portion of the exam. However, for PET/MRI, calculating the attenuation correction from the MRI exam is not straightforward and work is underway to find better solutions [94]. Also, the PET hardware must be MRI compatible and must not create inhomogeneities in the MRI scanner's main magnetic field [99]. Even though several challenges face PET/MRI technology, the potential benefits are great. The multiple contrasts generated by MRI, along with the possibility to generate functional MRI data and spectroscopic data, can provide even more information than a PET/CT scan.

Image fusion is another multimodal strategy that can be used to merge images from complementary modalities when integrated systems are not available. Image fusion of MR and CT images has been studied for cardiac functional imaging and brain imaging in patients with Parkinson's disease [100, 101]. This technique eliminates the need to have onsite integrated multi-modal systems, thus allowing scanning sessions to be conducted separately. This method also allows for more than two forms of image data to be fused together. Image fusion has also proven beneficial in the study of Hodgkin's lymphoma patients, where fused PET/CT and PET/MRI images were more accurate than each modality alone or in side-by-side image analysis [102]. While the registration methods employed by this technique are most accurate for the brain, that locale lacks a high level of accuracy generated in other areas of the body. There may also be discrepancies in fused images if the patient is not positioned the same on both scans.

Along with the development of multimodality imaging techniques, multimodal contrast agents are also being investigated. These agents could be used to track cell migration, location, and function as well as provide morphological information simultaneously in several modalities. Their use has been demonstrated by several studies of MRI/Optical and PET/MRI agents. For example, an MRI/optical agent created by McCarthy et al. used a targeted strategy to detect thrombosis [103]; similarly, the PET/MRI agent manufactured by Choi et al. was able to demonstrate sentinel lymph node detection [104]. Still yet, Lee et al. detected implanted tumors with a targeted PET/MRI nanoparticle [105]. Many of these agents are constructed using nanoparticles or large molecules that provide contrast in one modality as a backbone; then, the molecule is conjugated with a contrast agent from a different modality to the original backbone [103–105]. Other methods involve doping nanoparticles with a compound, thus allowing them to generate contrast in several modalities; this strategy is similar to the method used to construct the particle described by Yong in 2009 for both MRI and optical imaging [106]. An MRI/CT agent has also been developed by Alric et al. that consists of a gold nanoparticle coated with a gadolinium chelate [19].

Summary and Conclusions

Molecular medicine is being advertised as the way of the future, and molecular and cellular imaging are key in helping to pave its way. While improvements to individual imaging modalities are ongoing, significant advances in molecular and cellular imaging are likely to be made in regards to developing multi-modal imaging techniques. As summarized in Fig. 18.1, each of the individual imaging modalities differs with respect to spatial image resolution and sensitivity. Improvements in hardware and imaging protocols can help to improve image resolution and sensitivity of any imaging technique. Likewise, the quality of contrast agents can be improved, thereby enhancing the sensitivity and specificity of imaging techniques. Contrast agents that support imaging by more than one imaging modality will also need to be developed in order to effectively take advantage of multimodal imaging scanners currently under development. The future of *in vivo* imaging holds the promise of being able to obtain detailed anatomical and spatial localization of therapeutic cells of interest *in vivo* over extended periods of time. In the future, properly designed contrast agents should be able to provide information about the functional state of

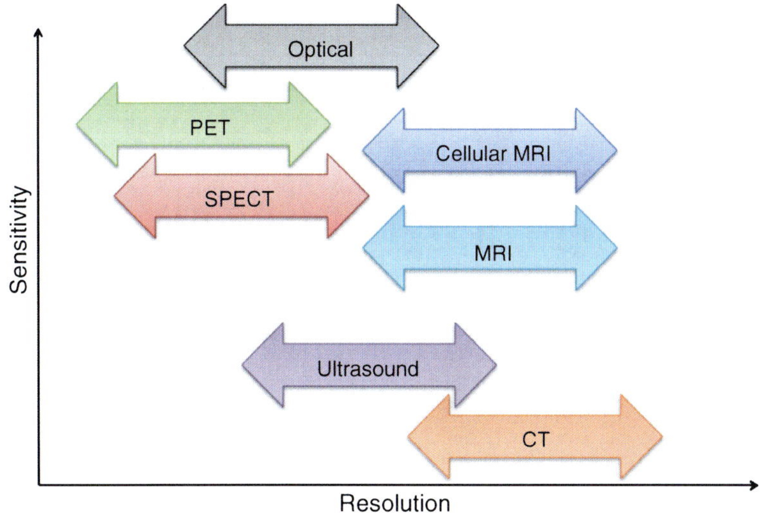

Fig. 18.1 Relative levels of sensitivity and resolution of multiple imaging modalities. *PET* positron emission tomography, *SPECT* single photon emission computed tomography, *MRI* magnetic resonance imaging, *CT* computer tomography

the cells of interest. With the development of appropriate multimodality image scanners, such information will be obtained ideally in a single scanning session. These multimodal imaging approaches have the potential to provide clinicians with invaluable information not only about the fate of transplanted therapeutic cells, but also information about their function *in vivo*. This information can be later correlated to the overall efficacy of the immunotherapy. By providing clinicians with the real-time, accurate diagnostic imaging information that permits rapid adjustments to an ongoing treatment, improvements in overall patient care and disease outcomes will be obtained.

References

1. Becker, Y. (2006). Molecular immunological approaches to biotherapy of human cancers – a review, hypothesis and implications. Anticancer Res **26**(2A): 1113–1134.
2. Chaudhuri, D., R. Suriano, et al. (2009). Targeting the immune system in cancer. Curr Pharm Biotechnol **10**(2): 166–184.
3. D'Elios, M. M., G. Del Prete, et al. (2009). New frontiers in cell-based immunotherapy of cancer. Expert Opin Ther Pat **19**(5): 623–641.
4. Nyolczas, N., S. Charwat, et al. (2009). Tracking the migration of cardially delivered therapeutic stem cells *in vivo*: state of the art. Regen Med **4**(3): 407–422.
5. Lee, Z., J. E. Dennis, et al. (2008). Imaging stem cell implant for cellular-based therapies. Exp Biol Med (Maywood) **233**(8): 930–940.
6. Roach, P. J., D. L. Bailey, et al. (2008). Enhancing lung scintigraphy with single-photon emission computed tomography. Semin Nucl Med **38**(6): 441–449.

7. Costelloe, C. M., E. M. Rohren, et al. (2009). Imaging bone metastases in breast cancer: techniques and recommendations for diagnosis. Lancet Oncol **10**(6): 606–614.
8. Strair, R. K., M. Towle, et al. (1990). Retroviral mediated gene transfer into bone marrow progenitor cells: use of beta-galactosidase as a selectable marker. Nucleic Acids Res **18**(16): 4759–4762.
9. Song, W., H. L. Kong, et al. (1997). Dendritic cells genetically modified with an adenovirus vector encoding the cDNA for a model antigen induce protective and therapeutic antitumor immunity. J Exp Med **186**(8): 1247–1256.
10. Limberis, M. P., C. L. Bell, et al. (2009). Identification of the murine firefly luciferase-specific CD8 T-cell epitopes. Gene Ther **16**(3): 441–447.
11. Vermandel, M. and X. Marchandise (2009). From a new kind of radiation to tomography: a history of the computed tomography. IRBM **30**(2): 33–39.
12. Hildebrandt, I. J. and S. S. Gambhir (2004). Molecular imaging applications for immunology. Clin Immunol **111**(2): 210–224.
13. Pichler, B. J., H. F. Wehrl, et al. (2008). Latest advances in molecular imaging instrumentation. J Nucl Med **49 Suppl 2**: 5 S–23 S.
14. Shih, C. M., W. H. Hsu, et al. (2003). Usefulness of chest single photon emission computed tomography with technetium-99 m methoxyisobutylisonitrile to predict taxol based chemotherapy response in advanced non-small cell lung cancer. Cancer Letters **199**(1): 99–105.
15. El-Baz, A., G. Gimel'farb, et al. (2009). Automatic analysis of 3D low dose CT images for early diagnosis of lung cancer. Pattern Recognition **42**(6): 1041–1051.
16. Schmidt, G. P., M. F. Reiser, et al. (2009). Whole-body imaging of bone marrow. Semin Musculoskelet Radiol **13**(2): 120–133.
17. Hainfeld, J. F., D. N. Slatkin, et al. (2006). Gold nanoparticles: a new x-ray contrast agent. Br J Radiol **79**(939): 248–253.
18. Kim, D., S. Park, et al. (2007). Antibiofouling polymer-coated gold nanoparticles as a contrast agent for *in vivo* x-ray computed tomography imaging. J Am Chem Soc **129**(24): 7661–7665.
19. Alric, C., J. Taleb, et al. (2008). Gadolinium chelate coated gold nanoparticles as contrast agents for both x-ray computed tomography and magnetic resonance imaging. J Am Chem Soc **130**(18): 5908–5915.
20. Rabin, O., J. Manuel Perez, et al. (2006). An x-ray computed tomography imaging agent based on long-circulating bismuth sulphide nanoparticles. Nat Mater **5**(2): 118–122.
21. Popovtzer, R., A. Agrawal, et al. (2008). Targeted gold nanoparticles enable molecular CT imaging of cancer. Nano Lett **8**(12): 4593–4596.
22. Piedra, M., A. Allroggen, et al. (2009). Molecular imaging with targeted contrast ultrasound. Cerebrovasc Dis **27 Suppl 2**: 66–74.
23. Turkbey, B., H. Kobayashi, et al. (2009). Imaging of tumor angiogenesis: functional or targeted? AJR Am J Roentgenol **193**(2): 304–313.
24. Arbab, A. S. and J. A. Frank (2008). Cellular MRI and its role in stem cell therapy. Regen Med **3**(2): 199–215.
25. Sengar, R. S., L. Spokauskiene, et al. (2009). Magnetic resonance imaging-guided adoptive cellular immunotherapy of central nervous system tumors with a T1 contrast agent. Magn Reson Med **62**(3): 599–606.
26. Ferrara, K., R. Pollard, et al. (2007). Ultrasound microbubble contrast agents: fundamentals and application to gene and drug delivery. Annu Rev Biomed Eng **9**: 415–447.
27. Kaur, T., R. A. Slavcev, et al. (2009). Addressing the challenge: current and future directions in ovarian cancer therapy. Curr Gene Ther **9**(6): 434–458.
28. Fisher, N. G., J. P. Christiansen, et al. (2002). Influence of microbubble surface charge on capillary transit and myocardial contrast enhancement. J Am Coll Cardiol **40**(4): 811–819.
29. Guo, D. P., X. Y. Li, et al. (2006). Ultrasound-targeted microbubble destruction improves the low density lipoprotein receptor gene expression in HepG2 cells. Biochem Biophys Res Commun **343**(2): 470–474.
30. Korpanty, G., J. G. Carbon, et al. (2007). Monitoring Response to Anticancer Therapy by Targeting Microbubbles to Tumor Vasculature. Clinical Cancer Research **13**(1): 323–330.

31. Leong-Poi, H. (2009). Molecular imaging using contrast-enhanced ultrasound: evaluation of angiogenesis and cell therapy. Cardiovasc Res **84** (2): 190–200.
32. Hoh, C. K., R. A. Hawkins, et al. (1993). Cancer detection with whole-body PET using 2-[18F]fluoro-2-deoxy-D-glucose. J Comput Assist Tomogr **17**(4): 582–589.
33. Cherry, S. R. (2006). The 2006 Henry N. Wagner lecture: of mice and men (and positrons)-advances in PET imaging technology. J Nucl Med **47**(11): 1735–1745.
34. Wu, A. M. (2009). Antibodies and antimatter: the resurgence of immuno-PET. J Nucl Med **50**(1): 2–5.
35. Zalutsky, M. R. (2006). Potential of immuno-positron emission tomography for tumor imaging and immunotherapy planning. Clin Cancer Res **12**(7 Pt 1): 1958–1960.
36. Tumeh, P. C., C. G. Radu, et al. (2008). PET imaging of cancer immunotherapy. J Nucl Med **49**(6): 865–868.
37. Kumar, R., A. Chauhan, et al. (2006). Role of PET/PET-CT in the management of sarcomas. Expert Rev Anticancer Ther **6**(8): 1241–1250.
38. Akins, E. J. and P. Dubey (2008). Noninvasive imaging of cell-mediated therapy for treatment of cancer. J Nucl Med **49 Suppl 2**: 180 S–195 S.
39. Serganova, I. and R. Blasberg (2005). Reporter gene imaging: potential impact on therapy. Nucl Med Biol **32**(7): 763–780.
40. Le, L. Q., J. H. Kabarowski, et al. (2002). Positron emission tomography imaging analysis of G2A as a negative modifier of lymphoid leukemogenesis initiated by the BCR-ABL oncogene. Cancer Cell **1**(4): 381–391.
41. Pantuck, A. J., J. Matherly, et al. (2002). Optimizing prostate cancer suicide gene therapy using herpes simplex virus thymidine kinase active site variants. Hum Gene Ther **13**(7): 777–789.
42. Dong, D., L. Dubeau, et al. (2004). Spontaneous and controllable activation of suicide gene expression driven by the stress-inducible grp78 promoter resulting in eradication of sizable human tumors. Hum Gene Ther **15**(6): 553–561.
43. Dobrenkov, K., M. Olszewska, et al. (2008). Monitoring the efficacy of adoptively transferred prostate cancer-targeted human T lymphocytes with PET and bioluminescence imaging. J Nucl Med **49**(7): 1162–1170.
44. Sato, T., A. Neschadim, et al. (2007). Engineered human tmpk/AZT as a novel enzyme/prodrug axis for suicide gene therapy. Mol Ther **15**(5): 962–970.
45. Khan, Z., W. Knecht, et al. (2010). Plant thymidine kinase 1: a novel efficient suicide gene for malignant glioma therapy. Neuro Oncol. **12**(6): 549–58.
46. Dubey, P., H. Su, et al. (2003). Quantitative imaging of the T cell antitumor response by positron-emission tomography. Proc Natl Acad Sci USA **100**(3): 1232–1237.
47. Koehne, G., M. Doubrovin, et al. (2003). Serial *in vivo* imaging of the targeted migration of human HSV-TK-transduced antigen-specific lymphocytes. Nat Biotechnol **21**(4): 405–413.
48. Herschman, H. R. (2004). Noninvasive imaging of reporter gene expression in living subjects. Adv Cancer Res **92**: 29–80.
49. Herschman, H. R. (2004). PET reporter genes for noninvasive imaging of gene therapy, cell tracking and transgenic analysis. Crit Rev Oncol Hematol **51**(3): 191–204.
50. Buursma, A. R., V. Rutgers, et al. (2006). 18F-FEAU as a radiotracer for herpes simplex virus thymidine kinase gene expression: in-vitro comparison with other PET tracers. Nucl Med Commun **27**(1): 25–30.
51. Saga, T., M. Koizumi, et al. (2009). Molecular imaging of cancer: evaluating characters of individual cancer by PET/SPECT imaging. Cancer Sci **100**(3): 375–381.
52. Barbet, J., F. Kraeber-Bodere, et al. (2008). What can be expected from nuclear medicine tomorrow? Cancer Biother Radiopharm **23**(4): 483–504.
53. Erasmus, J. J., E. Rohren, et al. (2009). Prognosis and reevaluation of lung cancer by positron emission tomography imaging. Proc Am Thorac Soc **6**(2): 171–179.
54. Lu, Y., H. Dang, et al. (2004). Bioluminescent monitoring of islet graft survival after transplantation. Mol Ther **9**(3): 428–435.
55. Fowler, M., J. Virostko, et al. (2005). Assessment of pancreatic islet mass after islet transplantation using *in vivo* bioluminescence imaging. Transplantation **79**(7): 768–776.

56. Lin, Y., R. Weissleder, et al. (2002). Novel near-infrared cyanine fluorochromes: synthesis, properties, and bioconjugation. Bioconjug Chem **13**(3): 605–610.
57. Ntziachristos, V., C. Bremer, et al. (2003). Fluorescence imaging with near-infrared light: new technological advances that enable *in vivo* molecular imaging. Eur Radiol **13**(1): 195–208.
58. Shah, K., E. Bureau, et al. (2005). Glioma therapy and real-time imaging of neural precursor cell migration and tumor regression. Ann Neurol **57**(1): 34–41.
59. Shah, K. and R. Weissleder (2005). Molecular optical imaging: applications leading to the development of present day therapeutics. NeuroRx **2**(2): 215–225.
60. Ntziachristos, V., C. H. Tung, et al. (2002). Fluorescence molecular tomography resolves protease activity *in vivo*. Nat Med **8**(7): 757–760.
61. Tung, C. H. (2004). Fluorescent peptide probes for *in vivo* diagnostic imaging. Biopolymers **76**(5): 391–403.
62. Giepmans, B. N., S. R. Adams, et al. (2006). The fluorescent toolbox for assessing protein location and function. Science **312**(5771): 217–224.
63. Sutton, E. J., T. D. Henning, et al. (2008). Cell tracking with optical imaging. Eur Radiol **18**(10): 2021–2032.
64. Bergers, G., K. Javaherian, et al. (1999). Effects of angiogenesis inhibitors on multistage carcinogenesis in mice. Science **284**(5415): 808–812.
65. Weissleder, R., C. H. Tung, et al. (1999). *In vivo* imaging of tumors with protease-activated near-infrared fluorescent probes. Nat Biotechnol **17**(4): 375–378.
66. Rochefort, H., M. Garcia, et al. (2000). Cathepsin D in breast cancer: mechanisms and clinical applications, a 1999 overview. Clin Chim Acta **291**(2): 157–170.
67. Tung, C. H., U. Mahmood, et al. (2000). *In vivo* imaging of proteolytic enzyme activity using a novel molecular reporter. Cancer Res **60**(17): 4953–4958.
68. Bogdanov, A. A., Jr., C. P. Lin, et al. (2002). Cellular activation of the self-quenched fluorescent reporter probe in tumor microenvironment. Neoplasia **4**(3): 228–236.
69. Mahmood, U., C. H. Tung, et al. (2002). Feasibility of *in vivo* multichannel optical imaging of gene expression: experimental study in mice. Radiology **224**(2): 446–451.
70. Walling, M. A., J. A. Novak, et al. (2009). Quantum dots for live cell and *in vivo* imaging. Int J Mol Sci **10**(2): 441–491.
71. Jaiswal, J. K., H. Mattoussi, et al. (2003). Long-term multiple color imaging of live cells using quantum dot bioconjugates. Nat Biotechnol **21**(1): 47–51.
72. Smith, A. M., G. Ruan, et al. (2006). Engineering luminescent quantum dots for *in vivo* molecular and cellular imaging. Ann Biomed Eng **34**(1): 3–14.
73. Zhang, H., D. Yee, et al. (2008). Quantum dots for cancer diagnosis and therapy: biological and clinical perspectives. Nanomedicine **3**(1): 83–91.
74. Dahan, M., S. Levi, et al. (2003). Diffusion dynamics of glycine receptors revealed by single-quantum dot tracking. Science **302**(5644): 442–445.
75. Michalet, X., F. F. Pinaud, et al. (2005). Quantum dots for live cells, *in vivo* imaging, and diagnostics. Science **307**(5709): 538–544.
76. Noh, Y. W., Y. T. Lim, et al. (2008). Noninvasive imaging of dendritic cell migration into lymph nodes using near-infrared fluorescent semiconductor nanocrystals. FASEB J **22**(11): 3908–3918.
77. Kraitchman, D. L., A. W. Heldman, et al. (2003). *In vivo* magnetic resonance imaging of mesenchymal stem cells in myocardial infarction. Circulation **107**(18): 2290–2293.
78. Bulte, J. W. and D. L. Kraitchman (2004). Iron oxide MR contrast agents for molecular and cellular imaging. NMR Biomed **17**(7): 484–499.
79. Sosnovik, D. and R. Weissleder (2005). Magnetic resonance and fluorescence based molecular imaging technologies. Prog Drug Res **62**: 83–115.
80. Kircher, M. F., J. R. Allport, et al. (2003). *In vivo* high resolution three-dimensional imaging of antigen-specific cytotoxic T-lymphocyte trafficking to tumors. Cancer Res **63**(20): 6838–6846.

81. Ponomarev, V., M. Doubrovin, et al. (2004). A novel triple-modality reporter gene for whole-body fluorescent, bioluminescent, and nuclear noninvasive imaging. Eur J Nucl Med Mol Imaging **31**(5): 740–751.
82. Li, C., W. Wang, et al. (2006). Dual optical and nuclear imaging in human melanoma xenografts using a single targeted imaging probe. Nucl Med Biol **33**(3): 349–358.
83. Thompson, M., D. M. Wall, et al. (2005). *In vivo* tracking for cell therapies. Q J Nucl Med Mol Imaging **49**(4): 339–348.
84. Damadian, R. (1971). Tumor detection by nuclear magnetic resonance. Science **171**(976): 1151–1153.
85. Magnitsky, S., R. M. Walton, et al. (2008). Magnetic resonance imaging detects differences in migration between primary and immortalized neural stem cells. Acad Radiol **15**(10): 1269–81.
86. Bos, C., Y. Delmas, et al. (2004). *In vivo* MR imaging of intravascularly injected magnetically labeled mesenchymal stem cells in rat kidney and liver. Radiology **233**(3): 781–9.
87. de Vries, I. J., W. J. Lesterhuis, et al. (2005). Magnetic resonance tracking of dendritic cells in melanoma patients for monitoring of cellular therapy. Nat Biotechnol **23**(11): 1407–1413.
88. Dekaban, G. A., J. Snir, et al. (2009). Semiquantitation of mouse dendritic cell migration *in vivo* using cellular MRI. J Immunother **32**(3): 240–251.
89. Valable, S., E. L. Barbier, et al. (2008). *In vivo* MRI tracking of exogenous monocytes/macrophages targeting brain tumors in a rat model of glioma. Neuroimage **40**(2): 973–983.
90. Verdijk, P., E. H. Aarntzen, et al. (2009). Limited amounts of dendritic cells migrate into the T-cell area of lymph nodes but have high immune activating potential in melanoma patients. Clin Cancer Res **15**(7): 2531–2540.
91. Helfer, B. M., A. Balducci, et al. (2010). Functional assessment of human dendritic cells labeled for *in vivo*(19)F magnetic resonance imaging cell tracking. Cytotherapy **12**(2): 238–50.
92. Martin, A. L., L. M. Bernas, et al. (2008). Enhanced cell uptake of superparamagnetic iron oxide nanoparticles functionalized with dendritic guanidines. Bioconjug Chem **19**(12): 2375–2384.
93. Wang, Y. X., H. H. Wang, et al. (2008). Pitfalls in employing superparamagnetic iron oxide particles for stem cell labelling and *in vivo* MRI tracking. Br J Radiol **81**(972): 987–988.
94. von Schulthess, G. and H.-P. Schlemmer (2009). A look ahead: PET/MR versus PET/CT. Eur J Nucl Med Mol Imaging **36**: 3–9.
95. Liu, W. and J. A. Frank (2009). Detection and quantification of magnetically labeled cells by cellular MRI. Eur J Radiol **70**(2): 258–264.
96. Brix, G., E. A. Nekolla, et al. (2009). Risks and safety aspects related to PET/MR examinations. Eur J Nucl Med Mol Imaging **36 Suppl 1**: S131–138.
97. Institute, N. E. (2003). Fact Sheet: Radiation and Human Health.
98. Lee, S.-W., P. Padmanabhan, et al. (2009). Stem cell-mediated accelerated bone healing observed with *in vivo* molecular and small animal imaging technologies in a model of skeletal injury. J Orthop Res **27**(3): 295–302.
99. Wehrl, H. F., M. S. Judenhofer, et al. (2009). Pre-clinical PET/MR: technological advances and new perspectives in biomedical research. Eur J Nucl Med Mol Imaging **36**(0 (supplement 1)): 56–68.
100. Cook, R. J., G. Fracchia, et al. (1998). Evolution of a surgical technique for posteroventral pallidotomy using CT/MR fusion and intraoperative macrostimulation. J Clin Neurosci **5**(1): 20–27.
101. Chenoune, Y., Y. Bouaoune, et al. (2007). MR/CT multimodal registration of short-axis slices in CT volumes. Conf Proc IEEE Eng Med Biol Soc **2007**: 4496–4499.
102. Furth, C., T. Denecke, et al. (2006). Correlative imaging strategies implementing CT, MRI, and PET for staging of childhood Hodgkin disease. J Pediatr Hematol Oncol **28**(8): 501–512.

103. McCarthy, J. R., P. Patel, et al. (2009). Multimodal manoagents for the detection of intravascular thrombi. Bioconjug Chem **20**(6): 1251–1255.
104. Choi, J. S., J. C. Park, et al. (2008). A hybrid nanoparticle probe for dual-modality positron emission tomography and magnetic resonance imaging. Angew Chem Int Ed Engl **47**(33): 6259–6262.
105. Lee, H.-Y., Z. Li, et al. (2008). PET/MR dual-modality tumor imaging using arginine-glycine-aspartic (RGD)-conjugated radiolabeled iron oxide nanoparticles. J Nucl Med **49**: 1371–1379.
106. Yong, K.-T. (2009). Mn-doped near-infrared quantum dots as multimodal targeted probes for pancreatic cancer imaging. Nanotechnology **20**: 015102 (015110 pp).

Part V
Transplantation

Chapter 19
Allogeneic and Autologous Transplantation Therapy of Cancer: Converging Themes

Daniel H. Fowler

Abstract Effective adoptive T-cell therapy of cancer occurs after allogeneic bone marrow transplantation, and is now increasingly observed after autologous transplantation. As the clinical practice of each field evolves and the biologic underpinnings of these two seemingly disparate approaches are elucidated, four converging themes have emerged. (1) T-cell antigen specificity. The alloreactive T-cell graft-versus-tumor response is difficult to dissociate from potentially lethal graft-versus-host disease (GVHD); as such, efforts are underway to enhance antigen-specificity after allogeneic transplantation. By comparison, the field of autologous T-cell therapy has now largely redefined itself by use of highly specific T-cell receptor reactivities, which may ultimately prove limiting in terms of tumor escape mechanisms; efforts are therefore underway to broaden antigenic reactivities after autologous transplantation. (2) Host conditioning. Myeloablative conditioning used in allogeneic transplantation causes significant morbidity and mortality even in young and healthy patients, thereby limiting broader application to the majority of cancer patients; in response, less intensive and better tolerated nonmyeloablative regimens are being developed. By comparison, autologous T-cell therapy has been primarily limited by lack of efficacy, and in response, investigators have increased host conditioning to myeloablative levels to create immune space that facilitates T-cell expansion and effectiveness *in vivo*. (3) T-cell function. The evolving discipline of T-cell biology will continue to enhance the efficacy of both autologous and allogeneic transplantation therapy. The quality of the T-cell response is of paramount importance, and is determined by T-cell differentiation status, apoptotic tendency, and cytokine phenotype vis-à-vis Th1, Th2, Treg, and Th17 balance; modulation of this balance for therapeutic gain will depend upon an ability to understand and control an emerging phenomenon termed T-cell plasticity. (4) T-cell-mediated immune pathology. GVHD after allogeneic transplantation has been an instrumental model system for understanding T-cell pathology; importantly, cellular and molecular mechanisms underlying GVHD may also contribute to toxicities observed after effective autologous T-cell therapy. As such,

D.H. Fowler (✉)
Bethesda, MD, USA
e-mail: dhfowler@helix.nih.gov

J. Medin and D. Fowler (eds.), *Experimental and Applied Immunotherapy*,
DOI 10.1007/978-1-60761-980-2_19,

both allogeneic and autologous immunotherapies are now confronted with an overall goal of maximizing T-cell efficacy while limiting T-cell toxicity. Attainment of such enhanced therapeutic windows may be facilitated by recent developments in immune cell modulation, including use of cell fate control genes and use of inhibitors of the mammalian target of rapamycin (mTOR) and JAK/STAT pathways. In conclusion, converging themes in autologous and allogeneic transplantation therapy indicate that a bright future will emerge for T-cell therapies, the success of which will be realized through advances in T-cell biology and T-cell engineering.

Keywords GVHD • JAK/STAT pathway • Pentostatin • Rapamycin • Th1/Th2

Transplantation: Paradigm Shift from Hematopoietic Reconstitution to T-Cell Biology

The rationale for performing allogeneic or autologous transplantation has in many cases shifted from reconstitution of hematopoiesis to delivery of therapeutic T-cell immunity. From the perspective of allogeneic bone marrow transplantation (BMT), the nidus of the paradigm shift can be traced to a publication in 1979 that demonstrated an association of GVHD, which was thought to be mediated by T cells, to potency of the transplant antileukemic effect observed after myeloablative host conditioning [1]. Anticancer effects observed after allogeneic BMT or hematopoietic stem cell transplantation (HSCT) are generally referred to as graft-versus-tumor (GVT) effects. In the future, it will be critical to determine whether allogeneic T-cell immunity might significantly expand the therapeutic application of allogeneic HSCT beyond the treatment of hematologic malignancy. Towards this end, in a multicenter analysis of over 100 patients treated in Europe published in 2006, antitumor responses against metastatic renal cell carcinoma after allogeneic HSCT were observed in more that 20% of patients [2]; and, in a pilot clinical trial, allogeneic HSCT yielded tumor regressions in patients with metastatic breast cancer [3].

By comparison, conventional autologous HSCT, which is still performed widely for therapy of chemotherapy-sensitive malignancies such as lymphoma and multiple myeloma, essentially represents a procedure to resuscitate hematopoiesis after high-dose chemotherapy. That is, a strong case has not been made that endogenous anticancer T cells naturally arise in the autologous setting, thereby placing the therapeutic emphasis upon chemotherapy. However, as detailed in Chaps. 1 and 12 of this book, recent advances in understanding of tumor antigens and in T-cell engineering have brought autologous T-cell therapy to the forefront of translational research. It is certainly possible to consider a future whereby advances in targeting the most relevant tumor antigens may allow autologous immunotherapy to match or surpass the curative effects of allogeneic HSCT. For example, autologous T-cell therapy might be capable of targeting antigens preferentially expressed on cancer stem cells; this approach was recently found to be successful in such an experimental model of brain cancer [4].

Harnessing this type of novel, tumor-specific T-cell effect may place anticancer T-cell therapy into a new dimension in terms of clinical efficacy.

The emphasis in autologous transplantation thus appears to be shifting away from chemotherapeutic anticancer effects to immune antitumor effects. However, as will be detailed below, it is important to note that high-dose chemotherapy is playing an emerging role in autologous immunotherapy efforts because of the beneficial impact of host immune depletion on the efficacy of adoptive T-cell therapy. Indeed, autologous T-cell therapy protocols are increasingly being performed in settings that require intensive chemotherapy and hematopoietic support in the form of growth factor or hematopoietic progenitor cell transplantation. As such, although both autologous and allogeneic approaches use chemotherapy, the therapeutic focus will be increasingly placed upon enhancement of the transplanted T cells. In this pursuit of T-cell therapy, it will nonetheless be important to more fully consider the potential effects of stem cell infusion on T-cell function: that is, in experimental murine models of autologous, tumor-specific T-cell therapy, coadministered hematopoietic stem cells could either promote or inhibit T-cell immunity depending upon infused cell ratios [5].

Antitumor Specificity

Allogeneic HSCT represents a mechanistically diffuse approach to the immune therapy of cancer that can involve B, NK, and T-cell effector pathways. The T-cell contribution to allogeneic GVT effects is perhaps the most potent; however, the T-cell biology that accounts for GVT effects closely mimics the pathogenesis of the most significant transplant complications, acute and chronic GVHD. For this reason, as has recently been reviewed [6], it has been difficult to separate beneficial GVT effects from detrimental GVHD clinically. With respect to antigen specificity and GVT effects, it will be important to determine the extent to which alloreactive T-cell responses contribute to GVT effects; and, by extension, it will be important to identify the success of allogeneic transplant approaches that seek to preferentially emphasize nonalloreactive, tumor-specific T-cell responses.

One such approach to limit alloreactivity consists of an *ex vivo* processing method whereby allogeneic T-cell-containing progenitor cell grafts are first stimulated with antigen presenting cells (APCs) from the host to activate the alloreactive T-cell receptor (TCR) repertoire, with subsequent photodepletion of activated T cells; this allograft engineering strategy is currently being tested in a pilot clinical trial [7]. In theory, the potential success of this and similar approaches is dependent upon a nonalloreactive (that is, tumor-reactive or tumor-specific) component of the donor TCR repertoire that can be activated independent of any adjuvant effect from inflammatory pathways promoted through alloreactivity. The extent to which GVT effects in the allogeneic setting can be attributed to alloreactive T-cell responses versus antigen-specific T-cell responses is currently unknown. The fields of allogeneic and autologous T-cell therapy are therefore converging at two important points.

First, tumor-reactive or tumor-specific T cells will increasingly be the focus of autologous transplantation therapy; at the same time, their role in allogeneic transplantation remains to be clarified and potentially harnessed. And, second, adjuvant effects of inflammatory pathways activated through microbial translocation and TLR4 activation are clearly related to GVT effects and GVHD in the allogeneic transplantation context [6]; at the same time, their role in autologous immunotherapy has also now been identified [8]. As such, issues of T-cell specificity and systemic host factors will be at the forefront of both autologous and allogeneic immunotherapy efforts.

A second method that attempts to harness tumor-specific immunity in the allogeneic setting relates to the adoptive transfer of T cells that are specific for minor histocompatibility antigens that are preferentially expressed on tumor cells relative to normal host tissue. However, it is not clear whether such preferential antigen expression can result in a favorable therapeutic window of potent antitumor responses with limited toxicity: for example, infusion of leukemia-associated minor histocompatibility antigen specific T cells was associated with pulmonary toxicity that may have been due to antigen expression in the lung [9]. Theoretically, this approach of targeting the donor T-cell response to minor histocompatibility antigens may also be limited by systemic toxicities that may occur when mediators such as fas ligand or TNF-α are leaked into the circulation [10]. Nonetheless, in experimental murine models, adoptive transfer of allogeneic T cells that are truly specific for tumor antigen (transgenic TCR specific for a melanoma antigen) can yield potent antitumor responses with limited GVHD [11]; as such, there remains a strong rationale for attempts to harness tumor-specific immunity in the allogeneic setting. Overall, a convergence of autologous and allogeneic immunotherapy fields is therefore now apparent, with each field attempting to find the optimal balance between parameters of tumor-specific T-cell responses and generalized adjuvant effects.

Immune Space

The concept of "immune space" is fundamental to current efforts in autologous and allogeneic T-cell therapy. Such space is not geographical in nature, but rather represents T-cell competition for limiting quantities of critical growth factors, including IL-7 and IL-15. It is important to emphasize that the current focus on IL-7 and IL-15 represents a dramatic shift away from IL-2, which has been used clinically for decades in an attempt to enhance the effects of adoptively transferred T cells. Indeed, clinical trials have demonstrated that IL-2 administration in humans can drive regulatory T-cell expansion [12], which overall appears to represent a detriment to T-cell therapeutic potency (reviewed in Chaps. 10 and 11). Without sufficient space, a particular adoptive T-cell transfer will have limited *in vivo* efficacy due to insufficient T-cell expansion within the host; in this scenario, adoptively transferred T cells simply are outnumbered by existing host T cells, which may contain immune suppressive populations such as Treg cells. The importance of homeostatic cytokine availability rather than physical space was clarified by experiments performed in a murine model of T-cell therapy of metastatic melanoma [13]. That is, intensive host preparation (with total body irradiation (TBI)) predictably and efficiently caused lymphocyte depletion prior to adoptive transfer; however, such lymphocyte depletion was not sufficient for optimal *in vivo* T-cell expansion if the host was genetically deficient in IL-7 or IL-15. In this manner, it was concluded that host T cells acted as "cytokine sinks" that reduced availability of cytokines critical for T-cell therapy.

It is interesting to consider this concept of immune space when evaluating current trends in allogeneic HSCT, in particular, transplantation using reduced-intensity or "nonmyeloablative" conditioning regimens. Reduced-intensity regimens are associated with reduced morbidity and mortality, thereby extending the applicability of transplantation to patients of advanced age or individuals with co-morbidities such as infection or organ dysfunction. With reduced conditioning, greater host immunity is present at the time of transplantation, with predictably lower levels of homeostatic cytokines available for donor T-cell expansion; thus, low levels of allogeneic T-cell engraftment can occur both due to increased host T-cell numbers and insufficient homeostatic cytokines available for donor T-cell expansion. Use of low-intensity conditioning indeed predisposes to mixed donor/host T-lymphoid chimerism; importantly, such mixed chimerism can be clinically stable with single-agent, posttransplant sirolimus administration and can lead to the amelioration of sickle cell anemia [14], for example. However, the relative immune tolerance associated with mixed chimerism may not be beneficial for allogeneic transplantation therapy of malignancy [15], and as such, one of our own research goals is to achieve more complete donor lymphoid engraftment in the low-intensity conditioning setting. One such approach is to evaluate new agents to deplete or suppress host immunity prior to transplantation. Specifically, we are evaluating pentostatin, which is a purine analog similar to fludarabine (which is most commonly used in reduced-intensity transplantation) [16]. Pentostatin has a unique mechanism of action that involves inhibition of adenosine deaminase, which is the enzyme deficient in a subset of patients with severe combined immunodeficiency that can be augmented by gene therapy [17]. Importantly, pentostatin combined with low-dose total body irradiation (TBI) has been shown to deplete host immunity and facilitate the engraftment of sibling donor and matched unrelated donor allografts [18]. However, it is currently not known whether the unique mechanism of action of pentostatin will confer superior alloengraftment results relative to agents such as fludarabine; it will therefore be important to compare and contrast the immune effects of fludarabine and pentostatin, including effects on homeostatic cytokines such as IL-7 and IL-15.

The creation of immune space in the setting of allogeneic HSCT may certainly be beneficial in terms of enhanced expansion of therapeutic donor T cells. That said, it should be noted, however, that patients with higher levels of serum IL-7 in the peritransplant period had an increased chance of developing acute GVHD after reduced-intensity transplantation [19]. It is important to note that GVHD can manifest in differential patterns after reduced-intensity allogeneic HSCT relative to myeloablative conditioning; however, the overall negative impact of GVHD in terms of morbidity and mortality has not been shown to be substantially reduced with reduced-intensity transplantation [20]. As such, elevated homeostatic cytokines that result from host conditioning prior to allogeneic HSCT represent both an opportunity for immune therapy and a risk in terms of GVHD. Therefore, attempts should be made to identify functionally defined T-cell subsets that possess a capacity to promote alloengraftment with reduced GVHD. Towards this aim, we have determined that infusion of donor CD4$^+$ T cells polarized to a Th2 cytokine phenotype effectively facilitated alloengraftment [21] in a murine model of graft rejection. Typically, donor T cells that facilitate engraftment do so through cytolytic pathways that are difficult to dissociate from GVHD [22]; however, in our recent study, we also found that donor Th2 cells operated by an IL-4/STAT6 dependent mechanism whereby donor Th2 cells activated STAT6 in host T cells, thereby transferring the tolerogenic Th2 state to host T cells by an "infectious" process [23]. As such, allogeneic immunotherapy can be facilitated by the creation of immune space, and also by the functional modulation of host T cells, such as in our Th2 cell strategy.

It is interesting to note that as efforts are underway to reduce immune space at the time of allogeneic HSCT (via reduced-intensity conditioning), efforts in the autologous immunotherapy setting seek to heighten immune space. Specifically, at the National Cancer Institute Surgery Branch, the conventional host preparation with combination fludarabine plus cyclophosphamide has been intensified by the further inclusion of up to 12 Gy TBI; in one study, inclusion of TBI yielded further increases in serum IL-7 and IL-15 and an associated increase in the complete and partial remission rate of patients with metastatic melanoma [24]. The beneficial effects of IL-7 and IL-15, in addition to driving adoptively transferred T-cell expansion, can also enhance T-cell function in ways that may benefit cancer immunotherapy: specifically, in a murine model, IL-7 and IL-15 broadened the CD8$^+$ T-cell response to subdominant epitopes [25].

The use of high-dose chemotherapy and TBI to create immune space has some limitation in terms of generalized applicability to the majority of cancer patients; that is, the majority of cancer patients are over the age of 50 and such regimens are associated with significant morbidity, particularly in individuals over the age of 50. As such, efforts are underway to mimic the effects of host conditioning through recombinant cytokine administration. For example, in a murine model of adoptive T-cell transfer for therapy of cancer, combination anti-CD25 antibody therapy (for regulatory T-cell depletion) and IL-7 administration improved adoptive T-cell therapy in lymphoreplete hosts. Most importantly, generation of immune space by this method actually increased resultant *in vivo* T-cell function (increased antigen-specific IFN-γ secretion; increased TCR epitope spreading) relative to TBI-based conditioning [26].

Further studies will most likely identify enhanced strategies for use of cytokines to induce immune space. For example, it was recently found that IL-7 has not only direct, positive effects on CD4$^+$ T-cell expansion but also a counter-regulatory effect that is mediated by IL-7 activation of plasmacytoid dendritic cells (DCs), which results in down-regulation of DC expression of MHC class II that is required for optimal CD4 cell expansion [27]. These findings suggest that IL-7 mediated immune space may be further enhanced through modulation of plasmacytoid DC biology. Therefore, it is possible that with a further understanding of and an ability to create immune space, adoptive T-cell therapy strategies might be defined that are not only safer but also more effective. Towards clinical translation of this concept, a phase I trial of IL-7 was recently completed in humans. As predicted from murine studies, effector T-cell responses were promoted whereas regulatory T cells (Tregs), which lack the IL-7 receptor alpha chain, were not expanded *in vivo* [28]. Clinical translation of IL-15 has been slower to progress largely due to commercial considerations; nonetheless, recent preclinical studies in nonhuman primates have confirmed the important role of IL-15 in creating immune space, particularly for $CD8^+$ T cells [29]. And finally, it should also be noted that additional cytokines may emerge as important for adoptive T-cell therapy: specifically, it was recently determined that IL-21, a cytokine closely related to IL-2, maintained cytotoxic T-cell function *in vivo* without inducing terminal T-cell differentiation, thereby leading to improved antitumor effects [30].

T-Cell Phenotype and Plasticity

The immunotherapeutic capacity of adoptively transferred T cells is not only dependent upon TCR specificity and sufficient immune space, but also on the T-cell cytokine phenotype. In the setting of murine allogeneic BMT, $CD4^+$, and $CD8^+$ T-cell subsets of the Th1/Tc1 phenotype (characterized in part by IFN-γ secretion and fas-based cytolysis) or Th2/Tc2 phenotype (characterized in part by IL-4 secretion and perforin-based cytolysis) differentially mediate GVHD and GVT effects [31]. In that study, Th1/Tc1 cells mediated potent GVT effects but were limited by GVHD, whereas Th2/Tc2 cells mediated minimal GVHD but induced modest GVT effects. These observations, as well as those from a multitude of additional studies in the literature, indicate that GVHD and GVT effects share a similar biology, thereby making their clinical separation difficult (reviewed in [6]). Given this biology, we have used murine models to develop an allogeneic transplantation strategy whereby a Th1/Tc1 response is first initiated for realization of GVT effects, with subsequent regulation of GVHD through delayed administration of Th2 cells generated *ex vivo* (sequential Th1 → Th2 therapy) [32]. An initial pilot clinical trial evaluating allograft augmentation with Th2 cells has been completed [33] by our group in 2006. Ongoing clinical trials are presently evaluating rapamycin-resistant Th2 cells, which we have shown in murine models to be more potent for regulation of established GVHD[32].

The research area related to T-cell cytokine phenotype has gained in complexity due to the characterization of at least two additional functional subsets of $CD4^+$ T cells, namely Th17 cells and regulatory T cells (Tregs). These subsets, which are both generated in the presence of TGF-β and either the presence of IL-6/STAT3 signaling (Th17 cells) or the absence of IL-6/STAT3 signaling (Tregs) (reviewed in [34]), help determine the outcome of allogeneic immune reactions, including GVHD and GVT effects. That is, in a murine allogeneic BMT model, coinfusion of donor Tregs inhibited GVHD while preserving GVT effects [35]. In a separate model, donor Tregs were found to impair both GVHD and GVT effects, but the GVT effect could be realized through the delayed, posttransplant infusion of Treg cells [36]. These observations in the allogeneic transplantation setting stand somewhat in contrast to the known effect of Tregs, which in general promote tumor growth (reviewed in Chaps. 10 and 11). Further research should be aimed at determining the mechanism(s) whereby allogeneic Tregs might modulate GVHD without impairing GVT effects. Potentially, it is interesting to speculate that these observations may reflect the anti-inflammatory properties of Tregs and the known link between chronic inflammation and cancer progression, which seems to be driven by the STAT3 activating cytokines IL-6 and IL-23 [37].

The complex role of T cell subsets in allogeneic transplantation has been further delineated through studies of the fourth major T-helper cell subset, Th17 cells. That is, a blockade of IL-6 posttransplant resulted in a shift away from Th17 cells and towards Treg cells, thereby reducing acute GVHD [38]. In addition, a deficiency of IL-23 (a cytokine known to propagate the Th17 lineage) in host APCs was shown to ameliorate GVHD [39]. Furthermore, in a murine model, chronic GVHD was characterized as a disease mediated by Th1 and Th17 subsets that avoided Treg cell inhibition [40]. Recent murine data using donor T cells that were rendered double-deficient for key polarizing cytokines further confirmed that Th1, Th17, and Th2 effector subsets can each contribute to GVHD but with different degrees of severity (Th1 and Th17 severity > Th2 severity) and different target tissue distribution (Th1 and Th17 cells with gut GVHD propensity; Th2 cells with pulmonary GVHD propensity) [41]. Therefore, a model can emerge whereby immune space generated during conditioning for allogeneic transplantation allows for the expansion of Th1, Th17, and Th2 cells; the relative T-helper cell subset expansion in this immune space will determine the type and severity of GVHD induced, with Tregs capable of modulating each type of T-helper cell pathology.

Antitumor effects induced through adoptive T-cell therapy in the autologous context, similar to the allogeneic context, are preferentially mediated through proinflammatory T-cells such as the Th1 and Th17 subsets (reviewed in [6]). Th1-type cells, which are driven in large part by IL-12 and IFN-α activation of T-cell STAT1 and STAT 4 pathways, have been clearly linked to antitumor efficacy [42]. Given this biology, gene therapy strategies are being developed to deliver high-levels of IL-12 in the tumor microenvironment for optimal induction of therapeutic Th1-type effector cells [43]. Recently in a murine model of melanoma, it was determined that adoptive transfer of Th1 or Th17 cells each mediated regression of established metastases. Of interest, Th17 cell transfer appeared to be more potent, and resulted in down-stream $CD8^+$ T-cell immunity and DC activation associated with effective antitumor responses [44]. Furthermore, murine $CD8^+$ T cells can also be induced to express high-levels of the Th17 signature cytokine (IL-17) and can mediate potent antitumor effects [45]. An interesting paradox therefore presents itself: chronic inflammation via IL-6 activates STAT3 for promotion of Th17 cells with resultant tumor progression; however, the adoptive transfer of highly purified Th17 cells represents a potentially potent strategy for cancer therapy. It is therefore clear that the role of T-helper cell subsets in the natural history of cancer (chronic setting) can stand in stark contrast with their role in adoptive T-cell therapy (acute setting). Further research will be needed to dissect the mechanisms accounting for this interesting dichotomy.

These observations suggest a bright future for an ability to use specialized subsets of T cells for optimization of antitumor effects (Th1 or Th17 cell therapy) or amelioration of GVHD or autoimmunity (Th2 or Treg cell therapy). However, such scenarios, which assume that adoptively transferred T cells can maintain their preferential cytokine polarization pattern *in vivo*, is threatened by an emerging body of evidence indicating that T-helper cells maintain a high-degree of plasticity in terms of cytokine polarity (reviewed in[34]). As this review details, current evidence suggests that the Th17 and Treg cell subsets are particularly susceptible to subset inter-conversion, whereas Th1 and Th2 subsets are more stable but still not fixed in their differentiation status. First, it should be acknowledged that there exists a high degree of cytokine promiscuity across the major T-helper subsets. For example, IFN-γ is markedly up-regulated in both Th1 and Th17 subsets [46]; IL-10, initially thought of as a Th2 cytokine, is now considered a counter-regulatory cytokine expressed by multiple lineages [47]; and IL-9, which was once considered a marker for a still further subset of T-helpers (Th9 cells), can be secreted by various T-helper subsets [48, 49]. Second, it was initially thought that T-helper cell expression of signature transcription factors (Th1 [T-bet]; Th2 [GATA-3]; Th17 [RORγt]; Treg [*FOXP3*]) might allow for clear dissection of the subset contribution. However, the literature is now replete with examples demonstrating that individual T-helper cells can coexpress the distinct, putatively exclusive transcription factors and also that T-helper cells can readily make a switch in transcription factor expression. As one example, the infidelity of Tregs has been documented now on several fronts: (1) Tregs may only transiently express *FOXP3* and can secrete inflammatory cytokines upon loss of *FOXP3* [50]; (2) individual Treg cells may express both *FOXP3* and the Th1 transcription factor, T-bet [51]; and (3) *FOXP3*-expressing Treg cells are particularly susceptible to conversion to RORγt-expressing Th17 cells after IL-6 induced STAT3 activation [52]. Such plasticity of adoptively transferred T cells can have deleterious consequences: for example, in a murine model of diabetes, *in vivo* conversion of adoptively transferred Tregs into effector T cells increased disease pathology [50]. Clearly, future advances in the field of adoptive T-cell therapy will require a better understanding of issues relating to T-cell differentiation plasticity and an ability to limit such plasticity for enhanced therapeutic effect. One such future direction relates to the role of epigenetic events: for example, DNA methylation status was found to dictate not only Treg cell *FOXP3* expression but also stability of expression. Importantly, in that study, the FDA-approved DNA hypomethylating agent azacytadine was capable of promoting *FOXP3* expression and stability [53]. In another study involving epigenetics, histone trimethylation maps helped explain the mixed differentiation features and inherent plasticity of Th1, Th2, Treg, and Th17 subsets [54]. A second area of ongoing investigation relates to a potential role of differential expression of microRNA (which are noncoding RNA species that regulate gene expression at a posttranscriptional level) on T-helper cell polarization (reviewed in [55]).

T-Cell Differentiation Status and Apoptotic Threshold

The efficacy of adoptive T-cell therapy is also dependent upon other functional parameters, the most important being T-cell differentiation status and apoptotic threshold. In murine models, autologous T cells with more limited differentiation status associate with improved antitumor effects upon adoptive transfer: that is, transfer of naïve T cells were more potent that T central memory cells, which were themselves more potent that terminally differentiated effector memory T cells [56]. These findings were at first somewhat counter-intuitive because effector memory T cells displayed enhanced IFN-γ secretion and more potent cytolytic capacity prior to adoptive transfer, however, the less differentiated T cells were able to sustain engraftment post-transfer, with the net result of heightened effector function *in vivo*. In our own murine studies involving rapamycin-resistant allogeneic Th2 cells, which express a T central memory phenotype, we found that the *in vivo* capacity to produce type II cytokines was increased by approximately tenfold relative to values obtained after infusion of control Th2 cells that expressed an effector memory phenotype [57]. The favorable effect of adoptive transfer of T central memory cells (in terms of prolonged enhanced of the memory T-cell pool) has also been demonstrated in nonhuman primates [58, 59]. Such observations have important implications for immuno-gene therapy: that is, how does one attain both a less-differentiated T-cell status and enrichment for antigen-specificity, which naturally requires multiple rounds of cell division? First, with advances in TCR and CAR gene transfer, it may be possible to genetically enforce antigen specificity into sorted, highly purified naïve T cells; however, it should be noted that naïve T cells are relatively rare in the adult population [58]. Alternatively, it is possible that pharmacologic maneuvers might promote effector T-cell expression of genes associated with less differentiated cells; for example, GSK3 inhibition and subsequent wnt signaling promoted a T-cell transcriptional program with stemlike characteristics that associated with enhanced antitumor immunity after adoptive T-cell transfer [60]. *Ex vivo* incorporation of rapamycin represents a second pharmacologic method for favorable modulation of T-cell differentiation status. Inhibition of T-cell mTOR signaling enhances T-cell expression of the T central memory markers, CD62 ligand and CCR7 [61]. In murine studies, we found that *ex vivo* rapamycin directly promoted Th2 cell expression of CD62L and CCR7 independent of Th2 cell division [31]; upon adoptive transfer, such rapamycin-generated Th2 cells manifested prolonged *in vivo* engraftment and mediated enhanced therapeutic effects in terms of preventing GVHD [32, 57] or graft rejection [23].

In addition to T-cell differentiation status, we have also found that the T-cell apoptotic threshold represents a critical determinant of the *in vivo* efficacy of adoptively transferred T cells. In initial studies, we determined that *ex vivo* administration of rapamycin, in addition to promoting T central memory differentiation markers, also yielded T cells with a multifaceted antiapoptotic phenotype [21]. In that study we determined that: (1) the antiapoptotic phenotype could be manifested in both Th1- and Th2-type T cells; (2) rapamycin-exposed T cells exhibited initial caspase activation but had markedly reduced activation of distal caspases, thereby indicating modulation of the intrinsic apoptotic pathway; (3) rapamycin-exposed T cells preferentially expressed antiapoptotic members of the bcl-2 family of genes relative to proapoptotic gene members; and (4) such modulation correlated with an enhanced capacity of T cells to accumulate *in vivo* with an advanced proliferative status. More recently [62], we have identified that the mechanism accounting for the antiapoptotic effect of *ex vivo* rapamycin relates to a cellular process known as autophagy. During conditions of increased cellular stress or signals of nutrient deprivation (such as mTOR inhibition via rapamycin), T cells may undergo autophagy as a means to self-digest cellular organelles such as mitochondria [63] in an attempt to diminish cellular energetic requirements to prolong cell survival [64]. In our experiments [62], we found that human T cells generated in the presence of rapamycin underwent autophagy as indicated by: (1) reduction in mitochondrial mass with associated improvement in mitochondrial membrane stability; and (2) alteration of autophagy-related genes, including Beclin-1 gene expression, which was required for realization of the antiapoptotic phenotype. And, most importantly, upon transfer of human rapamycin-resistant and apoptotic-resistant Th1 cells into immune-deficient murine hosts, such Th1 cells stably engrafted and mediated severe xenogeneic GVHD. In sum, these data indicate that autophagy can be harnessed *ex vivo* for the manufacture of T cells with enhanced function via attainment of an antiapoptotic phenotype. These results extend other recent findings in the literature supporting a role for autophagy in immune protection; specifically, autophagy improves host defense to tuberculosis [65] and viral infections [66]. Furthermore, autophagy of tumor cells can increase cross-presentation of relevant tumor antigens to dendritic cells *in vitro* and *in vivo* [66]. As such, future research should focus on improving an understanding of autophagy and methods to modulate this process, both from the *ex vivo* standpoint of enhancing T-cell function and the *in vivo* standpoint of favorably influencing tumor biology in the setting of T-cell therapy.

Autologous and Allogeneic Immunotherapy: Therapeutic Index

A favorable therapeutic index of autologous or allogeneic transplantation therapy for cancer therapy will require consistent attainment of complete remissions while simultaneously limiting morbidity and mortality associated with the T-cell therapy. With respect to allogeneic HSCT, both sides of this therapeutic index can be improved upon. That is, malignant disease relapse is a common cause of posttransplant mortality, particularly in more aggressive or advanced hematologic malignancies, solid tumors, or transplants involving reduced-intensity conditioning [67]; and, significant morbidity and mortality still persists due to direct toxicity from host conditioning and to down-stream effects from GVHD. Because the molecular and cellular mechanisms accounting for GVHD and GVT effects in general share a similar biology (reviewed in [6]), enhancement of GVT effects while limiting GVHD remains a lofty goal. Promising areas of future research, as previously detailed, include the delayed administration of cellular populations capable of modulating ongoing GVHD, including Treg cells [68] and Th2 cells [32]. It will be important to identify the molecular mechanism(s) of action of such cellular products in the event that more specific modulators of immune function might be used. As an example, we have recently found that human Treg cells inhibit human-into-mouse xenogeneic GVHD by a mechanism involving up-regulation of programmed death-ligand 1 (PD-L1) [69]; these data indicate that PD-L1, either as a drug therapy or a gene therapy, might be utilized to treat GVHD. On the other hand, cellular therapies typically operate by complex mechanisms that may not be amenable to such a "reductionist" approach; for example, in a murine model of GVHD, donor Th2 cells effectively treated established GVHD by a mechanism that involved Th2 cell secretion of IL-4 and IL-10, along with consumption of IL-2 that would otherwise be available for effector T cells responsible for GVHD propagation [32].

Nonetheless, there exists hope that such a targeted approach may yield a significant increase in our capacity to modulate the complex biology of GVHD, particularly with single molecule inhibitors of TNF-α (which contributes to the cytokine storm phase of GVHD). Recent results from a phase III trial, however, did not identify an overall beneficial effect of TNF inhibition posttransplant [70]. Because IL-6 has been shown in murine models to inhibit Treg cells while promoting Th17 cells [38], perhaps a single molecule inhibitory approach against IL-6 would yield dividends in terms of antiGVHD effects. Of note, an antibody to IL-6 has just been approved by the FDA for therapy of autoimmune disease, and as such, it is possible that clinical trials in the allogeneic HSCT setting with this reagent might be feasible [71]. Finally, it should be noted that the therapeutic index of allogeneic HSCT is restricted by the current usage of potent immunosuppressive agents, in particular, T cell calcineurin inhibitors such as cyclosporine A and tacrolimus. Paradoxically, allogeneic HSCT is the only form of immunotherapy that attempts to harness potent T-cell anticancer effects while simultaneously inhibiting T-cell clonal expansion and effector function. One potential solution to this clinical practice is the short-term administration of posttransplant cyclophosphamide without administration of other pharmacologic GVHD inhibitors [72]. With this approach, *in vivo* activated alloreactive T cells are clonally deleted, therefore presumably retaining only the tumor-specific repertoire that does not require systemic immune inhibition for control of GVHD. Importantly, this transplant strategy yielded satisfactory levels of GVHD control in a study involving 117 recipients of either HLA-matched sibling or matched-unrelated donor transplantation [72].

With respect to autologous transplantation, the consideration of therapeutic index has recently moved to the forefront of translational research. Previously, autologous transplantation was relatively safe but not associated with dramatic antitumor benefits. However, this situation has appeared to change dramatically because of the increased potential efficacy of autologous strategies that incorporate highly functional costimulated T cells or highly antigen-specific T cells. However, at the same time, the potential toxicities associated with autologous transplantation therapy have increased due to T-cell cytokine storm events, toxicities associated with intensive host preparation, and T-cell-mediated autoimmune events. In particular, gene therapy employing recombinant TCR genes or chimeric antigen receptors has offered great hope in terms of efficacy, but also has been associated with an apparent increase in toxicity, including: (1) development of vitiligo and uveitis in recipients of antimelanoma, antigen-specific T cells [73, 74]; (2) development of liver inflammation in recipients with metastatic renal cell carcinoma who received T cells specific for carbonic anhydrase IX [75]; and (3) most importantly, treatment-related deaths in a recipient of T cells expressing an anti-CD19 chimeric antigen receptor (CAR) [76] and in a recipient of T cells expressing an anti-ERBB2 CAR [77]. These toxicities can most likely be considered "on target" effects that are anticipated due to the specificity of the T cells and the expression of target antigen on normal tissue. T-cell-mediated toxicity in the autologous transplantation setting has also recently been described using host conditioning followed by the infusion of costimulated, polyclonal T cells for therapy of multiple myeloma [78]. In that study, a subset of patients developed "engraftment syndrome," which is more commonly observed after allogeneic transplantation; engraftment syndrome is thought to be mediated by a cytokine storm and is often characterized by high fever, hypoxia with pulmonary infiltration, and vascular leak syndrome [79]. As such, as the autologous T-cell effector mechanisms become more potent and as T-cell receptor reactivities gain in specificity, great caution must continue to be exerted in the clinical translation of these modalities. It is certainly possible that humans are particularly sensitive to the deleterious effects of dramatic and specific T-cell responses; it is also important to note that such responses in the past have not been predicted from animal modeling, as was observed in the uniform development of severe cytokine storm in a phase I trial of an anti-CD28 monoclonal antibody [80].

In sum, these data indicate that both autologous and allogeneic immunotherapy efforts might benefit from novel, improved methods to regulate T-cell responses in an attempt to harness antitumor responses with subsequent protection against autoimmunity or alloimmunity. Certainly, there exists a multitude of potential new methods of regulating immunity, but for the purpose of this text, I will focus on three areas that may have particular relevance to adoptive T-cell therapy: (1) use of suicide or "cell fate control" genes; (2) use of more specific modulators of T-cell biology such as postreceptor, Janus Kinase (JAK) inhibition; and (3) use of inhibitors of the mammalian target of rapamycin (mTOR inhibitors). First, auto- and allo-immunity generated by adoptively transferred T cells may be curtailed efficiently before serious adverse events occur if such T cells are forced to express a gene product such as herpes-simplex virus thymidine kinase (HSVTK), which uniquely allows such T cells to phosphorylate the prodrug ganciclovir, with subsequent T-cell sensitivity to apoptosis. This HSVTK strategy has been evaluated extensively in the clinic for prevention of GVHD after haplo-identical allogeneic HSCT. Recently it was reported that GVHD in this setting was amenable to ganciclovir therapy; it was concluded that this gene therapy method accelerated immune reconstitution, which is typically greatly diminished after HLA-mismatched transplantation [81]. Currently, this HSVTK suicide gene approach is being tested in a phase III clinical trial [81]. In spite of these potentially encouraging results with this HSVTK approach, several limitations may exist, including: (1) the transgene appears to be immunogenic [82];

(2) the gene-expressing T cells may not possess optimal effector function after *ex vivo* manipulation; and (3) the enzyme is relatively inefficient in terms of prodrug activation. In an attempt to overcome these potential obstacles, we have developed a novel cell fate control axis that incorporates an optimized TMPK enzyme that is of human origin (and therefore predictably nonimmunogenic) and highly efficient for activation of a novel and relatively nontoxic prodrug, AZT [83]. Ongoing research seeks to use this cell fate control strategy in adoptive T-cell therapies involving highly functional, rapamycin-resistant T cells.

Janus kinase inhibitors represent a second novel approach for the *in vivo* modulation of T-cell-mediated responses. The clinical development of such JAK inhibitors has occurred in parallel with the development of other tyrosine kinase inhibitors (TKI), most notably the use of imatinib for the targeted therapy of chronic myelogenous leukemia (reviewed in [84]). It should be noted that the JAK pathways are the distal mediators of the more proximal, postreceptor STAT signaling pathways [85]. As such, JAK inhibitors are not highly specific; that is, inhibition of a single JAK pathway can result in the inhibition of more than one STAT pathway, which therefore can manifest across cytokine receptor families [86]. At the current time, there are no reports relating to the successful manufacture of STAT pathway inhibitors, which would represent a highly selective tool for T-cell subset control. Nonetheless, it is believed that JAK inhibition will result in more selective immune suppressive effects relative to currently available medications such as corticosteroids or calcineurin inhibitors. Importantly, late-stage clinical trials involving a JAK3 inhibitor have shown promise for the control of auto-immune disease [87], thereby offering hope that more selective modulators of *in vivo* T-cell responses will soon become a reality in the clinic.

And finally, rapamycin treatment represents a third strategy deserving of further investigation for modulation of T-cell responses. Modulation of immunity through rapamycin and other mTOR inhibitors is a particularly attractive approach for cancer therapy because such inhibitors have displayed activity as anticancer agents [88]. Most notable in this regard is the rapamycin analog temsirolimus, which is FDA-approved for the therapy of refractory, metastatic renal cell carcinoma [89]. As previously detailed, we have used *ex vivo* administered rapamycin to generate both Th1- and Th2-type T cells [31], as well as regulatory T cells [69], that manifest an antiapoptotic phenotype [21] that occurs via autophagy [62]. These data indicate that *ex vivo* rapamycin administration, when used in combination with varying types of input T cell populations, varying recombinant cytokine cocktails, and varying methods of costimulation can be used to manufacture apoptosis-resistant T cells manifesting a wide array of cytokine phenotypes. It should be noted that these conclusions stand somewhat in contrast to other investigations, such as findings that: (1) rapamycin, in the absence of Th1/Th2 polarizing cytokines, preferentially promotes the generation of regulatory T cells [90]; and (2) the mTOR pathway differentially influenced Th1/Th2 differentiation in studies using mTORC1- and mTORC2-deficient T cells [91]. Finally, it should be noted that the mTOR pathway (recently reviewed in [92]) is critical for the regulation of a multitude of cell surface signaling events in both hematopoietic and nonhematopoietic cell types; as such, *in vivo* therapy with mTOR inhibitors can result in a wide-range of *in vivo* effects distinct from its known effects on the modulation of T cells . As one example, rapamycin is known to modulate dendritic cell biology, thereby resulting in immune suppression through inhibition of antigen-presentation and IL-12 production [93]; as such, one might predict that *in vivo* rapamycin therapy may be detrimental to adoptive T-cell therapeutic efforts. However, on the other hand, *in vivo* rapamycin therapy has been shown in a murine model to greatly enhance the generation of memory T-cell responses [94]. And in clinical trials, use of rapamycin for posttransplant GVHD prophylaxis resulted in improved overall survival and disease-free survival in patients with lymphoma [95], thereby indicating that mTOR inhibition yielded a direct antitumor effect that complemented the allogeneic GVT effect.

Conclusion

The fields of autologous and allogeneic transplantation have different origins, yet current research identifies converging themes that will continue to drive future translational efforts. (1) With the discovery of a multitude of tumor antigens relevant to nearly every neoplasm, it is anticipated that vaccine strategies and genetic engineering approaches such as the use of CARs will represent an important interface with adoptive T-cell therapies. (2) Beyond issues of T-cell specificity, a substantial research focus will seek to identify T cells with long-term engraftment capacity and incorporate an optimal mix of Th1, Th2, Th17, and Treg functionalities. (3) The role of host conditioning in such T-cell therapies will undoubtedly be an active area of investigation, with the goal of identifying approaches that create optimal immune space with minimal host toxicity. Because of the aging population and the increasing incidence of cancer with age, such low-intensity transplant therapies will be instrumental for T-cell therapy integration into the mainstream of anticancer therapies. (4) Ongoing research will need to focus on novel methods to control T-cell-mediated toxicities. Such toxicities are well-known to occur in the allogeneic setting (GVHD) and increasingly observed in the autologous setting (on-target tissue destruction after CAR-expressing T-cell therapy; engraftment syndrome after infusion of costimulated T cells). New methods of immune modulation may be advantageous for both autologous and allogeneic transplantation, including gene therapy using cell fate control cassettes or pharmaceuticals targeting mTOR or JAK pathways. Integration of these various disciplines holds great promise for the ultimate realization of T-cell therapy of cancer and will clearly require a substantial investment for expansion of clinical trials to evaluate the various approaches through an iterative approach.

References

1. Weiden PL, Flournoy N, Thomas ED, et al. Antileukemic effect of graft-versus-host disease in human recipients of allogeneic-marrow grafts. N Engl J Med. 1979;300:1068–1073.

2. Barkholt L, Bregni M, Remberger M, et al. Allogeneic haematopoietic stem cell transplantation for metastatic renal carcinoma in Europe. Ann Oncol. 2006;17:1134–1140.
3. Bishop MR, Fowler DH, Marchigiani D, et al. Allogeneic lymphocytes induce tumor regression of advanced metastatic breast cancer. J Clin Oncol. 2004;22:3886–3892.
4. Brown CE, Starr R, Martinez C, et al. Recognition and killing of brain tumor stem-like initiating cells by $CD8^+$ cytolytic T cells. Cancer Res. 2009;69:8886–8893.
5. Wrzesinski C, Paulos CM, Gattinoni L, et al. Hematopoietic stem cells promote the expansion and function of adoptively transferred antitumor CD8 T cells. J Clin Invest. 2007; 117:492–501.
6. Fowler DH. Shared biology of GVHD and GVT effects: potential methods of separation. Crit Rev Oncol Hematol. 2006;57:225–244.
7. Mielke S, Nunes R, Rezvani K, et al. A clinical-scale selective allodepletion approach for the treatment of HLA-mismatched and matched donor-recipient pairs using expanded T lymphocytes as antigen-presenting cells and a TH9402-based photodepletion technique. Blood. 2008;111:4392–4402.
8. Paulos CM, Wrzesinski C, Kaiser A, et al. Microbial translocation augments the function of adoptively transferred self/tumor-specific $CD8^+$ T cells via TLR4 signaling. J Clin Invest. 2007;117:2197–2204.
9. Warrren EH, Fujii N, Akatsuka Y, et al. Therapy of relapsed leukemia after allogeneic hematopoietic cell transplant with T cells specific for minor histocompatibility antigens. Blood. 2010;115(19):3869–3878.
10. Ksontini R, Colagiovanni DB, Josephs MD, et al. Disparate roles for TNF-alpha and Fas ligand in concanavalin A-induced hepatitis. J Immunol. 1998;160:4082–4089.
11. Boni A, Muranski P, Cassard L, et al. Adoptive transfer of allogeneic tumor-specific T cells mediates effective regression of large tumors across major histocompatibility barriers. Blood. 2008;112:4746–4754.
12. Zhang H, Chua KS, Guimond M, et al. Lymphopenia and interleukin-2 therapy alter homeostasis of $CD4^+CD25^+$ regulatory T cells. Nat Med. 2005;11:1238–1243.
13. Gattinoni L, Finkelstein SE, Klebanoff CA, et al. Removal of homeostatic cytokine sinks by lymphodepletion enhances the efficacy of adoptively transferred tumor-specific $CD8^+$ T cells. J Exp Med. 2005;202:907–912.
14. Hsieh MM, Kang EM, Fitzhugh CD, et al. Allogeneic hematopoietic stem-cell transplantation for sickle cell disease. N Engl J Med. 2009;361:2309–2317.
15. Bishop MR, Hou JW, Wilson WH, et al. Establishment of early donor engraftment after reduced-intensity allogeneic hematopoietic stem cell transplantation to potentiate the graft-versus-lymphoma effect against refractory lymphomas. Biol Blood Marrow Transplant. 2003;9:162–169.
16. Robak T, Korycka A, Lech-Maranda E, Robak P. Current status of older and new purine nucleoside analogues in the treatment of lymphoproliferative diseases. Molecules. 2009;14:1183–1226.
17. Aiuti A, Cattaneo F, Galimberti S, et al. Gene therapy for immunodeficiency due to adenosine deaminase deficiency. N Engl J Med. 2009;360:447–458.
18. Pavletic SZ, Bociek RG, Foran JM, et al. Lymphodepleting effects and safety of pentostatin for nonmyeloablative allogeneic stem-cell transplantation. Transplantation. 2003;76: 877–881.
19. Dean RM, Fry T, Mackall C, et al. Association of serum interleukin-7 levels with the development of acute graft-versus-host disease. J Clin Oncol. 2008;26:5735–5741.
20. Mielcarek M, Martin PJ, Leisenring W, et al. Graft-versus-host disease after nonmyeloablative versus conventional hematopoietic stem cell transplantation. Blood. 2003;102:756–762.
21. Mariotti J, Foley J, Jung U, et al. *Ex vivo* rapamycin generates apoptosis-resistant donor Th2 cells that persist *in vivo* and prevent hemopoietic stem cell graft rejection. J Immunol. 2008;180:89–105.
22. Martin PJ, Akatsuka Y, Hahne M, Sale G. Involvement of donor T-cell cytotoxic effector mechanisms in preventing allogeneic marrow graft rejection. Blood. 1998;92:2177–2181.
23. Mariotti J, Foley J, Ryan K, et al. Graft rejection as a Th1-type process amenable to regulation by donor Th2-type cells through an interleukin-4/STAT6 pathway. Blood. 2008;112:4765–4775.

24. Dudley ME, Yang JC, Sherry R, et al. Adoptive cell therapy for patients with metastatic melanoma: evaluation of intensive myeloablative chemoradiation preparative regimens. J Clin Oncol. 2008;26:5233–5239.
25. Melchionda F, Fry TJ, Milliron MJ, McKirdy MA, Tagaya Y, Mackall CL. Adjuvant IL-7 or IL-15 overcomes immunodominance and improves survival of the $CD8^+$ memory cell pool. J Clin Invest. 2005;115:1177–1187.
26. Cui Y, Zhang H, Meadors J, Poon R, Guimond M, Mackall CL. Harnessing the physiology of lymphopenia to support adoptive immunotherapy in lymphoreplete hosts. Blood. 2009; 114:3831–3840.
27. Guimond M, Veenstra RG, Grindler DJ, et al. Interleukin 7 signaling in dendritic cells regulates the homeostatic proliferation and niche size of $CD4^+$ T cells. Nat Immunol. 2009;10:149–157.
28. Sportes C, Hakim FT, Memon SA, et al. Administration of rhIL-7 in humans increases *in vivo* TCR repertoire diversity by preferential expansion of naive T cell subsets. J Exp Med. 2008;205:1701–1714.
29. Berger C, Berger M, Hackman RC, et al. Safety and immunologic effects of IL-15 administration in nonhuman primates. Blood. 2009;114:2417–2426.
30. Hinrichs CS, Spolski R, Paulos CM, et al. IL-2 and IL-21 confer opposing differentiation programs to $CD8^+$ T cells for adoptive immunotherapy. Blood. 2008;111:5326–5333.
31. Jung U, Foley JE, Erdmann AA, et al. *Ex vivo* rapamycin generates Th1/Tc1 or Th2/Tc2 Effector T cells with enhanced *in vivo* function and differential sensitivity to post-transplant rapamycin therapy. Biol Blood Marrow Transplant. 2006;12:905–918.
32. Foley JE, Mariotti J, Ryan K, Eckhaus M, Fowler DH. Th2 cell therapy of established acute graft-versus-host disease requires IL-4 and IL-10 and is abrogated by IL-2 or host-type antigen-presenting cells. Biol Blood Marrow Transplant. 2008;14:959–972.
33. Fowler DH, Odom J, Steinberg SM, et al. Phase I clinical trial of costimulated, IL-4 polarized donor $CD4^+$ T cells as augmentation of allogeneic hematopoietic cell transplantation. Biol Blood Marrow Transplant. 2006;12:1150–1160.
34. O'Shea JJ, Paul WE. Mechanisms underlying lineage commitment and plasticity of helper $CD4^+$ T cells. Science.;327:1098–1102.
35. Edinger M, Hoffmann P, Ermann J, et al. $CD4^+CD25^+$ regulatory T cells preserve graft-versus-tumor activity while inhibiting graft-versus-host disease after bone marrow transplantation. Nat Med. 2003;9:1144–1150.
36. Jones SC, Murphy GF, Korngold R. Post-hematopoietic cell transplantation control of graft-versus-host disease by donor CD425 T cells to allow an effective graft-versus-leukemia response. Biol Blood Marrow Transplant. 2003;9:243–256.
37. Langowski JL, Zhang X, Wu L, et al. IL-23 promotes tumour incidence and growth. Nature. 2006;442:461–465.
38. Chen X, Das R, Komorowski R, et al. Blockade of interleukin-6 signaling augments regulatory T-cell reconstitution and attenuates the severity of graft-versus-host disease. Blood. 2009;114:891–900.
39. Das R, Chen X, Komorowski R, Hessner MJ, Drobyski WR. Interleukin-23 secretion by donor antigen-presenting cells is critical for organ-specific pathology in graft-versus-host disease. Blood. 2009;113:2352–2362.
40. Chen X, Vodanovic-Jankovic S, Johnson B, Keller M, Komorowski R, Drobyski WR. Absence of regulatory T-cell control of TH1 and TH17 cells is responsible for the autoimmune-mediated pathology in chronic graft-versus-host disease. Blood. 2007;110:3804–3813.
41. Yi T, Chen Y, Wang L, et al. Reciprocal differentiation and tissue-specific pathogenesis of Th1, Th2, and Th17 cells in graft-versus-host disease. Blood. 2009;114:3101–3112.
42. Dobrzanski MJ, Reome JB, Hollenbaugh JA, Dutton RW. Tc1 and Tc2 effector cell therapy elicit long-term tumor immunity by contrasting mechanisms that result in complementary endogenous type 1 antitumor responses. J Immunol. 2004;172:1380–1390.
43. Labbe A, Nelles M, Walia J, et al. IL-12 immunotherapy of murine leukaemia: comparison of systemic versus gene modified cell therapy. J Cell Mol Med. 2009;13:1962–1976.
44. Martin-Orozco N, Muranski P, Chung Y, et al. T helper 17 cells promote cytotoxic T cell activation in tumor immunity. Immunity. 2009;31:787–798.

45. Hinrichs CS, Kaiser A, Paulos CM, et al. Type 17 CD8+ T cells display enhanced antitumor immunity. Blood. 2009;114:596–599.
46. Damsker JM, Hansen AM, Caspi RR. Th1 and Th17 cells: adversaries and collaborators. Ann N Y Acad Sci;1183:211–221.
47. Saraiva M, O'Garra A. The regulation of IL-10 production by immune cells. Nat Rev Immunol;10:170–181.
48. Lee YK, Mukasa R, Hatton RD, Weaver CT. Developmental plasticity of Th17 and Treg cells. Curr Opin Immunol. 2009;21:274–280.
49. Zhou L, Chong MM, Littman DR. Plasticity of CD4+ T cell lineage differentiation. Immunity. 2009;30:646–655.
50. Zhou X, Bailey-Bucktrout SL, Jeker LT, et al. Instability of the transcription factor *FOXP3* leads to the generation of pathogenic memory T cells *in vivo*. Nat Immunol. 2009;10:1000–1007.
51. Koch MA, Tucker-Heard G, Perdue NR, Killebrew JR, Urdahl KB, Campbell DJ. The transcription factor T-bet controls regulatory T cell homeostasis and function during type 1 inflammation. Nat Immunol. 2009;10:595–602.
52. Yang XO, Nurieva R, Martinez GJ, et al. Molecular antagonism and plasticity of regulatory and inflammatory T cell programs. Immunity. 2008;29:44–56.
53. Polansky JK, Kretschmer K, Freyer J, et al. DNA methylation controls *FOXP3* gene expression. Eur J Immunol. 2008;38:1654–1663.
54. Wei G, Wei L, Zhu J, et al. Global mapping of H3K4me3 and H3K27me3 reveals specificity and plasticity in lineage fate determination of differentiating CD4+ T cells. Immunity. 2009;30:155–167.
55. Merkenschlager M, Wilson CB. RNAi and chromatin in T cell development and function. Curr Opin Immunol. 2008;20:131–138.
56. Hinrichs CS, Borman ZA, Cassard L, et al. Adoptively transferred effector cells derived from naive rather than central memory CD8+ T cells mediate superior antitumor immunity. Proc Natl Acad Sci USA. 2009;106:17469–17474.
57. Foley JE, Jung U, Miera A, et al. *Ex vivo* rapamycin generates donor Th2 cells that potently inhibit graft-versus-host disease and graft-versus-tumor effects via an IL-4-dependent mechanism. J Immunol. 2005;175:5732–5743.
58. Berger C, Jensen MC, Lansdorp PM, Gough M, Elliott C, Riddell SR. Adoptive transfer of effector CD8+ T cells derived from central memory cells establishes persistent T cell memory in primates. J Clin Invest. 2008;118:294–305.
59. Hakim FT, Memon SA, Cepeda R, et al. Age-dependent incidence, time course, and consequences of thymic renewal in adults. J Clin Invest. 2005;115:930–939.
60. Gattinoni L, Zhong XS, Palmer DC, et al. Wnt signaling arrests effector T cell differentiation and generates CD8+ memory stem cells. Nat Med. 2009;15:808–813.
61. Sinclair LV, Finlay D, Feijoo C, et al. Phosphatidylinositol-3-OH kinase and nutrient-sensing mTOR pathways control T lymphocyte trafficking. Nat Immunol. 2008;9:513–521.
62. Amarnath S, Flomerfelt FA, Costanzo CM, et al. Rapamycin generates anti-apoptotic human Th1/Tc1 cells via autophagy for induction of xenogeneic GVHD. Autophagy. 2010;6(4):523–541.
63. Pua HH, Guo J, Komatsu M, He YW. Autophagy is essential for mitochondrial clearance in mature T lymphocytes. J Immunol. 2009;182:4046–4055.
64. Virgin HW, Levine B. Autophagy genes in immunity. Nat Immunol. 2009;10:461–470.
65. Gutierrez MG, Master SS, Singh SB, Taylor GA, Colombo MI, Deretic V. Autophagy is a defense mechanism inhibiting BCG and *Mycobacterium tuberculosis* survival in infected macrophages. Cell. 2004;119:753–766.
66. Li Y, Wang LX, Yang G, Hao F, Urba WJ, Hu HM. Efficient cross-presentation depends on autophagy in tumor cells. Cancer Res. 2008;68:6889–6895.
67. Bishop MR, Alyea EP, 3rd, Cairo MS, et al. Introduction to the reports from the National Cancer Institute first international workshop on the biology, prevention, and treatment of relapse after allogeneic hematopoietic stem cell transplantation. Biol Blood Marrow Transplant;16:563–564.

68. Taylor PA, Panoskaltsis-Mortari A, Swedin JM, et al. L-Selectin(hi) but not the L-selectin(lo) CD4+25+ T-regulatory cells are potent inhibitors of GVHD and BM graft rejection. Blood. 2004;104:3804–3812.
69. Amarnath S, Costanzo CM, Mariotti J, et al. Regulatory T cells and human myeloid dendritic cells promote tolerance via programmed death ligand-1. PLoS Biol;8:e1000302.
70. Couriel DR, Saliba R, de Lima M, et al. A phase III study of infliximab and corticosteroids for the initial treatment of acute graft-versus-host disease. Biol Blood Marrow Transplant. 2009;15:1555–1562.
71. Venkiteshwaran A. Tocilizumab. MAbs. 2009;1:432–438.
72. Luznik L, Bolanos-Meade J, Zahurak M, et al. High-dose cyclophosphamide as single agent, short-course prophylaxis of graft-versus-host disease. Blood. 2010;115(16):3224–3230.
73. Morgan RA, Dudley ME, Wunderlich JR, et al. Cancer regression in patients after transfer of genetically engineered lymphocytes. Science. 2006;314:126–129.
74. Johnson LA, Morgan RA, Dudley ME, et al. Gene therapy with human and mouse T-cell receptors mediates cancer regression and targets normal tissues expressing cognate antigen. Blood. 2009;114:535–546.
75. Lamers CH, Sleijfer S, Vulto AG, et al. Treatment of metastatic renal cell carcinoma with autologous T-lymphocytes genetically retargeted against carbonic anhydrase IX: first clinical experience. J Clin Oncol. 2006;24:e20–e22.
76. Brentjens R, Yeh R, Bernal Y, Riviere I, Sadelain M. Treatment of chronic lymphocytic leukemia with genetically targeted autologous T cells: case report of an unforeseen adverse event in a phase I clinical trial. Mol Ther. 2010;18:666–668.
77. Morgan RA, Yang JC, Kitano M, Dudley ME, Laurencot CM, Rosenberg SA. Case report of a serious adverse event following the administration of T cells transduced with a chimeric antigen receptor recognizing ERBB2. Mol Ther. 2010;18:843–851.
78. Rapoport AP, Stadtmauer EA, Aqui N, et al. Rapid immune recovery and graft-versus-host disease-like engraftment syndrome following adoptive transfer of Costimulated autologous T cells. Clin Cancer Res. 2009;15:4499–4507.
79. Spitzer TR. Engraftment syndrome following hematopoietic stem cell transplantation. Bone Marrow Transplant. 2001;27:893–898.
80. Suntharalingam G, Perry MR, Ward S, et al. Cytokine storm in a phase 1 trial of the anti-CD28 monoclonal antibody TGN1412. N Engl J Med. 2006;355:1018–1028.
81. Ciceri F, Bonini C, Stanghellini MT, et al. Infusion of suicide-gene-engineered donor lymphocytes after family haploidentical haemopoietic stem-cell transplantation for leukaemia (the TK007 trial): a non-randomised phase I-II study. Lancet Oncol. 2009;10:489–500.
82. Berger C, Flowers ME, Warren EH, Riddell SR. Analysis of transgene-specific immune responses that limit the *in vivo* persistence of adoptively transferred HSV-TK-modified donor T cells after allogeneic hematopoietic cell transplantation. Blood. 2006;107:2294–2302.
83. Sato T, Neschadim A, Konrad M, Fowler DH, Lavie A, Medin JA. Engineered human tmpk/AZT as a novel enzyme/prodrug axis for suicide gene therapy. Mol Ther. 2007;15:962–970.
84. Ghoreschi K, Laurence A, O'Shea JJ. Selectivity and therapeutic inhibition of kinases: to be or not to be? Nat Immunol. 2009;10:356–360.
85. Adamson AS, Collins K, Laurence A, O'Shea JJ. The Current STATus of lymphocyte signaling: new roles for old players. Curr Opin Immunol. 2009;21:161–166.
86. Karaman MW, Herrgard S, Treiber DK, et al. A quantitative analysis of kinase inhibitor selectivity. Nat Biotechnol. 2008;26:127–132.
87. van Gurp E, Weimar W, Gaston R, et al. Phase 1 dose-escalation study of CP-690 550 in stable renal allograft recipients: preliminary findings of safety, tolerability, effects on lymphocyte subsets and pharmacokinetics. Am J Transplant. 2008;8:1711–1718.
88. Ciuffreda L, Di Sanza C, Milella M. The mTOR pathway: a new target in cancer therapy. Curr Cancer Drug Targets. 2010;10(5):484–495.
89. Hudes G, Carducci M, Tomczak P, et al. Temsirolimus, interferon alfa, or both for advanced renal-cell carcinoma. N Engl J Med. 2007;356:2271–2281.

90. Battaglia M, Stabilini A, Roncarolo MG. Rapamycin selectively expands $CD4^+CD25^+$ *FOXP3*$^+$ regulatory T cells. Blood. 2005;105:4743–4748.
91. Delgoffe GM, Kole TP, Zheng Y, et al. The mTOR kinase differentially regulates effector and regulatory T cell lineage commitment. Immunity. 2009;30:832–844.
92. Delgoffe GM, Powell JD. mTOR: taking cues from the immune microenvironment. Immunology. 2009;127:459–465.
93. Hackstein H, Taner T, Zahorchak AF, et al. Rapamycin inhibits IL-4 – induced dendritic cell maturation *in vitro* and dendritic cell mobilization and function *in vivo*. Blood. 2003;101:4457–4463.
94. Araki K, Turner AP, Shaffer VO, et al. mTOR regulates memory CD8 T-cell differentiation. Nature. 2009;460:108–112.
95. Armand P, Gannamaneni S, Kim HT, et al. Improved survival in lymphoma patients receiving sirolimus for graft-versus-host disease prophylaxis after allogeneic hematopoietic stem-cell transplantation with reduced-intensity conditioning. J Clin Oncol. 2008;26:5767–5774.

Index

J. Medin and D. Fowler (eds.), *Experimental and Applied Immunotherapy*,
DOI 10.1007/978-1-60761-980-2, © Springer Science+Business Media, LLC 2011

G

H

I

N

O

P